Physics of Functional Materials

Physics of Functional Materials

Hasse Fredriksson

KTH Stockholm, Sweden

and

Ulla Åkerlind

University of Stockholm, Sweden

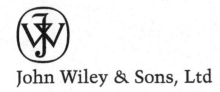

John Wiley & Sons, Ltd

Other Wiley Editorial Offices

John Wiley & Sons Inc., 111 River Street, Hoboken, NJ 07030, USA

Jossey-Bass, 989 Market Street, San Francisco, CA 94103-1741, USA

Wiley-VCH Verlag GmbH, Boschstr. 12, D-69469 Weinheim, Germany

John Wiley & Sons Australia Ltd, 42 McDougall Street, Milton, Queensland 4064, Australia

John Wiley & Sons (Asia) Pte Ltd, 2 Clementi Loop #02-01, Jin Xing Distripark, Singapore 129809

John Wiley & Sons Ltd, 6045 Freemont Blvd, Mississauga, Ontario L5R 4J3, Canada

Wiley also publishes its books in a variety of electronic formats. Some content that appears in print may not be available in electronic books.

Library of Congress Cataloging in Publication Data

Fredriksson, Hasse.
   Physics of Functional Materials / Hasse Fredriksson and Ulla Åkerlind.
      p.   cm.
   Includes bibliographical references and index.
   ISBN 978-0-470-51757-4 (cloth) — ISBN 978-0-470-51758-1 (pbk. : alk. paper)
   1. Physics.   2. Materials.   I. Akerlind, Ulla.   II. Title.
   QC21.3.F74 2008
   530.4—dc22

                                        2007050191

British Library Cataloguing in Publication Data

A catalogue record for this book is available from the British Library

ISBN 978-0-470-51757-4 (HB)
ISBN 978-0-470-51758-1 (PB)

Typeset in 10/12pt Times by Integra Software Services Pvt. Ltd, Pondicherry, India

Frontpage figures Reproduced with permission from

1. Posten Frimärken AB, Sweden, Nobel Prize stamps celebrating Schrödinger and de Broglie.

2. Parmanand Sharma, TMR Tohuku University, Sendai Magnetic Domains observed by MFM technology.

Contents

Preface   vii

1   Structures of Melts and Solids   1
  1.1   Introduction   1
  1.2   X-ray Analysis   2
  1.3   The Hard Sphere Model of Atoms   10
  1.4   Crystal Structure   12
  1.5   Crystal Structures of Solid Metals   19
  1.6   Crystal Defects in Pure Metals   24
  1.7   Structures of Alloy Melts and Solids   30
        Summary   39
        Exercises   42

2   Theory of Atoms and Molecules   45
  2.1   Introduction   46
  2.2   The Bohr Model of Atomic Structure   46
  2.3   The Quantum Mechanical Model of Atomic Structure   48
  2.4   Solution of the Schrödinger Equation for Atoms   57
  2.5   Quantum Mechanics and Probability: Selection Rules   65
  2.6   The Quantum Mechanical Model of Molecular Structure   71
  2.7   Diatomic Molecules   75
  2.8   Polyatomic Molecules   88
        Summary   89
        Exercises   94

3   Theory of Solids   97
  3.1   Introduction   97
  3.2   Bonds in Molecules and Solids: Some Definitions   98
  3.3   Bonds in Molecules and Nonmetallic Solids   100
  3.4   Metallic Bonds   112
  3.5   Band Theory of Solids   125
  3.6   Elastic Vibrations in Solids   146
  3.7   Influence of Lattice Defects on Electronic Structures in Crystals 151
        Summary   157
        Exercises   163

4   Properties of Gases   169
  4.1   Introduction   170
  4.2   Kinetic Theory of Gases   170
  4.3   Energy Distribution in Particle Systems: Maxwell–Boltzmann Distribution Law   174
  4.4   Gas Laws   178
  4.5   Heat Capacity   185
  4.6   Mean Free Path   192
  4.7   Viscosity   196
  4.8   Thermal Conduction   198
  4.9   Diffusion   199

4.10   Molecular Sizes   204
4.11   Properties of Gas Mixtures   205
4.12   Plasma – The Fourth State of Matter   208
       Summary   213
       Exercises   216

5   Transformation Kinetics: Diffusion in Solids   219
5.1   Introduction   220
5.2   Thermodynamics   220
5.3   Transformation Kinetics   236
5.4   Reaction Rates   242
5.5   Kinetics of Homogeneous Reactions in Gases   245
5.6   Diffusion in Solids   253
       Summary   276
       Exercises   283

6   Mechanical, Thermal and Magnetic Properties of Solids   287
6.1   Introduction   287
6.2   Total Energy of Metallic Crystals   288
6.3   Elasticity and Compressibility   290
6.4   Expansion   296
6.5   Heat Capacity   303
6.6   Magnetism   317
       Summary   333
       Exercises   337

7   Transport Properties of Solids. Optical Properties of Solids   341
7.1   Introduction   342
7.2   Thermal Conduction   342
7.3   Electrical Conduction   347
7.4   Metallic Conductors   350
7.5   Insulators   357
7.6   Semiconductors   362
7.7   Optical Properties of Solids   375
       Summary   388
       Exercises   394

8   Properties of Liquids and Melts   399
8.1   Introduction   400
8.2   X-ray Spectra of Liquids and Melts   400
8.3   Models of Pure Liquids and Melts   402
8.4   Melting Points of Solid Metals   406
8.5   Density and Volume   407
8.6   Thermal Expansion   409
8.7   Heat Capacity   412
8.8   Transport Properties of Liquids   415
8.9   Diffusion   416
8.10   Viscosity   425
8.11   Thermal Conduction   438
8.12   Electrical Conduction   439
       Summary   443
       Exercises   449

Answers to Exercises   455

Index   475

Preface

The basic idea and the initial aim of this book, *Physics of Functional Materials*, was to provide the necessary knowledge in physics for a deeper interpretation of many of the solidification and crystallization processes that are treated in the book *Materials Processing during Casting*, written for the university undergraduate level and published in March 2006. The present book fulfils this requirement and is at such a mathematical level that a basic knowledge of mathematics at university level is sufficient.

However, the book *Physics of Functional Materials* has a very wide and general character. It is by no means designed only for the purpose described above. On the contrary, this book may be useful and suitable for students in various sciences at the Master's and PhD level, who have not taken mathematics and physics as major subjects. Examples of such sciences are materials science, chemistry, metallurgy and many other scientific and technical fields where a basic knowledge of the foundations of modern physics and/or properties of materials is necessary or desirable as a basis and a background for higher studies.

Fundamental properties of different materials such as diffusion, viscosity, heat capacity and thermal and electrical conduction are examined more extensively in the present book than in available physics books of today. This book will fill a gap between the demand for and supply of such knowledge.

The atomistic view of matter requires a genuine background of modern physics from atoms via molecules to solid-state physics, primarily the modern band theory of solids, and the nature of bonds and crystal structure in solids. These topics are treated in Chapters 1–3 and are applied in later chapters, particularly in Chapter 7.

The three first chapters on modern physics are followed by applications of classical physics of material properties. Basic thermodynamics, properties of gases, including the kinetic theory of gases and the Boltzmann distributions of velocity and energy of molecules, transformation kinetics including chemical reactions and diffusion in solids, mechanical, thermal and magnetic properties of matter including ferromagnetism, are treated in Chapters 4–6.

Chapter 7 deals with thermal and electrical conduction in solids and their optical properties, particularly electrical conduction in metals and semiconductors. Polarization phenomena in crystals and optical activity in solids and liquids are discussed. In the last chapter, a short survey of the material properties of liquids is given.

Physics of Functional Materials contains solved examples in the text and exercises for students at the end of each chapter. Answers to all the exercises are given at the end of the book.

Acknowledgements

We want to express our sincere thanks to Dr Gunnar Benediktsson (KTH, Stockholm) for many long and fruitful discussions and support on solid-state physics. Dr Ulf Ringström (KTH, Stockholm) gave many valuable points of view on atomic and molecular physics and optics. We also owe our gratitude to Dr Göran Grimvall and Dr Ragnar Lundell (KTH, Stockholm) for valuable help concerning aspects of solid-state theory and solid mechanics, respectively.

Dr Jonas Åberg, Thomas Bergström and Hani Nassar (Department of Casting of Metals, KTH, Stockholm) gave us practical support concerning the ever-lasting computer problems throughout the years. We thank them gratefully for their patience and unfailing help. We also owe our gratitude to Dr Gunnar Edvinsson (University of Stockholm) and Dr Thomas Antonsson (Department of Casting of Metals, KTH, Stockholm) for valuable computer help. Colleges at the Institutes of Physics at the Universities of Uppsala and Lund and KTH, Stockholm, kindly allowed us to use freely their problem collections in physics. We thank them for this generous offer. Some of their exercises have been included in this book. We also thank Dr Olof Beckman, University of Uppsala, for permission to use some polarization figures.

We are most grateful for financial support from The Iron Masters Association in Sweden. Finally and in particular we want to express our sincere gratitude to Karin Fredriksson and Lars Åkerlind. Without their constant support and great patience through the years this book would never have been written.

Hasse Fredriksson
Ulla Åkerlind
Stockholm, Sweden
March 2008

1

Structures of Melts and Solids

1.1 Introduction	1
1.2 X-ray Analysis	2
1.2.1 Methods of X-ray Examination of Solid Materials	2
1.2.2 X-ray Examination and Structures of Metal Melts	4
1.3 The Hard Sphere Model of Atoms	10
1.3.1 Atomic Sizes	10
1.3.2 The Hard Sphere Model of Liquids	11
1.4 Crystal Structure	12
1.4.1 Crystal Structure of Solids	12
1.4.2 Types of Crystal Structures	13
1.4.3 Lattice Directions and Planes	15
1.4.4 Intensities of X-ray Diffractions in Crystal Planes	18
1.5 Crystal Structures of Solid Metals	19
1.5.1 Coordination Numbers	19
1.5.2 Nearest Neighbour Distances of Atoms	20
1.5.3 BCC, FCC and HCP Structures in Metals	21
1.6 Crystal Defects in Pure Metals	24
1.6.1 Vacancies and Other Point Defects	25
1.6.2 Line Defects	26
1.6.3 Interfacial Defects	28
1.7 Structures of Alloy Melts and Solids	30
1.7.1 Basic Concepts	30
1.7.2 Structures of Liquid Alloys	31
1.7.3 Structures of Solid Alloys	33
Summary	39
Exercises	42

1.1 Introduction

Crystallization is the process of transferring a material from a liquid or a gas phase into a solid state of regular order. The three aggregation states have widely different atomic structures and properties. In order to understand crystallization processes, it is essential to have a thorough knowledge of the atomic structure of both the melt or gas and of the new solid phase.

Physics of Functional Materials   Hasse Fredriksson and Ulla Åkerlind
© 2008 John Wiley & Sons, Ltd

In this chapter, the atomic structures of melts and various solid phases will be examined. The structures of pure elements and of alloys and chemical compounds will be discussed. The chapter starts with a short introduction to X-ray analysis, which is a very important tool for investigating atomic structures.

1.2 X-ray Analysis

The energetic X-radiation was discovered in 1901 by the German physicist W. K. Roentgen. The source was an evacuated tube where electrons, accelerated by an electric field, hit a metal target (Figure 1.1). When the electrons suddenly lost their kinetic energy, continous X-ray radiation was emitted together with a few discrete X-ray wavelengths characteristic of the target atoms (Figure 1.2).

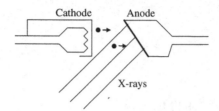

Figure 1.1   The principle of an X-ray tube. Electrons are accelerated in a strong electric field between an anode and an indirectly heated cathode, hit the anode and lose their kinetic energy successively during numerous collisions. Their kinetic energies are transformed into continuous X-radiation. The whole equipment is included in a highly evacuated tube.

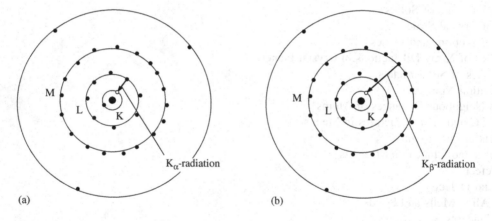

Figure 1.2   (a) The origin of the characteristic X-radiation: When a high-speed electron hits an electron in an inner shell of a target atom, both electrons leave the atom and a vacancy is left in its inner shell. The rest of the process is described in (b).
(b) The vacancy is filled by an electron from an outer shell of the atom and an X-ray photon with a wavelength characteristic for the anode material is emitted. There are several alternative wavelengths, depending on the shell from which the jumping electron emanates. The X-ray lines K_α and K_β shown in (a) and (b) appear simultaneously as numerous such processes occur at the same time.

X-ray analysis of materials has been used since the beginning of the 20th century to investigate the construction and structure of materials. Measurements, calculations and interpretations of the results were made by hand for many decades. Today, completely automatic computer-based X-ray spectrometers, which perform both the experiments and the analysis of the results, are commercially available. However, it is necessary to understand the principles of their function, which are described briefly below.

1.2.1 Methods of X-ray Examination of Solid Materials

A very important method to provide information on the structure of materials is X-ray diffraction measurements. In solid crystalline materials, the diffraction pattern is caused by the atoms in the crystal lattice (Figure 1.3). If a crystal surface is exposed to parallel monoenergetic X-radiation with a known angle of incidence, Bragg's law gives the condition for coherence (in phase) of the diffracted waves:

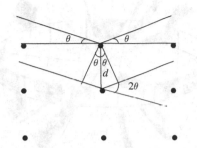

Figure 1.3 X-ray diffraction in a crystal.

$$2d \sin \theta = p\lambda \tag{1.1}$$

where

d = distance between the atomic planes
θ = angle between the incident ray and the atomic plane
λ = wavelength of the monoenergetic X-ray
p = an integer $\neq 0$.

If λ is known and θ can be measured, Bragg's law can be used to determine the distance d between the atomic planes in the crystal lattice.

An X-ray crystallographic examination of a single crystal will in principle be performed as described in Figure 1.4.

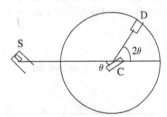

Figure 1.4 The principle of an X-ray spectrometer.

The radiation from an X-ray tube S falls on a turnable single crystal C. The angle between the incident and diffracted rays is 2θ. The crystal is tilted stepwise. For every angle of incidence θ the detector D is placed at the corresponding angle of diffraction 2θ and the intensity of the diffracted radiation is measured. In this way, the intensity as a function of θ is obtained and the whole X-ray spectrum of the solid material can be obtained.

The method described above is time consuming. If the single crystal is replaced with a powder compound, consisting of small crystals of the solid material, it is no longer necessary to tilt the specimen to find high-intensity angles. All possible crystallographic directions are present in the powder compound.

In practice, the Debye–Scherrer method of X-ray crystallographic investigations on metallic or other crystalline materials is the most common one. The principle is the same as that described in Figure 1.4. The apparatus used is shown in Figure 1.5. The detector consists of a cylinder-shaped photographic film. The monoenergetic radiation from the X-ray tube passes through a narrow vertical slit and then the specimen, a thin powder compound, placed at the centre of the cylinder.

The diffracted radiation with the highest intensity has the angle of scattering 2θ (Figure 1.7). The diffracted radiation has a cone-shaped form. The radiation is registered on the photographic film as four slightly curved lines. These lines are characteristic of the solid material and will appear symmetrically around the direct beam when the film is developed. Figure 1.6b shows the appearence of the diffraction pattern of zinc.

The positions of the lines are measured. The pattern on the film is analysed and the various distances between the atomic planes in the crystals are calculated. In this way, the structure of the solid material can be derived.

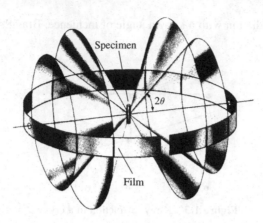

Figure 1.5 X-ray examination according to the Debye–Sherrer method. Reproduced with permission from B. D. Cullity, *Elements of X-Ray Diffraction*, © Addison-Wesley Publishing Company, Inc. (now under Pearson Education).

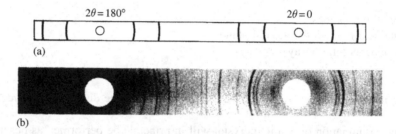

Figure 1.6 (a) The appearence of a Debye–Sherrer pattern. (b) Debye–Sherrer pattern of zinc. Reproduced with permission from B. D. Cullity, *Elements of X-Ray Diffraction*, © Addison-Wesley Publishing Company, Inc. (now under Pearson Education).

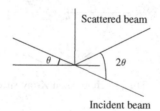

Figure 1.7 The angle 2θ of scattering is the angle between the incident and scattered beams.

1.2.2 X-ray Examination and Structures of Metal Melts

By replacing the powder compound by a metal melt at constant temperature, the corresponding Debye–Sherrer measurements can be made on a liquid metal.

The results deviate very much from those obtained for crystalline powders. Instead of a few sharp, well-defined lines one obtains several wide, unsharp maxima and minima. The resulting patterns can be analysed and fairly detailed information on the structures of the melts can be obtained. This is illustrated below by two examples.

Structures of Pure Metal Melts

The X-ray pattern of liquid gold at 1100 °C is shown in Figure 1.8. The intensity of the diffracted radiation is plotted versus $\sin \theta / \lambda$, where θ is *half* the scattering angle and λ is the X-ray wavelength. In Figure 1.8, the main peak is found to the left and a series of subsidary peaks to the right.

The short vertical lines in Figure 1.8 show the corresponding X-ray pattern of crystalline gold. The intensities of the diffracted waves are indicated by the heights of the lines. There is a clear correspondence between the positions along the horizontal axis of the main peaks of liquid gold and the main lines of crystalline gold.

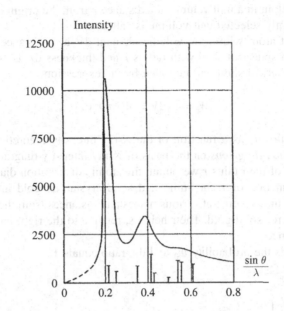

Figure 1.8 The intensity of scattered X-radiation as a function of sin θ/λ. The curve originates from scattering in liquid gold at 1100 °C. The vertical lines originate from scattering in crystalline gold. Reproduced with permission from N. S. Gingrich, The diffraction of X-rays by liquid elements, *Rev. Mod. Phys.* **15**, 90–100. © 1943 The American Physical Society.

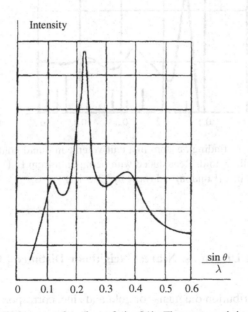

Figure 1.9 The intensity of scattered X-radiation as a function of sin θ/λ. The curve originates from scattering in liquid zinc at 460 °C. Reproduced with permission from N. S. Gingrich, The diffraction of X-rays by liquid elements, *Rev. Mod. Phys.* **15**, 90–100. © 1943 The American Physical Society.

An X-ray intensity spectrum such as those in Figures 1.8 and 1.9 supply the necessary basis to obtain information concerning the structure of a liquid or solid metal. However, to interpret an intensity diagram and obtain information from it, such as type of structure and atomic distances, it is necessary to transform the X-ray spectrum into another type of diagram. The background of this is given in next section.

From X-ray Plots to Atomic Distribution Diagrams

We consider the distribution of atoms around an arbitrary atom anywhere in the liquid and choose this atom as the origin.

The probability of finding another atom in a unit volume at a distance r from the origin is called w_r. The average probability of finding another atom in any randomly selected unit volume is called w_0.

The probability of finding the next atom within the volume element dV at a distance r from the origin atom equals the product of dV and w_r. If we choose a spherical shell with radius r and thickness dr as volume element, the probability dW_r of finding another atom within the spherical shell will be given by the expression

$$dW_r = w_r dV = w_r 4\pi r^2 dr \tag{1.2}$$

It is possible to derive the probability w_r as a function of the X-ray intensity theoretically. When this function is known, it is possible to draw *atomic distribution diagrams* on the basis of X-ray intensity diagrams such as those in Figures 1.8 and 1.9. By plotting $4\pi r^2 w_r$ as a function of the radius r we obtain the atomic distribution diagram of the metal melt in question.

The ratio w_r/w_0 is plotted as a function of r for liquid and solid crystalline gold in Figure 1.10. It shows the relative probability of finding gold atoms in a unit volume at various places and distances from the origin. In analogy with Figure 1.8, the vertical lines represent the values for solid gold. Their heights, relative to the right-hand scale, give the number of atoms per unit volume at the indicated distance.

At infinity, the probability w_r equals the probability w_0 and the ratio equals 1.

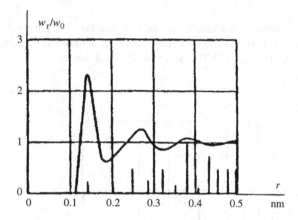

Figure 1.10 The relative probability w_r/w_0 of finding an atom in a unit volume in liquid gold at a distance r from the origin is a function of r. The vertical peaks correspond to crystalline gold. Reproduced with permission from F. D. Richardson, *Physical Chemistry of Melts in Metallurgy*, Vol. 1. © 1974 Academic Press Inc. (London) Ltd (now by Elsevier).

Interpretation of Atomic Distribution Diagrams. Nearest Neighbour Distances, Coordination Shells and Coordination Numbers

Figures 1.11 and 1.12 are the atomic distribution diagrams for gold and zinc, corresponding to Figures 1.8 and 1.9, respectively. Each curve can be regarded as a product of two functions of r, the probability w_r (Figure 1.10) and $4\pi r^2$, which is a parabola. Both figures show the same characteristics.

The first peak of the atomic distribution function gives the most probable distance between nearest neigbour atoms and is called the *nearest neighbour distance*. The peak appears close to the same position as the nearest neighbour distance in the solid crystalline metal.

The subsiding peaks at larger distances also show some correspondence with peaks in the solid metals but this decreases with increasing distance. A very likely explanation is that the melt still possesses some of its short-range order of the solid crystal but has lost its long-range regularity.

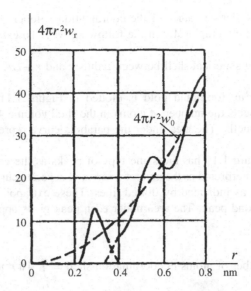

Figure 1.11 Atomic distribution diagrams of liquid gold at 1100 °C and of gold crystals. The probability of finding an atom within a spherical shell with the radius r is a function of r. The curve corresponds to liquid gold. Reproduced with permission from F. D. Richardson, *Physical Chemistry of Melts in Metallurgy*, Vol. 1. © 1974 Academic Press Inc. (London) Ltd (now by Elsevier).

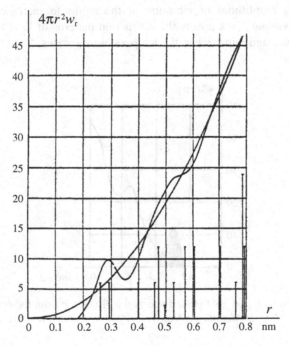

Figure 1.12 Atomic distribution diagrams of liquid zinc at 460 °C and of zinc crystals. The probability of finding an atom within a spherical shell with the radius r is a function of r. The curve corresponds to liquid zinc and the vertical peaks to crystalline zinc. Reproduced with permission from F. D. Richardson, *Physical Chemistry of Melts in Metallurgy*, Vol. 1. © 1974 Academic Press Inc. (London) Ltd (now by Elsevier).

The curves in Figures 1.11 and 1.12 drop steeply to the left and become zero for a characteristic small r value in both cases. This value corresponds to the nearest possible spacing between a pair of atoms owing to their mutual resistance to interpenetration. The characteristic distance is 0.22 nm for liquid gold.

The vertical peaks for solid gold and zinc are interpreted as *coordination shells* in the crystal structure. The maximum number of atoms, that can be included within a coordination shell is called the *coordination number* of the shell. These atomic concepts can be applied to both liquids and solids.

From diffraction data, it is possible to derive values of the coordination numbers for the first and sometimes also for the second coordination shell. This can be done graphically in the following way. To explain the method, we concentrate on the atomic distribution diagram for gold.

We consider the number of atoms in a spherical shell between radius r and $r+\Delta r$. It equals $w_r \times 4\pi r^2 \Delta r$ and is called *the radial distribution function.*

The radial distribution function $4\pi r^2 w_r$ for liquid gold is plotted in Figure 1.11 as a function of r together with the function $4\pi r^2 w_0$. The latter curve represents the number of atoms in the shell volume $4\pi r^2 \Delta r$ in a fictional space with equally distributed atoms, i.e. no coordination shells. The area under the parabola curve represents the total number of atoms within a sphere with radius r.

The real curve for liquid gold in Figure 1.11 has the same type of peaks as the corresponding curve in Figure 1.10. The liquid gold curve in Figure 1.11 can be interpreted as the sum of a series of separate curves, which arise from each coordination shell and start from the horizontal axis as indicated by dotted lines. These extrapolations can be done fairly accurately for the first and sometimes also for the second peak. The area under each peak gives approximately the number of atoms in the coordination shell:

$$\text{Number of atoms in coordination shell} = \int_{r_0}^{r_{min}} 4\pi \, r^2 w_r \mathrm{d}r \tag{1.3}$$

where r_0 and r_{min} are the limiting r values of the peak.

Estimations of the area under the first peak in Figure 1.13 indicate that there are about 8.5 atoms in the first coordination shell of liquid gold. The values for the second peak are too uncertain for any worthwhile calculations.

Increasing temperature causes the amplitudes of vibration of the atoms to increase and the fluctuations around their equilibrium positions become more violent. This affects the diffraction patterns of liquid metals and the peaks of the curves in Figures 1.11 and 1.12 become lower and wider when the temperature increases.

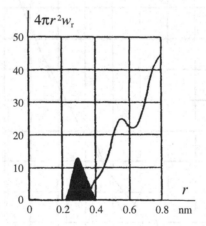

Figure 1.13 Radial distribution function of liquid gold plotted versus the distance r from the origin. Reproduced with permission from F. D. Richardson, *Physical Chemistry of Melts in Metallurgy*, Vol. 1. © 1974 Academic Press Inc. (London) Ltd (now by Elsevier).

Example 1.1

The atomic distribution diagram of zinc is given in Figure 1.12 on page 7. Use this diagram to answer the following questions:

(a) Does the smooth parabola have any physical significance?
(b) What is the smallest possible distance between two atoms in liquid zinc at 460 °C?
(c) What is the nearest neighbour distance in crystalline zinc?
(d) What is the reason for claiming that there is a short-range order in molten zinc?

(e) What is the coordination number in the first shell of solid crystalline zinc?

(f) What is the coordination number in the first shell of liquid zinc at 460 °C?

Solution:

(a) The smooth curve represents the number of atoms in a spherical shell at a distance r from the origin atom provided that the probability of finding an atom in unit volume is equal everywhere. This is not the case and therefore the answer to the question is 'no'.

(b) The distance is found at the intersection of the curve and the r axis. $r = 0.18$ nm.

(c) The nearest neighbour distance is the r value of the first 'crystalline' peak. $r = 0.26$ nm.

(d) The r values of the first two 'crystalline' peaks are very close to the most probable distance in molten zinc, the r value of the first top.

(e) The coordination number of the first shell is the sum of the number of the atoms in the first two sub-shells, i.e. $6 + 6 = 12$.

(f) The probability of finding an atom within a spherical shell with the radius r can be written as

$$dW_r = 4\pi r^2 w_r dr \tag{1'}$$

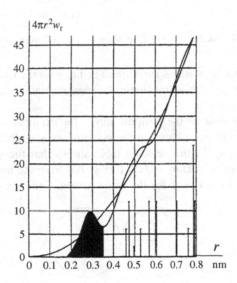

We obtain the number of atoms by integration of the function (1'):

$$W_r = \int_{r_0}^{r_{min}} 4\pi r^2 w_r dr$$

where $r_0 = 0.18$ nm and r_{min} equals the r value of the first minimum of the curve. It is 0.36 nm as shown in the figure.

As we have no analytical function of w_r, we have to integrate graphically. The dark area under the curve represents the demanded number of atoms. Its area is calculated as 11.

Answer:

(a) No.

(b) $r = 0.18$ nm.

(c) $r = 0.26$ nm.

(d) The r values of the first two 'crystalline' peaks are very close to the most probable distance in molten zinc.

(e) 12.

(f) 11.

1.3 The Hard Sphere Model of Atoms

The radius of an atom is not a well-defined quantity. An atom consists of a tiny core surrounded by an electron cloud of position-dependent density with no sharp 'surface' limit. However, in crystallography and metallurgy, the simple model of the atom as a hard sphere with a well-defined radius is in many cases most useful and satisfactory. Figure 1.10 on page 6 shows that this is a reasonable model.

1.3.1 Atomic Sizes

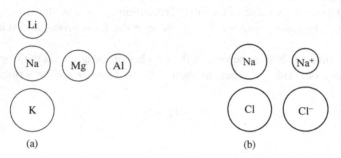

(a)   (b)

Figure 1.14   Sketch of 'radii' of some atoms and ions (not to scale).

The size of an atom in a liquid or a solid varies with its surroundings, i.e. with the external forces acting on the electron cloud. Hence the atom diameter varies with the type of binding (ionic, covalent, metallic, van der Waals) to neighbouring atoms, the coordination numbers and the state of ionization.

The atomic radii of metals within a *period* of the periodic table, for example from Na to Al, decrease with increasing nuclear charge because of the increasing attraction between the nucleus and the electrons.

A comparison of the radii within a *group* in the periodic table shows that the radii increase downwards because filled electron shells force the outermost electrons to occupy a shell further out from the nucleus. Supply or removal of one or more electrons, i.e. ionization, may change the atom radius for the same reason. Some examples are given in Figure 1.14.

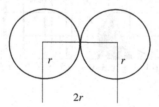

Figure 1.15   Distance between atoms.

The hard sphere model of atoms is used to define the normal diameter of a metal atom from a practical point of view. In a crystal the atoms are supposed to be spheres in contact with each other (Figure 1.15).

Atomic radius = half the distance of the closest approach of atomic centres in a crystal of the pure element

Figure 1.16   Crystal structure of solid zinc. Reproduced with permission from B. D. Cullity, *Elements of X-Ray Diffraction*, © Addison-Wesley Publishing Company, Inc. (now under Pearson Education).

Coordination numbers and coordination shells and distances between nearest neighbour atoms in melts and solids will be discussed in Section 1.5.2. Some data of the most common metals are given in Table 1.9 on page 21. It is reasonable to assume from Figure 1.16 that space is an important parameter.

However, it is impossible to illustrate crystal structures by drawing hard spheres. Instead, the atoms are often represented by points. This will be done to a great extent below. However, one should keep in mind that the hard sphere model is much closer to reality than the point model.

1.3.2 The Hard Sphere Model of Liquids

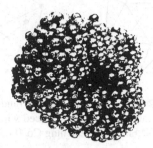

Figure 1.17 Random distribution of close-packed atoms as a model of a metal melt. Reproduced with permission from J. D. Bernal, in *Liquids: Structure, Properties, Solid Interactions*, edited by T. J. Hughel. © Elsevier Publishing Co 1965.

It is difficult to define a unit cell in a liquid in the same way as in a solid. The reason is that the atoms change their positions relative to each other incessantly. However, in the early 1960s Bernal made attempts to interpret the structure of a melt as a crowd of irregularly orientated atoms that does not provide enough space to include one more atom among the others (Figure 1.17). His approach was generally in good agreement with the results of X-ray examinations of metal melts.

By tedious measurements and calculations (statistical geometry), he and his co-workers found that the structure of a liquid can be successfully described by the following geometric model.

The atoms are represented by hard spheres situated at the corners of five different types of rigid polyhedra with the following properties:

1. The edges of the polyhedra must all be of approximately equal length.
2. It is impossible to introduce one more sphere into the centre of the liquid without stretching the distances between the spheres.

The five types of polyhedra, the so-called canonical holes or Bernal polyhedra, are

- tetrahedron
- octahedron
- dodecahedron
- trigonal prism
- archimedean antiprism.

The liquid is considered to consist of a mixture of the polyhedra in Figure 1.18. By performing a statistical treatment of the composition of the liquid, Bernal and co-workers calculated the relative numbers of polyhedra:

tetrahedra	73%
half-octahedra	20%
tetragonal dodecahedra	3%
trigonal prisms	3%
archimedean antiprisms	1%

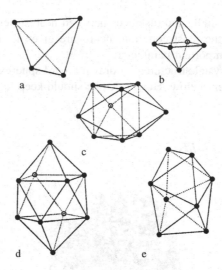

Figure 1.18 Unit blocks in Bernal's liquid model. The five canonical holes of random close packing are (a) tetrahedron, (b) octahedron, (c) dodecahedron, (d) trigonal prism and (e) archimedean antiprism. Reproduced with permission from J. D. Bernal, in *Liquids: Structure, Properties, Solid Interactions*, edited by T. J. Hughel. © Elsevier Publishing Co 1965.

Bernal's liquid model is obviously a hybrid between a close-packed random heap of atoms and atoms arranged in an ordered state. It is interesting to note that by rapid quenching of a melt, one finds solids consisting of the polyhedra in question. It is denoted an icosahedral phase and has fivefold symmetry.

The hard sphere model of liquids is an excellent illustration of the statement that short-range order exists in liquids.

1.4 Crystal Structure

1.4.1 Crystal Structure of Solids

Most solids, including all metals, have a crystalline structure. The strength and nature of the forces between the atoms of the crystal lattice and also the electron clouds around the nuclei determine the macroscopic properties of the solid, for example strength, elasticity and conduction of heat and charge. Hence it is very important to examine the crystalline structure of matter in order to understand the background of its behaviour.

A striking example of the influence of the forces between the atoms on the mechanical strength of the solid is a comparison between graphite and diamond, both of which consist of carbon (Figure 1.19).

The graphite structure consists of parallel layers of planar six-membered rings of carbon atoms. The carbon atoms within the rings are held together by strong covalent bonds. The interaction between the layers is weak and due to van der Waals bonds. This is the reason for the weakness of graphite towards shearing forces and its use as a lubricant.

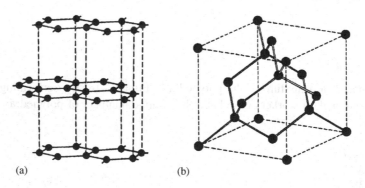

(a)   (b)

Figure 1.19 Crystalline structure of (a) graphite and (b) diamond. (a) Adapted with permission from W. Hume-Rothey, R. E. Smallman and C. W. Haworth, *The Structure of Metals and Alloys*. (Published by the Institute of Metals, 1969) now © W. S. Maney & Son Ltd. (b) Reproduced with permission from C. Kittel, *Introduction to Solid State Physics*, 6th edn, P. 19. © 1986 John Wiley & Sons, Inc.

Diamond is one of the hardest elements in nature. This mechanical strength is due to the non-planar covalent bonds between the carbon atoms.

In the following we will define some basic concepts used in crystallography and give a review of the different types of existing crystallographic structures. These features will often arise in the discussions in this and later chapters.

1.4.2 Types of Crystal Structures

The skeleton of a crystallographic structure is a *space lattice*. A space lattice is defined as *a collection of 'mathematical' points, arranged in such a way that each of the points is surrounded by precisely the same configuration*. The view will always be the same, independent of which point we chose as observation site. Each lattice point always lies at a centre of symmetry. Every lattice point on one side of the observation site always has a corresponding lattice point in an identical position but situated on the opposite side.

Crystals are build of *unit cells* or *primitive cells*. Each mathematical point in the lattice is replaced by an atom or group of atoms. The crystal is built of rows of such unit cells. Even if the lattice structure is basically simple, the crystal structure can be very complicated as the unit cell might consist of tens of thousands of atoms. This is the case in proteins and other organic structures.

The crystal structure is hence determined by two factors:

1. the lattice structure (lattice)
2. the configuration of the unit cell (basis).

Crystal structure = lattice + basis

As early as 1848, long before X-ray methods of crystallography were known, the French crystallographer A. Bravais stated that the atomic arrangements in all crystalline solids can be referred to 14 fundamental crystal classes, consisting of four types of space lattice in combination with seven systems of unit cells. These 14 crystal classes are listed below together with a short characteristic in each case. The 14 crystal classes are the only ones which can be packed to fill space, which is a necessary condition.

The name crystal comes from the Greek word *krystallos*, which means ice. The structure of ice is normally hexagonal (see below).

The names and characteristics of the seven systems of unit cells are given in Table 1.1 together with drawings of the different crystal types. In Table 1.2, the common lattice symbols are listed. Table 1.3 gives more detailed information on the 14 types of crystal classes in a condensed form.

The classification of the unit cells is based on *the degree of symmetry of the various crystal types*. The number of symmetry axes, symmetry planes and symmetry centres are used as a measure of the degree of symmetry.

Table 1.1 Bravais' 14 fundamental types of crystal lattices.

Type of unit cell	Type of space lattice			
	Simple	Base-centred	Body-centred	Face-centred
Cubic 3 axes at right-angles, all of equal length		Non-existent		
Hexagonal 2 equal axes at 120° angle, each at right-angles to the third axis of different length		Non-existent	Non-existent	Non-existent

(continued overleaf)

Table 1.1 (*continued*)

Type of unit cell	Type of space lattice			
	Simple	Base-centred	Body-centred	Face-centred
Tetragonal 3 axes at right-angles, 2 of them of equal length		Non-existent		Non-existent
Trigonal (rhombohedral) 3 equally inclined axes, not at right angles, all of equal length		Non-existent	Non-existent	Non-existent
Orthorombic 3 axes at right-angles, all of different lengths				
Monoclinic 3 axes, one pair not at right-angles, all of different lengths			Non-existent	Non-existent
Triclinic 3 axes, all at different non-right-angles, all of different length		Non-existent	Non-existent	Non-existent

Diagrams reproduced with permission from *Understanding Science*, No. 15. © 1966/1967 Sampson Low, Marston & Searle, London.

Table 1.2 Space lattice terms.

Term	Characteristics	Lattice symbol
Simple	Atoms in the corners of the unit cell only	S (simple) or P (primitive)
Body-centred	Atoms in the corners of the unit cell + an extra atom in the centre of the unit cell	BC or I (interior) (an extra atom inside)
Base-centred	Atoms in the corners of the unit cell + an extra atom at the centres of the base and top surfaces of the unit cell	C (extra atoms in the parallel C-faces)
Face-centred	Atoms in the corners of the unit cell + an extra atom at the centre of each surface of the unit cell	FC

Table 1.3 Crystal systems and Bravais lattices.

Symbol	Axial lengths and angles	Bravais lattice	Lattice symbol
Cubic	Three equal axes at right-angles $a = b = c \quad \alpha = \beta = \gamma = 90°$	Simple Body-centred Face-centred	S BC FC
Hexagonal	Two equal coplanar axes at 120°, third axis at right-angles $a = b \neq c \quad \alpha = \beta = 90° \; \gamma = 120°$	Simple	S
Tetragonal	Three equal axes at right-angles, two equal $a = b \neq c \quad \alpha = \beta = \gamma = 90°$	Simple Body-centred	S BC
Rhombohedral or trigonal	Three equal axes, equally inclined $a = b = c \quad \alpha = \beta = \gamma \neq 90°$	Simple	S
Orthorhombic	Three unequal axes at right-angles $a \neq b \neq c \quad \alpha = \beta = \gamma = 90°$	Simple Body-centred Base-centred Face-centred	S BC C FC
Monoclinic	Three unequal axes, one pair not at right-angles $a \neq b \neq c \quad \alpha = \gamma = 90° \neq \beta$	Simple Base-centred	S C
Triclinic	Three unequal axes unequally inclined, none at right-angles $a \neq b \neq c \quad \alpha \neq \beta \neq \gamma \neq 90°$	Simple	S

The lattice symbols have been explained in Table 1.2. The number of atoms per unit cell is given by

$$N_{\text{unit cell}} = N_{\text{interior}} + \frac{N_{\text{face}}}{2} + \frac{N_{\text{corner}}}{8} \tag{1.4}$$

A *primitive* cell is a unit cell which contains one atom per unit cell. If the unit cell contains more than one atom, other symbols are used.

1.4.3 Lattice Directions and Planes

The British crystallographer Miller introduced a nomenclature for directions and planes in crystals, which is generally accepted. We will briefly explain the significance of his system, as it will be used in some of the following chapters.

Lattice Directions

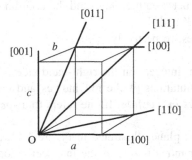

Figure 1.20 Some different crystal directions through the origin in a simple cubic crystal. The units lengths on the axes are a, b and c, respectively.

A *direction* in a lattice, is characterized by three integers, u, v and w. The directions are generally written <uvw>, which means that all permutations of the uvw indices and combination of signs occur. If one refers to a *specific line with the given direction*, square brackets are used, [uvw].

A line with the given direction is drawn through the origin. The integers u, v and w represent the coordinates of the lattice points along the line. It is customary to choose the smallest possible integers. Negative numbers are indicated by a bar above the figure instead of a minus sign; u, v and w are called *the indices of the direction*.

Some examples of special lines are given in Figure 1.20.

Lattice Planes. Miller Indices

The orientation of the normal to a crystal plane in a crystal is completely defined by the three intercepts between the plane and the *x*, *y* and *z* axes in a coordinate system. However, it is not convenient to use these intercepts directly to describe the direction of the normal of the plane, as infinity will be involved if a plane is parallel to any of the coordinate axes. Instead, the Miller suggested the following definition of a crystal plane (Figure 1.21):

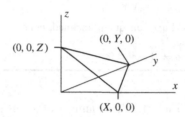

Figure 1.21 The intersections between the coordinate axes and the plane are *X*, *Y* and *Z*.

$$(hkl) = \left(\frac{D}{X}\frac{D}{Y}\frac{D}{Z}\right) \tag{1.5a}$$

where

$$X = \frac{D}{h}; \quad Y = \frac{D}{k}; \quad Z = \frac{D}{l} \tag{1.5b}$$

and

hkl = *integers* called the *Miller indices*
X, *Y*, *Z* = coordinates of the intersections between the plane and the coordinate axes
D = smallest common denominator of *X*, *Y* and *Z*
a, *b*, *c* = unit lengths on the *x*, *y* and *z* axes, respectively.

The constant *D* makes it possible to obtain integers on the right-hand side of Equation (1.5a). The *Miller indices*, written within curly brackets {}, mean that all permutations of the hkl indices and combinations of signs occur. They define the *directions of the normals to the type of planes in question*. If one refers to a *specific plane with the given direction*, ordinary parentheses are used, (hkl).

If *X* = *Y* = *Z* = 1, the Miller indices of the plane, illustrated in Figure 1.21, are (111). If the plane is parallel with one of the coordinate axes, the corresponding Miller index is zero as the intersection occurs at infinity.

Figure 1.22 gives some examples. A bar over an index indicates a negative value of the intercept.

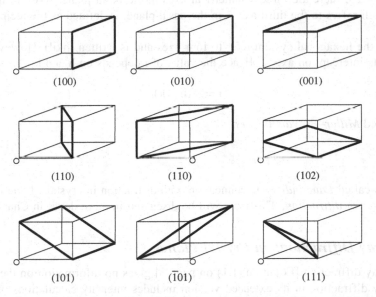

Figure 1.22 Miller indices of some lattice planes. The origin of the coordinate system is marked by a small circle in the figures.

Example 1.2

The units on the crystallographic axes in a crystal are a, b and c. Calculate the Miller indices of a plane with the intercepts $2a$, $3b$ and $1.5c$ at the coordinate axes.

Solution:

We obtain the plane intercepts by dividing the intercepts with the length units of the respectively axes.
   The plane intercepts are 2, 3 and 1.5.
   The reciprocals of the intercepts are 1/2, 1/3 and 2/3.
   If these numbers are multiplied by 6, the smallest common denominator, we obtain the Miller indices.

Answer:

The Miller indices are (324).

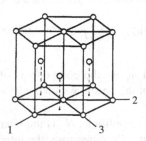

Figure 1.23 Hexagonal close-packed structure. Reproduced with permission from B. D. Cullity, *Elements of X-Ray Diffraction*, © Addison-Wesley Publishing Company, Inc. (now under Pearson Education).

Sometimes one finds sets of *four* Miller indices instead of three in the literature. They are used in the *hexagonal system*.

As can be seen in Figure 1.23, there are three symmetrical axes in the basal plane. Axes 1 and 2 are the axes of the unit cell. The fourth index i, which refers to the third axis 3 in the basal plane, is defined as the reciprocal of the intercept on the third axis.

The indices of a plane in the hexagonal system refer to four axes and is written (hkil). However, the intercepts of a plane on axes 1 and 2 determine its intercept on axis 3. Hence the value of i depends on h and k:

$$i = -(h+k) \tag{1.6}$$

This type of indices are called *Miller–Bravais indices*.

Laue Indices

In Chapter 3, we will use so-called *Laue indices* in connection with diffraction in crystals. Laue indices are defined as *Miller indices multiplied by the order of diffraction*. This topic will be discussed in more detail in Chapter 3 (page 133).

1.4.4 Intensities of X-ray Diffractions in Crystal Planes

The Bragg condition of X-ray diffraction [Equation (1.1) on page 3] gives no information on the intensities of the diffracted X-rays. The theory of X-ray diffraction in its extended version includes intensity calculations of the diffracted radiation. A detailed treatment is beyond the scope of this book, but a brief outline of the extended theory is given below.

It can be shown that in the special case of a *cubic* crystal structure, Bragg's law is replaced by the simple condition

$$\sin^2 \theta = \frac{\lambda^2}{4a^2} \left(h^2 + k^2 + l^2 \right) \tag{1.7}$$

This generalized form of Bragg's law for a cubic structure will be used in the intensity discussion below.

Intensity of the Diffracted Radiation

If the incident X-ray beam has an intensity I_0, the intensity I of the diffracted beam can be written as

$$I = constant \times I_0 S\,(\mathrm{hkl})^2 \left(N^3\right)^2 n_{\mathrm{hkl}} \tag{1.8}$$

where
  $S(\mathrm{hkl}) =$ geometric structure factor
  $I, I_0 =$ intensities
  $N =$ number of unit cells
  $n_{\mathrm{hkl}} =$ occurrence number.

The $S(\mathrm{hkl})$ factor is a measure of the resulting amplitude of the diffracted X-ray beam. It can be written as

$$S(\mathrm{hkl}) = \sum_{\mathrm{j}} f_{\mathrm{j}}\, e^{-2\pi i \left(u_{\mathrm{j}} h + v_{\mathrm{j}} k + w_{\mathrm{j}} l \right)} \tag{1.9}$$

where (hkl) are the Miller indices and u_{j}, v_{j} and w_{j} are the components of a position vector r_{j} successively describing the different basic positions of atoms in the unit cell of the crystal which contribute to the sum Σ; f_{j} is called the *atomic scattering factor* and $i = \sqrt{-1}$.

The intensity is proportional to the square of the amplitude. N^3 is the *number of unit cells*. The amplitude is the vector sum of the single coherent amplitudes from all the unit cells. Therefore, the intensity is proportional to $(N^3)^2$.

The *occurrence number* n_{hkl} has the following significance. Suppose that a certain angle θ, measured for diffraction in a polycrystalline material with a cubic structure, corresponds to the condition $(h^2 + k^2 + l^2 = 4\pi a^2 \sin^2 \theta / \lambda^2)$ [Equation (1.7)]. If $h^2 + k^2 + l^2 = 1$ it is fulfilled by six sets of crystal planes $[(100), (010), (001), (\bar{1}00), (0\bar{1}0)$ and $(00\bar{1})]$ or shorter $\{100\}$. The resulting diffraction pattern is the sum of these six sets of independent crystal reflections. Hence the intensity increases by an occurence factor $n_{hkl} = 6$ compared with reflection in a single set of planes, for example (010). Similarly, all permutations of positive and negative integers hkl in combination with the condition $h^2 + k^2 + l^2 = 2$ give the occurrence number $n_{hkl} = 12$.

Intensities of Diffraction Lines

The geometric structure factor $S(hkl)$ can be calculated by applying Equations (1.8) and (1.9) to various cubic crystal structures, for example FCC and BCC. The results can be summarized in the simple conditions given in Tables 1.4 and 1.5.

Table 1.4 FCC structure.

$S(hkl)$	Condition
$4f$	hkl = odd integers (unmixed indices)
$4f$	hkl = even integers (unmixed indices)
0	For all other hkl combinations (mixed indices)

The unit cell in an FCC structure contains four atoms (page 22). The basic positions have the coordinates (0, 0, 0), (0, $1/2$, $1/2$), ($1/2$, 0, $1/2$) and ($1/2$, $1/2$, 0). These values are inserted into Equation (1.9), which gives the values $4f$ or 0, depending on the values of hkl.

For example, the plane sets {200} and {111} give X-ray reflections, whereas the plane sets {210} do not reflect the incident X-radiation.

Table 1.5 BCC structure.

$S(hkl)$	Condition
$2f$	$h + k + l$ = even integers
0	$h + k + l$ = odd integers

The unit cell in a BCC structure contains two atoms (page 22). The basic positions have the coordinates (0, 0, 0) and ($1/2$, $1/2$, $1/2$). These values inserted into Equation (1.9) give the conditions in Table 1.5.

Hence the plane sets {110} give X-ray reflections, but not the plane sets {111}.

1.5 Crystal Structures of Solid Metals

1.5.1 Coordination Numbers

As indicated in Section 1.2.2 (page 6), diffraction data can be used to obtain information about the structure of solid materials, distances in the crystal lattice and number of atoms in the coordination shells and nearest neighbour distances. Measurements and calculations have been performed for many pure metals in the same way as for gold and zinc (page 7).

A collection of values for the most probable nearest neighbour distances and the coordination numbers for some common liquid and solid metals is given in Table 1.9 on page 21.

Many metals have a *close-packed structure*, which means that the number of atoms per unit cell is higher than 1, the number of atoms in a simple crystal structure. The most common close packed structures are given in Table 1.6.

Table 1.6 The most common close-packed metal structures.

Name	Structure	No. of atoms per unit cell
BCC	Body-centred cubic	2
FCC	Face-centred cubic	4
HCP	Hexagonal close-packed	2

1.5.2 Nearest Neighbour Distances of Atoms

The nearest neighbour distances d of the atoms in a lattice can be calculated from the lattice parameters. This has been done for the three most common metal structures in Figure 1.24.

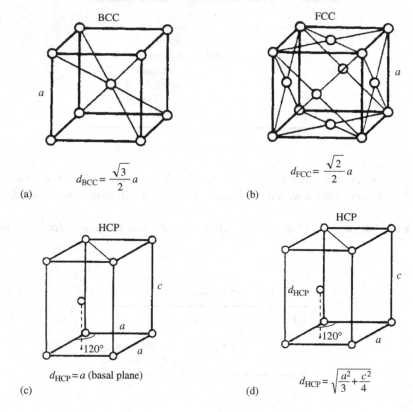

$$d_{BCC} = \frac{\sqrt{3}}{2} a$$

(a)

$$d_{FCC} = \frac{\sqrt{2}}{2} a$$

(b)

$$d_{HCP} = a \text{ (basal plane)}$$

(c)

$$d_{HCP} = \sqrt{\frac{a^2}{3} + \frac{c^2}{4}}$$

(d)

Figure 1.24 Relationships between nearest neighbour distances and lattice parameters for BCC, FCC and HCP structures. Reproduced with permission from B. D. Cullity, *Elements of X-Ray Diffraction*, © Addison-Wesley Publishing Company, Inc. (now under Pearson Education).

If the atoms are regarded as hard spheres in contact with each other, it is possible to derive the radii of the atoms from the nearest neighbour distances.

It has been pointed out in Section 1.3.1 that the radii of the atoms depend on several factors. Tables 1.7 and 1.8 give some examples of the dependence of atomic diameter on ionization and coordination number.

Table 1.7 Change of radius due to ionization.

Atom/ion	Diameter of atom/ion (nm)
Fe	0.248
Fe^{2+}	0.166
Fe^{3+}	0.134

Table 1.8 Change of coordination number due to size contraction.

Change of coordination number	Size contraction of atom diameter (%)
12 → 8	3
12 → 6	4
12 → 4	12

A change of coordination number means a change of crystal structure. The lattice constant changes when the crystal changes from one structure to another and is a function of the packing of the atoms.

Table 1.9 Interatomic distances and coordination numbers of some common metals at their melting points. Reproduced with permission from F. D. Richardson, *Physical Chemistry of Melts in Metallurgy*, Vol. 1. © 1974 Academic Press Inc. (London) Ltd (now by Elsevier).

Metal	Solid properties		Liquid properties	
	Coordination number	Nearest neighbour distance (nm)	Coordination number	Nearest neighbour distance (nm)
Na	8	0.372	9.5	0.370
K	8	0.452	9.5	0.470
Mg	12	0.320	10.0	0.335
Al	12	0.286	10.6	0.296
Ge	4	0.245	8.0	0.270
Sn	4 or 2[a]	0.302 or 0.318[a]	8.5	0.327
Pb	12	0.350	8.0	0.340
Sb	3 or 3[a]	0.291	6.1	0.312
Bi	3 or 3[a]	0.309 or 0.353[a]	7–8	0.332
Cu	12	0.256	11.5	0.257
Ag	12	0.289	10.0	0.286
Au	12	0.288	8.5	0.285
Hg	6 or 6[a]	0.301 or 0.347[a]	10.0	0.307

[a]Different directions. The unit cell is asymmetric.

Table 1.9 and other available data show that Na, K, Cu, Ag and Au have similar nearest neighbour distances and first coordination numbers in the liquid and solid states. They all have close-packed structures (HCP 12, FCC 12, BCC 8) and high coordination numbers in the solid state.

Another related group of metals is Ge, Sn and Bi, which have complicated crystal structures and low coordination numbers in the solid state. This type of metals changes considerably during melting. Both their first coordination numbers and their nearest neighbour distances increase substantially.

1.5.3 BCC, FCC and HCP Structures in Metals

BCC Structure

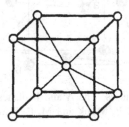

Figure 1.25 Unit cell of the BCC structure. Reproduced with permission from B. D. Cullity, *Elements of X-Ray Diffraction*, © Addison-Wesley Publishing Company, Inc. (now under Pearson Education).

This structure is a simple cubic lattice with an additional atom in the centre of the cube (Figure 1.25).

$$\text{Number of atoms per unit cell} = (8 \times 1/8) + 1 = 2$$

$$\text{Number of nearest neighbours} = 8$$

Examples of metals with a BCC structure are **α-Fe, δ-Fe, β-Ti, Li** and **V**.

FCC Structure

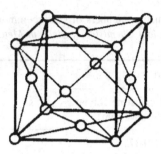

Figure 1.26 Unit cell of the FCC structure. Reproduced with permission from B. D. Cullity, *Elements of X-Ray Diffraction*, © Addison-Wesley Publishing Company, Inc. (now under Pearson Education).

This structure is a simple cubic lattice with an additional atom in the centre of each of the six faces of the cube (Figure 1.26).

$$\text{Number of atoms per unit cell} = (8 \times 1/8) + (6 \times 1/2) = 4$$

$$\text{Number of nearest neighbours} = 12$$

Examples of metals with an FCC structure are **γ-Fe, Al, Cu, Pb** and **β-Co**.

In order to facilitate the study of the stacking sequence of atomic planes in an FCC structure, it has to be drawn in a different way than in Figure 1.26. A better representation is shown in Figure 1.28a and b, where the crystal is rotated in such a way that the parallel (111) planes in Figure 1.27 are horizontal and the perpendicular principal diagonal of the cube has a vertical direction. Every atomic plane contains numerous atoms. For clarity, only a few of them are drawn in each plane.

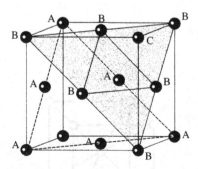

Figure 1.27 FCC structure with some {111} atomic planes. The A and B planes are, of course, parallel. Only one atom of the parallel C plane is shown. Reproduced with permission from J. R. Mook and M. E. Hall, *Solid State Physics*, 2nd edn. © 1991 John Wiley & Sons, Ltd.

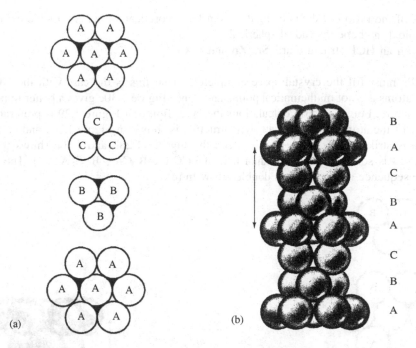

Figure 1.28   (a) Stacking sequence of some parallel atomic (111) planes in an FCC structure. The planes are seen from above. The interstitial sites where the following upper atoms rest are marked with black.
(b) Stacking sequence of the parallel atomic (111) planes in the FCC structure in (a) seen from the side. The atomic positions of the A, B and C layers are all different and repeat periodically. The stacking sequence in (a) corresponds to the part of (b) marked with a vertical double arrow in (b). The sequence in (a) is repeated about twice in (b). Reproduced with permission from J. S. Blakemore, *Solid State Physics*, 1st edn. © 1969 W. B. Saunders Company (now Elsevier).

HCP Structure

The BCC and FCC structures are simple in the sense that atoms are directly placed in the corners of the Bravais point lattice. The *hexagonal close-packed structure* is more complicated as every lattice corner is occupied by an atom associated with an additional atom outside the lattice corners like a dumb-bell. All dumb-bells are parallel in space. The structure can be regarded as two lattices connected rigidly with one another. The unit cell is shown in Figure 1.29a and b.

$$\text{Number of atoms per unit cell} = (8 \times 1/8) + 1 = 2$$

$$\text{Number of nearest neighbours} = 12$$

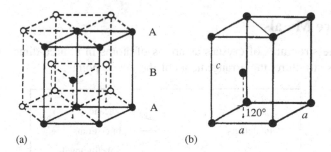

Figure 1.29   (a) HCP structure and its unit cell. Only one of the three B atoms is enclosed within the unit cell.
(b) Single unit cell of the HCP structure. Reproduced with permission from B. D. Cullity, *Elements of X-Ray Diffraction*, © Addison-Wesley Publishing Company Inc. (now under Pearson Education).

The ratio of the distances c/a (Figure 1.29b) in an ideal HCP structure formed of hard spheres in contact with each other is

$$\frac{c}{a} = \sqrt{\frac{8}{3}} = 1.633 \tag{1.10}$$

The experimental values of the ratio c/a deviate slightly from the theoretical value. A possible explanation could be that the atoms are slightly ellipsoidal in shape instead of spherical.

Examples of metals with an HCP structure are. **Sn**, **Zn** and **α-Co**.

The arrays of unit cells must fill the crystal space completely. That this is the case with the HCP structure is shown in Figure 1.30a and c . The atoms are not mathematical points and hence Figure 1.30c gives a better impression than Figure 1.30b of the 'close-packed' structure. For geometric calculations the latter figure or Figure 1.29 is preferable, however.

The stacking sequence of the atomic planes in an HCP structure is shown in Figure 1.30a and c. The resemblance between Figures 1.30a and 1.28a is striking. A comparison between the Figures 1.28b and 1.30c shows that the stacking sequence of HCP (A–B–A–B–A . . .) is somewhat simpler than that of FCC (A–B–C–A–B–C–A . . .). The stacking sequence in (a) corresponds to twice the sequence marked with a double arrow in (c).

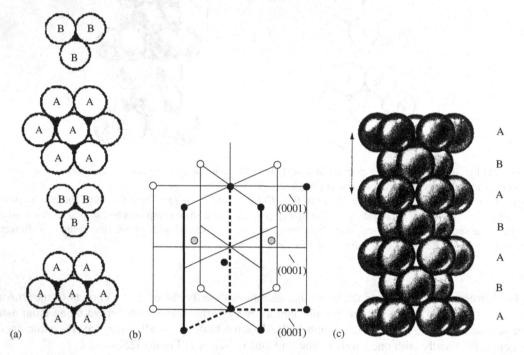

Figure 1.30 (a) Each plane in an HCP structure is shown from above. A and B planes alternate.
(b) Stacking sequence of the atomic planes in an HCP structure. The black atoms form a unit cell. (Figure 1.30 a and b, reproduced with permission from B. D. Cullity, *Elements of X-Ray Diffraction*, © Addison-Wesley Publishing Company, Inc. (now under Pearson Education).)
(c) Stacking sequence of the atomic planes in an HCP structure shown from the side.

1.6 Crystal Defects in Pure Metals

In Section 1.4.2 we discussed the structures of crystals as arrays of atoms in a three-dimensional regular pattern. The *shape* of a crystal is not the same as its structure, the arrangements of the lattice points.

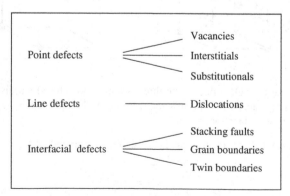

Real crystals are never perfect. They always contain crystal defects, which give a distorted lattice. Even if the fraction is not more than one atom out of place in 10 000, this is enough to have a strong influence on many important properties of the material, for example mechanical strength, diffusion and crystal growth.

The imperfections, i.e. all sorts of deviations from the regular crystal pattern, can be divided into three main classes:
The most important subgroups are listed in the box and have the following features.

- The *point defects* are local. They concern just a few atoms close to the defect.
- The *line defects* are more long-range. Even atoms far from the centre of the defect are influenced.
- The *interfacial defects* are three-dimensional phenomena. They will be discussed in a later chapter.

1.6.1 Vacancies and Other Point Defects

- A vacancy = a missing atom in a lattice site.

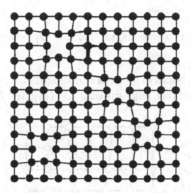

Figure 1.31 Vacancies in a crystal lattice. Reproduced with permission from R. Cotterill, *The Cambridge Guide to the Material World*, 1st edn. © 1985 Cambridge University Press.

Vacancies are the most important point defects in metals (Figure 1.31). They will be discussed further in later chapters in connection with density, diffusion phenomena and thermodynamic relationships.

Impurity atoms also occur in crystal lattices and cause distortion. Impurity atoms, added on purpose, give doped materials (semiconductors). Impurity atoms occur either as *interstitials* or *substitutionals*:

- an interstitial atom = an atom between the ordinary lattice sites
- a substitutional atom = an atom instead of a lattice atom in an ordinary site.

Interstitials and substitutionals (Figure 1.32a and b) will be treated in Section 1.7.3 on page 33.

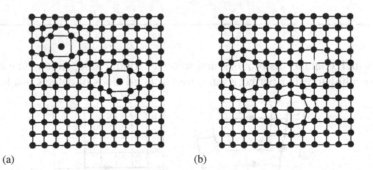

(a) (b)

Figure 1.32 (a) Interstitial atoms in a crystal lattice. (b) Substitutional atoms and a vacancy in a crystal lattice. Reproduced with permission from R. Cotterill, *The Cambridge Guide to the Material World*, 1st edn. © 1985 Cambridge University Press.

Note the deformation of the equilibrium positions of the nearest neighbours in all three cases.

All point defects can move within the crystal lattice. In metals near the melting point, the vacancy frequence is about 1:1000 and the jumping rate for an atom is of the order 10^9 per second.

1.6.2 Line Defects

Dislocations

Dislocations are *line defects* in the crystal lattice. To illustrate them we use a model in which the atoms are represented by hard spheres and the bonds between them by elastic springs (Figure 1.33b). This model is used on a simple cubic lattice.

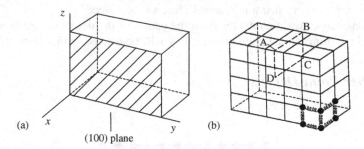

Figure 1.33 Model of a simple cubic lattice in two versions, (a) and (b). Reproduced with permission from D. Hull, *Introduction to Dislocations*. © 1965 Pergamon Press Ltd.

A virtual cut in the plane ABCD inhibits the bonds in the area concerned. This approach is used to describe the basic geometry of dislocations. There are two types of dislocations, *edge* dislocations (Figures 1.34) and *screw* dislocations (Figure 1.35).

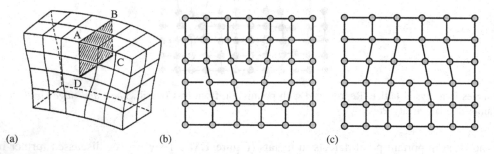

Figure 1.34 (a) Edge dislocations are formed by inserting or removal of an extra atom plane. Reproduced with permission from D. Hull, *Introduction to Dislocations*. © 1965 Pergamon Press Ltd (now with Elsevier).
(b) Positive edge dislocation. A new row appears in the crystal lattice. Reproduced with permission from J. Weertman and J. R. Weertman, *Elementary Dislocation Theory*. © Oxford University Press.
(c) Negative edge dislocation. A row disappears in the crystal lattice. Reproduced with permission from J. Weertman and J. R. Weertman, *Elementary Dislocation Theory*. © Oxford University Press.

A useful tool to characterize dislocations is a concept called *Burgers vector*, which is described in the box below.

Burgers Vector

Consider a defect crystal, which contains dislocations, and select an arbitrary loop around one of them, for example an edge dislocation (a). The corresponding loop in a perfect crystal, free from all kinds of defects inside the loop, is shown in (b).

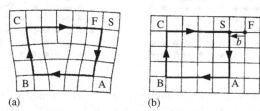

Figure Box 1.a and b (a) Closed loop: S(start)ABCF(finish). (b) Closed loop: SABCF. Reproduced with permission from D. Hull, *Introduction to Dislocations*. © 1965 Pergamon Press Ltd (now with Elsevier).

The two loops differ because of the dislocation. Burgers vector is defined as the vector b where $|b| = FS$ (figure(b)).

Burgers vector lies in the same plane as the loop in case of edge dislocations. A *positive edge dislocation* (Figure 1.34b and figure (b) in the box) arises when an extra layer of atoms is inserted in the slot caused by the virtual cut ABCD. In this case, Burgers vector is positive. A dislocation is *negative* when a row disappears (Figure 1.34c), which corresponds to a negative Burgers vector. These types of faults occur owing to stacking faults arising during condensation of vacancies.

Burgers vector is perpendicular to the (perfect) crystal plane in case of screw dislocations. A *screw dislocation* is produced by adapting a shear stress along the line AB (Figure 1.35) in the absence of atomic bonds within the area ABCD. The shear stress produces a permanent displacement of the crystal on one side of AB relative to the other (Figure 1.35). The displacement is one lattice spacing. There are two possible directions of Burgers vector in case of screw dislocations.

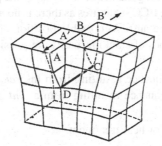

Figure 1.35   Screw dislocation. Reproduced with permission from D. Hull, *Introduction to Dislocations*. © 1965 Pergamon Press Ltd (now with Elsevier).

This final state can also be achieved if we use another approach. Let a vector AD, with the end D fixed along the line DC (Figure 1.35), rotate in the vertical (100) plane in the clockwise direction, seen from D towards C, around the axis DC and simultaneously move forward one lattice spacing, AA' or BB', per revolution around DC. The point A then describes a right-hand helix (Figure 1.36a), which corresponds to a positive Burgers vector. The point B moves in the same way and the dislocation is called a *right-hand screw dislocation*.

If the helix, on the other hand, retreats one lattice spacing per clockwise revolution or advances one lattice spacing when rotating one counter-clockwise revolution, the dislocation is called a *left-hand screw dislocation* (Figure 1.36b), which corresponds to a negative Burgers vector.

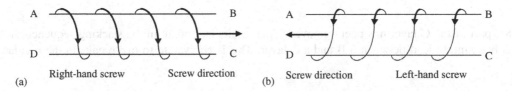

Figure 1.36   (a) Right-hand screw. (b) Left-hand screw.

A screw dislocation converts a pile of crystal planes into a single continuous helix. When the helix intersects the surface a step is formed (Figure 1.37a), which cannot be eliminated by adding further atoms. The crystal grows as a never-ending spiral (Figure 1.37b).

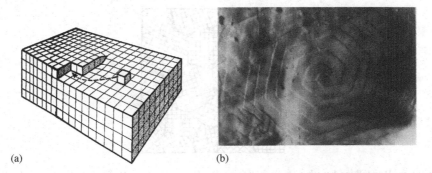

Figure 1.37   (a) Evidence of a screw dislocation on the crystal surface. (b) Dislocation spiral on the surface of a polypropylene crystal. Reproduced with permission from R. Cotterill, *The Cambridge Guide to the Material World*, 1st edn. © 1985 Cambridge University Press.

1.6.3 Interfacial Defects

Stacking Faults

A crystal lattice can be regarded as large number of crystal planes piled on one another in a regular way. The normal stacking sequences in FCC and HCP structures have been given in Figure 1.28b on page 23 and Figure 1.30c on page 24, respectively.

The normal stacking sequences of an HCP structure can be denoted ABAB and so on, where the letters are simply a short designation of the atomic configurations in the respective crystal planes. Similarly, the normal stacking sequences in an FCC structure can be described as ABCABC and so on.

A stacking fault is an interfacial defect where the regular sequence order of the planes is broken. Stacking faults are not expected in crystals with ABAB sequences in, for example, BCC structures as there is no alternative for an A layer resting on a B layer.

For structures with sequences ABCABC, there are two possible positions for an A layer. It can rest on either a B or a C layer. In crystals with an FCC structure, two types of stacking faults are possible, intrinsic and extrinsic stacking faults:

- An *intrinsic* stacking fault is the change in sequence resulting from the *removal* of a layer.
- An *extrinsic* stacking fault is the change in sequence resulting from an *introduction* of an extra layer.

Stacking faults can arise during solidification or heat treatment.

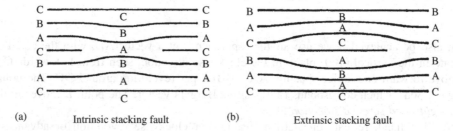

| (a) | Intrinsic stacking fault | (b) | Extrinsic stacking fault |

Figure 1.38 Stacking faults in the (111) plane of an FCC lattice: (a) intrinsic and (b) extrinsic stacking faults. Reproduced with permission from D. Hull, *Introduction to Dislocations*. © 1965 Pergamon Press Ltd (now with Elsevier).

In Figure 1.38a, part of the C layer has been removed. This results in a break in the stacking sequence. In Figure 1.38b, an extra A layer has been introduced between a B and a C layer. This is equivalent to *two* breaks in the regular sequence order.

Grain Boundaries

Crystalline metals consist of aggregates of small crystals with mutually different orientations. The interfaces between these grains are called *grain boundaries* (Figure 1.39). They can be observed in an optical microscope using polarized light or by etching the sample.

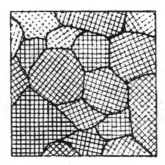

Figure 1.39 Grain boundaries. Reproduced with permission from W. Hume-Rothey, R. E. Smallman and C. W. Haworth, *The Structure of Metals and Alloys*. (Published by the Institute of Metals, 1969) now © W. S. Maney & Son Ltd.

A grain boundary is equivalent to a dense array of static dislocations. It works like a high barrier to moving dislocations. The grain boundaries promote hardening. The mechanical strength of a crystal is inversely proportional to the average grain diameter. The smaller the grains are, the better will be the mechanical properties of the material.

Twinned Crystals

Twinned crystals is a common term for crystals which consist of two parts showing mutual symmetry. By some sort of a symmetry operation, one part can be brought to coincide with the other. The two main kinds of symmetry operations are

- 180° rotation about an axis, called the twin axis
- reflection across a plane, called the twin plane.

The border plane between the two parts of a twinned crystal is called the *composition plane*. In the case of reflection twins, the composition plane and the reflection plane may or may not coincide.

Owing to the origin, two kinds of twinned crystals are of special interest:

- Annealing twins in FCC metals and alloys (for example, Cu, Ni, brass and Al) that have been cold-worked and then annealed to cause recrystallization. They also form during crystallization directly from melts.
- Deformation twins occuring in deformed HCP metals (for example, Zn and Mg) and BCC metals (for example, α-Fe and W).

Annealing twins in FCC metals are *rotation twins* as the parts can be brought to coincidence by rotation. The two parts are related by a fictive 180° rotation about a twin axis of the direction <111>. The real cause is stacking faults during growth.

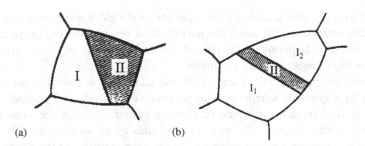

Figure 1.40 FCC annealing twins consisting of (a) two and (b) three parts. Reproduced with permission from B. D. Cullity, *Elements of X-Ray Diffraction*, © Addison-Wesley Publishing Company Inc. (now under Pearson Education).

Annealing twins occasionally appear under a microscope as in Figure 1.40a. Part II is twinned with respect to part I. The two parts are in contact along the composition plane (111), which makes a visible linear trace.

More often, the grains consist of three parts (Figure 1.40b). The two parts are separated by a *twin band*. Such FCC annealing twins are formed by changes in the normal growth mechanism. During growth two stacking faults are created, which are shown schematically in Figure 1.41.

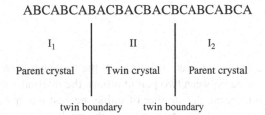

Figure 1.41 The origin of twin bands in FCC annealing twins in the case illustrated in Figure 1.40b.

In real materials, the number of crystal layers A, C, B in the twin band, shown in Figure 1.41, is thousands of times greater. In a single grain, several twin bands of different directions can be found.

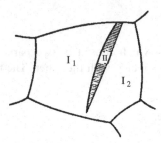

Figure 1.42 HCP deformation twins. Reproduced with permission from B. D. Cullity, *Elements of X-Ray Diffraction*, © Addison-Wesley Publishing Company Inc. (now under Pearson Education).

Deformation twins are caused by deformation and occur in both BCC and HCP lattices. Figure 1.42 shows the most common form of a twin band in HCP metals. The composition 'plane' is curved.

In general, a crystal may twin repeatedly and produce several new orientations. If crystal I_1 twins to form II, which twins to form I_2, etc., then II and I_2 are said to be the first-order, second-order, etc., twins of parent crystal I_1.

1.7 Structures of Alloy Melts and Solids

An alloy can be characterized as a metallic liquid or solid, consisting of a close combination of two or more elements. Often one metal occurs in a high concentration. It is called the *parent metal* or *solvent*. Any chemical element can be used as an *alloying element* or *solute*. Examples of nonmetal alloying elements are carbon, oxygen, nitrogen and sulfur. However, the only alloying elements used in high concentrations are metals.

Alloys are normally prepared by melting the parent metal and adding an accurately measured amount of each alloying element to the melt, followed by vigorously stirring in order to obtain a homogeneous product.

The experimental methods to determine the structure of alloy melts and solids are the same as those used for pure metals and described at the beginning of this chapter. The types of structures of alloys are generally the same as for pure metals. However, it is worth spending a few pages on discussing briefly the characteristics of the structures of alloy melts and solids.

1.7.1 Basic Concepts

All systems achieve a stable equilibrium by spontaneously going over to their lowest possible energy levels. Solid metal solutions are no exceptions from this basic law. Owing to the complexity of solid-state systems, it is difficult to perform quantitative energy calculations in these cases.

All effects that occur when an alloying metal is dissolved in a parent metal are consequences of a transition of the system in order to reach its minimum energy level. Atoms become displaced in the crystal lattice, they change their sizes as a function of composition of the solid solution, new phases appear and chemical compounds, clusters or superlattices are formed.

In order to understand these effects qualitatively, we will discuss the structures of various types of solid alloys.

Random Solid Solutions

The structure of an alloy depends on the forces between the atoms. If the force between an alloying atom and a parent atom is equal to, larger than or smaller than the force between two parent atoms, the internal energy is constant, increases or decreases when the atoms rearrange themselves to increase the number of unlike nearest neighbours.

If the interactions between any atoms are equal, the atoms mix easily and homogeneously in any proportions. An solid alloy of this kind is called a *random solid solution*. Many alloys approximately belong to this type.

Types of Solid Solutions

A crystal lattice can be regarded as positive ions in a lattice surrounded by free electrons moving between them. This type of bonding is fairly indifferent to the exact proportions of the two types of atoms in the lattice and their distribution in the lattice. Random solutions of alloying elements in a parent metal over a wide range of composition are thus possible.

It is possible to change the composition of such alloys almost continuously. At any composition the alloy is fully homogeneous and its structure differs only slightly from those of alloys with neighbouring compositions. A few alloys have completely miscible components, i.e. all concentrations of the alloying element from 0 to 100% are possible. Equal crystal structure of the two components is a necessary condition.

Normally the composition of homogeneous alloys is possible only within certain limits. When the limited range includes one of the pure components of the alloy, the solution is said to be a *primary solid solution* with this component as solvent and the other as solute. If no pure components are included in a solid solution, it is said to be a *secondary solid solution*. The two components have compositions in the intermediate range in this case. Secondary solid solutions often have different crystal structures than their components. The term *intermediate phases* is a common name for *secondary solid solutions* and *intermetallic compounds*.

Solid solutions are either *substitutional* or *interstitial* (Figure 1.43 and 1.44). The type of solution has a great influence on the properties of the alloy.

Figure 1.43 Substitutional solid solution. Reproduced with permission from A. G. Guy, Elements of Physical Metallurgy, 2nd edn, 1980. © Addison-Wesley Publishing Co, Inc (now under Pearson Education).

Figure 1.44 Interstitial solid solution. Reproduced with permission from A. G. Guy, Elements of Physical Metallurgy, 2nd edn, 1980. © Addison-Wesley Publishing Co, Inc (now under Pearson Education).

In a *substitutional* solid solution, the alloying atoms replace some of the parent atoms in the crystal lattice. When the alloying element is another metal, a substitutional solution is often formed. In an *interstitial* solution, the alloying atoms are small enough to take sites between the normal parent atoms filling the lattice. Common examples of this type of solution are carbon or nitrogen dissolved in a metal.

Alloys can consist of both interstitial and substitutional solid solutions at the same time. Examples are stainless steels, which contain interstitially dissolved carbon together with substitionally dissolved chromium, nickel and/or other metals.

1.7.2 Structures of Liquid Alloys

The structures of liquid alloys normally do not differ very much from those of pure metal melts. The X-ray diffraction patterns show that there exists short-range order both in liquid alloys and in pure metal melts. This is especially the case for alloys with low melting points. In alloys with high melting points, the short-range order with preferred nearest neighbour associations tends to become lost when the temperature increases, owing to increasing thermal agitation (internal kinetic energy) of the atoms.

In the uncomplicated cases of simple binary liquid alloys, the position of the first diffraction peak gradually changes from the position that it has in one of the pure metals to the position of the other metal when the composition changes from 0 to 100% of the alloying metal. The position of the first peak of the alloy can be predicted by simple interpolation when the composition is known (Figure 1.45).

In some cases the single peak is replaced by a double peak on the way from 0 to 100%. This can be explained by the formation of an intermediate chemical compound with a special interatomic distance instead of the interpolated liquid solution distance. Figure 1.46 illustrates this. The X-ray diffraction patterns of pure Au, 50:50 Au–Sn, pure Sn and solutions with intermediate compositions are shown. In this case, a short-range order compound AuSn is formed in the liquid with an Au–Sn distance of 0.285 nm. This agrees closely with the distance 0.284 nm found in solid crystalline AuSn.

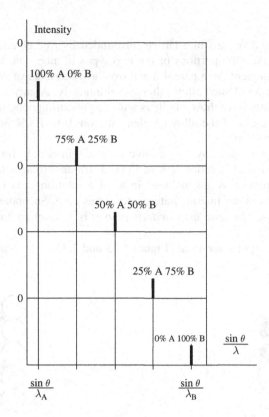

Figure 1.45 Schematic X-ray diffraction patterns in liquid binary alloys of various compositions. No attention should be paid to the intensities, only to the positions of the lines. Reproduced from F. D. Richardson, *Physical Chemistry of Melts in Metallurgy*, Vol. 1. © 1974 Academic Press Inc. (London) Ltd (now by Elsevier).

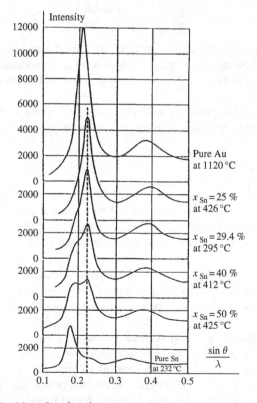

Figure 1.46 X-ray diffraction pattern in liquid AuSn of various compositions. Reproduced from F. D. Richardson, *Physical Chemistry of Melts in Metallurgy*, Vol. 1. © 1974 Academic Press Inc. (London) Ltd (now by Elsevier).

1.7.3 Structures of Solid Alloys

Substitutional Solid Solutions. Hume-Rothery's Rules

Most metals for practical use are alloys. They are developed for special purposes such as better strength, better hardness, better chemical resistance against corrosion, better heat resistance or other better properties than the pure metal.

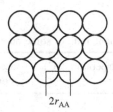

$2r_{AA}$

Figure 1.47 Distance between the atoms in a pure metal A.

The composition of an alloy cannot be chosen arbitrarily. The metals must form a solid solution and intermediate phases must be avoided. As a result of the work of Hume-Rothery in the early 1930s it is possible to judge whether the metals in the planned proportions can form a solid substitutional solution or not.

Hume-Rothery's results can be summarized in four rules. The first rule concerns the sizes of the atoms. The atoms are regarded as spheres, touching each other and filling the space. As a measure of the atomic radius of a metal we use half the distance between two nuclei in the pure element (Figure 1.47).

1. The Relative Size Rule

The more the solute atom differs in size from the solvent, the lower will be the solubility of the metal. This is described with the aid of a relative size factor, defined as

$$\text{Size factor} = 1 + \frac{r_{\text{solute}} - r_{\text{solvent}}}{r_{\text{solvent}}} \tag{1.11}$$

If this factor is larger than 1.14, it is unlikely that a solid solution can form and the solubility will be low. A complete solid solution is obtained only if the size factor is less than 1.08.

2. The Electrochemical Rule

The more electropositive one of the metals is and the more electronegative the other one is, the lower will be the solubility of the two metals. If the difference in chemical affinity of the two metals is large, the two atoms form a compound instead of a solid solution.

3. The Relative Valence Rule

If the alloying metal and the basic metal differ in valence, *the electron ratio*, i.e. the average number of valence electrons per atom, will be changed by alloying. Crystal structures are more sensitive to a decrease in the number of electrons than to an increase. This is the reason why *a high-valence metal dissolves a low-valence metal poorly, whereas a low-valence metal may dissolve a high-valence metal well.*

4. The Lattice Type Rule

Only metals with identical lattice structures are completely miscible, i.e. can form solid solutions of any proportions.

There are exceptions to Hume-Rothery's rules, but overall they are very useful in predicting qualitative solubilities of metals.

The first rule is a necessary but not sufficient condition. If the *relative size factor* is disadvantageous, the solubility of the metals will be poor, even if the other conditions are fulfilled.

The relative size rule is an effect of the strain around a misfit atom (see Figure 1.32b on page 25). The atoms of the solvent are displaced from their normal values in the crystal lattice because of the smaller or larger solute atom. These changes need some extra energy, which limits the solubility.

The sizes of the metal atoms are obtained from the lattice constants of the pure element. However, the sizes of the atoms change as a function of the composition of an alloy, owing to changes in its degree of ionization, electron concentration or crystal structure.

An example of the *relative valence rule* is the CuSi system. Substitutionally dissolved silicon atoms can occur in copper as they increase the number of free electrons. If a copper atom replaces a silicon atom in a silicon crystal, the single valence electron of coppar is not enough to form four covalent bonds to neighbouring silicon atoms and the crystal structure is seriously changed.

Table 1.10   Lattice constants and atomic radii of some pure metals.

Metal	Crystal structure	Temperature (°C)	Lattice constant (nm)	Atomic radius (hard sphere model) (nm)
Cu	FCC	18	0.361	0.255
Ag	FCC	25	0.408	0.288
Al	FCC	25	0.404	0.285
Mg	HCP	25	0.320	0.319
Zn	HCP	25	0.266	0.265
Fe	BCC	25	0.286	0.248

Hume-Rothery and co-workers found that if the size factor was advantageous and the relative valence rule was fulfilled, the limit of solubility was set by the *electron concentration*. The solubility limit was about 1.4 free electrons per atom in all cases. Divalent zinc dissolves in copper up to 40 at-% whereas trivalent aluminium dissolves up to about 20 at-% and tetravalent germanium up to 13 at-%.

Most alloys are substitutionally solid solutions. Copper and the other metals in the same group in the periodic table are excellent solvents for many metals. Copper lies in the middle with respect to atomic radius and electrochemical factor. It dissolves considerable amounts of many metals and forms the basis of many commercially important alloys such as brasses, bronzes and Cu–Ni alloys.

The transition metals, iron and other metals in the same group, also form many solid solutions. Iron has a advantageous size factor for many metals such as Al, Co, Cr, Cu, Mn, Mo, Ni, Pt, V and W. These metals dissolve extensively in FCC-iron (austenite) and in BCC-iron (ferrite).

Interstitial Solid Solutions

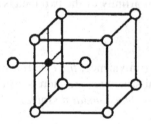

Figure 1.48   An interstitial carbon atom (black) in a (white) α-Fe lattice (BCC structure).

The only alloying elements which are small enough to form interstitial solid solutions are H, C, N and B. The interstitial solutions are genuine alloys with metallic properties. The lattice rule plays an important role in forming interstitial solid solutions. There is, for example, more room for carbon atoms in austenite (γ-Fe, FCC structure) than in ferrite (δ-Fe, BCC structure). The solubility of carbon is therefore about eight times higher in austenite than in ferrite (1.7 and 0.2 wt-%, respectively).

The difference in solubility of carbon between austenite and ferrite depends of the available space for C atoms. An FCC structure offers much more space for interstitials than a BCC structure (Figure 1.48). The difference is very important in the heat treatment of carbon steels.

Most alloys consist of substitutional solid solutions. There are few interstitial solid solutions compared with the substitutional ones, but some of them are *very* important. The interstitial solid solution of carbon in iron is the basis of steel hardening.

The hydrides, nitrides, carbides and borides of the transition metals are important groups of interstitial solid solutions. Their structures are simple when the interstitial atom has a radius < 59% of that of the metal atom. Most of them have the FCC or HCP structure and occasionally BCC. When the ratio exceeds 59%, the nonmetal atom is too large for the interstitial sites and a more complex structure is formed. In carbon steel the ratio is 0.63. The carbon atom is too large to fit into the BCC crystal structure and cementite, an intermediate phase Fe_3C with a complex structure, is formed. Interstitial nitrogen in 18Cr–8Ni stainless steel is important to maintain the iron in its austenitic form.

Intermediate Phases

When an alloying element is added to a base metal in such quantities that the limit of solid solubility is exceeded, a secondary or *intermediate phase* appears. The secondary phase can be another solid solution, a chemical compound or a phase with a structure other than the one of the primary solid solution.

According to their structure, the intermediate phases in alloys can roughly be classified as

- electrochemical or valence compounds
- size factor compounds
- electron compounds.

Electrochemical Compounds

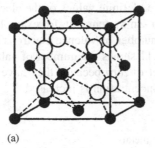 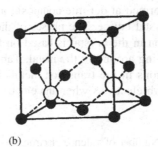

(a)   (b)

Figure 1.49   Some examples of electrochemical compounds. (a) The compound Mg_2Si has the same structure as CaF_2. (b) Structure of ZnS. Reproduced with permission from W. Hume-Rothey, R. E. Smallman and C. W. Haworth, *The Structure of Metals and Alloys*. (Published by the Institute of Metals, 1969) now © W. S. Maney & Son Ltd.

Electrochemical compounds obey the valence law. They are formed by electropositive and electronegative elements. Mg_2Si and ZnS are typical examples (Figure 1.49). Many of them have simple structures of the types found in ionic crystals. They often have high melting points, indicating strong ionic or covalent bonds. Their properties are mainly nonmetallic. They are brittle and have poor electrical conductivities.

Size Factor Compounds

The size factor compounds have compositions and structures that correspond to the lowest possible energies, lower than the sum of the energies of the separate components. The component atoms in size factor compounds are closely packed and have often high coordination numbers.

Examples are the so-called *sigma phases* such as FeCr, CoCr, FeV, FeW, Mn_3Cr and Mn_3V, which might form in high-alloy steels and heat-resistant alloys. They are brittle and harmful because they cause cracks in the material when cold-worked. Sigma phases in steel are normally formed owing to segregation during the solidification process. Methods for decreasing the segregation have been developed and are used in industry.

An important group of size factor compounds are *Laves phases*. They are sometimes used for special applications in electronic components. In other cases they are destructive and must be avoided, for example as precipitation during hardening of high-temperature materials.

Laves phases appear at comparatively large differences between the component atoms, i.e. the A atoms are about 22.5% larger than the B atoms. Their compositions are of the type AB_2.

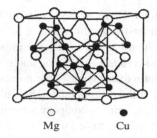

Mg   Cu

Figure 1.50 Structure of MgCu$_2$, a Laves phase with a diamond lattice. It has a complicated structure with very high coordination number. Reproduced with permission from W. Hume-Rothey, R. E. Smallman and C. W. Haworth, *The Structure of Metals and Alloys*. (Published by the Institute of Metals, 1969) now © W. S. Maney & Son Ltd.

Three different structures are possible, one cubic and two hexagonal, depending on the number of free electrons per atom. The cubic structure is stable in the range 1.33–1.8 free electrons per atom. Above 1.8 electrons per atom, the hexagonal structures are the stable ones. Examples are MgCu$_2$ (Figure 1.50), AgBe$_2$, MgZn$_2$, CaMg$_2$, TiFe$_2$ and MgNi$_2$. Striking properties of the Laves phases are dense packing and high coordination numbers. Each A atom has 16 neighbours, 4 A and 12 B, whereas each B atom has 12 neighbours. In a normal close-packed structures the coordination number is 12.

Electron Compounds

The appearance of intermediate phases is very sensitive to the electron ratio, i.e. the *number of valence electrons per atom* in the crystal lattice. They are formed at definite compositions, i.e. definite values of the electron ratio of the alloy, and vary also with the structure of the crystal lattice. The phases of the Cu–Zn system can, for example, be explained in this way.

The free electron concentration in the crystal is based on the number of valence electrons which each atom contributes to the common electron cloud moving around in the metal. Table 1.11 shows the number of valence electrons in some metals.

The number of valence electrons in the transition metals is set to *zero* because they have partly unfilled d states in their electronic structures which absorb electrons when the electron concentration is increased.

Table 1.11 Number of valence electrons in some metals.

Type of metal	Metal	Valence
First group of the periodic table	Cu, Ag, Au	1
Second group of the periodic table	Mg, Zn, Cd, Be	2
Third group of the periodic table	Al	3
Fourth group of the periodic table	Si, Ge, Sn	4
Fifth group of the periodic table	Sb	5
Transition metals. Seventh group of the periodic table	Fe, Co, Ni, Pt, V, W	0

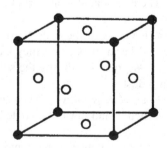

Figure 1.51 Structure of β brass. Its chemical composition is CuZn. The average electron ratio of CuZn is 1.5.

Copper is monovalent and its structure is FCC. The initial electron ratio is therefore 1. If zinc is added to the base metal, the electron ratio increases. Adding, for example, 50 at-% of zinc, i.e. 50 Zn atoms are added to 100 Cu atoms, gives the average electron ratio $[(100 \times 1) + (50 \times 2)]/150$ or 1.33. At increasing Zn concentration a BCC β phase appears (Figure 1.51). If the Zn content is raised further, a complex cubic γ phase appears at a ratio of 1.62 and a close-packed hexagonal ε phase at a ratio of 1.75. Each phase exists over a range of compositions, as shown in the phase diagram of the Cu–Zn system (Figure 1.52). The α and η phases are the primary solid solutions.

The unit cell of complex γ brass (Cu_5Zn_8) consists of 27 unit cells of BCC β brass minus two atoms. Using Table 1.11, we obtain the value of the electron ratio of γ brass as $(5 \times 1 + 8 \times 2)/(5 + 8) = 21 : 13$. A large number of compounds have the same electron ratio value.

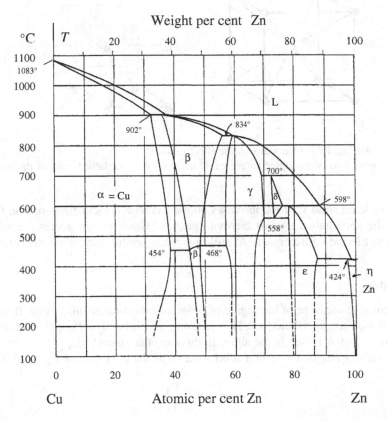

Figure 1.52 Phase diagram of the system Cu–Zn. Reproduced with permission from M. Hansen and K. Anderko, *Constitution of Binary Alloys*, 2nd edn. © 1958 McGraw-Hill Book Company, Inc.

Metal carbides are good and important examples of electron compounds. The carbides of the transition metals are extremely hard. They are used in cutting tools and other applications where mechanical strength at high temperatures is needed. The carbides of some transition metals are extremely stable with a simple cubic structure. Others (MoC, Mo_2C, WC, W_2C and Ta_2C) have a hexagonal structure. Chromium carbides are important and occur in alloy steels. Chromium is soluble in cementite and forms three types of carbides: $Cr_{23}C_6$, $Cr_{27}C_3$ and Cr_3C_2.

Ordered Solid Solutions. Superlattices

If each atom in a solid metal is preferably surrounded by atoms of the same kind, a *clustered state* may be obtained.

If unlike atoms attract each other *more* than like atoms, the structure of the resulting alloy varies considerably with the nature of the forces involved. If the unlike atoms differ electrochemically, the bond between them become partly ionic and the structure is characterized as *intermetallic*. If one component is strongly electronegative, e.g. O, Cl and S, a true chemical compound is formed (*intermetallic compound*) and the material is no longer metallic. If both the unlike atoms are metals, the material will form an alloy and remain metallic. This case will be discussed more in detail.

In a stable solid solution of two metals A and B, the internal energy is smaller than the sum of the separate internal energies of the two metals. Hence an ordered structure is favoured and attraction forces exist between the two types of atoms.

X-ray examination and other methods show that a *long-range order* exists and the two types of atoms are arranged in a regular, alternating pattern at low or moderate temperatures through the entire crystal. This longe-range order of the crystal can be regarded as two or more interpenetrating lattices, firmly attached to each other. Such an *ordered solid solution* is called a *superlattice*. A perfect superlattice can be formed only if the two kinds of atoms occur in simple proportions. Both primary and secondary solutions develop superlattices. An example of a simple superlattice is given in Figure 1.53a.

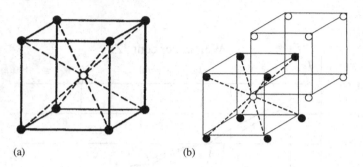

(a)   (b)

Figure 1.53   (a) Superlattice of $AuCu_3$. Its is identical with the BCC structure.
(b) The BCC structure can be regarded as two coupled simple cubic lattices. Only one lattice point of the second lattice is seen in (a).

If the temperature of the ordered solid solution increases above a certain *critical temperature*, the attractive forces are not strong enough to maintain the superlattice against thermal agitation. However, the forces are still there and a *short-range order* may exist locally but not in the entire crystal. At still higher temperatures the atom distribution becomes random.

Degree of Short-range Order

It is possible to obtain a quantative measure of the degree of order in a short-range order state. If the alloying B atoms and the parent A atoms in a solid solution are distributed totally at random, the probability P_A that a certain neighbour of a B atom is an A atom equals the fraction x_A of A atoms in the alloy. In this case the ratio $P_A/x_A = 1$.

We define the *short range order coefficient* α of a solid solution as the deviation of P_A/x_A from 1:

$$\alpha = 1 - \frac{P_A}{x_A} \tag{1.12}$$

The short-range order coefficient is small and varies with the composition of the solid solution. An example is given in Figure 1.54.

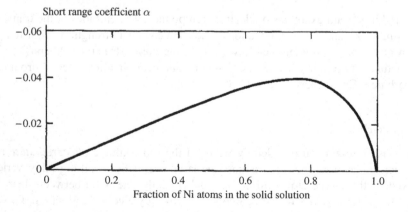

Figure 1.54   The short-range order coefficient of AuNi solid solutions as a function of composition. Reproduced with permission from A. G. Guy, Elements of Physical Metallurgy, 2nd edn, 1980. © Addison-Wesley Publishing Co., Inc (now under Pearson Education).

Three different cases can be distinguished:

- $P_A/x_A = 1$:

  If $P_A/x_A = 1$ there is no preference of neighbours and $\alpha = 0$ which gives a disordered or *random solid solution*. This is the case in all solid solutions at *high* temperatures. At lower temperatures there are two alternatives:
- $P_A/x_A < 1$:

  If the B atoms are surrounded preferably by B atoms then $P_A < x_A$ and $\alpha > 0$. Hence α is positive in a *clustered state*.
- $P_A/x_A > 1$:

  If the B atoms are surrounded preferably by A atoms then $P_A > x_A$ and $\alpha < 0$. In this case α is negative and a short-range order exists in the solid solution.

The three cases are summarized in Table 1.12.

Table 1.12 State of solid solutions as a function of temperature and short range order coefficient.

T	Low	High	Low
$\alpha < 0$	Ordered solid solution Superlattice		
$\alpha = 0$		Random order	
$\alpha > 0$			Clustered state

Summary

■ *X-ray Studies of Solids and Melts*

X-ray studies of solid metals shows the existence of a regular structure. The vertical lines in X-ray diagrams refer to the solid and are interpreted as coordination shells.

The coordination number is the number of atoms which can be included in a coordination shell.
The curves with broad peaks in the diagrams originate from the corresponding melt. They show the existence of a short-range order in the liquid.

■ *X-ray Analysis of Melts*

X-ray data can be transformed into atomic distribution diagrams. From them it is possible to derive

- the nearest neighbour distance (r value of first maximum)
- the nearest possible distance between two atoms (intersection with r axis)
- the coordination number (area under first peak).

■ *Crystal Structure*

The crystal structure is of the utmost importance for the properties of solids.
Crystals are built of unit cells arranged in a space lattice. The crystal structure is determined by two factors

- the lattice structure (lattice)
- the configuration of the unit cell (basis).

$$\text{Crystal structure} = \text{lattice} + \text{basis}$$

There are 14 fundamental types of crystal classes. Three of them, BCC, FCC and HCP, are common in metals.

■ *Lattice Directions and Lattice Planes*

Lattice Directions

A *direction* in a lattice is characterized by three integers, u, v and w. The directions are generally written <uvw>, which means that all permutations of the uvw indices and combination of signs occur. If one refers to a *specific line with the given direction*, square brackets are used, [uvw].

Lattice Planes

Miller indices (hkl) are used to define crystal planes; h, k and l are integers proportional to the reciprocal intersections between the plane and the coordinate axes:

$$(hkl) = \left(\frac{D}{X} \frac{D}{Y} \frac{D}{Z}\right)$$

The constant D makes it possible to obtain integers on the right-hand side of the equation. The *Miller indices*, written within curly brackets {}, mean that all permutations of the hkl indices and combination of signs occur. They define the *directions of the normals to the type of planes in question*. If one refers to a *specific plane with the given direction*, ordinary brackets are used, (hkl).

In hexagonal systems, the four Miller–Bravais indices (hkil) are used. The index i is given by the relationship $i = -(h + k)$.

Intensities of X-ray Diffraction Lines in Cubic Structures

FCC Structures

hkl = odd integers (unmixed indices) give diffraction lines with intensities > 0 ($S > 0$).
hkl = even integers (unmixed indices) give diffraction lines with intensities > 0 ($S > 0$).
All other hkl combinations (mixed indices) give diffraction lines with intensities zero ($S = 0$).

BCC Structures

$h + k + l$ = even integers give diffraction lines with intensities > 0 ($S > 0$).
$h + k + l$ = odd integers give diffraction lines with intensities zero ($S = 0$).

■ *Crystal Defects*

Point Defects

Vacancies, substitutionals and interstitials.

Line Defects

Edge dialocations, screw dialocations.

Interfacial Defects

Stacking faults, grain boundaries and twin boundaries.

■ *Structures of Liquid Alloys*

There are no significant structural differences between liquid alloys and pure metal melts.

■ *Structures of Solid Alloys*

Basic Concepts

A primary solid solution includes one of the pure components.
A secondary solid solution does not include any of the pure components.
In a random solid solution the atoms mix easily and in any proportions.

Hume-Rothery's Rules

Substitutional Solid Solutions

The composition of a substitutional solid solution cannot be chosen arbitrarily. Four conditions limits the possibilities of forming a solid solution of two metals:

1. *The Relative Size Rule*:
   The more the solute and solvent atoms differ in size the lower will be the solubility.

   $$\text{Size factor} = 1 + \frac{r_{\text{solute}} - r_{\text{solvent}}}{r_{\text{solvent}}}$$

   If it is > 1.14 the solubility is low; if it is < 1.08 the solubility is complete.
2. *The Electrochemical Rule*:
   The more electropositive one of the metals is and the more electronegative the other one is, the lower will be the solubility of the two metals. If the difference in chemical affinities of the two metals is large, the two atoms form a compound instead of a solid solution.
3. *The Relative Valence Rule*:
   A high-valence metal dissolves a low-valence metal poorly, whereas a low-valence metal dissolve a high-valence metal well.
4. *The Lattice Type Rule*:
   Only metals with identical lattice structures are completely miscible, i.e. can form solid solutions of any proportions.

Interstitial Solid Solutions

The only alloying elements which are small enough to form interstitial solid solutions are small atoms such as H, C, N and B.
   The interstitial phases are genuine alloys with metallic properties.
   Hume-Rothery's rules are also valid for interstitial solid solutions.
   A solid solution can be interstitial and substitutional at the same time. Stainless steel is an example.

Intermediate Phases

When an alloying element is added to a base metal in such quantities that the limit of solid solubility is exceeded, a secondary or intermediate phase appears.
   The secondary phase can be another solid solution, a chemical compound or a phase with a structure other than the one of the primary solid solution.
   According to their structure, the intermediate phases in alloys can roughly be classified as

- electrochemical or valence compounds
- size factor compounds
- electron compounds.

Ordered Solid Solutions

If each atom in a solid metal is preferably surrounded by atoms of the same kind, a *clustered state* might be obtained.
   If there is no preference in attraction between alloying B atoms and parent A atoms, both types of atoms are distributed totally at random.
   If unlike atoms attract each other *more* than like atoms, the structure of the resulting alloy varies considerably with the nature of the forces. If the two types of atoms are arranged in a regular, alternating pattern, the alloy is characterized as an *ordered solid solution*, either as a short-range order solution or, at low temperature, as a *superlattice* with a regular alternation of unlike atoms through the entire crystal.

$$\text{Short-range order coefficient } \alpha = 1 - \frac{P_A}{x_A}$$

$\alpha > 0$: clustered state.
$\alpha = 0$: random order
$\alpha < 0$: ordered solid solution; superlattice

Exercises

1.1 A high voltage is applied on an X-ray tube.

   (a) Describe the emitted X-ray spectrum from the tube.
   (b) The X-radiation hits a target of copper. What is the minimum voltage for emission of the K series? The wavelength of the K_α line is 0.154 nm.
   (c) The K_β photons have a higher energy than the K_α photons. Is it possible to choose the voltage over the X-ray tube is such a way that K_α photons but no K_β photons are emitted?

1.2 A powder specimen that consists of small crystals is exposed to a beam of parallel X-rays of wavelength 0.112 nm. A photographic plate is placed 10 cm behind the specimen perpendicular to the X-ray beam.

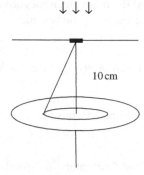

   Depending on the random orientation of the crystals in the specimen, a system of rings appears on the plate when developed. The radii of the rings were measured as 4.26 and 10.80 cm.
   Calculate the distance between the atomic planes that cause this ring system.

1.3 The characteristic X-ray spectrum of a solid metal consists of sharp spectral lines. The corresponding spectrum of a metal melt has a number of wide maxima instead. Why?

1.4 Radiation from an X-ray tube with a target of copper is diffracted in an X-ray spectrograph. A beam of the radiation enters the spectrograph and a strong intensity maximum (first-order diffraction) is found when the angle between the entering and diffracted beam is 31.7°. The crystal in the spectrograph consists of NaCl (FCC structure). Its density is $2.16 \times 10^3 \, \text{kg/m}^3$.
   Calculate the wavelength of the X-ray line in question and try to identify it.

1.5 Consider Figures 1.10 and 1.11.

   (a) What is the sense of the quantities w_r and w_0 in Figure 1.10? What is the relationship between the curves in Figures 1.10 and 1.11?
   (b) Determine the shortest possible distance between two gold atoms in liquid gold.
   (c) Determine the nearest neighbour distance for liquid gold.
   (d) Make a graphical estimation of the coordination number of liquid gold at 1100 °C.

1.6 What fraction of the crystal volume is filled with atoms in the cubic structures

   (a) SC (simple cubic)
   (b) BCC
   (c) FCC

if the atoms are regarded as hard spheres, which touch each other?

1.7 (a) Show that $c/a = \sqrt{8/3}$ for an ideal HCP crystal.
   (b) Calculate the maximum fraction of the available volume which can be filled if the crystal structure is HCP.
   Hint: The unit cell of HCP is shown in Figure 1.29b on page 23.

1.8 The crystal structure of Fe is BCC at temperatures below 637 °C and FCC above this temperature. Derive the percentage density change when an iron specimen is transferred from a BCC to an FCC structure. Assume that the distances between nearest neighbours do not change at the transition.

1.9 (a) Zinc oxide has a hexagonal close-packed structure (HCP) with two formula units per unit cell. The density of ZnO in this phase has been determined experimentally as $5.60 \times 10^3 \, kg/m^3$. Calculate a theoretical value of the density with the aid of known constants and the following data: $a = 0.3243 \, nm$ and $c = 0.5195 \, nm$.

     (b) Discuss possible reasons for the discrepancy between the experimental and theoretical values of the density. Check whether ZnO is an ideal HCP crystal or not.

     (c) Zinc oxide also appears in an FCC structure. The density of ZnO with this structure has been measured as $5.47 \times 10^3 \, kg/m^3$. Calculate the corresponding length of the unit cell.

1.10 Copper has an FCC crystal structure. Use standard table values to calculate

     (a) the lattice constant, i.e. the length of the edge in the conventional unit cell

     (b) the distance to the nearest neighbour atom

     (c) the number of nearest neighbours to each atom.

1.11 (a) Define the geometric structure factor and the rules for the structure type (BCC and FCC) when all the atoms are of the same kind.

     (b) Which of the following planes give lines in the X-ray spectrum of a BCC crystal: (100), (110), (111), (200), (210), (211), (220), (221), (222), (300), (310), (311), (320), (321).

     (c) Which planes in (b) give X-ray lines if the crystal has FCC structure?

1.12 In an electron diffraction experiment on MgO (cubic structure), the following values of the diameters of the rings were obtained:

$$19.9 \, mm \quad 37.8 \, mm \quad 49.9 \, mm$$
$$22.9 \, mm \quad 39.7 \, mm \quad 51.1 \, mm$$
$$32.4 \, mm \quad 45.8 \, mm \quad 56.0 \, mm$$

The acceleration voltage of the electrons was 90.0 kV. The distance L between the specimen and the target was 609 mm.

     (a) Determine the type of cubic lattice and identify the reflections.

     (b) Calculate the length of the edge in the conventional unit cell.

1.13 Describe briefly the different types of point defects in crystal lattices.

1.14 Give some examples of line defects and interfacial defects in crystals.

1.15 Discuss the two types of solid solutions.

     (a) Which type occurs preferably in alloys? Why?

     (b) Describe Hume-Rothery's rules for such solid solutions.

     (c) Give examples of atoms which appear in the other type of solid solutions. Why do they belong to this type?

1.16 What is an intermediate phase and when does it appear in an alloy?

1.17 Explain the following concepts:

     (a) random solid solution

     (b) ordered solid solution

     (c) superlattice

     (d) short-range order

     (e) short-range order coefficient

     (f) cluster.

2

Theory of Atoms and Molecules

2.1 Introduction		46
2.2 The Bohr Model of Atomic Structure		46
	2.2.1 The Hydrogen Atom	46
	2.2.2 Many-electron Atoms	48
2.3 The Quantum Mechanical Model of Atomic Structure		48
	2.3.1 Blackbody Radiation and Photoelectric Effect	48
	2.3.2 Matter Waves	50
	2.3.3 Wave Mechanics and Quantum Mechanics	51
	2.3.4 The Schrödinger Equation	52
	2.3.5 Physical Interpretation of the Wave Function Ψ	54
	2.3.6 The Heisenberg Uncertainty Principle	54
	2.3.7 Solutions of the Schrödinger Equation. Particle in a Box	55
2.4 Solution of the Schrödinger Equation for Atoms		57
	2.4.1 The Hydrogen Atom	57
	2.4.2 Quantum Numbers and Their Interpretation	58
	2.4.3 Quantum Numbers of Many-electron Atoms	62
2.5 Quantum Mechanics and Probability: Selection Rules		65
	2.5.1 Electron Density Distribution	66
	2.5.2 Selection Rules for Electronic Transitions	68
	2.5.3 Zeeman Effect	69
2.6 The Quantum Mechanical Model of Molecular Structure		71
	2.6.1 The H_2^+ and H_2 Molecules	72
2.7 Diatomic Molecules		75
	2.7.1 Classification of Electronic States	75
	2.7.2 The Rigid Rotator	76
	2.7.3 The Harmonic Oscillator	80
	2.7.4 Eigenfunctions and Probability. Density Distribution of the Harmonic Oscillator	81
	2.7.5 Spectra of Diatomic Molecules	82
	2.7.6 Selection Rules	84
2.8 Polyatomic Molecules		88
Summary		89
Exercises		94

Physics of Functional Materials   Hasse Fredriksson and Ulla Åkerlind
© 2008 John Wiley & Sons, Ltd

2.1 Introduction

Modern theory of atomic and molecular physics is the fundamental basis of all materials science and a key to understanding material structure and processes. This chapter gives a brief survey of atomic and molecular structure, based on quantum mechanics, as a preparation and basis for the following chapters, especially Chapters 3–6.

In Chapters 4, 5 and 6, a survey of the most important properties of gases, solids and liquids is given. This knowledge is essential for materials science when various aspects of crystallization processes are discussed.

2.2 The Bohr Model of Atomic Structure

2.2.1 The Hydrogen Atom

Line spectra, for example emission line spectra specific for hydrogen and other elements and the Fraunhofer absorption lines in the continuous spectrum of the Sun, were discovered at the beginning of the 19th century. The origin of the spectral lines could not be explained.

In 1885, the Swiss schoolteacher J. J. Balmer found an empirical equation for some of the strongest lines in the visible part of the spectrum of hydrogen (Figure 2.1):

$$\frac{1}{\lambda} = constant \times \left(\frac{1}{2^2} - \frac{1}{n^2} \right) \quad \text{Balmer series} \tag{2.1}$$

where
λ = wavelength of the spectral line
n = positive integer >2.

The origin of the lines remained a puzzling and unanswered question for a long time.

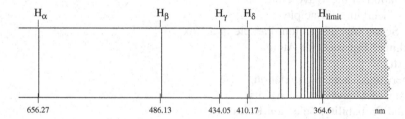

Figure 2.1 Balmer series for hydrogen. The line spectrum is emitted, for example, from an electric arc in a discharge tube filled with hydrogen gas at low pressure. Many H_2 molecules dissociate into excited H atoms, which emit a line spectrum, including the Balmer series. Reproduced with permission from A. Beiser, *Modern Physics. An Introductory Survey.* © 1968 Addison-Wesley Publishing Company (now under Pearson Education).

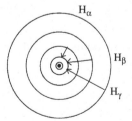

Figure 2.2 When an H atom becomes excited, it absorbs exactly the amount of energy necessary to move the electron from the ground state $n = 1$ to a higher orbit. The average lifetime of the electron in the excited state is of magnitude 10^{-8} s. Then the electron returns to a lower orbit, for example one of those indicated, and emits a photon $h\nu$: $h\nu = E_n - E_m$. This is the explanation of the Balmer series and other spectral lines characteristic for hydrogen.

On the basis of experiments on scattering α particles by a thin gold foil, Rutherford introduced his 'classical' atomic model in 1911. According to his model, a hydrogen atom consists of small nucleus (a proton) surrounded by an electron which rotates in a circular orbit around the nucleus like a planet around the Sun. This model could not explain the origin of the line spectrum of the hydrogen atom.

The first explanation was given by Bohr when he published his atomic 'semiclassical' model in 1913 (Figure 2.2). He combined Rutherford's classical model with four postulates:

1. The dynamic equilibrium of the system nucleus and electron conforms with Newton's mechanics.
2. Electrons can only stay in certain stationary orbits or stationary states around the nucleus.
3. The angular momentum of every electron round the centre of its orbit will in the stationary state of the system be equal to an integer multiple of $h/2\pi$, where h is Planck's constant:

$$L = n\frac{h}{2\pi} \tag{2.2}$$

4. Only when an electron transition occurs from one stationary state to another is radiation emitted or absorbed with a frequency ν which is given by the relationship

$$\Delta E = h\nu$$

where ΔE is the energy difference between the two states.

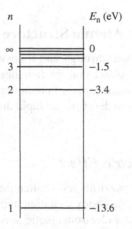

Figure 2.3 Energy levels of the H atom.

The potential and kinetic energies of the electron can be calculated by classical mechanics. The energy of a free electron, i.e. the electron at rest at an infinite distance from the nucleus ($n = \infty$), is chosen as zero on the energy scale. The sum is equal to the total energy of the atom. It can be shown that the energy levels of the atom are (Figure 2.3)

$$E_n = -\frac{13.6\,\text{eV}}{n^2} \tag{2.3}$$

where
$\quad n$ = integer quantum number ≥ 1
$\quad E_n$ = energy of a hydrogen atom in the orbit n.

The lowest value of n is 1, which corresponds to the ground state, the lowest possible energy level of the hydrogen atom. For the total energy $E > 0$, all energies are allowed as the kinetic energy of a free electron is continuous.

Single-electron Atoms

Bohr's fourth postulate in combination with Equation (2.3) led to excellent agreement with the measured wavelengths of the Balmer series and the empirical Equation (2.1). A modified formula is valid also for other *single-electron atoms*:

$$E_n = -13.6 \text{ eV} \times \frac{Z^2}{n^2} \tag{2.4}$$

where Z is the number of protons in the nucleus. An example is He^+ with $Z = 2$.

2.2.2 Many-electron Atoms

Bohr initially assumed circular electron orbits. Later, Sommerfeld introduced elliptical orbits for the electrons and three additional quantum numbers (denoted l, m_1 and m_s today) were defined to describe the orbit of each electron for better agreement with experiments.

Even with this addition, Bohr's theory could not successfully explain the structure of more complicated atoms and their spectra, e.g. the spectra of the alkali metals, which are similar to the hydrogen spectrum. Many spectral lines show a fine structure, i.e. instead of a single monochromatic line they consist of a narrow group of lines with slightly different wavelengths, due to splitting of the energy levels involved. The Bohr model was unable to explain this phenomenon.

A new theory, *quantum mechanics*, was introduced in the middle of the 1920s. It is a mathematically very advanced theory and much more abstract and difficult to grasp and visualize than Bohr's theory.

To overcome this complexity, some of the terminology of the Bohr model is still used, which is somewhat inadequate but helps to give an comprehensible picture of the abstract model.

2.3 The Quantum Mechanical Model of Atomic Structure

The Bohr model of the atom represented a great breakthrough and it answered the old question of the origin of the spectral lines. Bohr abandoned classical physics when he stated that no radiation is emitted by the electron in its stationary orbit around the nucleus of the atom. He introduced quantum aspects in atomic physics with his fourth postulate. To understand the development of quantum mechanics, it is necessary to describe the rapid and exciting development of physics at the beginning of the 20th century.

2.3.1 Blackbody Radiation and Photoelectric Effect

At the end of the 19th century, there were several puzzling and challenging experimental results in physics that could not be explained by classical physics. Two examples are blackbody radiation and photoelectric effect.

At any temperature $T > 0$ K, every object radiates electromagnetic waves. The radiation depends on the temperature and on the character of the radiating surface. An object which absorbs all incident radiation, independent of wavelength, is called a blackbody. An object at thermal equilibrium with its surroundings radiates as much energy per unit time as it absorbs. A blackbody is a perfect emitter and absorber. No real body is a blackbody. A cavity with black walls and a small opening is the best approximation of a blackbody.

The intensity distribution of the radiation emitted from such a cavity at a constant temperature T was studied by Kirchhoff experimentally in 1860 and the intensity was plotted as a function of the frequency of the radiation (Figure 2.4). Rayleigh and Jeans derived a theoretical expression for the energy density inside the cavity. They assumed that the energy density was a continuous function of the frequency ν and calculated the energy distribution using classical Boltzmann statistics (Chapter 4, Section 4.3) in analogy with the distribution of kinetic energy of the molecules in a gas. The average energy of the 'radiation gas' was calculated as $k_B T$.

In Figure 2.4b, the Rayleigh–Jeans radiation law curve is included. The agreement with experimental values is good at low frequencies but the disagreement at high frequencies is known as 'the ultraviolet disaster'.

Max Planck proposed in 1900 that the energy is *not* a continuous function of the frequency but only *discrete* energy states are allowed (Figure 2.5):

$$E = n \times h\nu \tag{2.5}$$

where

n = a positive integer 1, 2, 3, . . .

h = Planck's constant

ν = frequency of the radiation.

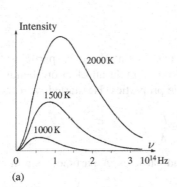

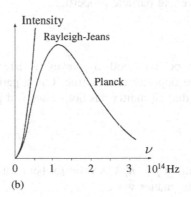

(a)   (b)

Figure 2.4 Experimental intensity distribution of the radiation emitted from a cavity at constant temperature T as a function of frequency. Planck's law: $E(\nu)\mathrm{d}\nu = \dfrac{8\pi\nu^2}{c^3}\dfrac{h\nu\mathrm{d}\nu}{\mathrm{e}^{\frac{h\nu}{k_{\mathrm{B}}T}}-1}$ (a) Experimental results. (b) Theoretical curves of Rayleigh–Jeans and Planck. Reproduced with permission from I. Lindgren, J. Nilsson, O. Beckman, E. Karlsson and T. Kivikas, *Fysik 3: Kvantfysik*. © 1971 Almqvist & Wiksell Forlag AV, Stockholm.

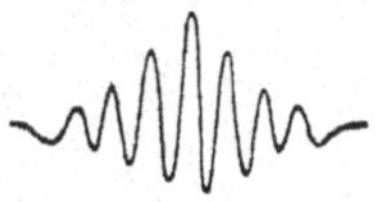

Figure 2.5 A quantum $h\nu$ is regarded as a small group of waves which amplify each other by constructive interference within a limited region and destroy each other by destructive interference elsewhere in space.

As can be seen in Figure 2.4, Planck's radiation law gives excellent agreement with the experimental curve. This was the first introduction of a quantum of radiation and a complete break with classical physics. Planck's hypothesis was strongly supported when Einstein in 1905 published a paper on the photoelectric effect, described in Figure 2.6. It could easily be explained in terms of Planck's quantum theory.

Figure 2.6 Photoelectric effect. Incident monochromatic light hits a metal surface. If the frequency ν of the light exceeds a particular frequency ν_0, electrons are emitted from the metal surface. The velocities of the electrons can be determined from the deviation in a known magnetic field.

Einstein assumed that the incident light behaved like a stream of light quanta or *photons*, each with energy $h\nu$, and not like a wave. He set the minimum energy ϕ, required to release an electron from a metal surface, equal to the energy $h\nu_0$, of the quantum, which is just able to knock out the electron, i.e. $\phi = h\nu_0$. Einstein applied the energy law to the process:

$$h\nu = \phi + \frac{mv^2}{2} = h\nu_0 + \frac{mv^2}{2} \qquad (2.6)$$

where v is the velocity of the electron and ϕ is called the *work function*. The last term represents the kinetic energy of the released electron, which equals $h\nu - h\nu_0$, where $\nu > \nu_0$.

If $v < v_0$, *no electrons are emitted* even if the intensity of the light is increased substantially. This is in excellent agreement with the experimental observations, which are impossible to explain by classical physics.

On the other hand, there are numerous experiments which show the wave character of light and other electromagnetic radiation, particularly all sorts of interference experiments. It was necessary to accept the duality of electromagnetic radiation. It has simultaneously wave and particle properties.

2.3.2 Matter Waves

Electromagnetic radiation behaves both as waves (interference experiments) and as particles (photoelectric effect). Next question to ask was if the opposite also is true. Can a particle, for example an electron, behave like a wave? In 1923, de Broglie boldly suggested that all matter has both wave and particle properties. He started with the momentum of a photon:

$$p = \frac{h\nu}{c} = \frac{h}{\lambda} \tag{2.7}$$

and assumed that the equation $p = h/\lambda$ is valid for both waves and particles. A particle has a momentum equal to mv, which gives the wavelength of the matter wave:

$$\lambda_{deB} = \frac{h}{mv} \tag{2.8}$$

where
   v = velocity of the particle
   λ_{deB} = de Broglie wavelength of the particle, which moves with the velocity v
   h = Planck's constant
   m = mass of the particle.

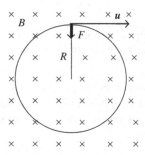

Figure 2.7   Deviation of a beam of electrons in a magnetic field. The electrons behave like particles. The radius of the circle is determined by the relationship $mv^2/R = Bev$.

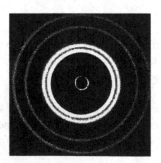

Figure 2.8   Diffraction pattern on passing a beam of electrons through an Al foil. The de Broglie wavelength of the electrons was the same as the wavelength of the X-rays in the experiment in Figure 2.9. Reproduced from F. Blatt, *Modern Physics*. © McGraw-Hill Inc. (1992).

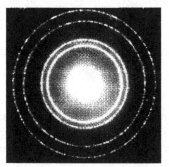

Figure 2.9   Diffraction pattern obtained by passing monochromatic X-rays through the same Al foil as in the experiment shown in Figure 2.8. Reproduced from F. Blatt, *Modern Physics*. © McGraw-Hill Inc. (1992).

The expression *mv* of the momentum is often denoted by *p*. It is valid in both classical and relativistic mechanics. In the latter case, *m* represents the relativistic mass of the particle.

de Broglie's hypothesis was confirmed by two famous experiments. The wave character of electrons was demonstrated by Davisson and Germer in the USA in 1927 and independently by G. P. Thomson (son of J. J. Thomson, who discovered the electron in 1897) in England in 1928 by electron diffraction through a thin metal foil (Figure 2.8). Their experiments showed excellent agreement with Equation (2.8) and the resemblance with X-ray diffraction (Figure 2.9) is striking.

The electron distribution behind the foil in Figure 2.8 can be calculated as a diffraction phenomenon if the wave properties of the moving electrons are considered. The result can be verified experimentally on a screen or a photographic plate.

Alternatively, the electrons can be regarded as particles and the electron distribution behind the foil can be found by statistical calculations. The electrons can be registered by a mobile particle detector at various points in the plane of the screen. The statistical distribution of a large number of electrons agrees with the diffraction pattern. The probability of finding electrons at the bright areas of the diffraction pattern is high whereas few electrons hit the detector at the dark positions of the diffraction pattern.

A beam of electrons shows wave properties during propagation. On interaction with matter, for example a photographic plate or particle detector, the electrons behave like particles. High-speed electrons are relativistic, as is shown in the box.

Relativistic Mechanics

According to the theory of special relativity, we have the following fundamental equations:

$$E_{\text{total}} = mc^2$$

$$E_{\text{kin}} = mc^2 - m_0\,c^2$$

$$p = mv$$

where

$$m = \frac{m_0}{\sqrt{1 - \dfrac{v^2}{c^2}}}$$

Classical mechanics is a special case of relativistic mechanics which is valid when $v \ll c$. Relativistic effects usually have to be considered when $v > 0.1c = 3 \times 10^7\,\text{m/s}$. The corresponding energy of an electron is $\sim 2.5\,\text{keV}$.

2.3.3 Wave Mechanics and Quantum Mechanics

The discovery of the particle–wave duality led to a fruitful period of physics and is closely associated with the development of two new theories, *wave mechanics* and *quantum mechanics*.

Schrödinger concentrated on de Broglie's matter waves and developed the theory of *wave mechanics*. In 1925 he proposed a general wave equation for the determination of the wavelength of the matter wave of a moving particle (page 53). This equation is fundamental in wave mechanics and can be applied to all sorts of waves and particles.

Heisenberg concentrated on the particle or quantum aspect and discussed the impossibility of the simultaneous, infinitely accurate determination of both the position and momentum of a moving particle. He published his fundamental uncertainty principle (page 54) in 1927.

Quantum mechanics concentrates on the quantization of various physical quantities of particles whereas wave mechanics describes the particle as a matter wave. In fact, it soon became clear that the two theories are two aspects of the same fundamental theory, nowadays generally called quantum mechanics. The two theories proved to be different mathematical formulations of the same physical theory.

Many other great physicists have contributed to the development of quantum mechanics, among them Born, Jordan, Dirac and Pauli. Born's interpretation of the wave function, which deals with the probability of finding the electron within a certain volume element (page 54), is the link between Heisenberg's uncertainty principle and Schrödinger's wave mechanics.

Quantum mechanics has been applied to photons, electrons, protons, neutrons, atoms, molecules and even macroscopic particles with great success. Classical mechanics is a special case of quantum mechanics just as geometric optics is a special case of wave optics. Classical mechanics holds for large particles, for example a moving ball, but fails for small particles in high-speed motion. The de Broglie wavelength is very small for a heavy particle [Equation (2.8)] and no deviations from classical mechanics can be observed.

Quantum mechanics is often considered to be a difficult subject. The main trouble, apart from mathematical difficulties, is our experience of the macroscopic world, which is difficult to abandon in the microscopic world, where the normal laws of mechanics fail.

2.3.4 The Schrödinger Equation

To understand the motion and distribution of the electrons in atoms and molecules, we have to consider their wave properties.

A beam of electrons passing a thin metal foil (Figure 2.8) is an example of a nonstationary process. This general case corresponds to travelling matter waves. In this book we intend to apply quantum mechanics only to electrons in stationary states in atoms and molecules. Stationary states correspond to standing waves. Hence electrons in *stationary states* in atoms and molecules will be treated as *standing matter waves*.

Classical Equations for a Standing Wave

The general differential equation of a standing mechanical or acoustic wave in one dimension can be written as

$$\frac{\partial^2 s}{\partial t^2} = v^2 \frac{\partial^2 s}{\partial x^2} \tag{2.9}$$

The solution of Equation (2.9) is

$$s = s_0 \sin \frac{\omega x}{v} \sin \omega t \tag{2.10}$$

where
s = displacement from the equilibrium position
s_0 = amplitude of the wave
t = time
v = velocity of the wave
x = coordinate
ν = frequency of the wave
λ = wavelength
ω = angular frequency of the wave = $2\pi\nu = 2\pi\dfrac{v}{\lambda}$.

Equation (2.10) can be written in a more general form as

$$s = \psi(x) \sin \omega t \tag{2.11}$$

where $\psi(x)$ is the amplitude of the wave at position x.

If we introduce this expression for s and its derivatives into Equation (2.9), we obtain after division by $\sin \omega t$

$$v^2 \frac{\partial^2 \psi}{\partial x^2} = -\omega^2 \psi \tag{2.12}$$

If ω is replaced by $2\pi v/\lambda$, Equation (2.12) can be divided by v and we obtain

$$\frac{\partial^2 \psi}{\partial x^2} + \frac{4\pi^2}{\lambda^2} \psi = 0 \tag{2.13}$$

where ψ is a function of position x.

The general equation of a standing wave in three dimensions can be written as

$$\frac{\partial^2 \psi}{\partial x^2} + \frac{\partial^2 \psi}{\partial y^2} + \frac{\partial^2 \psi}{\partial z^2} + \frac{4\pi^2}{\lambda^2} \psi = 0 \tag{2.14}$$

where the amplitude ψ of the wave is a function of x, y and z.

The Schrödinger Equation

In order to determine the stationary energy states of the system, we have to set up the wave equation in analogy with the equation of a vibrating string. The wave function Ψ is analogous to the displacement s of a point on the vibrating string from its equilibrium position but it has a different significance. Ψ is a function of both time and position.

The wavelength λ is in this case the matter wavelength of the electron, i.e. the de Broglie wavelength defined as in Equation (2.8). If we introduce λ_{deB} instead of λ into the time-independent Equation (2.14), we obtain

$$\frac{\partial^2 \psi}{\partial x^2} + \frac{\partial^2 \psi}{\partial y^2} + \frac{\partial^2 \psi}{\partial z^2} + \frac{4\pi^2 m^2 v^2}{h^2} \psi = 0$$

If we substitute the kinetic energy $mv^2/2$ by the total energy E minus the potential energy E_{pot} and introduce the symbol $\hbar = h/2\pi$, we obtain

$$\frac{\partial^2 \psi}{\partial x^2} + \frac{\partial^2 \psi}{\partial y^2} + \frac{\partial^2 \psi}{\partial z^2} + \frac{2m}{\hbar^2} \left(E - E_{\text{pot}} \right) \psi = 0 \tag{2.15}$$

Equation (2.15) is the differential equation of *amplitude* of the matter wave. E_{pot} is a function of x, y and z. When Equation (2.15) is solved, the amplitude ψ of the matter wave is obtained as a function of x, y and z.

The displacement of a standing wave also depends on time. In Equations (2.10) and (2.11), we used a sine function. The total time-dependent wave function Ψ of a matter wave also varies periodically with time. Instead of a sine function for the time dependence, we will use an exponential function, which facilitates calculations, and obtain

$$\Psi = \psi e^{-2\pi i v t} \tag{2.16}$$

where

Ψ = total wave function
ψ = amplitude of the wave function
v = frequency of the matter wave
t = time.

If we introduce the expression for Ψ in Equation (2.16) and its partial derivatives with respect to t and x into the three-dimensional form of Equation (2.9), we obtain the general differential equation for Ψ:

$$-\frac{\hbar^2}{2m} \left(\frac{\partial^2 \Psi}{\partial x^2} + \frac{\partial^2 \Psi}{\partial y^2} + \frac{\partial^2 \Psi}{\partial z^2} \right) + \left(E - E_{\text{pot}} \right) \Psi = i\hbar \frac{\partial \Psi}{\partial t} \tag{2.17}$$

This is the *Schrödinger equation*, published in 1925, the fundamental equation in quantum mechanics which replaces the equations of classical mechanics for atomic systems.

Equation (2.17) is to be solved for atomic systems. According to de Broglie, the frequency v of the matter wave is related to the total energy E of the system by the relationship

$$E = hv \tag{2.18}$$

Hence we can rewrite Equation (2.16) as

$$\Psi = \psi \, e^{-i\frac{E}{\hbar}t} \tag{2.19}$$

Often it is sufficient to solve Equation (2.15) and obtain the amplitude ψ of the matter wave. This is the case for standing matter waves.

Equation (2.17) is often called *the time-dependent Schrödinger equation* and Equation (2.15) as the *time-independent Schrödinger equation*.

2.3.5 Physical Interpretation of the Wave Function Ψ

The discussion on pages 50–51 indicates that the position of an electron is closely related to statistics and probability. In quantum mechanics it is impossible to tell the exact position of an electron. The best one can do is to predict the probability of finding the electron within a given volume element. This is an essential feature of quantum mechanics, which was indicated in the discussion earlier so far but has no concrete expression.

It is known from elementary wave theory that the square of the amplitude is proportional to the intensity of a wave. In 1926, Born suggested, in analogy with this, that the value of the wave function Ψ within a volume element $dxdydz$ at position (x, y, z) is related to the probability of finding the particle within the volume element:

- The probability of finding a particle within a volume element dxdydz is

$$|\Psi|^2\, dxdydz = \Psi\Psi^*dxdydz \tag{2.20}$$

where Ψ^* is the complex conjugate of Ψ.

This definition can be used to normalize the wave function:

$$\iiint\limits_{\text{space}} \Psi\Psi^*dxdydz = 1 \tag{2.21}$$

In the special case of standing waves, $\Psi\Psi^*$ can be replaced by $\psi\psi^*$ because we have, according to Equation (2.16),

$$\Psi\Psi^* = \psi e^{-2\pi i \nu t}\psi^*e^{+2\pi i \nu t} = \psi\psi^* \tag{2.22}$$

Hence it is sufficient in the case of stationary states to deal with the amplitude function ψ instead of Ψ and use the normalization equation:

$$\iiint\limits_{\text{space}} \psi\psi^*dxdydz = 1 \tag{2.23}$$

The probability concept is a most useful interpretation of the wave function. The concept has already been used in Chapter 1 on page 6. As we will see later, it can be used to find the electron probability distribution around the nucleus of the atom and to find selection rules for transition from one orbital to another in an atom or molecule.

In all atomic and molecular applications, only stationary states are considered. Hence from now on we will only discuss the solutions of the time-independent Schrödinger equation and use the expression $\psi\psi^*dxdydz$ to calculate probabilities in atomic and molecular systems.

2.3.6 The Heisenberg Uncertainty Principle

Independently of Schrödinger's theory, Heisenberg analysed the possibilities of determining the position and momentum of a moving particle simultaneously. His result, published in 1927, is an essential part of quantum mechanics, by no means in contradiction with wave mechanics. Born's interpretation of the wave function represents a link between Schrödinger's wave mechanics and Heisenberg's quantum mechanics. The outlines of his theory will be discussed briefly below.

Consider a particle with mass m which moves with the velocity v. According to de Broglie (page 50), its momentum can be written as $mv = h/\lambda_{\text{deB}}$, where λ_{deB} is the de Broglie wavelength of the moving particle.

Accurate measurement of the wavelength requires an extended wave. If we want an exact value of the wavelength or the momentum of the particle, then the position of the particle is completely undetermined according to wave mechanics. The probability of finding the particle within a given volume element $d\tau$ is $\Psi\Psi^*d\tau$ or $\psi\psi d\tau$. It is the same everywhere as the de Broglie wave has an infinite extension.

Similarly, if we wish to define the position of a particle very accurately, the wave function must differ from zero only at a given point. This can only be achieved by overlapping of many sine waves of all wavelengths from zero up to infinity. This makes the wavelength and hence the momentum completely uncertain:

- Position and momentum cannot be measured exactly simultaneously.

This is a consequence of *Heisenberg's uncertainty principle*, which is closely related to wave mechanics:

$$\Delta y\Delta p_y \geq \frac{h}{4\pi} \tag{2.24}$$

where

h = Planck's constant

Δy = uncertainty in position (standard deviation)

Δp_y = uncertainty in momentum (standard deviation).

The Heisenberg uncertainty principle is also valid for another pair of quantities, energy and time:

$$\Delta E \Delta t \geq \frac{h}{4\pi} \tag{2.25}$$

Because it is impossible to determine the position and momentum of an electron simultaneously, well-defined electron orbitals are impossible. The only possibility is to calculate the probability of finding an electron within a certain volume element as a function of position.

2.3.7 Solutions of the Schrödinger Equation. Particle in a Box

The Schrödinger equation is soluble if the following conditions are fulfilled:

$$\psi \text{ must everywhere be} \tag{2.26}$$

- single-valued
- finite
- continuous (both ψ and its derivative)
- and vanish at infinity (if the particle position is finite).

If and only if these conditions are fulfilled is the Schrödinger equation soluble, not for arbitrary values of E, but for *special* values of E only. These values, which are called *eigenvalues,* can be derived from the conditions given above.

The corresponding wave functions are called the *eigenfunctions.* They represent the stationary states of the wave motion, for which the waves do not cancel owing to destructive interference.

In order to demonstrate clearly that the conditions (2.26), in spite of their nonmathematical form, lead to concrete values of the eigenvalues E, we will discuss an electron in a one-dimensional box as a simple example.

Example 2.1

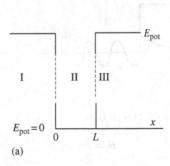

(a)

An electron is trapped in a rectangular well and can move back and forth within the well. The potential energy E_{pot} is zero within the well and very high outside the well.

Use quantum mechanical calculations to find the eigenvalues of the electron, i.e. find the possible energy levels E of the electron in the box.

Solution:

The electron can move in the limited region $0 \leq x \leq L$. According to classical physics the electron is trapped within the box. The probability of finding the electron within regions I and III is zero.

To obtain the quantum mechanical probability of finding the electron within an interval dx we have to solve the Schrödinger equation:

$$\frac{\partial^2 \psi}{\partial x^2} + \frac{2m}{\hbar^2}\left(E - E_{pot}\right)\psi = 0 \tag{1'}$$

for each of the three regions.

For *region II* we have

$$\frac{\partial^2 \psi}{\partial x^2} + \frac{2m}{\hbar^2}\left(E - 0\right)\psi = 0 \tag{2'}$$

The solution of Equation (2') is

$$\psi_{II} = A\sin\left(\sqrt{\frac{2mE}{\hbar^2}}x + \varphi\right) \tag{3'}$$

where A and φ are two constants, which will be determined by boundary conditions.

Inside *region I*, the potential energy E_{pot} is larger than the total energy E. In this case we write the Schrödinger equation as

$$\frac{\partial^2 \psi}{\partial x^2} - \frac{2m}{\hbar^2}\left(E_{pot} - E\right)\psi = 0 \tag{4'}$$

Equation (4') has a solution which is the sum of two exponential functions:

$$\psi = B\exp\left[\sqrt{\frac{2m(E_{pot} - E)}{\hbar^2}}x\right] + C\exp\left[-\sqrt{\frac{2m(E_{pot} - E)}{\hbar^2}}x\right] \tag{5'}$$

where B and C are constants. At $x = -\infty$ the function must vanish. For this reason, the constant C must be zero.

$$\psi_I = B\exp\left[\sqrt{\frac{2m(E_{pot} - E)}{\hbar^2}}x\right] \tag{6'}$$

ψ must be continuous and single-valued everywhere and even at the junction between regions I and II. This condition is sketched graphically in the figure below on the right.

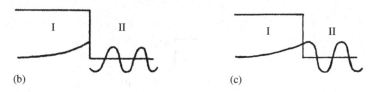

(b)   (c)

This condition results in the relationship $\psi_I = \psi_{II}$ at $x = 0$ or

$$B = A\sin\varphi \tag{7'}$$

It can be seen from Equation (6') that if E_{pot} is very large the amplitude $\psi_I \to 0$ when x approaches zero from the negative side. The continuity condition requires that $\psi_I(0) = \psi_{II}(0)$ or $0 = A\sin\varphi$. As $A \neq 0$, φ must be zero and we obtain from (3')

$$\psi_{II} = A\sin\left(\sqrt{\frac{2mE}{\hbar^2}}x\right) \tag{8'}$$

Equation (7′) shows that if $\varphi = 0$, B also becomes zero. This means that if $E_{pot} = \infty$ the probability of finding the electron outside the box is zero.

(d)   (e)

Analogously, there must not be a discontinuity at $x = L$ which gives the relationship $\psi_{II} = \psi_{III} = 0$ for $x = L$:

$$\psi_{II} = A \sin \left(\sqrt{\frac{2mE}{\hbar^2}} L \right) = 0 \tag{9'}$$

Hence the standing wave within the rectangular well has nodal points at $x = 0$ and $x = L$ (figure above on the right). This condition can be written as

$$\sqrt{\frac{2mE}{\hbar^2}} L = n\pi \tag{10'}$$

where n is an integer >0. From Equation (10′), we obtain the resulting quantization condition for the energy.

Answer:

The only possible energy levels of the electron are

$$E = \frac{h^2}{8m} \frac{n^2}{L^2}$$

where n is a positive integer.

Example 2.1 shows that the eigenvalues, i.e. the discrete energy values, are a natural consequence of the solution of the Schrödinger equation. Similarly, quantum numbers appear in the solutions of ψ in atomic and molecular systems. The mathematical calculations are far more complicated in these cases than in Example 2.1 and will only be described briefly in the text.

2.4 Solution of the Schrödinger Equation for Atoms

2.4.1 The Hydrogen Atom

The principles of solving the Schrödinger equation for single-electron atoms has been sketched on pages 53 and 55. Here the solution will be discussed for the concrete case of the hydrogen atom and hydrogen-like ions.

The basis of the calculations is Equation (2.15) on page 53. In this case, the potential energy of the electron is equal to

$$E_{pot} = -\frac{1}{4\pi\varepsilon_0} \frac{e \times Ze}{r} \tag{2.27}$$

Inserting this value into Equation (2.15), we obtain

$$\frac{\partial\psi^2}{\partial x^2} + \frac{\partial\psi^2}{\partial y^2} + \frac{\partial\psi^2}{\partial z^2} + \frac{2m}{\hbar^2}\left(E + \frac{Ze^2}{4\pi\varepsilon_0 r}\right)\psi = 0 \tag{2.28}$$

This differential equation can be solved and gives values of ψ, which are single valued, continuous and finite, for all positive values of E but only for the following negative values of E:

$$E = -\frac{me^4}{(4\pi\varepsilon_0)^2 \times 2\hbar^2} \frac{Z^2}{n^2} \tag{2.29}$$

where

m = mass of the electron
e = charge of the electron
ℏ = Planck's constant/2π
Z = number of protons in the nucleus of the atom
n = positive integer.

2.4.2 Quantum Numbers and Their Interpretation

The Principal Quantum Number

The *principal quantum number is represented by n*. If the values of the constants are inserted into Equation (2.29), it becomes identical with the old Bohr equation [Equation (2.4) on page 48]. In this case, the Bohr model and the quantum mechanical model give the same result.

For each eigenvalue of the Schrödinger equation there is normally more than one eigenfunction. Each eigenfunction represents an orbital. These eigenfunctions are distinguished by two additional quantum numbers, l and m_l, which appear in the solutions of ψ.

The Azimuthal Quantum Number

The quantum number l is the *azimuthal quantum number*. It can have the values

$$l = 0, 1, 2, \ldots, (n-1) \tag{2.30}$$

Apart from the hydrogen-like atoms, there are small differences in energy between states with different l values and equal n.

The value of l influences the shape of the orbital and its angular momentum. All orbitals with $l = 0$ are spherically symmetrical. Orbitals with higher l values are more complicated. A simple example ($l = 1$) is given on page 68.

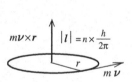

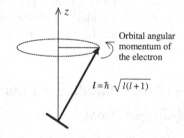

Figure 2.10 The classical angular momentum of a particle with mass m in a circular orbital with radius r is $m\nu \times r$. According to Bohr's postulate 3 on page 47 the angular momentum of the electron in its orbit (fat circle) is a vector of length $|l| = n \times h/2\pi$.

Figure 2.11 The quantum mechanical angular momentum vector l has the length $|l| = \hbar\sqrt{l(l+1)}$. This l vector precesses incessantly around an axis of constant direction. Its maximum projection on the axis is $l \times \hbar$.

The azimuthal quantum number can be given a physical interpretation, which is easy to imagine concretely. It can be interpreted as *a measure of the quantum mechanical angular momentum of the electron in its orbital*. The quantum mechanical angular momentum is equal to $\hbar\sqrt{l(l+1)}$ (Figure 2.11).

The classical angular momentum $n \times h/2\pi$ (Bohr's postulate 3, page 47) cannot represent the angular momentum because this would contradict Heisenberg's uncertainty principle. Both direction and size cannot have exact values simultaneously. The quantum mechanical angular momentum has an exact value, whereas its *direction* is totally undetermined, owing to the precession motion of the l vector. This is in agreement with the uncertainty principle.

The l values are obviously very characteristic of the electron orbitals in the atom. A special terminology has been introduced to represent them as single letters, shown in Table 2.1.

Table 2.1 Nomenclature for electron orbitals.

l	0	1	2	3
Name	s	p	d	f

The Magnetic Quantum Number

The quantum number m_l is called the *magnetic quantum number*. It can have the values

$$m_l = 0, \pm 1, \pm 2, \ldots, \pm l \tag{2.31}$$

which means $2l + 1$ different values. It can be interpreted in the following way. A result of quantum mechanics is that

- The orbital angular momentum in atoms is space quantized.

The quantization is defined by the magnetic quantum number. Allowed orientations of the orbital angular momentum vector are those which give a projection on a fixed direction equal to the magnetic quantum number m_l multiplied by $\hbar$, i.e. $h/2\pi$ (Figure 2.12).

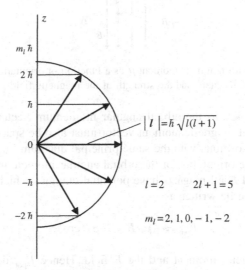

Figure 2.12 Each energy level splits into $2l + 1$ different energy levels owing to different directions of the angular momentum vector. The components of l in the direction of the magnetic field is quantized and has the value $m_l \times \hbar$. The directions correspond to orbitals with slighly different energies. Hence each l level splits into $2l + 1$ slightly different energy levels in a magnetic field.

Figure 2.13 Magnetic moment of a single coil with current I.

The name of the third quantum number indicates that magnetic effects must be involved. This topic will be discussed below.

An electron in its orbit corresponds to a circular current (a coil with one round). The circular current is equivalent to a small magnet or rather a magnetic plate with a magnetic moment, proportional to the current (Figure 2.13).

In analogy with this, an orbital in an atom can be regarded as a single coil with a negative current and consequently as a small magnetic plate (Figure 2.14) with a magnetic moment in the opposite direction to that in Figure 2.13.

Each orbital in an atom is characterized by its principal quantum number n and its azimuthal quantum number l. As explained on page 58, the latter defines the size of the orbital angular momentum vector. In analogy with Figure 2.10, the

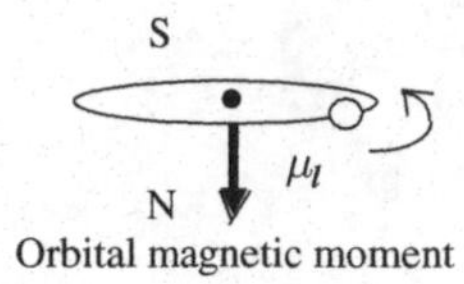

Figure 2.14 Orbital magnetic moment of an electron.

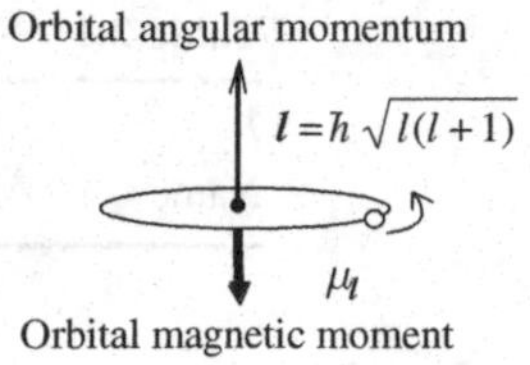

Figure 2.15 Orbital angular momentum and orbital magnetic moment of an electron.

direction is that shown in Figure 2.15. It is also important to note that the orbital angular momentum and the orbital magnetic moment always have opposite directions.

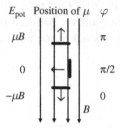

Figure 2.16 Potential energy of a magnet with magnetic moment μ as a function of orientation. The magnetic potential energy is defined as the negative scalar product of the magnetic moment and the strength of the magnetic field.

Because the $\boldsymbol{\mu}_l$ vector is rigidly coupled to the orbital angular momentum vector (Figure 2.15) and the orbital angular momentum is space quantized, the orbital magnetic moment vector must also be space quantized in a magnetic field.

In the absence of a magnetic field, all orbitals with the same principal quantum number n and azimutal quantum number l have the same energy, independent of the orientation of the orbital angular momentum.

On the other hand, in a magnetic field $\boldsymbol{B}$ all magnets have potential energies which depend on their orientation relative to the magnetic field (Figure 2.16). This can be written as

$$E_{\text{pot}} = -\boldsymbol{\mu} \cdot \boldsymbol{B} = -\mu B \cos \varphi \qquad (2.32)$$

where φ is the angle between the magnetic moment and the $\boldsymbol{B}$ field. Hence E_{pot} depends on the direction in space of $\boldsymbol{\mu}$. Figure 2.16 shows the values of E_{pot} for some special positions of $\boldsymbol{\mu}$.

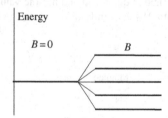

Figure 2.17 Splitting of degenerate energy levels in a magnetic field. The figure shows the energy levels of the five l orbitals ($m_l = 0$, ± 1, ± 2) indicated in Figure 2.12, without and with a magnetic field of strength B.

If the magnet is free to move, it orients itself in the direction of the magnetic field as this position correspond to the minimum potential energy $[-\mu B(\varphi = 0)]$. A maximum effort has to be made to turn the magnet in the opposite position $[+\mu B(\varphi = \pi)]$. This is also valid for the orbital magnet moment with $\mu = \mu_l$. The potential energies of the three positions in Figure 2.16 will be $E_{\text{pot}} = 0, \pm \mu_l B$.

Hence orbitals with different values of their magnetic quantum numbers have different energies in a magnetic field. In the presence of an external magnetic field B, the $2l + 1$ orbitals with different orientations in Figure 2.12 have slightly different energies. The single energy level splits into $2l + 1$ energy levels in a magnetic field (Figure 2.17). The equidistant splittings between successive levels are proportional to the strength of the magnetic field B. If the magnetic field is zero, all the $2l + 1$ energy levels coincide. They are said to be degenerate.

The introduction of the magnetic quantum number m_l makes it possible to explain the very puzzling fact that spectral lines split into narrow multiplets in a magnetic field. This phenomenon is called the *Zeeman effect*, which is treated on pages 69–71.

The Spin Quantum Number

The fourth quantum number, which characterizes the orbital electrons in the atoms, is the *spin quantum number*, postulated and introduced by Uhlenbeck and Goudsmit. By use of the electron spin, several unexplained phenomena could be explained, among them the Zeeman effect and the periodic table of the elements.

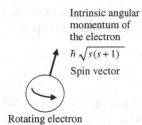

Figure 2.18   Intrinsic angular momentum of the electron.

Each electron in an orbital behaves as if it had an intrinsic rotation (Figure 2.18). Hence the rotating electron must have an intrinsic angular momentum. In analogy with the orbital angular momentum $|l| = h/2\pi \times \sqrt{l(l+1)}$ or $\hbar\sqrt{l(l+1)}$, the intrinsic angular momentum of the electron can be described by a so-called spin vector: $|s| = h/2\pi \times \sqrt{s(s+1)}$ or $\hbar\sqrt{s(s+1)}$, where s is the spin quantum number.[1] The value of the spin s is a half integer:

$$s = \tfrac{1}{2} \tag{2.33}$$

Like the orbital angular momentum, the spin vector is space quantized in a magnetic field. Two rotation directions are possible, as shown in Figures 2.19 and 2.20, which correspond to

$$m_s = \pm\tfrac{1}{2} \tag{2.34}$$

The spinning of the electron corresponds to electrical charge in motion and causes an intrinsic magnetic moment $\boldsymbol{\mu}_s$ of the electron.

The spin vector is coupled to the orbital angular momentum vector, either parallel or antiparallel. Hence each (n, l, m_l) orbital splits into two levels with slightly different energy. $m_s = \pm\tfrac{1}{2}$ is often symbolized by arrows: spin up $= \uparrow$ and spin down $= \downarrow$.

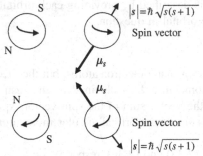

Figure 2.19   Intrinsic angular momentum vector (electron spin) and magnetic moment vector of the electron.

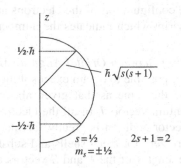

Figure 2.20   Space quantization of the spin vector.

---

[1] The spin quantum number s (italic) should not be confused with the letter s (roman) in Table 2.1 on page 59. The latter is merely a letter designation of a special type of electron orbital.

2.4.3 Quantum Numbers of Many-electron Atoms

The Pauli Principle

The number of possible orbitals in a hydrogen atom and in hydrogen-like ions is determined by the four quantum numbers n, l, m_l and m_s. Another consequence of quantum mechanics is that

- In an atom no more than one electron can have the same set of values of the four quantum numbers n, l, m_l and m_s.

This is the *Pauli exclusion principle*. This important principle makes it possible to understand the theoretical background of the periodic table of the elements and the existence of shells and subshells of the electrons around the atom nucleus.

Table 2.2 gives the theoretical background to the periodic system of the elements, which was primarily published in 1869 by Mendeleev, based on an empirical knowledge of the chemical properties of the elements.

Quantum Numbers of Many-electron Atoms

The chemical properties of atoms are determined by the outer electrons. When the atomic number Z increases, the electrons successively fill the empty orbitals. The orbitals with the lowest energy are filled primarily as equilibrium corresponds to lowest possible energy of the atom.

Electron Configuration

As a simple example of an electron orbital description of an atom, we choose sodium, which has the atomic number 11 and hence 11 orbital electrons. The electron configuration of Na (Figure 2.21) can be written as $1s^2 2s^2 2p^6 3s$. The *first* figures are the principal quantum numbers and the letters s and p represent the value of the azimuthal quantum number according to Table 2.1 on page 59.

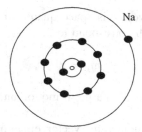

Figure 2.21 Configuration of the Na atom.

The configuration of the electrons around the nucleus of an atom is described by providing each orbital designation with a superscript which indicates the number of electrons in the respective orbital in question.

Coupling Between Orbital Angular Momentum and Electron Spin

It is not possible to obtain exact solutions of the Schrödinger equation for many-electron atoms, but the theoretical background will be the same as that given above. The total orbital angular momentum L is obtained as the resultant of the angular momentum vectors l_i of all the electrons. The total spin vector S is the vector sum of the spin vectors s_i of all the electrons. The vector addition is greatly facilitated by the fact that the vector sum of the orbital angular momentum and the resultant spin are zero for filled shells and subshells.

The lengths of the L and S vectors are calculated using Equations (2.35) and (2.36), respectively:

$$|L| = \hbar\sqrt{L(L+1)} \tag{2.35}$$

$$|S| = \hbar\sqrt{S(S+1)} \tag{2.36}$$

where $\hbar = h/2\pi$.

Table 2.2 Maximum numbers of electrons in orbitals according to the Pauli principle.

n	1	2		3			4			
l	0	0	1	0	1	2	0	1	2	3
Type of orbital	s	s	p	s	p	d	s	p	d	f
m_l	0	0	1 0 −1	0	1 0 −1	2 1 0 −1 −2	0	1 0 −1	2 1 0 −1 −2	3 2 1 0 −1 −2 −3
m_s	$\pm\frac{1}{2}$	$\pm\frac{1}{2}$	$\pm\frac{1}{2}$ $\pm\frac{1}{2}$ $\pm\frac{1}{2}$	$\pm\frac{1}{2}$	$\pm\frac{1}{2}$ $\pm\frac{1}{2}$ $\pm\frac{1}{2}$	$\pm\frac{1}{2}$ $\pm\frac{1}{2}$ $\pm\frac{1}{2}$ $\pm\frac{1}{2}$ $\pm\frac{1}{2}$	$\pm\frac{1}{2}$	$\pm\frac{1}{2}$ $\pm\frac{1}{2}$ $\pm\frac{1}{2}$	$\pm\frac{1}{2}$ $\pm\frac{1}{2}$ $\pm\frac{1}{2}$ $\pm\frac{1}{2}$ $\pm\frac{1}{2}$	$\pm\frac{1}{2}$ $\pm\frac{1}{2}$ $\pm\frac{1}{2}$ $\pm\frac{1}{2}$ $\pm\frac{1}{2}$ $\pm\frac{1}{2}$ $\pm\frac{1}{2}$
No. of electrons	2	8		18			32			
Name of shell	K	L		M			N			

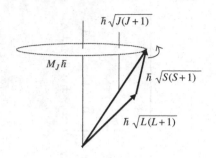

Figure 2.22 *L–S* coupling in atoms. The L and S vectors precess around the resultant J vector. J is the total angular momentum. The vector J precesses around a given axis.

The vectors L and S are coupled to each other in analogy with the l and s vectors in a single electron orbital (page 61). The coupling occurs in such a way that the vectors L and S precess around the resultant vector J (Figure 2.22). The quantum number J is an integer or a half integer depending on whether the quantum number S is an integer or half integer:

$$J = L + S, \; L + S - 1, \; L + S - 2, \; \ldots, \; |L - S| \tag{2.37}$$

The vector J also performs a precession motion.

Hence there are $2S + 1$ alternatives owing to the electron spin, which is the *multiplicity* of the energy level. The resultant precesses around a fixed direction and does not contradict Heisenberg's uncertainty priciple. $2S + 1$ is the *multiplicity* of the energy level.

In a magnetic field, the resultant J vector is space quantized in analogy with the space quantization of the l vector of a single electron orbital. The projection of the vector J equals $M_J\hbar$ where the magnetic quantum number M_J can have the values

$$M_J = J, J - 1, \ldots, 0, \ldots, -J \tag{2.38}$$

Nomenclature

To describe the energy state of an atom, a terminology corresponding to s, p, d, f, $\ldots$, for an orbital electron has been introduced to express the values of L and S. An energy state is called S, P, D, F, $\ldots$ when $L = 0, 1, 2, 3, \ldots$. The multiplicity $2S + 1$ is used as a superscript on the S, P, D, F, $\ldots$ states to indicate their multiplicity. Some common examples are given in Table 2.3.

Table 2.3 Electron configurations and electronic states of some atoms.

Atom	Electron configuration	L	S	Possible electronic states
H	$1s$	0	$1/2$	^{2}S
He	$1s^2$	0	0	^{1}S
O	$1s^2\,2s^2\,2p^4$	0, 1, 2	0, 1	^{1}S, ^{3}P, ^{1}D
Na	$1s^2\,2s^2\,2p^6\,3s$	0	$1/2$	^{2}S
Al	$1s^2\,2s^2\,2p^6\,3s^2\,3p$	1	$1/2$	^{2}P
Cl	$1s^2\,2s^2\,2p^6\,3s^2\,3p^5$	1	$1/2$	^{2}P

The resultant J is often given as a subscript after the capital letter, for example the doublet ^{2}P consists of the levels $^2P_{3/2}$ and $^2P_{1/2}$.

Energy Levels

The energy levels of the orbitals can be calculated by accurate measurements of the spectral lines of the atoms and identification of their upper and lower levels. The behaviour of the lines in a magnetic field supports the analysis (page 69).

The energy difference of an 'electron jump' from an upper energy level to a lower energy level, which defines the frequency of the emitted spectral line, is simply

$$h\nu = E' - E'' \tag{2.39}$$

where

ν = frequency of the photon

E', E'' = energy of the upper and lower energy level, respectively.

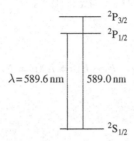

Figure 2.23   Splitting of the yellow doublet line of Na.

The multiplet levels E' and E'' (doublets, triplets, . . .) are often so close that they appear as a single energy state. The spectral lines that involve such states appear as single lines in low resolution but consist of several lines. An example is the yellow line of sodium (Figure 2.23). The lines are said to have a fine structure.

2.5 Quantum Mechanics and Probability: Selection Rules

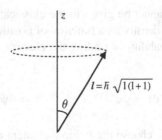

Figure 2.24   The precession motion of the *l* vector of an electron illustrates Heisenberg's uncertainty principle.

Some quantities in quantum mechanics have exact values without violating Heisenberg's uncertainty principle. The orbital angular momentum of an electron or an atom has an exact value. Consequently, it is impossible to determine the direction in space of the vector (page 58). The angle θ in Figure 2.24 is exact but the vector precesses around the z axis and its direction cannot be determined.

Another example is the quantum mechanical width ΔE of energy states in the atom. In the ground state the mean lifetime Δt is infinite, which allows $\Delta E = 0$ according to Heisenberg's uncertainty principle (pages 54–55). All excited states, however, have a finite quantum mechanical width.

Example 2.2

An atom is excited when a valence electron in its lowest orbital is moved to a higher orbital. The average lifetime in the excited state is of the magnitude 10^{-8} s. After this time it returns to the lowest orbital and emits a photon. Calculate the quantum mechanical uncertainty in energy of the upper orbital and the uncertainty in the frequency of the emitted photon.

Solution:

In this case, Heisenberg's uncertainty principle can be written as [Equation (2.25) on page 55]

$$\Delta E' \geq \frac{h}{4\pi\Delta t} = \frac{6.6 \times 10^{-34}\,\text{J s}}{4\pi \times 10^{-8}\,\text{s}} \approx 5 \times 10^{-27}\,\text{J}$$

We also have

$$h\nu = E' - E'' \Rightarrow h\Delta\nu = \Delta E'$$

which gives

$$h\Delta\nu = \frac{h}{4\pi\Delta t} \quad \Rightarrow \quad \Delta\nu = \frac{1}{4\pi \times \Delta t} = \frac{1}{4\pi \times 10^{-8}} \approx 10^7\,\text{s}^{-1}$$

Answer:

The quantum mechanical uncertainty in the energy level is of magnitude 5×10^{-27} J. The corresponding uncertainty in the frequency of the emitted photon is of magnitude $10^7\,\text{s}^{-1}$.

2.5.1 Electron Density Distribution

The exact position of an electron in an orbital cannot be given in the classical sense. Instead, the electron distribution can be described by a 'probability cloud' or probability density as a function of position around the nucleus of the atom. The electron density distribution can be described as the probability of finding the electron inside a given volume element by studying the function [compare Equation (2.16) on page 53]:

$$\Psi\Psi^*\mathrm{d}\tau = \psi e^{-2\pi i t}\psi^* e^{+2\pi i t}\mathrm{d}\tau = \psi\psi^*\mathrm{d}\tau \tag{2.40}$$

This function can be illustrated graphically. If we choose the volume element $\mathrm{d}\tau$ as a spherical shell $4\pi r^2\mathrm{d}r$, we can illustrate the probability of finding the electron in a special orbital in any direction at a distance from the nucleus within the interval r to $r + dr$ graphically as a function of r. Figure 2.25 give some simple examples. If $l = 0$ the electron density distribution shows spherical symmetry.

The electron clouds are three-dimensional. Figure 2.26 shows a sketch of a cross-section of the electron clouds for the s orbitals given in Figure 2.25.

The values of r in Figure 2.25, which corresponds to the most likely probability, has approximately the same magnitude as the major semi-axis of the corresponding elliptical Bohr orbit.

The electron distribution is shown in two different ways in Figures 2.25 and 2.26. A third method, which has the advantage of space visualization, is used in Figure 2.27a. This shows schematically the wave functions of an s electron and three p-electrons, i.e. the configuration sp^3. Figure 2.27b shows the corresponding values of $\psi\psi^*$.

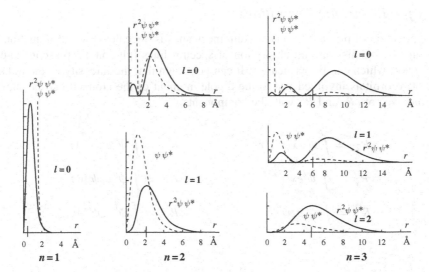

Figure 2.25   Probability density distribution of the electron of the H atom for $n = 1, 2, 3$ as a function of r. $1\,\text{Å} = 0.1\,\text{nm}$ or $10^{-10}\,\text{m}$. The vertical axis is double. It represents the values of $r^2\psi\psi^*$ (full curves) and $\psi\psi^*$ (dotted curves). The vertical marks on the r axis indicate the major semi-axis of the corresponding Bohr orbits for $n = 1, 2, 3$. Reproduced from G. Hertzberg, *Atomic Spectra and Atomic Structure*, Dover, 1944.

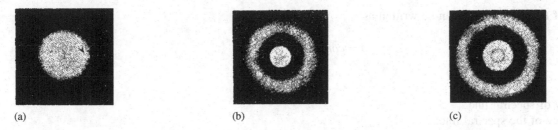

(a)   (b)   (c)

Figure 2.26   Electron clouds for (a) $n = 1$, (b) $n = 2$ and $l = 0$ and (c) $n = 3$ and $l = 0$. Reproduced from G. Hertzberg, *Atomic Spectra and Atomic Structure*, Dover, 1944.

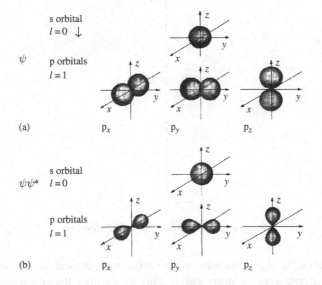

Figure 2.27   (a) Wave functions of an s electron and three p electrons. The figure shows that the wave functions of the p electrons are perpendicular to each other.

(b) $\psi\psi^*$ of the s and p electrons. The three p wave functions in (a) and (b) overlap partly but have been separated into three figures for better visualization. © 1971 Ingvar Lindgren, Jan Nilsson, Olof Beckman, Erik Karlsson, Toivelemb Kivikas (Published by Almqvist & Wiksell Forlag AV, Stockholm).

2.5.2 *Selection Rules for Electronic Transitions*

Another very important application of probability in quantum mechanics is the calculation of transition probabilities from one orbital or energy state to another. Emission and absorption of spectral lines (photons) represents so-called dipole radiation. The dipole moment is a vector, which can be written $e\,\mathbf{r}$. It can be shown that the intensity of the radiation is proportional to the wave functions of the two orbitals involved and to the dipole moment of the atom. It can be shown that the intensity is proportional to the amplitude vectors $\mathbf{R}'$ and $\mathbf{R}''$ with the components

$$R_x^{'/''} = \int \psi' ex\psi''^* d\tau \qquad\qquad R_x^{''/'} = \int \psi'' ex\psi'^* d\tau \qquad\qquad (2.41)$$

$$R_y^{'/''} = \int \psi' ey\psi''^* d\tau \qquad\qquad R_y^{''/'} = \int \psi'' ey\psi'^* d\tau \qquad\qquad (2.42)$$

$$R_z^{'/''} = \int \psi'' ez\psi''^* d\tau \qquad\qquad R_z^{''/'} = \int \psi'' ez\psi'^* d\tau \qquad\qquad (2.43)$$

where
  $\mathbf{R}$ = amplitude vector of a spectral line
  x, y, z = coordinates
  ψ', ψ'' = amplitudes of the wave functions of the orbitals involved.

The intensity of the spectral line can be written as

$$I = constant \times \nu^3 R^{'/''} R^{''/'} \qquad\qquad (2.44)$$

where
  I = intensity of spectral line
  ν = frequency of the spectral line.

Figure 2.28 Emission or absorption line.

When the $\mathbf{R}$ vectors, which involve the eigenfunctions of the two orbitals, are calculated for all transitions between arbitrary levels in the atoms, the values become zero in many cases. This means that the transition is forbidden and cannot occur. Simple *selection rules,* i.e. changes of quantum numbers for allowed transitions, can be listed (Table 2.4).

An s electron in an H atom can only jump to a p orbital and a p electron can jump to an s orbital or a d orbital. An electron can jump from a ^{3}P state to another ^{3}P state or to a ^{3}S or a ^{3}D state.

Another example is the normal Zeeman effect, which can be explained by application of selection rules.

Table 2.4 Examples of selection rules.

Hydrogen atom	Many-electron atoms
Δn = any value	Δn = any value
	$\Delta J = 0, \pm 1$
	except $J' = 0$ to $J'' = 0$
$\Delta l = \pm 1$	$\Delta L = 0, \pm 1$
	simultaneously $\Delta l = \pm 1$
	for the 'jumping' electron
$\Delta m_l = +1$ or -1	$\Delta M_J = 0, \pm 1$
$\Delta s = 0$	$\Delta S = 0$

2.5.3 Zeeman Effect

On pages 59, we found that each orbital energy level, characterized by the quantum number l, splits into $2l+1$ energy levels in a magnetic field. The same phenomenon occurs in many-electron atoms where the resultant $\boldsymbol{J}$ vector is space quantized (Figure 2.29). The magnetic moment μ_J in its different space positions opposite to the space quantized $\boldsymbol{J}$ vector in a magnetic field is the cause of the splitting of the energy levels (page 60).

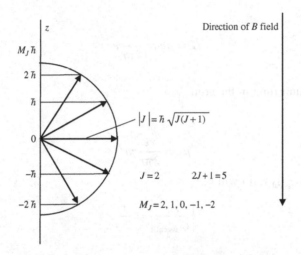

Figure 2.29 Provided that $S = 0$ each energy levels splits into $2J+1$ different energy levels owing to different directions of the $\boldsymbol{J}$ vector. Its component in the direction of the magnetic field is quantized and has the values $M_l\hbar$. The directions correspond to orbitals with slightly different energies. Hence each J level splits into $2J+1$ slightly different energy levels in a magnetic field.

If we apply Equation (2.32) (page 60) it is possible to calculate the intervals between successive energy levels when the atoms (light source) are (is) placed in a magnetic field $\boldsymbol{B}$. The energies of the split levels is expressed as in Equation (2.32), $E_{\text{pot}} = \mu B \cos\theta$ where φ is the angle between $\boldsymbol{\mu}$ and $\boldsymbol{B}$ as the B field is directed in the opposite z direction (Figure 2.16 on page 60).

On page 60, we found that the magnetic and angular momentums are rigidly coupled to each other. Therefore, $|\boldsymbol{\mu}|$ is proportional to $|\boldsymbol{J}|$ with the magnitude $\hbar\sqrt{J(J+1)}$ and we can write

$$E_{\text{pot}} = -constant \times \sqrt{J(J+1)}B\cos(\pi - \varphi) \tag{2.45}$$

where $\pi - \varphi$ is the angle between $\boldsymbol{J}$ and $\boldsymbol{B}$. Figure 2.29 and Equation (2.45) give the following expression for $-\cos(\pi - \varphi)$:

$$-\cos(\pi - \varphi) = \cos\varphi = \frac{M_J}{\sqrt{J(J+1)}} \tag{2.46}$$

The constant in Equation (2.45) is the so-called *Bohr magneton*. Its value is derived in the box below.

Magnetic Moment of an Orbital Electron. Calculation of the Bohr Magneton

Classical Theory

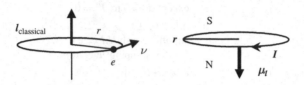

Consider an electron in its orbit around its nucleus (left figure). Its angular momentum $l_{classical}$ is defined as

$$|l_{classical}| = |r \times mv| = mvr \tag{1'}$$

The electron in its orbit is equivalent to an electric current in a coil with radius r (right figure). Such a a magnetic plate (page 59) has the magnetic moment

$$|\mu_l| = I\,\pi r^2 \tag{2'}$$

In this case, the current I will be

$$I = ne = \frac{v}{2\pi r}e \tag{3'}$$

where
   n = the number of rounds per unit time in the orbit.

Equations (2') and (3') give

$$\mu_l = \frac{ve}{2\pi r}\pi r^2 \tag{4'}$$

After reduction of μ_l, the $\mu_l/l_{classical}$ ratio will be

$$\left|\frac{\mu_l}{l_{classical}}\right| = \frac{e}{2m} \tag{5'}$$

Quantum Mechanical Theory

If we replace the classical angular momentum with the quantum mechanical expression for l in Equation (5'), we obtain

$$\mu_l = \frac{e}{2m}\hbar\sqrt{l(l+1)} = constant \times \sqrt{l(l+1)} \tag{6'}$$

or

$$\mu_l = \mu_B\sqrt{l(l+1)} \tag{7'}$$

Identification of equations (6') and (7') shows that the constant μ_B is

$$\mu_B = \frac{e\hbar}{2m} \tag{8'}$$

The constant μ_B is called the *Bohr magneton*. It has the value $9.274 \times 10^{-24}\,\text{A m}^2$.

Equation (7') is valid for all the orbital electrons in an atom and also for their resultant L. Provided that the resultant $S = 0$ it is also valid for the total angular momentum J in this case as $J = L$ when $S = 0$.

However, it should be pointed out that L can *not* be replaced by J when $S \neq 0$ in Equation (7').

Inserting the value of the Bohr magneton:

$$\mu_B = \frac{e\hbar}{2m} \tag{2.47}$$

into Equation (2.45), we obtain with the aid of Equation (2.46)

$$E_{pot} = \mu_B B M_J \tag{2.48}$$

The intervals between two successive energy levels (Figure 2.30) are

$$\Delta E_{pot} = \mu_B B \Delta M_J \tag{2.49}$$

As $\Delta M_J = 1$, we finally obtain

$$\Delta E_{pot} = \mu_B B = \frac{e\hbar}{2m} B \tag{2.50}$$

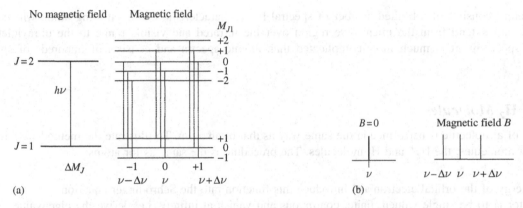

Figure 2.30 Normal Zeeman effect. (a) Sketch of electron jumping from a d orbital to a p orbital in the absence and the presence of a magnetic field.
(b) A single spectral line splits into three lines in a magnetic field. The splitting is proportional to the strength of the magnetic field.

The intervals ΔE_{pot} between the successive levels are proportional to the strength of the magnetic field and *equal* in the upper and lower states (Figure 2.30). Owing to the selection rules in Table 2.4 on page 69, the spectral line without a magnetic field is split into three narrow lines and no more in a magnetic field (Figure 2.30).

The normal Zeeman effect described above is the simplest splitting in a magnetic field. It is valid only when $S = 0$. The more complicated anomalous Zeeman effect [$S \neq 0$ and Equation (2.48) is *not* valid] is beyond the scope of this book but it may be mentioned as an illustration that the two yellow Na lines (Figure 2.23 on page 65) split into four and six lines, respectively, in a magnetic field. The reason for the anomalous Zeeman effect will be explained in Chapter 6 (page 319).

Spectral lines emitted by atoms in electrical fields also split into several components. This so-called Stark effect is not as simple to interpret in general terms as the normal Zeeman effect.

2.6 The Quantum Mechanical Model of Molecular Structure

The most powerful tool for the experimental investigation of atomic and molecular structure is the analysis of the spectra of atoms and molecules. The structures of molecules are much more complicated than those of atoms.

The inner filled electron shells change very little when a molecule is formed but the electrons in the outer orbital, the valence electrons, change from atomic orbitals into molecular orbitals. To characterize the molecular orbitals we will introduce the terms σ, π and δ in analogy with s, p and d orbitals (Table 2.5). These designations have a physical significance.

Table 2.5 Nomenclature for molecular orbitals.

m_l	0	1	2	3
Name	σ	π	δ	ϕ

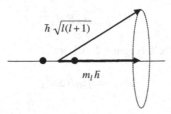

Figure 2.31 Projection of the l vector on the nuclear axis.

They tell us the value of the projection of the l vector, in terms of $\hbar$, on the axis of the nuclei of a diatomic molecule (Figure 2.31).

The spectrum of an atom consists of a limited number of spectral lines, characteristic of the emitting atom, whereas the spectrum of a molecule can extend from the microwave region over the infrared and visible range to the ultraviolet part of the electromagnetic spectrum. It is much more complicated than atomic spectra and consists of hundreds of spectral lines.

2.6.1 The H$_2^+$ and H$_2$ Molecules

The theoretical analysis of a molecule is performed in the same way as that of an atom. To illustrate the method we will first discuss the two simplest molecules, the H_2^+ and H_2 molecules. The procedure is the same as for atoms:

1. Find the potential energy of the orbital electron and introduce this function into the Schrödinger equation.
2. Find the conditions for ψ to be single valued, finite, continuous and vanish at infinity, i.e. derive the eigenvalues of the orbital.
3. Solve the Schrödinger equation, i.e. derive the eigenfunctions corresponding to each eigenvalue.

The H$_2^+$ Molecule

The potential energy of the electron can be written as (Figure 2.32)

$$E_{\text{pot}} = \frac{e^2}{4\pi\varepsilon_0}\left(-\frac{1}{r_1} - \frac{1}{r_2} + \frac{1}{r}\right) \tag{2.51}$$

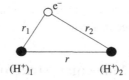

Figure 2.32 Classical picture of the H_2^+ ion.

The solution of the Schrödinger equation is rather complicated in this case and will be omitted here. Instead, the deduction of the eigenvalues will be based on the following qualitative discussion and illustrated graphically.

Consider a hydrogen atom together with its eigenfunction and a proton (H$^+$) at a large distance from each other. The two alternatives are sketched in Figures 2.33a and b. The two undisturbed wave functions ψ_1 and ψ_2 of the electron are equally probable. ψ_1 and ψ_2 are the amplitudes of the wave functions of H atoms 1 and 2, respectively.

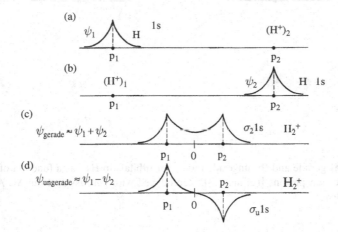

Figure 2.33   (a) The electron belongs to proton 1. (b) The electron belongs to proton 2. (c) The 'gerade' wave function corresponds to a stable H$_2^+$ molecule. (d) The 'ungerade' wave function corresponds to an unstable H$_2^+$ molecule. Reproduced with permission from M. Alonso and E. Finn, *Fundamental University Physics*. © Addison-Wesley.

When the nuclei approach each other, the orbital of the electron changes. The electron no longer belongs to the initial H nucleus. It belongs to either nucleus and is shared between them.

$$H^+ + H \leftrightarrow H + H^+ \tag{2.52}$$

The resulting eigenfunction of the molecule H$_2^+$ will neither be ψ_1 nor ψ_2 but a combination of the two wave functions. There are two alternatives:

$$\psi_{\text{gerade}} = \psi_1 + \psi_2 \tag{2.53}$$

or

$$\psi_{\text{ungerade}} = \psi_1 - \psi_2 \tag{2.54}$$

The subscripts gerade (German for even) and ungerade (German for odd) of the resulting molecular wave function amplitudes means that the function is symmetrical or antisymmetrical, respectively, in space.

The symmetrical wave function ψ_{gerade} does not change sign when indices 1 and 2 are exchanged in Equation (2.53). ψ_{ungerade} is antisymmetrical as the wave function changes sign when the indices 1 and 2 are exchanged in Equation (2.54).

The amplitudes of the molecular wave functions can be used to calculate the probability of finding the electron at various positions in the vicinity of the nuclei. This is shown in Figure 2.33.

The electron cloud in the region between the protons helps to keep the molecule together owing to the attraction between the protons and the negative charge in spite of the repulsion between the two nuclei (Figures 2.33c and 2.34a). The electron cloud on either side of the nuclei will help to pull the nuclei apart. Obviously the *gerade* molecular orbital acts as 'glue' between the nuclei and results in a stable H$_2^+$ molecule. The *ungerade* molecular orbital has a very low electron density between the protons and corresponds to an instable H$_2^+$ molecule (Figures 2.33d and 2.34b).

Each eigenfunction ψ has its own eigenvalue E. In this case we have one *gerade* and one *ungerade* orbital with corresponding eigenvalues, i.e. one stable and one unstable energy state. If the energy of each H atom initially is zero, we have two energy states of the molecule with very different energies. The energy difference between the states is called the *exchange energy*, which only can be explained by quantum mechanics. Exchange energy is a very common and important phenomenon in connection with molecule formation.

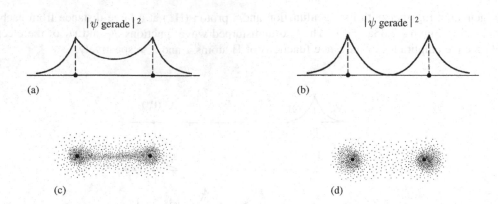

Figure 2.34   Electron density for (a) gerade and (b) ungerade molecular orbitals in H_2^+ as a function of position along the proton–proton axis and in a plane that contains the two protons [(c) and (d)]. Reproduced with permission from M. Alonso and E. Finn, *Fundamental University Physics*. © Addison-Wesley.

The difference in energy of the two states depends on the distance between the two protons. The energy of the two states as a function of the distance between the two protons is given in Figure 2.35. The lower curve represents the ground state of the H_2^+ molecule. The upper state is unstable. ψ_{gerade} is said to be a *bonding wave function*, whereas $\psi_{ungerade}$ is an *antibonding wave function*.

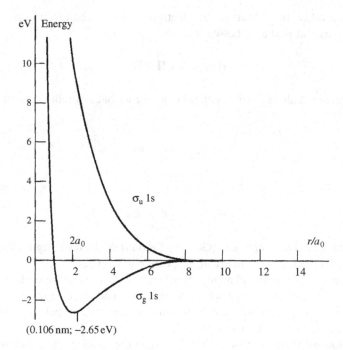

Figure 2.35   The energy of the two lowest energy states in the H_2^+ molecule as a function of the distance between the protons. The equilibrium distance between the two nuclei $= 2a_0$, where a_0 is the Bohr radius, i.e. the radius of the $n = 1$ orbital in the H atom. As $a_0 = 0.053\,\mathrm{nm}$, the equilibrium distance corresponds to $0.106\,\mathrm{nm}$. Reproduced with permission from M. Alonso and E. Finn, *Fundamental University Physics*. © Addison-Wesley.

The H_2 Molecule

The simplest neutral molecule is the H_2 molecule, which consists of two protons and two electrons. The potential energy of this four-body system consists of six terms corresponding to the interaction between electron 1 and the two protons, electron 2 and the two protons and an interaction term for the electrons and one for the protons. The solution of the Schrödinger equation for the H_2 molecule will therefore be much more complicated than that of H_2^+.

Instead, we will use the results for the H_2^+ molecule and combine them with the knowledge that each orbital can accomodate two electrons with anti-parallel electron spin vectors (Table 2.2 on page 63). Both electrons have bonding wave functions, which results in a very stable molecule. Similarly, there is an unstable state where the two electrons have parallel electron spin vectors (Figure 2.36). Figure 2.37 shows the energy of the molecule as a function of the internuclear distance.

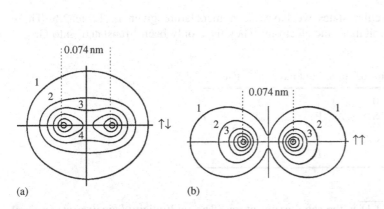

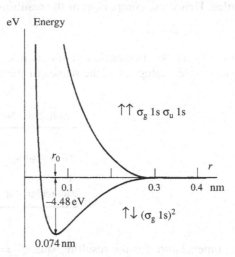

Figure 2.36 Probability density distribution of the electrons in (a) the ground state and (b) the lowest repulsive state of the H_2 molecule. The figures 1, 2, 3, ... represent the relative probability densities. Reproduced with permission from G. Hertzberg, *Atomic Spectra and Atomic Structure*, Dover, 1944.

Figure 2.37 Energy of the H_2 molecule as a function of the distance between the two protons. Reproduced with permission from M. Alonso and E. Finn, *Fundamental University Physics*. © Addison-Wesley.

Two binding electrons represent a stronger bond in the H_2 molecule than the single electron in the H_2^+ molecule. This results in a shorter internuclear distance between the protons at equilibrium and a more stable molecule, i.e. a deeper minimum of the ground state, than the corresponding quantities found in the H_2^+ molecule. The calculated curve in Figure 2.37 agrees well but not completely with the experimental results. We will come back to this discrepancy in Section 3.3.3 in the next chapter, where the bond of the H_2 molecule is further discussed (pages 108–109).

At infinite distance between the two protons there are no longer any forces acting between them. Then the H_2 molecule has dissociated into two H atoms in their ground states. The concepts of bonding energy and dissociation energy will be discussed in Chapter 3.

2.7 Diatomic Molecules

2.7.1 Classification of Electronic States

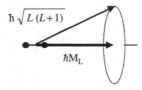

Figure 2.38 Precession of the orbital angular momentum vector L around the internuclear axis.

In Section 2.4.3, we have seen that the resulting orbital angular momentum $|L| = \hbar\sqrt{L(L+1)}$ is constant (Figure 2.38). In a diatomic molecule, the L vector precesses around the internuclear axis and its component M_L can only have integer values owing to the same type of space quantization as the orbital angular momentum vector l in atoms (page 58 and Figure 2.12 on page 59):

$$M_L = 0, \pm 1, \pm 2, \ldots, \pm L \tag{2.55}$$

The orbital electrons move in the electrostatic field of the two nuclei, which is very strong. The consequence of this is that the energy states with different M_L values have widely different energies and the internuclear axis is a very special direction. The change of sign of M_L corresponds classically to a reversed rotation direction of the electrons, which does not change the energy. Hence the energy levels which correspond to $\pm M_L$ are degenerate and coincide. The L vector is no longer important, only the component along the nuclear axis makes sense.

In order to distinguish between atomic and molecular quantities, ordinary letters are used for atoms and Greek letters for molecules. Hence the component of the resulting orbital angular momentum of a diatomic molecule is called Λ:

$$\Lambda = 0, 1, 2, \ldots, L \tag{2.56}$$

In analogy with the nomenclature for atoms, for molecular states we have the nomenclature given in Table 2.6. These designations are analogous to the names of the atomic orbitals of the electrons. They have only been 'translated' into Greek.

Table 2.6   Nomenclature for atomic and molecular states

L	0	1	2	3
Atomic state	S	P	D	F
Λ	0	1	2	3
Molecular state	Σ	Π	Δ	Φ

The nomenclature for the resulting spin $|S| = \hbar\sqrt{S(S+1)}$ is the same as for atoms. The multiplicity (superscript) is equal to $2S+1$. For example, if $S = \frac{1}{2}$ and $\Lambda = 1$, the state is called a $^2\Pi$ state.

2.7.2 The Rigid Rotator

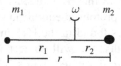

Figure 2.39   Model of a diatomic molecule. The interatomic distance is $r = r_1 + r_2$.

The simplest possible model of a rotating diatomic molecule is the dumb-bell model. The two nuclei are assumed to be point-like with masses m_1 and m_2. They are rigidly joined at a distance r at the ends of a weightless rod (Figure 2.39).

The dumb-bell rotates around its centre of mass. The classical expression of its rotation energy is

$$E_{\text{rot}} = \frac{1}{2}I\omega^2 \tag{2.57}$$

where
I = moment of inertia of the dumb-bell
ω = angular frequency $= 2\pi\nu_{\text{rot}}$
ν_{rot} = rotation frequency.

The moment of inertia of the dumb-bell in Figure 2.39 is

$$I = m_1 r_1^2 + m_2 r_2^2 \tag{2.58}$$

It can be shown that

$$I = \frac{m_1 m_2}{m_1 + m_2} r^2 \tag{2.59}$$

where

$m_{1,2}$ = masses of the two nuclei
r = internuclear distance.

If we introduce the so-called reduced mass

$$\mu = \frac{m_1 m_2}{m_1 + m_2} \tag{2.60}$$

into Equation (2.59), we obtain

$$I = \mu r^2 \tag{2.61}$$

i.e. the dumb-bell can be replaced by a single mass μ at distance r from the origin which rotates around an axis through origin perpendicular to the plane of rotation. Such a system is called a *simple rigid rotator* (Figure 2.40).

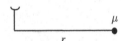

Figure 2.40 Simple rigid rotator.

Quantum Mechanical Theory of the Rigid Rotator

In order to determine the possible energy states of a rigid rotator according to quantum mechanics, we have to solve the Schrödinger equation. No potential energy is associated with the rotation as long as the rotator is completely rigid. Hence we obtain

$$\frac{\partial \psi^2}{\partial x^2} + \frac{\partial \psi^2}{\partial y^2} + \frac{\partial \psi^2}{\partial z^2} + \frac{2\mu}{\hbar^2}(E - 0)\psi = 0 \tag{2.62}$$

Single-valued, finite and continuous solutions of eigenfunctions ψ are obtained if the eigenvalues are

$$E_{\text{rot}} = \frac{\hbar^2 J(J+1)}{2\mu r^2} = \frac{\hbar^2 J(J+1)}{2I} \tag{2.63}$$

where J is the *rotation quantum number*, which is an integer, $J = 0, 1, 2, 3, \ldots$.

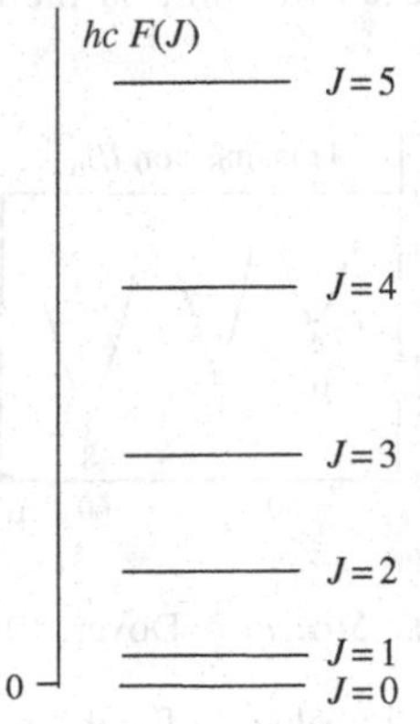

$hc\, F(J)$

——— $J=5$

——— $J=4$

——— $J=3$

——— $J=2$

——— $J=1$
0 — ——— $J=0$

Figure 2.41 Energy levels of a rigid rotator. $E_{\text{rot}} = 0$ at $J = 0$.

The rotational energy levels (Figure 2.41) are written by tradition as

$$E_{\text{rot}} = hcF(J) = hcBJ(J+1) \qquad (2.64)$$

B is is obtained from Equations (2.63) and (2.64):

$$B = \frac{\hbar^2}{2hc\mu r^2} \qquad (2.65)$$

and is called the *rotation constant*. $F(J)$ and B are measured in m^{-1}.

Combination of Equation (2.63) and the classical expression

$$E_{\text{rot}} = \frac{P^2}{2I}$$

gives an expression for the angular momentum P:

$$P = \hbar\sqrt{J(J+1)} \qquad (2.66)$$

Equation (2.66) agrees with the general result for the angular momentum vectors in atoms.

We introduce the vector $\boldsymbol{J}$, which represents the *angular momentum* vector and has the magnitude

$$|\boldsymbol{J}| = \hbar\sqrt{J(J+1)} \qquad (2.67)$$

Combining Equation (2.66) with $P = I\omega = I \times 2\pi\nu_{\text{rot}}$, we obtain the *rotational frequency*:

$$\nu_{\text{rot}} = \frac{h}{4\pi^2 I}\sqrt{J(J+1)} \qquad (2.68)$$

Example 2.3

In the far-infrared part of spectrum, a pure rotational spectrum of HCl has been found. When continuous infrared radiation above $30\,\mu m$ passes through a tube containing gaseous HCl molecules and is allowed to hit a detector, a number of nearly equidistant absorption maxima are found. The figure shows some of them.

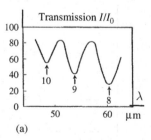

(a)

Reproduced from G. Hertzberg, *Atomic Spectra and Atomic Structure*, Dover, 1944.

Calculate the distance between the atoms in the HCl molecule for the most abundant Cl isotope with the aid of the measured wavenumbers $(83.0, 104.1, 124.30, 145.0, 165.5, 185.9, 206.4, 226.5) \times 10^2\,m^{-1}$.

Solution:

(b)

According to Equation (2.64), the rotational energy levels can be written as

$$E_{rot} = hcF(J) = hcBJ(J+1) \tag{1'}$$

The differences between the energy levels correspond to the absorption maxima:

$$\Delta E_{rot} = hc[F(J) - F(J-1)] = hcB[J(J+1) - (J-1)J]$$

or

$$\Delta E_{rot} = hc \times 2BJ$$

If we calculate the difference between the absorption maxima, we obtain

$$\Delta^2 E_{rot} = hc \times 2B \tag{2'}$$

The wavenumber difference between the absorption maxima is obviously equal to $2B$. It can be calculated from the experimental values given in the text: $(21.1, 20.2, 20.73, 20.48, 20.35, 20.52, 20.12) \times 10^2 \, m^{-1}$. Using the method of least squares, the mean value can be calculated as $20.7 \times 10^2 \, m^{-1}$ and we obtain $2B = 20.7 \times 10^2 \, m^{-1}$.

From Equations (2.63) and (2.64), we obtain

$$hcB = \frac{\hbar^2}{2I} = \frac{\hbar^2}{2} \frac{1}{m_1 m_2 r^2 / (m_1 + m_2)}$$

or

$$r = \sqrt{\frac{\hbar^2}{2} \frac{m_1 + m_2}{Bhcm_1 m_2}} = \sqrt{\frac{h}{8\pi^2 c} \frac{1+35}{10.35 \times 10^2 \times 1 \times 35 \times 1.66 \times 10^{-27}}}$$

$$r = 1.29 \times 10^{-10} \, m$$

Answer:

The distance between the H and Cl atoms in the ground state of the $H^{35}Cl$ molecule is 0.13 nm.

2.7.3 The Harmonic Oscillator

The simplest possible model of a vibrating diatomic molecule is the harmonic oscillator. The two atoms oscillate towards and away from each other around an equilibrium distance. It can be shown that this system is equivalent to a spring with force constant k and equilibrium length r_e acting on a mass point of mass μ, where μ is the reduced mass of the two atom masses m_1 and m_2 (Figure 2.42).

$$\mu = \frac{m_1 m_2}{m_1 + m_2} \tag{2.60}$$

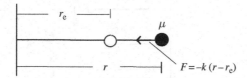

Figure 2.42 Harmonic oscillator.

The force F acting on the mass is proportional to the deviation from equilibrium:

$$F = -k(r - r_e) = \mu \frac{d^2(r - r_e)}{dt^2} \tag{2.69}$$

where
  r = distance between the atoms at time t
  r_e = equilibrium distance between the atoms
  μ = reduced mass of the two atoms.

If for simplicity we introduce $x = r - r_e$, we obtain

$$-kx = \mu \frac{d^2 x}{dt^2} \tag{2.70}$$

Integration of Equation (2.70) gives

$$x = x_0 \sin\left(\sqrt{\frac{k}{\mu}} t + \varphi_0\right) = x_0 \sin(2\pi \nu_{osc} t + \varphi_0) \tag{2.71}$$

where the frequency of oscillations is

$$\nu_{osc} = \frac{1}{2\pi} \sqrt{\frac{k}{\mu}} \tag{2.72}$$

Quantum Mechanical Theory of the Harmonic Oscillator

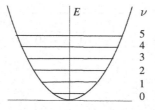

Figure 2.43 Energy levels of a harmonic oscillator in a potential curve $kx^2/2$.

The potential energy of the harmonic oscillator is

$$E_{pot} = \frac{kx^2}{2} \qquad (2.73)$$

where $x = r - r_e$. This expression is introduced into the Schrödinger equation:

$$\frac{\partial\psi^2}{\partial x^2} + \frac{2\mu}{\hbar^2}\left(E - \frac{kx^2}{2}\right)\psi = 0 \qquad (2.74)$$

The condition for the existence of single-valued, finite and continuous solutions of ψ is

$$E_{vibr} = \hbar\sqrt{\frac{k}{\mu}}\left(v + \frac{1}{2}\right) \qquad (2.75)$$

where v is the *vibration quantum number*. It can only have integer values, $v = 0, 1, 2, 3, \ldots$.

The equidistant energy levels of the harmonic oscillator are shown in Figure 2.43. Even if the vibration quantum number is zero, the molecule is vibrating and not at rest at the bottom of the potential curve. This is in complete agreement with Heisenberg's uncertainty principle. *If* there were no vibrations, both *position* (equilibrium distance between the nuclei) and *momentum* (zero) would have exact values, which is forbidden according to quantum mechanics. However, this is not the case and Heisenberg's uncertainty principle is *not* contradicted.

By combining Equations (2.72) and (2.74) we obtain the following expression for the eigenvalues, i.e. the vibration energies:

$$E_{vibr} = hcG(v) = hc\frac{\nu_{osc}}{c}\left(v + \frac{1}{2}\right)$$

or

$$E_{vibr} = hc\omega_e\left(v + \frac{1}{2}\right) \qquad (2.76)$$

where the *vibrational constant* ω_e is

$$\omega_e = \frac{\nu_{osc}}{c} = \frac{1}{2\pi c}\sqrt{\frac{k}{\mu}} \qquad (2.77)$$

As can be seen from Equation (2.76), the SI unit of $G(v)$ and ω_e is m^{-1}. ω_e is the traditional designation of the vibration constant. It should *not* be confused with the angular frequency ω, which has the dimension s^{-1}.

2.7.4 Eigenfunctions and Probability. Density Distribution of the Harmonic Oscillator

To each eigenvalue of the harmonic oscillator there is one eigenfunction. The square of an eigenfunction represent the probability of finding the oscillator at a particular distance r between the atoms. The probability densities as a function of the deviation from the equibrilium position of the oscillator are illustrated graphically for some values of v in Figure 2.44.

Figure 2.44 shows that the probability is reminiscent of a classical oscillator for high values of v. A pendulum, for example, spends more time at the end points than at the lowest point of its orbit where its velocity has a maximum. Consequenly, the probability is higher at the end points than elsewhere.

The lower the vibrational quantum number is, the higher will be the quantum mechanical probability deviation from the classical result. This is best shown for $v = 0$. Here the probability of finding the oscillator in either end position or turning point is small, which is in sharp contradiction with everyday experience.

So far we have treated the rigid rotator and the harmonic oscillator separately. In reality, vibrational and rotational transitions occur simultaneously in molecules. If an electron transition also occurs simultaneously, a band spectrum is obtained in the visible part of the electromagnetic spectrum. This will be discussed briefly below.

If no electronic transition occurs, simultaneous vibrational and rotational transitions give a rotational–vibrational spectrum, i.e. a number of spectral lines in the near–infrared part of spectrum.

If no electronic transition or vibrational transition occurs, only rotational transitions occur and a rotational spectrum appears in the far-infrared part of spectrum.

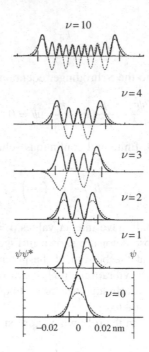

Figure 2.44 Probability density distributions of the harmonic oscillator for different values of the vibrational quantum number. The dotted curves represent the eigenfunctions. The continuous curves are the squared eigenfunctions. Reproduced from G. Hertzberg, *Atomic Spectra and Atomic Structure*, Dover, 1944.

2.7.5 Spectra of Diatomic Molecules

So far it has been possible to solve the Schrödinger equation only for a few of the simplest molecules. Most of the present knowledge of energy levels, internuclear distances and other quantities of molecules is based on accurate measurements and analysis of molecular spectra. Lines are classified and their quantum numbers determined with the aid of the selection rules, derived from quantum mechanics. If the analysis has been successful, the molecular constants can be calculated.

A spectral line, emitted by an atom, originates from a transition of an electron from one orbital to another. An electronic transition between two orbitals in a molecule may be accompanied by a simultaneous change of the vibrational and rotational state of the molecule:

$$h\nu = \Delta E_e + \Delta E_{vibr} + \Delta E_{rot} \tag{2.78}$$

The contributions are ranked in decreasing order. Pure rotational spectra appear in the microwave or far-infrared region and involve no change of the outermost electron or the vibration state. A necessary condition is that the molecule has an electrical dipole moment.

Combined vibrational and rotational spectra of diatomic molecules appear in the infrared region. Each value of ΔE_{vibr} is accompanied by many alternative values of ΔE_{rot}.

An electronic transition ΔE_e is accompanied by a number of alternative changes ΔE_{vibr}. Each combination of $\Delta E_e + \Delta E_{vibr}$ is accompanied by many alternative values of ΔE_{rot}. The spectrum is a band system consisting of a number of rotation bands, one for each vibrational transition.

Anharmonic Oscillator

Comparison between theory and experiments shows that the harmonic oscillator and the rigid rotator are models which have to be improved to give better agreement. Figure 2.45 shows that the energy curve is not symmetrical like that in Figure 2.43 on page 80. The average internuclear distance increases with increasing v values. Therefore, the effective ω_e [Equation (2.77) on page 80] decreases with increasing v. Better agreement with experiments is obtained if correction terms are added to the vibrational energy in Equation (2.76), which gives

$$\Delta E_{vibr} = hc\left[\omega_e'\left(v'+\frac{1}{2}\right)-\omega_e'x_e'\left(v'+\frac{1}{2}\right)^2\right]-hc\left[\omega_e''\left(v''+\frac{1}{2}\right)-\omega_e''x_e''\left(v''+\frac{1}{2}\right)^2\right] \tag{2.79}$$

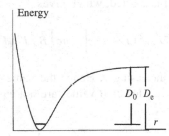

Figure 2.45 Potential curve of a diatomic molecule.

The potential curve of an anharmonic oscillator is shown in Figure 2.45. It is no longer symmetrical. On the left-hand side it is very steep, which corresponds to strong repulsive forces which appear when the electron clouds of the two nuclei starts to overlap. On the right-hand side the attraction forces becomes successively weaker when the internuclear distance increases. Finally the molecule dissociates into two free atoms. The *dissociation energy* D_e is defined as the depth of the potential curve. It can be calculated with the aid of the vibrational constants:

$$D_e = \frac{\omega_e^2}{4\omega_e x_e} \quad \text{or} \quad D_0 = \frac{\omega_e^2}{4\omega_e x_e}-\frac{\omega_e}{2} \tag{2.80}$$

If the depth of the potential curve is measured from the energy level $v = 0$ instead of the bottom of the potential curve, the dissociation energy D_0 is obtained.

In many cases, the so-called *Morse function* is a good approximation of the true potential energy curve of molecules. It is frequently used as a potential energy function:

$$E_{pot} = D_e\left[1-e^{-\beta(r-r_e)}\right]^2 \tag{2.81}$$

where β is a constant which depends of the reduced mass and the dissociation energy of the molecule:

$$\beta = constant \times \omega_e\sqrt{\frac{\mu}{D_e}} \tag{2.82}$$

The Morse function for the H_2 molecule is shown in Figure 2.46 as a dotted curve. It is very convenient to use the Morse function for most diatomic molecules.

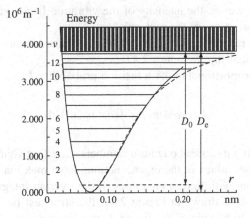

Figure 2.46 The potential curve of the ground state of the H_2 molecule. The energy is given in $10^6\,m^{-1}$, which is a multiple of the unit of wavenumber. If figures on the energy axis are multiplied by hc, the energy expressed in joules is obtained. A convenient unit for dissociation energies is electron volts. Reproduced from G. Hertzberg, *Atomic Spectra and Atomic Structure*, Dover, 1944.

Nonrigid Rotator

The molecules become stretched for high J values, owing to the increasing centrifugal force, when the rotation becomes more rapid. The moment of inertia increases and the value of the constant B in Equation (2.64) on page 78 decreases. Small correction terms are added when Equation (2.64) is applied, which gives

$$\Delta E_{\text{rot}} = hc\left[B'_{v'}J'\left(J'+1\right) - D'_{v'}J'^2\left(J'+1\right)^2 \right] - hc\left[B''_{v''}J''\left(J''+1\right) - D''_{v''}J''^2\left(J''+1\right)^2 \right] \tag{2.83}$$

The internuclear distance also increases with increasing v. Hence the values of B and D change with v. This is indicated with a subscript v in Equation (2.83). The upper and lower B values are not equal.

2.7.6 Selection Rules

Selection Rules for Electronic Transitions

The *selection rules for electronic transitions* are much more complicated in molecules than in atoms and will not be discussed in detail here. The molecular orbital angular momentum couples to the rotation of the molecule and the quantum number J represents the total angular momentum.

For the most common coupling cases, the following selection rule for the electronic transition is valid:

$$\Delta\Lambda = 0, \pm 1 \tag{2.84}$$

In analogy with atoms the multiplicity of the molecular states do not change:

$$\Delta S = 0 \tag{2.85}$$

Hence $^2\Sigma - {}^2\Sigma$ and $^3\Pi - {}^3\Sigma$ transitions may occur whereas $^3\Sigma - {}^1\Sigma$ and $^2\Delta - {}^2\Sigma$ are both forbidden, i.e. have zero intensity.

Selection Rules and Band Intensities for Vibrational Transitions

For *vibrational transitions* in the near-infrared region, the selection rule for the vibrational quantum number is

$$\Delta v = \pm 1 \tag{2.86}$$

If an electronic transition occurs simultaneously with the change of the vibration state of the molecule, there is *no restriction for the vibrational quantum number*. However, the intensity of the vibration bands varies widely. This can be understood by considering the electronic and vibrational states of the molecule.

The electronic transition occurs very rapidly and the internuclear distance changes very little during this time. Hence the transition must be represented by a vertical line in Figure 2.47.

The intensity of a vibration band is proportional to the transition probability I:

$$I = constant \times \int \psi(v')\psi(v'')^* \mathrm{d}r \tag{2.87}$$

Vibrational bands are denoted by their vibrational quantum numbers. The first figure is v' and the second is v''.

Figure 2.44 on page 82 shows that the values of the eigenfunctions have maxima at the turning points for high values of v and in the middle for $v = 0$. Transitions with maximum values of the overlap integral in Equation (2.87) correspond to the strongest vibrational transitions. In the case drawn in Figure 2.47, the strongest bands are the 4–0 band, the 0–2 band and the 1–5 band. The transition on the left-hand side of the figure does not appear at all as the molecule dissociates before the transition occurs.

Figure 2.48 shows two typical band spectra in the ultraviolet and visible regions of the spectrum.

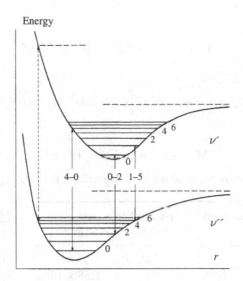

Figure 2.47   The most intense vibration bands of a band system are found by studying the transition probability of the tentative bands. Reproduced with permission from M. Alonso and E. Finn, *Fundamental University Physics*. © Addison-Wesley.

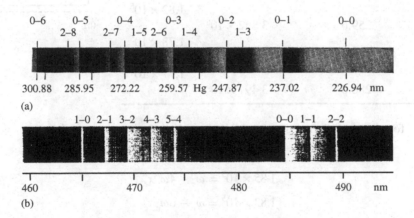

Figure 2.48   (a) The γ bands of the NO molecule in emission. The strongest bands originate from $v' = 0$.
(b) Part of the band spectrum of the AlO molecule.

Example 2.4

Figure 2.48a shows the ultraviolet band spectrum of NO, a $^2\Sigma - ^2\Pi$ transition where $^2\Pi$ is the ground state of the molecule. Use the information in the picture to make a rough estimation of the dissociation energy D_0 of the NO molecule in its ground state, expressed in eV.

Solution:

The dissociation energy can be calculated by use of Equation (2.80) on page 83 if the ω_e'' and $\omega_e''' x_e''$ values are known. These values can be estimated from the vibration bands given in the figure.

We assume that the origin of each band is approximately equal to its sharp band head and choose the six most intense bands, related to the lower energy state, for our calculation.

The difference between the lower vibration levels $v'' + 1$ and v'' can be written as

$$\Delta G_{v''+1/2} = G(v'' + 1) - G(v'') \tag{1'}$$

or, in terms of ω_e'' and $\omega_e'' x_e''$:

$$\Delta G_{v''+1/2} = \omega_e'' (v''+1.5) - \omega_e'' x_e'' (v''+1.5)^2 - \left[\omega_e'' (v''+0.5) - \omega_e'' x_e'' (v''+0.5)^2\right]$$

After reduction, we obtain

$$\Delta G_{v''+1/2} = \omega_e'' - \omega_e'' x_e'' \left[(v''+1.5)^2 - (v''+0.5)^2\right]$$

which can be reduced to

$$\Delta G_{v''+1/2} = \omega_e'' - 2\omega_e'' x_e'' (v''+1) \tag{2'}$$

The wavelengths of the strongest band heads are read from the figure and the $\Delta G_{v''+1/2}$ values are calculated.

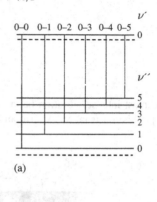

(a)

Vibration band $v'-v''$	λ of band head (nm)	Wavenumber (m^{-1})	$\Delta G_{v''+1/2}$ (m^{-1})
0–0	226.9	4.407×10^6	
			1.88×10^5
0–1	237.0	4.219×10^6	
			1.85×10^5
0–2	247.9	4.034×10^6	
			1.82×10^5
0–3	259.6	3.852×10^6	
			1.78×10^5
0–4	272.2	3.674×10^6	
			1.77×10^5
0–5	286.0	3.497×10^6	

Equation (2′) is applied for $v'' = 0, 1, 2, 3, 4$ and 5, which gives the equation system

$$1.88 \times 10^5 = \omega_e'' - 2\omega_e'' x_e''$$
$$1.85 \times 10^5 = \omega_e'' - 4\omega_e'' x_e''$$
$$1.82 \times 10^5 = \omega_e'' - 6\omega_e'' x_e''$$
$$1.78 \times 10^5 = \omega_e'' - 8\omega_e'' x_e''$$
$$1.77 \times 10^5 = \omega_e'' - 10\omega_e'' x_e''$$

This equation system consists of five equations and has only two unknown quantities. In order to minimize the errors and use all the information, we solve the system by the method of least squares. We multiply each equation with the coefficient of the first variable ω_e'' and then add the equations. Then we repeat the same procedure for the second variable and obtain the following equations:

$$9.10 \times 10^5 = 5\omega_e'' - 30\omega_e'' x_e'' \tag{3'}$$
$$54.02 \times 10^5 = 30\omega_e'' - 220\omega_e'' x_e'' \tag{4'}$$

The solution of this equation system is

$$\omega_e'' = 1.91 \times 10^5 \text{m}^{-1} \quad \omega_e'' x_e'' = 0.0145 \times 10^5 \text{ m}^{-1}$$

The dissociation energy is obtained from the equation

$$D_0 = \frac{\omega_e^2}{4\omega_e x_e} - \frac{\omega_e}{2} = \frac{(1.91 \times 10^5)^2}{4 \times 0.0145 \times 10^5} - \frac{1.91 \times 10^5}{2} \text{ m}^{-1} = 61.9 \times 10^5 \text{ m}^{-1}$$

$$D_0 = \frac{hc}{e} \times 61.9 \times 10^5 = 7.7 \text{ eV}$$

Answer:

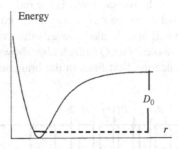

The dissociation energy of NO in its ground state is calculated as 7–8 eV.
This value is higher than the table values 5.3 eV (D_0) and 6.5 eV (D_e) for the 2Π ground state of NO. The relative error in $\omega_e'' x_e''$ is much larger than that in ω_e''. The original equation system shows that $\Delta G_{v''+1/2}$ in the last equation with the highest statistical weight of $\omega_e'' x_e''$ is too high. This results in a too small $\omega_e'' x_e''$ value and a too high value of the dissociation energy. The discrepancy between the band origins and the band heads is another source of error.

Selection Rules for Rotational Transitions

The selection rules for the quantum number J are, in case of a simultaneous electron jump,

$$\Delta J = 0, \pm 1 \qquad \text{for } \Delta\Lambda = \pm 1 \tag{2.88}$$

$$\Delta J = \pm 1 \quad \text{for } \Delta\Lambda = 0 \text{ (and no jump)} \tag{2.89}$$

The condition $\Delta J = \pm 1$ is also valid for pure rotational transitions in the far-infrared region and for rotation–vibration bands in the near-infrared region where $\Delta\Lambda = 0$ because there is no electron jump at all.

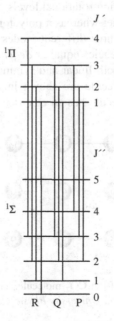

Figure 2.49 Rotational lines of a $^1\Pi - ^1\Sigma$ transition.

The *rotational structure* of each vibration band has two or three branches corresponding to the selection rule (2.88) and (2.89). If $\Delta J = J' - J'' = 0$ we have a Q branch. For an R branch $\Delta J = +1$. If $\Delta J = -1$ the branch is called a P branch (Figure 2.49).

Both the R and P branches start at the band origin. The R branch initially 'walks' towards the violet side of spectrum, turns and 'walks' towards the red if $B'' > B'$. The P branch 'walks' towards the red all the time. Such bands are said to have a band head and to be shaded to the red. This is the most common case because r_e'' normally is smaller than r_e'.

If $B' > B''$, the P branch 'walks' to the red, turns and 'walks' towards the violet. The R branch 'walks' towards the violet all the time and the band is shaded towards the violet. The Q branch also shows a head, i.e. is shaded.

The lines are named after their J'' values. Hence the first lines in the branches in Figure 2.49 will be R(0), Q(1) and P(2).

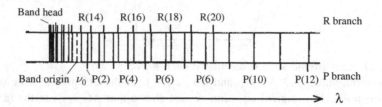

Figure 2.50 Rotational structure of one of the bands in an $^1\Sigma - {}^1\Sigma$ transition. The bands are shaded towards the red. Reproduced with permission from G. Hertzberg, *Atomic Spectra and Atomic Structure*, Dover, 1944.

Characteristic of a rotation band is the maximum or minimum frequency of one (R or P branch) or two (Q branch and R or P branch) of the branches, which gives the band a 'shaded' appearance with a sharp edge, normally towards the red (Figure 2.50). This is shown in Figure 2.48 on page 85. The explanation is the turn of the branches explained above.

2.8 Polyatomic Molecules

The energy levels and the spectra of polyatomic molecules are even more complicated than those of diatomic molecules. Polyatomic molecules in general have several sets of vibration energy levels as the atoms can vibrate with different vibration frequencies between pairs of atoms. This gives a manifold of vibration bands which may overlap and belong to different vibrational systems.

Each vibration level is associated with corresponding rotational levels. A diatomic molecule has only one moment of inertia for an axis perpendicular to an axis through the nuclei whereas a polyatomic molecule in general has three different moments of inertia in the principal directions x, y and z and three different modes of rotation. The result is a complicated spectrum.

Symmetry may make some of the vibration frequencies equal or *degenerate*, which makes the spectrum somewhat simpler. An example of this is the CO_2 molecule, which is both linear and symmetrical.

The normal modes of vibration of the CO_2 molecule are shown in Figure 2.51. The condition for a vibration–rotation spectrum in the near-infrared region is that the molecule has an electrical dipole moment.

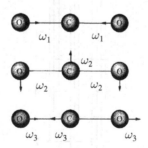

Figure 2.51 The three alternative modes of vibration of the CO_2 molecule. The sizes of the atoms are not drawn to scale. Reproduced with permission from M. Alonso and E. Finn, *Fundamental University Physics*. © Addison-Wesley.

The CO_2 molecule has no permanent electrical dipole moment. In the vibration mode ω_1 there is no induced dipole moment either. In the case of ω_2 and ω_3 electrical dipole moments are induced during the vibrations. Hence infrared bands corresponding to the frequencies ω_2 and ω_3 appear but not the bands corresponding to ω_1.

Most polyatomic molecules are nonlinear and asymmetric. Vibrations in such a molecule result in a rapidly varying electrical dipole moment of the molecule.

Summary

■ *Quantum Mechanics*

Quantum mechanics is the basis of modern atomic and molecular physics. The basic tool is the Schrödinger equation:

$$-\frac{\hbar^2}{2m}\left(\frac{\partial^2 \Psi}{\partial x^2} + \frac{\partial^2 \Psi}{\partial y^2} + \frac{\partial^2 \Psi}{\partial z^2}\right) + \left(E - E_{\text{pot}}\right)\Psi = \mathrm{i}\hbar\frac{\partial \Psi}{\partial t}$$

The relationship between the time-dependent wave function and the amplitude of the matter wave is

$$\Psi = \psi\mathrm{e}^{-\mathrm{i}\frac{E}{\hbar}t}$$

Matter Waves

Wavelength of matter wave:

$$\lambda_{\text{deB}} = \frac{h}{mv}$$

Differential equation of the *amplitude* of the matter wave:

$$\frac{\partial^2 \psi}{\partial x^2} + \frac{\partial^2 \psi}{\partial y^2} + \frac{\partial^2 \psi}{\partial z^2} + \frac{2m}{\hbar^2}\left(E - E_{\text{pot}}\right)\psi = 0$$

This equation is valid under stationary conditions (standing waves). It is often referred to as the time-independent Schrödinger equation. ψ is a function of x, y and z.

Physical Interpretation of the Wave Function

The probability of finding a particle within a volume element $\mathrm{d}x\mathrm{d}y\mathrm{d}z$ at stationary conditions is

$$|\Psi|^2\,\mathrm{d}x\mathrm{d}y\mathrm{d}z = \Psi\Psi^*\mathrm{d}x\mathrm{d}y\mathrm{d}z = \psi\psi^*\mathrm{d}x\mathrm{d}y\mathrm{d}z$$

Heisenberg's Uncertainty Principle

$$\Delta y\Delta p_y \geq \frac{h}{4\pi} \quad \text{or} \quad \Delta E\Delta t \geq \frac{h}{4\pi}$$

■ *Quantum Numbers in Atoms and Their Intepretation*

Solution of the Schrödinger equation for an orbital electron of an H atom gives expressions for the eigenfunction ψ (standing waves) and the eigenvalue E.

The four quantum numbers that characterize the orbitals of the electrons appear in the solutions:

- n = principal quantum number = number of 'shell'
  $n = 1, 2, 3$ for the K, L and M shells, respectively
- l = azimuthal quantum number
  $l = 0, 1, 2, 3, \ldots, (n-1)$

Length of orbital angular momentum vector of the electron:

$$|l| = \hbar\sqrt{l(l+1)}$$

- m_l = magnetic quantum number
  $m_l = 0, \pm1, \pm2, \pm3, \ldots, \pm l$

The projection of the orbital angular momentum vector in the direction of a magnetic field $= m_l\hbar$.

- s = spin quantum number = $1/2$

Length of spin vector of the electron:

$$|s| = \hbar\sqrt{s(s+1)}$$

The projection of the spin vector on the direction of a magnetic field $= m_s\hbar$.

$$m_s = \pm\tfrac{1}{2}$$

Nomenclature for Atomic Orbitals

l	0	1	2	3
Name	s	p	d	f

■ *Many-electron Atoms*

The Pauli Principle
In an atom no more than one electron can have the same set of values of the four quantum numbers n, l, m_l and m_s.

Quantum Numbers of Many-electron Atoms
The total orbital angular momentum L and the total spin vector S are obtained as the resultants of the angular momentum vectors l_i and s_i, respectively, of all the electrons:

$$|L| = \hbar\sqrt{L(L+1)}$$
$$|S| = \hbar\sqrt{S(S+1)}$$

The vector addition is greatly facilitated by the fact that filled shells and subshell have zero L and S resultants.

L and S are coupled in such a way that the quantum number J defines the resultant vector J. The quantum number J is an integer or half-integer when the quantum number S is an integer or half- integer, respectively:

$$J = L+S, L+S-1, L+S-2, \ldots, L, \ldots, |L-S|$$

The multiplicity of the energy state $= 2S+1$.

In a magnetic field, the resultant J vector is space quantized:

$$M_J = J, J-1, \ldots, 0, \ldots, -J$$

Nomenclature

L	0	1	2	3
Atomic state	S	P	D	F

The multiplicity of an electronic is indicated by a superscript before the symbol. ^{2}P means that $L = 1$ and $S = \tfrac{1}{2}$.

■ *Selection Rules for Electronic Transitions*

Spectral lines are emitted when an electron 'jumps' from an upper to a lower orbital. The intensities of the lines depend on the electron density distributions in the two electronic states and can be calculated from the two wave functions. In some cases the probability of a transition is zero. Simple transition rules have been found.

Important selection rules:

Hydrogen atom	Many-electron atoms
$\Delta n =$ any value	$\Delta n =$ any value
	$\Delta J = 0, \pm 1$
	except $J' = 0$ to $J'' = 0$
$\Delta l = \pm 1$	$\Delta L = 0, \pm 1$
	simultaneously $\Delta l = \pm 1$
	for the 'jumping' electron
$\Delta m_l = +1$ or -1	$\Delta M_J = 0, \pm 1$
$\Delta s = 0$	$\Delta S = 0$

■ Zeeman Effect

If a light source is placed in a magnetic field, the spectral lines split into components owing to splitting of the energy levels of the atoms:

$$\Delta E_{pot} = \mu_B B \Delta M_J$$

In the case of the normal Zeeman effect (singlet lines, e.g. $\Delta S = 0$), the energy difference between successive energy levels is the same for all energy levels ($\Delta M_J = 1$):

$$\Delta E_{pot} = \mu_B B = \frac{e\hbar}{2m} B$$

where the Bohr magneton $\mu_B = e\hbar/2m$.

Each line is split from a single line with wavenumber ν into three lines with wavenumbers $\nu \pm \Delta \nu$, owing to the selection rules.

In the case of the anomalous Zeeman effect ($\Delta S \neq 0$), the lines are split into many more components than three and the simple theory is *not* valid.

■ Molecular Orbitals in Diatomic Molecules

The structures of molecules are more complicated than those of atoms. An exact solution of the Schrödinger equation is possible only for the simplest molecules (H_2^+ and H_2).

Electrons in molecular orbitals have the same quantum numbers, which describe spin and orbital angular momentum, as in atoms. More important than the *l* vector is its projection on the axis between the two nuclei of a diatomic molecule.

Nomenclature for Molecular Orbitals

m_l	0	1	2	3
Name	σ	π	δ	ϕ

The H_2^+ Molecule

When the protons are far apart, the electron in an H_2^+ molecule can belong to either proton:

$$H(\text{wave function } \psi_1) + p_2 \quad \text{or} \quad p_1 + H(\text{wave function } \psi_2)$$

When the protons approach, two 'composed' wave functions are formed:

$$\psi_{gerade} = \psi_1 + \psi_2 \quad \text{and} \quad \psi_{ungerade} = \psi_1 - \psi_2$$

The *gerade* symmetrical wave function corresponds to a stable state and high electron density between the nuclei. The electron cloud act as 'glue'. The *ungerade* asymmetric wave function corresponds to an unstable state and low electron density between the nuclei.

The phenomenon when two states with highly different energies are formed from two initial states with equal energies is called the *exchange energy*. It is an common and important quantum mechanical effect.

Classification of Electronic States

In a diatomic molecule, the resulting angular momentum vector L precesses around the internuclear axis. Its component along the axis is space quantized:

$$M_L = 0, \pm1, \pm2, \ldots, \pm L$$

The orbital electrons move in the electrostatic field of the two nuclei, which is very strong.

Energy states with different M_L values, independent of sign, have widely different energies. The vector L is no longer important, only the component Λ along the nuclear axis makes sense.

Nomenclature for Molecular States

Λ	0	1	2	3
Molecular state	Σ	Π	Δ	Φ

The multiplicity of a molecular electronic state is the same as for atoms. If $S = \frac{1}{2}$ and $\Lambda = 0$, the electronic state is $^2\Sigma$.

■ *Rigid and Nonrigid Rotators*

The simplest possible model of a rotating diatomic molecule is the dumb-bell model. The dumb-bell rotates around its centre of mass:

$$E_{\text{rot}} = \frac{1}{2}I\omega^2$$

$$I = \mu r^2 = \frac{m_1 m_2}{m_1 + m_2} r^2$$

The solution of the Schrödinger equation gives the possible energy states for the rigid rotator:

$$E_{\text{rot}} = \frac{\hbar^2 J(J+1)}{2\mu r^2} = \frac{\hbar^2 J(J+1)}{2I}$$

or

$$E_{\text{rot}} = hcF(J) = hcBJ(J+1)$$

$$B = \frac{\hbar^2}{2hc\mu r^2}$$

The rotator becomes slightly stretched during rotation. In such a nonrigid rotator, the increase in r is considered by a correction term and the rotational energy can be written as

$$E_{\text{rot}} = hcF(J) = hc\left[BJ(J+1) - DJ^2(J+1)^2\right]$$

■ *Harmonic and Anharmonic Oscillators*

The simplest possible model of a vibrating diatomic molecule is the harmonic oscillator. The two atoms oscillate toward and away from each other around an equilibrium distance. The solution of the Scrödinger equation give the energy levels:

$$E_{vibr} = \hbar \sqrt{\frac{k}{\mu}} \left(v + \frac{1}{2} \right) \quad \mu = \frac{m_1 m_2}{m_1 + m_2}$$

or

$$E_{vibr} = hcG(v) = hc\omega_e \left(v + \frac{1}{2} \right) \quad \omega_e = \frac{1}{2\pi c} \sqrt{\frac{k}{\mu}}$$

As the potential well of a diatomic molecule is not symmetrical, the oscillator is anharmonic. This is considered by a correction term and the vibrational energy can be written as

$$E_{vibr} = hc \left[\omega_e \left(v + \frac{1}{2} \right) - \omega_e x_e \left(v + \frac{1}{2} \right)^2 \right]$$

Even if $v = 0$, the molecule is always vibrating.

The dissociation of a diatomic molecule is approximately

$$D_e = \frac{\omega_e^2}{4\omega_e x_e}$$

or

$$D_0 = \frac{\omega_e^2}{4\omega_e x_e} - \frac{\omega_e}{2}$$

■ *Selection Rules in Molecular Transitions*

Electron jumps from one orbital to another in molecules are usually accompanied by simultaneous changes of the vibrational and rotational energy states. Changes of the vibrational and rotational energy states with no electronic transition also occur in addition to rotational transitions alone.

Electronic Transitions

$$\Delta\Lambda = 0, \pm 1 \quad \text{and} \quad \Delta S = 0$$

Vibrational Transitions

No restrictions in combination with an electronic transition:

$$\Delta v = \pm 1 \text{ in the absence of an electronic transition}$$

Rotational Transitions

In combination with an electronic transition:

$$\Delta J = 0, \pm 1 \text{ for } \Delta\Lambda = \pm 1$$

$$\Delta J = \pm 1 \text{ for } \Delta\Lambda = 0$$

$$\Delta J = \pm 1 \text{ in the absence of an electronic transition}$$

Exercises

2.1 Calculate the ionization energy of the hydrogen atom (eV) when the wavelength of the H_α line in the Balmer series is known to be 656.3 nm. The charge of the electron, Planck's constant and the speed of light are supposed to be known and can be found in standard tables.

2.2 Calculate the minimum energy that has to be supplied to an unexcited hydrogen atom to enable it to emit the H_β line in the Balmer series. The ionization potential of hydrogen is 13.6 eV.

2.3 (a) Calculate the maximum wavelength of light that is able to release photoelectrons from a sodium electrode if the work function of sodium is 2.3 eV.

(b) Photons of wavelength 200 nm hit the electrode. Calculate the maximum kinetic energy of the released photoelectrons.

2.4 A certain photocell is made of a quartz tube and equipped with two electrodes. One of them is covered with a layer of an alkali metal. The electrodes are connected with a sensitive amperometer and a variable-voltage source. When the alkali metal electrode is exposed to light of wavelength 435.8 nm the current decreases to zero at a certain voltage between the electrodes. If the electrode is exposed to photons of wavelength 253.7 nm the voltage has to be changed by 2.04 V to maintain zero current. Calculate Planck's constant, with $e = 1.601 \times 10^{-19}$ As and $c = 3.00 \times 10^8$ m/s.

2.5 An X-ray tube has a constant voltage. The most energetic radiation emitted by the tube has the wavelength 0.300 nm. Calculate the smallest de Broglie wavelength of the electrons in the tube. The relativistic effects on the electrons can be neglected.

2.6 X-radiation of wavelength 0.0496 nm hits a crystal lattice under the Bragg angle (first order). In another experiment, a beam of neutrons hits the lattice under the same angle. The neutrons are reflected in the same way as the photons. What is the kinetic energy (eV) of the neutrons?

2.7 Describe the fundamental way of using quantum mechanics to find energy levels and quantum numbers of atomic and molecular systems.

2.8 The wave function of the hydrogen atom in its ground state is

$$\psi = \left(\pi a_0{}^3\right)^{-\frac{1}{2}} e^{-\frac{r}{a_0}}$$

where a_0 is the radius of the smallest Bohr orbit. According to the statistical interpretation of the wave function, the charge density at the point (x, y, z) in space is $\rho(x, y, z) = e|\psi|^2$.

(a) Show that the probability distribution has a maximum for $x = a_0$, i.e. the probability that the electron will be found at this distance from the nucleus is larger than for any other distance.

(b) Calculate the average values of r and r^2.

2.9 (a) Describe the quantum numbers that characterize the orbitals in the atoms and the designations of the electrons in the various orbitals.

(b) What is the signification of the Pauli principle?

(c) Describe the nomenclature and background of the energy states in atoms.

(d) List the selection rules for electronic transitions from one orbital to another (electron jumps) in atoms.

(e) An atom can have both singlet and triplet states. Is this in agreement with the selection rules?

2.10 One of the strong spectral lines of Hg is a transition between two energy levels with an energy difference 4.892 eV.

(a) What wavelength has the spectral line? (Check your answer in a standard table.)

(b) Calculate the number of quanta that are emitted per second by a 100 W Hg lamp if 5% of the supplied energy to the lamp is transformed into radiation of the given wavelength.

2.11 Calculate the energy required to ionize a neutral He atom completely in its ground state, by removing the two electrons one by one. In the series 1s ^{1}S–np ^{1}P the wavelength approaches 50.43 nm for large values of n. It is also known that the ionization energy of hydrogen is 13.6 eV.

2.12 Find the electron configuration and type of electronic state for unexcited H, He, Li, Be, B, F, Ne, Na, Cl, Ar and K atoms.

Hints:

1. Filled shells and subshells do not contribute to the L and S vectors.
2. Shells or subshells with one missing electron can be treated as shells or subshells with only one electron when the L and S vectors are derived.

2.13 The first excited configuration of the neutral Be atom is 2s2p with the terms $^3P_0$, $^3P_1$, $^3P_2$ and $^1P_1$. If the ground term energy is chosen as zero level the wavenumbers $(1/\lambda)$ (in the customary unit cm^{-1}) are $^3P_0$ 21 978.28 cm^{-1}, $^3P_1$ 21 978.92 cm^{-1} and $^3P_2$ 219 81.27 cm^{-1}. In the beryllium spectrum, another close multiplet with the following wavelengths and wavenumbers has been found:

1	2	3	4	5	6
265.0760	265.0694	265.0619	265.0596	265.0550	265.0454 nm
37 713.80	37 714.74	37 715.81	37 716.14	37 716.79	37 718.16 cm^{-1}

It has been found that these lines represent transitions from the excited triplet 2p^2 $^3P_2$, $^3P_1$, $^3P_0$ down to the first mentioned triplet.

(a) Make a diagram and draw the allowed transitions, according to the selection rules.
(b) Identify the measured values in the above table with these transitions.

2.14 Describe the normal Zeeman effect and give an explanation for a simple case.

2.15 The red line in the Cd spectrum $(1/\lambda = 15\,530.00\,\text{cm}^{-1})$ is emitted at the transition 5s^{1}5d^1 $^1D_2 \rightarrow$ 5s^{1}5p^1 $^1P_1$. If the light source is placed in a strong magnetic field of 1.0 T the lines split into several components. Illustrate this in a diagram and calculate the wavenumbers for the allowed transitions. All singlet states obey the rules of the normal Zeeman effect.

2.16 The H$_2{}^+$ molecule has in its ground state an energy that is 2.65 eV lower than the energy of a system that consists of a hydrogen atom in its ground state and a proton at infinite distance from each other. Energy of 4.48 eV has to be supplied to transfer the H$_2$ molecule in its ground state into two hydrogen atoms at infinite distance from each other. Calculate:

(a) the energy of H$_2{}^+$ relative to H$^+$ + H$^+$ + e$^-$ at infinite distance from each other;
(b) the energy of H^{2+} + e$^-$, at infinite distance from each other, relative to the energy of two neutral atoms H + H, at infinite distance from each other, in their ground state;
(c) the ionization energy of H$_2$.

2.17 (a) Describe the nomenclature and background of the electron energy states in diatomic molecules.
(b) List the selection rules for electron transitions from one orbital to another (electron jumps) in the molecules.
(c) Give the L, S and J values for a $^2\Delta$ state and name its components. Are transitions to a $^2\Sigma$ state, a $^2\Pi$ state and another $^2\Delta$ state possible?

2.18 Show that the energy values of the rigid rotator can be derived easily by combining the general quantization rule of angular momentum and the classical relationship between rotational energy and angular momentum.

2.19 Give a survey of the selection rules in diatomic molecules in the case of

(a) pure rotation of the molecule, vibration + rotation and vibration + rotation when an electronic transition occurs simultaneously, respectively.
(b) Discuss briefly the intensity of vibration + rotation bands in the latter case.

2.20 In the near-infrared part of the spectrum the CO molecule has an absorption spectrum at 2144 cm^{-1}. It corresponds to the transition from $v'' = 0$ to $v' = 1$. Calculate

(a) the basic frequency of the corresponding vibration in the molecule
(b) the zero level of the vibrational energy.

2.21 The Morse function:

$$E_{\text{pot}} = D_e \left[1 - e^{-(r-r_e)}\right]^2$$

is often used as an approximation of the potential curve of diatomic molecules. As the vibrational energy levels are not equidistant, the model of an anharmonic oscillator is used to describe their positions:

$$E_{\text{vibr}} = hcG(v) = hc\left[\omega_e\left(v + \tfrac{1}{2}\right) - \omega_e x_e\left(v + \tfrac{1}{2}\right)^2\right]$$

From an analysis of $^1\Pi{-}^1\Sigma$ bands in the MgO molecule, the following values of the vibrational constants have been derived:

$$\omega_e'' = 785\,\text{cm}^{-1}\text{and } \omega_e'' x_e'' = 5.1\,\text{cm}^{-1}.$$

Calculate the dissociation energies D_e and D_0 of the ground state $^1\Sigma$ of the molecule. Give the answers in eV.

2.22 Two consecutive lines in the rotational spectrum of the HCl molecule have a wavenumber difference of $21.2\,\text{cm}^{-1}$. Calculate:

(a) the moment of inertia of the molecule with respect to an axis through the centre of mass and perpendicular to the axis of the molecule
(b) the distance between the atoms.

2.23 An absorption line in the rotational spectrum of the CO molecule has been observed to absorb microwaves of the frequency 1.153×10^{11} Hz. Calculate the distance between the nuclei.

2.24 Why are the spectra of polyatomic molecules in most cases much more complicated than those of diatomic molecules?

2.25 From the spectrum of the water molecule, H_2O, the moments of inertia through the centre of mass of the molecule with respect to the axes 1 and 2 in the plane of the molecule have been calculated. The values $I_1 = 1.92 \times 10^{-47}\,\text{kg m}^2$ and $I_2 = 1.02 \times 10^{-47}\,\text{kg m}^2$ were derived.

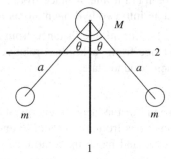

Calculate the distance between the O and H nuclei and the angle H–O–H.

3

Theory of Solids

3.1 Introduction   97
3.2 Bonds in Molecules and Solids: Some Definitions   98
    3.2.1 Binding Energy and Dissociation Energy. Ionization Energy. Electron Affinity   98
    3.2.2 Sublimation Energy and Condensation Energy. Cohesive Energy. Lattice Energy   100
3.3 Bonds in Molecules and Nonmetallic Solids   100
    3.3.1 Molecular Bonds   101
    3.3.2 Ionic Bonds   102
    3.3.3 Covalent Bonds   108
3.4 Metallic Bonds   112
    3.4.1 Free Electron Model of a Metal   112
    3.4.2 Classical Model of the Electron Gas   112
    3.4.3 Quantum Mechanical Model of the Electron Gas   114
3.5 Band Theory of Solids   125
    3.5.1 Origin of Energy Bands in Solids   125
    3.5.2 Energy Bands in Solids. Brillouin Zones   126
3.6 Elastic Vibrations in Solids   146
    3.6.1 Phonons   146
3.7 Influence of Lattice Defects on Electronic Structures in Crystals   151
    3.7.1 Influence of Lattice Defects on Electronic Structures in Nonmetallic Crystals   151
    3.7.2 Influence of Lattice Defects on Electronic Structures in Metals and Semiconductors   153
Summary   157
Exercises   163

3.1 Introduction

Solid-state physics is based on modern atomic and molecular physics, especially quantum mechanics. These theories have been briefly discussed in Chapter 2. In this chapter, the outlines of the modern theory of solids are given. Most of the theory presented here is the basis for understanding different types of crystallization processes discussed in later chapters. It is also the basis for the models which are used for the analysis and control of such processes.

Physics of Functional Materials   Hasse Fredriksson and Ulla Åkerlind
© 2008 John Wiley & Sons, Ltd

3.2 Bonds in Molecules and Solids: Some Definitions

In Sections 3.3 and 3.4, we will discuss bonds between atoms in molecules and solids. As an introduction to this field, we will start here with some general definitions.

The general definition of potential energy, which corresponds to a force, is

$$E_{pot} = \int_a^x -F dx \tag{3.1}$$

Attractive forces are positive and the corresponding potential energy is negative. Repulsive forces are negative and the corresponding potential energy is positive. The zero level of the potential energy is determined by the constant a in Equation (1).

These concepts can be applied to the interaction between two particles, which may be two atoms or a nuclei and an electron.

3.2.1 Binding Energy and Dissociation Energy. Ionization Energy. Electron Affinity.

Figure 3.1 Attractive force between two atoms at a distance r in a coordinate system x. In the case of atomic interaction the constant a in Equation (1) is chosen in such a way that the potential energy is zero when the two atoms are at an infinite distance from each other. Reproduced with permission from M. Alonso and E. Finn, *Fundamental University Physics*. © Addison-Wesley.

The electrons in their orbits around the nucleii are mainly responsible for the *forces* between the atoms in molecules and solids.

As we shall see below, the origin of the attractive forces varies, but they are always caused by the electrons in the outermost orbitals of the atoms. The potential energy of these forces is negative if we use the coordinate system suggested in Figure 3.1 and choose zero potential energy at infinity ($a = \infty$).

Strong repulsive forces appear when the distance between two atoms is so small that their outer filled electron orbitals begin to overlap. The potential energy is positive and the potential energy increases rapidly when the distance is further decreased.

The minimum potential energy of a system corresponds to its equilibrium state. Both the attractive and repulsive forces depend on the interatomic distance. At the equilibrium distance the net force is zero as the attractive and repulsive forces balance each other and the total energy has a minimum. The general shape of the energy curve is the same as in Figures 2.35 and 2.37 on pages 74 and 75, respectively.

Binding Energy and Dissociation Energy

Consider a stable diatomic molecule AB. The two atoms A and B are bound to each other and the total energy of the system is negative. The two nucleii vibrate relative to each other.

The *binding energy* E_B of the molecule is defined as

> $E_B = $ *the energy released when the two atoms A and B are moved from infinity to their*
>
> *equilibrium distance and form a stable molecule AB.*

The equilibrium distance is the average distance between the vibrating nuclei A and B. Binding energies of some diatomic molecules are given in Table 3.1.

The binding energy is equal to the *dissociation energy*, D_0, of the molecule or the energy required to separate the atoms A and B and move them to infinite distance from each other (Figure 3.2):

$$E_B = D_0 \tag{3.2}$$

These definitions can be generalized to many-particle systems, for example polyatomic molecules and solids, where all the components initially are at an infinite distance from each other.

Table 3.1 Binding energies of some diatomic molecules.

Molecule	Binding energy (eV)
H_2	4.5
Cl_2	2.5
Na_2	0.72
O_2	5.1

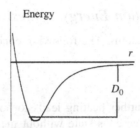

Figure 3.2 Total energy of a diatomic molecule as a function of distance. The ground state is marked. Dissociation energies are discussed on page 83 in Chapter 2. Reproduced with permission from M. Alonso and E. Finn, *Fundamental University Physics*. © Addison-Wesley.

Ionization Energy and Electron Affinity

The *ionization energy*, E_{ion}, of an electron in a stable orbit of an atom is defined as

> $E_{ion} =$ *the energy required to move an electron from its orbit in the atom to infinity.*

Each electron has its own ionization energy in an atom. The electrons in the highest energy level become ionized first. Some examples of ionization energies are given in Table 3.2.

Table 3.2 Ionization energies of some metals.

Atom	Ionization energy (eV)
Na	5.1
K	4.3
Al	6.0
Cu	7.7
Fe	7.9

When an electron is removed from the atom, the latter becomes a positive ion. Addition of an extra electron in an empty orbit of an atom gives a negative ion. The *electron affinity*, E_{aff}, is defined as

> $E_{aff} =$ *the energy which is released when an electron is moved from infinity to lowest possible*
>
> *orbit in an atom forming a stable negative ion.*

The electron affinity is a measure of the ability of an atom to attract electrons from outside. Table 3.3 gives some examples. Energy is released when an electron is added to a halide atom.

Table 3.3 Electron affinities of some atoms.

Atom	Electron affinity (eV)
F	3.5
Cl	3.7
Br	3.4
I	3.1
O	1.5

3.2.2 Sublimation Energy and Condensation Energy. Cohesive Energy. Lattice Energy

The definitions given above refer to molecules or atoms. The following concepts refer to crystals.

Sublimation Energy and Condensation Energy

When a solid is heated, it melts in most cases. Further heating leads to boiling, i.e. a vapour is formed. Sometimes atoms may be transferred directly from the solid to the gaseous state without the intermediate liquid state. This process is called *sublimation*. The sublimation energy of an atom is defined as

E_S = *the energy required to move an atom from its position in the solid to infinity.*

The reverse process is the *condensation* of a vapour directly into a solid. The released *condensation energy* per atom is equal to the sublimation energy:

$$E_S = E_{cond} \tag{3.3}$$

Cohesive Energy. Lattice Energy

Cohesive energy = *the energy which has to be added to one stoichiometric unit of a crystal to separate its components into neutral free atoms at rest at infinite distance from each other.*

Lattice energy is a concept which is used in connection with ionic crystals. It is defined as

Lattice energy = *the energy which has to be added to one stoichiometric unit of a crystal to separate its component ions into free ions at rest at an infinite distance from each other.*

3.3 Bonds in Molecules and Nonmetallic Solids

Solids are classified as crystalline or amorphous. In crystalline matter, the atoms has a periodic ordered structure whereas the atoms in amorphous matter show a random order. The dominant part of solids has a crystalline structure.

In Chapter 1, we described different crystal structures and the methods for studying them. Here we will discuss the theory of bonds between the atoms in molecules and solids using the results of quantum mechanics. On the basis of this theory, the properties of crystalline nonmetallic solids can be understood.

Crystalline solids can be classified according to the predominant type of bonds between the atoms in the crystal lattice. No crystal materials belong 100% to any of the pure types of bonds listed below, but to a mixture of several types. The main types of bonds are

- molecular bonds
- ionic bonds
- covalent bonds
- metallic bonds.

The first three types of bonds listed above will be discussed in Section 3.3. The last type of bond is very important in metals and responsible for their particular properties. Metallic bonds will be discussed in Section 3.4.

3.3.1 Molecular Bonds

Molecular bonds are much weaker than the other three types of bonds listed above. The origin of molecular bonds is dipole interaction between molecules.

Dipole Interaction Between Molecules

The electrical dipole moment of a dipole is a vector defined as

$$p = qr \tag{3.4}$$

where q is the electrical charge and r a vector directed from the negative towards the positive charge (Figure 3.3).

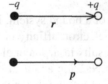

Figure 3.3   Dipole moment.

When two molecules at infinite distance are brought closer to each other, they will be affected by electrostatic forces. Even if a molecule has a zero dipole moment it may be a fluctuating dipole owing to vibrations in the molecule, for instance. Examples of *fluctuating dipoles* are mentioned on page 106. In the vicinity of a permanent or temporary dipole, the orbitals of other neighbouring molecules become slightly displaced and the molecules become *induced dipoles*.

This mutual interaction between dipoles is the origin of the *van der Waals forces* between the molecules. The weak, attractive van der Waals forces are inversely proportional to the seventh power of the distance between the molecules (see page 106). They will be discussed more extensively in connection with real gases in Chapter 4.

Molecular Bonds in Solids

Molecular bonds also occur in solids which consist of molecules and not of ions. The forces between the molecules are the weak van der Waals forces between permanent, fluctuating or induced dipoles.

The best-known example of this type of bonding is graphite, which is discussed on page 111. The graphite lattice consists of layers of hexagonal carbon rings held together by covalent bonds. Adjacent layers act as macromolecules and are held together by weak van der Waals forces, which account for the fragile, flaky and slippery nature of graphite.

Molecular bonds are very weak compared with ionic and covalent bonds.

3.3.2 Ionic Bonds

Ionic Bonds in Molecules

A simple and typical example of an ionic bond in a free molecule is the formation of a free Na^+Cl^- molecule from a free Na atom and a free Cl atom. The molecule formation occurs in practice in a mixture of sodium vapour and chlorine gas. The molecule formation can be described as follows.

An Na atom and a Cl atom at infinite distance from each other approach gradually, owing to the thermal kinetic motion in the gas. When their outermost orbits are close to each other, the outer electron clouds of the atoms rearrange and an Na^+Cl^- molecule is formed. The K and L shells in the Na atom are filled and there is a single valence electron in the 3s orbital (configuration $1s^2\ 2s^2\ 2p^6\ 3s$). The Cl atom has the configuration $1s^2\ 2s^2\ 2p^6\ 3s^2\ 3p^5$. The 3s electron of Na leaves its orbit and forms the eighth and missing electron in the M subshell of Cl. Then both ions have filled shells and subshells. There is a strong electrostatic attractive force between the ions.

The distance between the ions decreases until it equals the equilibrium distance for the Na^+Cl^- molecule, when the electrostatic attraction force balances the strong repulsive force, which appears when the electron clouds of the two subshells begin to overlap. Owing to the Pauli exclusion principle, no electrons can have four equal quantum numbers. Hence some electrons must be excited to higher energy levels. This process requires much energy and results in a steep energy curve. The equilibrium corresponds to the lowest possible total energy of the system and results in a stable ionic molecule.

The energy of the Na^+Cl^- molecule can be calculated theoretically by studying the energy of the system as a function of the interionic distance r.

Energy of a Free Ionic Molecule

As shown above, the molecule formation in the case of Na^+Cl^- can be described schematically to occur in three steps:

1. The Na atom at infinite distance is ionized.
2. The electron is absorbed by a Cl atom at infinite distance.
3. The Na^+ ion and the Cl^- ion at infinite distance are brought together until their interionic distance equals the equilibrium distance r_e.

The total energy is negative, which is characteristic of a bound system, i.e. a stable molecule. The first step requires addition of the ionization energy. During the second step, the electron affinity energy is released. The balance between the attractive electrostatic and the repulsive electron shell forces results in a potential curve similar to those of other diatomic molecules (Figure 3.4). The potential curve ends at infinity in an Na^+ ion and a Cl^- ion. The energy scale in Figure 3.4 is chosen in the normal way, i.e. $E = 0$ for an Na atom and a Cl atom at infinite distance from each other.

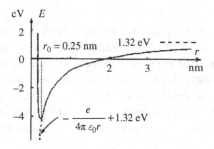

Figure 3.4 Potential energy of a free Na^+Cl^- molecule as a function of the interionic distance r. At $r = \infty$ the molecule is separated into an Na^+ ion and a Cl^- ion. Reproduced with permission. © E. Lindholm (Deceased).

The energy required to transfer an electron from the Na atom to the Cl atom can be calculated from the ionization energy of Na and the electron affinity of Cl (see pages 99–100):

$$E_{ion}^{Na} - E_{aff}^{Cl} = 5.14 - 3.82\,eV = 1.32\,eV$$

The total energy of the Na^+Cl^- system consists of the sum of four energy terms in addition to the energy required for the electron transfer:

$$E_{total} = E_{attr} + E_{rep} + E_{covalent} + E_{vib} + 1.32\,eV \tag{3.5}$$

The *first* term corresponds to the normal potential energy $-q^2/4\pi\,\varepsilon_0 re$ (eV) of the attractive electrostatic force between two point charges of different signs, where q is the charge of the ion.

The *second* term is the electrostatic potential energy, which corresponds to the strong repulsive force which appears at small interionic distances. Born assumed that the repulsive energy could be written empirically as a constant times $1/r^n$, where n is a number specific to each type of crystal. An alternative is an exponential function, discussed in connection with ionic crystals.

The *third* term is the covalent energy. As pointed out on page 101, no bond belongs to a single type of bond. Even in a typical ionic crystal such as Na^+Cl^- the bond is not 100% ionic. The wave function of the electron can be written as

$$\psi_{total} = f_{ion}\psi_{ion} + f_{covalent}\psi_{covalent} \tag{3.6}$$

where the fractions f_{ion} and $f_{covalent}$ are constants. In typical ionic molecules the fraction $f_{covalent}$ is small but never zero. Hence the bond is to a minor extent covalent. This matter will be discussed further in connection with covalent bonds below. In the case of Na^+Cl^-, the covalent energy $E_{covalent}$ can be neglected in comparison with the ionic energy.

The *fourth* term represents the vibrational energy. Owing to the Heisenberg uncertainty principle (Chapter 2, page 54) the ions cannot be at rest relative to each other. The two ions perform harmonic vibrations around their common centre of mass (Chapter 2, page 80). The vibrational energy is small in this case and can be neglected in comparison with the other terms.

Hence the total energy of a free Na^+Cl^- molecule can be written as

$$E_{total} = -\frac{1}{4\pi\varepsilon_0}\frac{e}{r} + \frac{constant}{r^n} + 1.32\,eV \tag{3.7}$$

The constant in Equation (3.7) can be determined from the equilibrium condition $dE/dr = 0$ for $r = r_e$.

During the third step, the binding energy of the Na^+Cl^- molecule is released. Figure 3.4 shows that it is equal to the dissociation energy D_0 (Chapter 2, page 83) of the ionic molecule. This quantity can be estimated experimentally.

The theory of ionic molecules, given above, is generally valid for free ionic molecules.

Ionic Bonds in Ionic Crystals

Ionic bonds occur in crystals which mainly consist of positive and negative ions. The attractive electrostatic forces between neighbouring, differently charged ions are strong and hold the crystal firmly together. As in free ionic molecules, the equilibrium interionic distance is determined by the balance between the electrostatic attractive forces between the ions and the electrostatic repulsive forces between the outer electron shells of the positive and negative ions.

Ionic crystals are hard and brittle and have high melting points. The valence electrons are firmly bound to the negative ions (Figure 3.5). Hence no transport of electrons through the crystal is possible. Ion crystals are insulators at room temperature. At high temperatures the ions become mobile and ion conduction is possible.

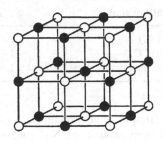

Figure 3.5 Na^+Cl^- crystal. The difference in size of the two types of ions is not considered.

Lattice Energy of Ionic Crystals

Total Energy

The theory of the lattice energy was developed by Born at the beginning of the 20th century. To describe it we choose the formation of an Na^+Cl^- crystal from a piece of Na metal and Cl_2 gas. The mechanism of the formation of Na^+ and Cl^- ions in a crystal lattice is the same as for a free Na^+Cl^- molecule. However, in the case of a crystal lattice the presence of other positive and negative ions, which surround each ion, has to be taken into consideration. Also, other effects are different in a crystal lattice than in a free molecule.

The total energy of an Na^+Cl^- unit or, in the general case, one stoichiometric unit, consists of a total of five contributions in addition to the transfer energy of an electron from Na to Cl. Four of them are analogous to the corresponding contributions in a free ionic molecule, but are modified in some cases. The fifth term, E_{pol}, which is due to interaction between electrical dipoles, has no equivalent in a free ionic molecule.

$$E_{total} = E_{attr} + E_{rep} + E_{covalent} + E_{vib} + E_{pol} + 1.32 \, eV \tag{3.8}$$

Each of the five other terms will be discussed shortly below. It should be noted that they must be expressed in eV.

Attractive Electrostatic Potential Energy

The electrostatic interaction between each ion and all the ions in the whole crystal lattice must be taken into consideration.

The Na^+Cl^- crystal has an FCC lattice. Each Na^+ ion is surrounded by six Cl^- ions at a distance R, by 12 Na^+ ions at a distance $R\sqrt{2}$ and by eight Cl^- ions at a distance $R\sqrt{3}$ and so on. The potential energy, expressed in eV, of the interaction between one Na^+ ion and all the other ions in the crystal will be

$$E_{attr} = \frac{-e}{4\pi\varepsilon_0 R}\left(\frac{6}{\sqrt{1}} - \frac{12}{\sqrt{2}} + \frac{8}{\sqrt{3}} - \frac{6}{\sqrt{4}} + \frac{24}{\sqrt{5}} \cdots\right) \tag{3.9a}$$

The same sum is obtained if we consider the interaction of a Cl^- ion with neighbouring ions. If we add the two sums, each bond is included twice. Hence the potential energy of the interaction of an Na^+Cl^- unit with the rest of the crystal lattice can be written (expressed in eV) as

$$E_{attr} = \frac{A}{4\pi\varepsilon_0}\frac{-e}{R_0} \, (eV) \tag{3.9b}$$

where

A = Madelung's constant for the crystal structure
R_0 = smallest possible distance between unequally charged ions in the crystal lattice
e = charge per ion.

Equations (3.9a) and (3.9b) show that the Madelung constant A in this case is the expression in the parentheses in Equation (3.9a).

In the general case, the charge is an integer multiple of the elementary charge and A is a pure number, determined entirely by the particular structure of the crystal. Table 3.4 give the values of A for some crystal structures.

Table 3.4 Values of Madelung's constant of some crystal structures.

Crystal/structure	A
CsCl (SC)	1.7627
NaCl (FCC)	1.7476
ZnS (diamond structure)	1.6381

Each Cs^+ ion is surrounded by eight Cl^- ions (simple cubic structure). For space reasons, Na^+Cl^- has a different structure to CsCl. There is not space enough for eight Cl^- ions around the Na^+ ion, which is considerably smaller than the Cs^+ ion. Instead, each Na^+ ion is surrounded by six Cl^- ions (FCC structure).

Repulsive Electrostatic Potential Energy

The repulsive potential energy has the same origin as that in a free ionic molecule. The repulsive forces are short-range forces and hence the repulsive potential energy depends only on the electron distribution in the outer electron shells of the nearest neighbour ions. Born's approximation is valid. It can be written (in the energy unit eV) as

$$E_{rep} = \frac{e}{4\pi\varepsilon_0} \frac{B}{R^n} \tag{3.10}$$

where n and B are constants. The value of n can be calculated from measurements of the compressibility of the crystal. The value of n for Na^+Cl^- has been found to be 9.1.

The attractive and repulsive electrostatic forces balance each other at equilibrium. The constant B can be calculated from the equilibrium condition

$$\left[d\left(E_{attr} + E_{rep} \right)/dR \right]_{R=R_e} = 0 \tag{3.11}$$

or

$$\frac{d}{dR}\left[\frac{e}{4\pi\varepsilon_0} \left(\frac{-A}{R} + \frac{B}{R^n} \right) \right]_{R=R_e} = \frac{e}{4\pi\varepsilon_0} \left(\frac{A}{R^2} + \frac{-Bn}{R^{n+1}} \right)_{R=R_e} = 0$$

which gives the relationship

$$B = \frac{A}{n} R_e^{n-1} \tag{3.12}$$

Consider a crystal consisting of N_0 positive and N_0 negative ions. If we neglect the surface effects, *the total electrostatic binding energy of the crystal* will be

$$E_{electrostatic} = N_0 \left(E_{attr} + E_{rep} \right)_{equilibrium} = N_0 \frac{A}{4\pi\varepsilon_0} \frac{-e}{R_e} \left(1 - \frac{1}{n} \right) \tag{3.13}$$

The total binding energy of $2N_0$ ions is equal to the expression in Equation (3.13). The sum seems to be double this amount but then each pair of ions would have been counted twice, which is not correct. The total electrostatic energy per ion pair is then

$$\frac{E_{electrostatic}}{N_0} = \frac{A}{4\pi\varepsilon_0} \frac{-e}{R_e} \left(1 - \frac{1}{n} \right) \tag{3.14}$$

A still better empirical approximation of the repulsive electrostatic energy than Equation (3.10) is an exponential function of the type

$$E_{rep} = Z\lambda e^{-\frac{R}{\rho}} \tag{3.15}$$

where

R = distance between nearest neighbour ions
Z = the number of nearest neighbours of any ion
λ, ρ = specific empirical parameters for each type of crystal.

If Equation (3.10) is replaced by Equation (3.15), analogous calculations give

$$E_{electrostatic} = N_0 \frac{A}{4\pi\varepsilon_0} \frac{-e}{R_e} \left(1 - \frac{\rho}{R_e} \right) \tag{3.16}$$

The electrostatic energy per ion pair (in eV) will then be

$$\frac{E_{electrostatic}}{N_0} = \frac{A}{4\pi\varepsilon_0} \frac{-e}{R_e} \left(1 - \frac{\rho}{R_e} \right) \tag{3.17}$$

Residual Covalent Binding Energy

Even in a typical ionic crystal such as Na^+Cl^- the bond is not 100% ionic but also to some extent covalent and molecular. This is taken into account by adding two energy terms, $E_{covalent}$ in analogy with the free ionic molecule and a term E_{pol} (see below).

In the case of an Na^+Cl^- crystal, the covalent binding energy is small and can be neglected.

Zero-point Vibrational Energy

In analogy with the free ionic molecule, the zero point vibrational energy must be taken into account. In a crystal with N_0 ions there are $3N_0 - 6$ different modes of vibration. The added zero point vibrational energies of these vibrations contribute to the total energy of the Na^+Cl^- unit.

The total vibrational energy, which is the sum of all the zero point vibrational energies, can be written in eV (compare page 83 in Chapter 2) as

$$E_{vibr} = \sum_i \frac{hc}{e} \omega_i \left(n^i_{vibr} + \frac{1}{2} \right) \tag{3.18}$$

where n^i_{vibr} are the vibrational quantum numbers.

Polarization Energy

A molecule is said to be polarized when the centre of its positive charges does not coincide with the centre of its negative charges. The molecule is a dipole. Temporary dipoles may arise in the crystal, owing to vibrations, and cause induced dipoles. We have seen on page 101 that the dipole–dipole interaction is the origin of molecular bonds. Such weak van der Waals forces also act within crystals. The interaction between dipoles results in polarization energy, which contributes to the lattice energy of the Na^+Cl^- unit. To find this energy, we use the definition of the electrical dipole moment p on page 101:

$$p = qx \tag{3.19}$$

A homogeneous electric field induces a dipole in an atom. The dipole moment is proportional to the electrical field:

$$p = \alpha E \tag{3.20}$$

where
  p = dipole moment of the induced atomic dipole
  α = polarizability of the atom
  E = electric field.

The energy of the atomic dipole in the electric field (in SI units) can be found by integration:

$$E_{pol} = -\int_0^x qE dx = -\int_0^x qE d\frac{\alpha E}{q} = -\frac{1}{2}\alpha E^2 \tag{3.21}$$

A crystal may be considered as a system of oscillating dipoles. If the electric field originates from a dipole, the electric field at a point at a distance R from the dipole varies as $1/R^3$. As the polarization energy is proportional to E^2, the polarization energy is proportional to $(1/R^3)^2$:

$$E_{pol} = -\frac{constant}{R^6} \tag{3.22}$$

The constant includes the factor $1/e$, which has to be introduced to express E_{pol} in eV.

Calculation of the Lattice Energy

According to the definition on page 100, the lattice energy of an ionic crystal can be written as

$$E_{lattice} = (E_{total})_{R=\infty} - (E_{total})_{R=R_0} \tag{3.23}$$

It can be calculated with the aid of Equation (3.8) in combination with expressions of the various energy contributions, i.e. Equations of the type (3.13) or (3.17), (3.18) and (3.22). The covalent energy contribution is small and can be been neglected.

In Table 3.5, the results of such calculations for an Na^+Cl^- unit are given. The table shows the relative magnitude of the different energy types.

Table 3.5 Lattice energy and energy types per Na^+Cl^- unit.

Type of energy	Energy (eV)
Attractive energy	−8.92
Repulsive energy	+1.03
Covalent energy	~0
Zero point vibrational energy	+0.08
Polarization energy	−0.13
Calculated lattice energy	−7.94

Experimental Determination of the Lattice Energy

The calculation of the lattice energy above is based on theory. Alternatively, it is possible to derive the lattice energy from experimentally determined quantities only. This can be done with the aid of the so-called Born–Haber cycle. As an example, we consider once again an Na^+Cl^- crystal.

The Born–Haber cycle starts with a piece of solid Na and Cl_2 gas and builds the Na^+Cl^- crystal step by step:

1. An Na atom is released from the Na lattice when sublimation energy is added.
2. The free Na atom is ionized by addition of ionization energy.
3. A Cl_2 molecule is dissociated. The formation of one Cl atom requires half the dissociation energy.
4. The Cl atom absorbs an electron. Electron affinity energy is released.

The above steps can be written as chemical formulae. The quantities refer to the formation of one pair of solid Na^+Cl^-.

$$Na_{solid} + E_S^{Na} \rightarrow Na_{vapour}$$

$$Na_{vapour} + E_{ion}^{Na} \rightarrow Na_{vapour}^+ + e^-$$

$$\tfrac{1}{2}Cl_2 + \tfrac{1}{2}D_{Cl_2} \rightarrow Cl$$

$$Cl + e^- \rightarrow Cl_{vapour}^- + E_{aff}^{Cl}$$

We can also describe the overall process in terms of lattice energy:

$$Na_{vapour}^+ + Cl_{vapour}^- \rightarrow (Na^+Cl^-)_{solid} + E_{lattice}$$

Na^+Cl^- is a very stable crystal. When it is formed, heat of formation, which we will call Q, is released. This process can be described by

$$Na_{solid} + \tfrac{1}{2}Cl_2 \rightarrow (Na^+Cl^-)_{solid} + Q$$

If we add the first five chemical formulae and subtract the sixth one, we obtain the relationship

$$E_{lattice} = E_S^{Na} + E_{ion}^{Na} + \tfrac{1}{2}D_{Cl_2} - E_{aff}^{Cl} + Q \tag{3.24}$$

All the quantities included in the right-hand side of Equation (3.24) have been determined experimentally. If we introduce their tabulated values we obtain

$$E_{lattice} = 1.1 + 5.1 + 1.2 - 3.8 + 4.3 = 7.9 \, eV$$

The agreement with the theoretical value in Table 3.5 is very good. Hence the theory is satisfactory and essentially correct.

The energy of an Na^+Cl^- unit of a crystal is shown in Figure 3.6. For comparison, the energies of both a free Na^+Cl^- molecule (Figure 3.4) and an Na^+Cl^- unit of a crystal are shown side by side. The energy scales in the two figures are the same. In both cases the energy well is deep, which means that the bonds are strong, especially in the crystal.

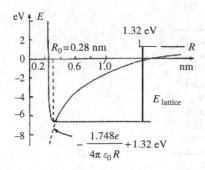

Figure 3.6　Potential energy of an Na^+Cl^- crystal as a function of the interionic distance R. $E_{lattice} = 7.9\,eV$. Reproduced with permission. © E. Lindholm (Deceased).

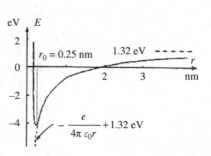

Figure 3.4　Potential energy of a free Na^+Cl^- molecule as a function of the interionic distance r. At $r = \infty$ the molecule is separated into an Na^+ ion and a Cl^- ion. Reproduced with permission. © E. Lindholm (Deceased).

3.3.3 Covalent Bonds

Cl atoms have filled K and L shells and seven electrons in the M shell. If two Cl atoms share two electrons they are able to form a stable Cl_2 molecule with both their M subshells filled:

$$: \ddot{C}l \cdot + \cdot \ddot{C}l : \rightarrow : \ddot{C}l : \ddot{C}l : + Q$$

The two atoms share the electron pair, which gives the lowest possible energy of the system.

This type of bond is called a *covalent bond* or *electron pair bond* or *homopolar bond*. The bonds are very strong and covalent solids are therefore characterized by high melting points and high mechanical strength. They are poor conductors of heat and electricity because there are no non-localized electrons which can carry energy or charge from one place to another. The electron excitation energies of covalent solids are high, of the magnitude of several eV. The excitation energy of diamond, for example, is 6 eV. As the thermal average energy k_BT at room temperature is of the magnitude 0.025 eV, covalent solids are normally in their electronic ground states.

A very important example of covalent bonds is the so-called hybrid formation of carbon in methane, which explains the special type of bonds which occur in carbon, silicon and germanium (page 110).

Bond in the H_2 Molecule

The free H_2 molecule is a typical example of a covalent bond and of great theoretical interest because it is the only molecule which allows exact theoretical calculations.

On pages 72–74 in Chapter 2, the ionic molecule H_2^+ was discussed. The bond between the equal H atoms was found to be strong and the probability of finding the electron between the nuclei comparatively high, which results in a deep potential well of the molecule. In the case of H_2 two electrons instead of one give an even stronger bond than that of H_2^+.

The strength of the bond in H_2^+ was found to be due to *exchange energy*. In the case of H_2, exchange energy appears for both electrons, but other types of so-called *resonance energy* are also involved. The interchange frequency of the electrons between the two nuclei is of the magnitude $10^{18}\,s^{-1}$.

The assumption by Fermi and Dirac that *electrons are indistinguishable* is closely related to quantum mechanics and the Pauli principle. This assumption is essential and very successful in connection with metallic bonds, as will be shown later. Here we also have to consider the fact that the two electrons in H_2 are indistinguishable.

When the two H atoms 1 and 2 approach each other and molecular orbitals are formed, it is impossible to know which one is electron A and which is electron B. The wave functions for the united molecule are combinations of the wave functions $\psi_1(A)$ and $\psi_1(B)$, $\psi_2(A)$ and $\psi_2(B)$. Calculations show that the lowest energy is given by the symmetrical, *gerade* wave function:

$$\psi_g = \psi_1(A)\psi_2(B) + \psi_1(B)\psi_2(A) \tag{3.25}$$

Both electrons are in the molecular orbital (σ_g 1s) and have anti-parallel spins. The energy changes with the internuclear distance as is shown in Figure 3.7 (Figure 2.35 on page 74 in Chapter 2).

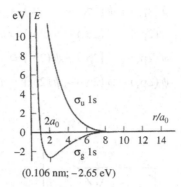

(0.106 nm; −2.65 eV)

Figure 3.7 Energy of the H_2^+ molecule as a function of the distance between the two protons. Reproduced with permission from M. Alonso and E. Finn, *Fundamental University Physics.* © Addison-Wesley.

The corresponding antibonding wave function is an antisymmetric *ungerade* wave function which corresponds to an instable molecular orbital (σ_u 1s). Two electrons with parallel spins fill the state.

The total energy varies with the internuclear distance. The ground state of the stable H_2 molecule has a deep minimum which corresponds to the binding energy. The pair of electrons results in a strong attractive force which balances the repulsion between the two H^+ nuclei. This is a typical covalent bond.

However, bonds in molecules never belong 100% to one type of bond, for example covalent or ionic bonds, but are always a mixture of two types or more.

In the case of H_2 there is no total agreement between the theoretical calculations of the energy curve and the experimental values. There is a possibility that both electrons are near one or the other nucleus, which results in an ionic molecule $H(1)^+H(2)^-$ or $H(2)^+H(1)^-$ and we obtain (page 103)

$$\psi_{\text{total}} = f_{\text{covalent}}\psi_{\text{covalent}} + f_{\text{ionic}}\psi_{\text{ionic}} \tag{3.6}$$

When this effect is taken into consideration, complete agreement between theory and experiment is achieved.

Covalent Bonds in Carbon

Graphite and transparent diamond crystals are well-known structures of carbon. Diamond is very hard and is used in cutting tools. Graphite is used in pencils and as a lubricant. Carbon powder and active carbon consist of ground graphite. Carbon often appears as a solute in metals and is therefore of particular importance in metallurgy.

Hybrid Formation of sp³ Orbitals in CH₄

Carbon normally has the valence 4. This is difficult to understand from the electron configuration of the carbon atom, $1s^2\,2s^2\,2p^2$. To explain why the valence is 4 we will consider the simplest of all symmetric free carbon molecules, CH_4.

When the four H atoms approach the carbon atom they become excited into a higher energy level than the ground state. One of the 2s electrons in carbon becomes excited up to a 2p orbital and the new electron configuration in carbon will be $1s^2\,2s\,2p^3$. The threshold energy, which has to be supplied, is a necessary contribution that makes it possible to achieve symmetry of the bonds, which corresponds to the lowest possible total energy, lower than the sum of the initial energies of the four H atoms and the C atom.

The wave functions that correspond to the 2p electrons are identical apart from their directions. For symmetry reasons they correspond to three perpendicular directions in free space (Figure 2.27 on page 68 in Chapter 2). For simplicity we call the wave functions $\psi(2p_x)$, $\psi(2p_y)$ and $\psi(2p_z)$.

Because the electrons are indistinguishable (as those in the H_2 molecule), the spherically symmetric wave function $\psi(2s)$ and the three perpendicular wave functions $\psi(2p_x)$, $\psi(2p_y)$ and $\psi(2p_z)$ combine and give four new wave functions, which are identical, apart from their directions. We call them $\psi(t_a)$, $\psi(t_b)$, $\psi(t_c)$ and $\psi(t_d)$. Their directions are symmetrical. If the C atom is located to the centre of a tetrahedron, the directions of the wave functions and the bonds coincide with the directions to the corners of the tetrahedron (Figure 3.8).

$$\psi\left(2s\right)+\psi\left(2p_x\right)+\psi\left(2p_y\right)+\psi\left(2p_z\right)=\psi\left(t_a\right) \tag{3.26}$$

$$\psi\left(2s\right)+\psi\left(2p_x\right)-\psi\left(2p_y\right)-\psi\left(2p_z\right)=\psi\left(t_b\right) \tag{3.27}$$

$$\psi\left(2s\right)-\psi\left(2p_x\right)+\psi\left(2p_y\right)-\psi\left(2p_z\right)=\psi\left(t_c\right) \tag{3.28}$$

$$\psi\left(2s\right)-\psi\left(2p_x\right)-\psi\left(2p_y\right)+\psi\left(2p_z\right)=\psi\left(t_d\right) \tag{3.29}$$

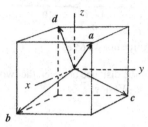

Figure 3.8   Tetrahedron directions. Reproduced with permission. © E. Lindholm (Deceased).

Hence there will be four t electrons in the tetrahedron directions instead of one 2s electron and three 2p electrons. The total energy that corresponds to this set of t wave functions is a minimum and lower than the energy of the 2s and three 2p electrons. Hence this symmetrical configuration is stable. The symmetric covalent tetrahedral bonds are characteristic of diamond (Figure 3.9).

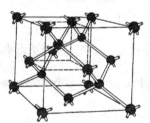

Figure 3.9   Tetrahedral bonds in diamond. Adapted with permission from M. J. Sinnott, *The Solid State for Engineers.* © 1958 John Wiley & Sons, Inc.

However, hybridization of carbon is only possible in connection with chemical binding, because the excitation of the 2s electron in carbon requires supply of energy from outside.

With the same wave function representation as in Figure 2.27 on page 68 in Chapter 2, we can illustrate the C–H bond as follows:

H   +   C   →   C–H

1s   +   2p   →   t

The bonds in methane, CH_4, are covalent. Each of the four t electrons of carbon forms an electron pair with the 1s electron of each of the four hydrogen atoms. The two electrons with anti-parallel spins are in a molecular σ orbital. The reason why hybridization occurs is that the hybrid wave functions lead to the lowest possible energy of the system.

For the same reason as above, the semiconductors Si and Ge have tetrahedral bonds analogous to those in carbon. We will come back to this topic in connection with semiconductors.

Hybridization is a general phenomenon. It is *not* specific to carbon and *not* restricted to sp³ orbitals.

Bonds in Graphite

Another type of hybrid carbon bonds appear in graphite and in aromatic organic compounds where stable planar C_6 rings, for example benzene (C_6H_6) and its derivatives, are formed (Figure 3.10).

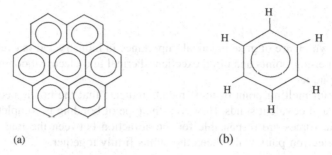

(a) (b)

Figure 3.10 (a) Planar layer of hexagonal carbon rings in graphite. (b) Benzene molecule. The C–C and C–H bonds are strong covalent σ bonds.

The types of hybrid formation are analogous in diamond and graphite (Figure 3.11). In the case of graphite and aromatic substances, the interaction between one 2s electron and two 2p electrons result in three localized σ bonds with sp^2 hybrid wave functions.

Each carbon atom in the ring has four valence electrons, which makes 24 electrons. Eighteen of them are used for the σ bonds and the remaining six π bonding electrons are not localized to any particular C atoms but are free to move along the planar layer of carbon rings but *not* perpendicularly to them. This explains the electrical conductivity of graphite, whereas diamond is an excellent insulator.

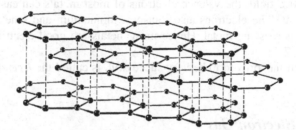

Figure 3.11 Structure of graphite. The consecutive layers of carbon rings act as macromolecules which are held together by van der Waals forces. Reproduced with permission from M. Alonso and E. Finn, *Fundamental University Physics*. © Addison-Wesley.

In Section 3.3.2, we discussed bonds in ionic crystals and the influence of the neighbouring ions in the crystal. In solids with mainly covalent bonds there is also multiple interaction between the atoms in the crystal. These topics will be discussed in Section 3.5. Of particular importance is a comparison between the energy levels in diamond and the semiconductors silicon and germanium, which have the same type of covalent bonds and the same crystalline structure.

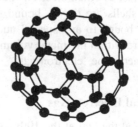

Figure 3.12 'Football' structure of C_{60}. Reproduced with permission from S. Lillieborg.

In 1985, a new type of carbon structure was found, when macromolecules of carbon, so-called fullerenes, were prepared. The best-known example is C_{60}, which consists of 20 hexagonal faces and 12 pentagonal faces and has the shape of a football. Most of the fullerenes crystallize in an FCC structure. The unit cells are kept together by van der Waals forces.

Figure 3.12 on page 111 shows that the same type of carbon rings appear in the fullerenes as in graphite.

3.4 Metallic Bonds

Metals are a special type of solids which are of great practical importance in industry and in everyday life. Metallic solids have good mechanical strengths, high melting points and often excellent thermal and electrical conductivities. They have relatively low ionization energies and are opaque.

The mechanical strength and high melting point of most metals indicate that the bonds between the atoms in a metal are strong just like the bonds in ionic and covalent solids. However, the type of bonding is completely different in all three cases. In ionic solids, strong electrostatic forces are responsible for the attraction between the ions. In covalent solids, the strong interatomic forces are caused by electron pairs which keep the atoms firmly together.

3.4.1 Free Electron Model of a Metal

We have seen that the valence electrons of the atoms are responsible for the nature of the strong bonds in ionic and covalent solids. In both cases the valence electrons are bound to the ions and atoms, respectively. In covalent solids, the strong bonds are caused by electron pairs.

Most metals have a few, weakly bound electrons in their outermost incomplete electron shells. In Chapter 1 we found that metals normally have high coordination numbers. Hence there are far from enough valence electrons to form electron pair bonds between all near neighbour atoms in a metal.

Instead, *all the valence electrons are assumed to be shared between all the metal atoms*. A metallic crystal can be regarded as a three-dimensional array of positive ions firmly held together by the attraction from a common electron cloud, consisting of all the valence electrons in the crystal. The electrons belong to the *whole* crystal lattice and *not* to any particular metal ion.

If the metal is exposed to an electric field, the valence electrons of most metals can easily move in a direction opposite to the field and carry the electric current. The electrons also transport momentum and kinetic energy in case of a temperature gradient instead of an electric field across the metal. Hence most metals are good electrical and thermal conductors. These properties will be treated in Chapter 7.

The theory suggested above, which has proved to be very successful, is called the *free electron model of a metal*. It will be treated extensively below.

3.4.2 Classical Model of the Electron Gas

The English physicist J. J. Thomson detected the electron in 1897 and measured e/m, the ratio of its charge and mass. By studying the motion of small charged oil drops in an electrical field, the American physicist R. A. Millikan made very careful measurements of the electron charge. Then the mass and charge of a free electron could be estimated separately:

$$e = -1.60 \times 10^{-19}\,\text{A s} \quad m = 9.11 \times 10^{-31}\,\text{kg}$$

The mobile electrons in a metal do not have the same mass as an electron in free space. For this reason, the concept *effective mass m^** has been introduced. The effective mass depends on the energy of the electron inside the metal. We will come back to this phenomenon later (pages 145–146).

A free metal atom in its ground state has a number of filled electron shells around the nucleus and one or several valence electrons in orbitals in the next shell. The filled shells are tightly bound to the nucleus whereas the valence electrons are supposed to be free in the sense that they are not bound to any special nucleus. As mentioned above, the valence electrons belong to all the ions and can easily move anywhere within the metal volume but not outside. They can be compared with the molecules in an ideal gas, which explains the former name 'electron gas'.

Thermal Distribution of Energies in the Classical Electron Gas

The kinetic theory of gases, which was introduced at the end at the 19th century by Maxwell, could successfully explain the properties of gases. Among other things, it was possible by simple means to calculate the relationship between the average value of the kinetic energy of a molecule in the gas and the temperature. As the number of molecules in a gas is very large, statistical methods could be applied. The so-called Maxwell–Boltzmann distribution gives the number of gas molecules per unit volume which have energies within an energy interval between E and $E + dE$, which corresponds to velocities between (v_x, v_y, v_z) and $(dv_x + dv_x, v_y + dv_y, v_z + dv_z)$. The Maxwell–Boltzmann distribution and the kinetic theory of gases are extensively discussed in Chapter 4.

As is shown in Chapter 4, it is possible to find the number of independent particles with kinetic energies within the interval E and $E + dE$ as a function of E. The shape of the curve depends on the temperature.

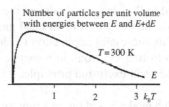

Figure 3.13 Maxwell–Boltzmann distribution of particle energies at room temperature and thermal equilibrium. Reproduced with permission from M. Alonso and E. Finn, *Fundamental University Physics.* © Addison-Wesley.

Figure 3.13 shows the Maxwell–Boltzmann energy distribution of particles as a function of particle energy at a temperature of 300 K. It can be concluded that few particles have energies equal to $3k_B T$ and still fewer have energies equal to $10k_B T$. Hence we can in practice regard $10k_B T$ as an upper limit of the kinetic energy of the particles.

This result was applied to an electron gas, i.e. to the valence electrons in a metal. If the Maxwell–Boltzmann distribution is valid, the most energetic valence electrons have energies $< 10k_B T$ at room temperature and thermal equilibrium, which corresponds to ~ 0.025 eV.

In the classical free electron theory, the electrons were regarded as free noninteracting classical particles in a potential well, which keeps the electrons inside the metal, trapped in a potential well.

The validity of the classical theory of the electron gas in a metal could be checked with the aid of the photoelectric effect [Equation (2.6)] discussed on page 49 in Chapter 2. When a photon with energy $h\nu$ hits a metal surface, an electron with velocity v_{ext} outside the metal may be emitted:

$$h\nu = \phi + \frac{m v_{\text{ext}}^2}{2} \tag{3.30}$$

where ϕ is a material constant called *work function*, which is the minimum energy required to release the most energetic trapped valence electrons from the metal surface (Figure 3.15).

Figure 3.14 Photoelectric effect. The released electron deviates in a magnetic field and its velocity can be calculated from the measured radius R and the known strength of the field B.

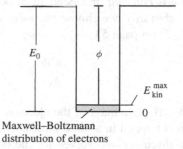

Maxwell–Boltzmann distribution of electrons

Figure 3.15 Potential well for trapped valence electrons at the surface of a metal. The figure also shows the kinetic energies of the valence electrons provided that the Maxwell–Boltzmann distribution (Chapter 4) was valid (which it is *not*). The calculations give $E_{\text{kin}}^{\text{max}} \approx 10k_B T \approx 0.025$ eV.

The released electron deviated in a circle in a known magnetic field B (Figure 3.14). The radius R of the circle was measured and the velocity of the electron could be calculated.

When the photon energy and the velocity of the released electron were known from experiments, ϕ could easily be calculated from Equation (3.30).

The experiments showed clearly that the valence electrons in metals have *much higher kinetic energies* than 0.025 eV at room temperature. One reason for the total failure of the classical model of the electron gas in a metal is that Maxwell–Boltzmann statistics is not valid for electrons.

3.4.3 Quantum Mechanical Model of the Electron Gas

As shown above strong objections can be raised to the classical model of the electron gas. It is not reasonable to regard the electrons as noninteracting classical particles like the molecules in an ideal gas.

The introduction of quantum mechanics and its successful application to electrons in atoms and molecules raised the idea that quantum mechanics could be applied also to free electrons in a metal. The valence electrons in atoms have energies calculated from quantum mechanics and obey the Pauli exclusion principle. It is highly unlikely that they behave like classical particles as free electrons in a metal.

Sommerfeld replaced the classical model with a model based on quantum mechanics. He published his quantum mechanical model in 1928. As we shall see below, his results deviated strongly from the results of the classical electron gas model.

The Schrödinger Equation of Free Electrons in a Metal

In order to solve the Schrödinger equation for a free electron in a metal, we must know the potential energy of the free electron as a function of position. It is no easy task to find this function as the electron is exposed to the electric field from all the nuclei with their inner electron shells in the metal and from all the other free electrons.

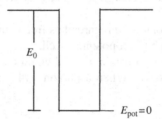

Figure 3.16   Potential well of the free valence electrons in a metal.

In the classical theory, the interactions between the free electron and the nuclei and between the free electrons were assumed to be zero. Sommerfeld did not neglect the interactions but he made the simplifying assumption that the potential energy of the free electron is *constant*, independent of position inside the metal (Figure 3.16). As the zero level of the potential energy can be chosen arbitrarily, we choose the value $E_{pot} = 0$ in analogy with Figure 3.15 on page 113. Hence we obtain, according to Equation (2.15) on page 53,

$$\frac{\partial^2 \psi}{\partial x^2} + \frac{\partial^2 \psi}{\partial y^2} + \frac{\partial^2 \psi}{\partial z^2} + \frac{2m^*}{\hbar^2}(E - 0)\psi = 0 \quad \left(\hbar = \frac{h}{2\pi}\right) \tag{3.31}$$

where m^* is the effective mass of the electron in the metal (page 112).

The electron is trapped in a three-dimensional box equal to the volume of the metal. In fact, the problem with a particle in a box has been discussed and solved for the one-dimensional case in Example 2.1 on page 55 in Section 2.3.7. The solution has to be extended to three dimensions in this case.

Solution of the Schrödinger Equation in One Dimension

We found in Example 2.1 in Chapter 2 (page 55) that the eigenvalue and eigenfunction of the electron in a one-dimensional box with length L_x are (some indices have been added here)

$$E = \frac{\hbar^2}{2m^*}\frac{n_x^2}{L_x^2} \quad \text{and} \quad \psi = A\sin\left(\sqrt{\frac{2m^*E}{\hbar^2}}x\right) \quad 0 < x < L \tag{3.32}$$

The calculations and equations will be much simpler in the following if we introduce the wavenumber k or rather the *wavevector* **k**, which is a vector in the direction of the wave motion:

$$\mathbf{k} = k_x\hat{x} + k_y\hat{y} + k_z\hat{z} \tag{3.33}$$

with the magnitude

$$|\mathbf{k}| = \frac{2\pi}{\lambda} \tag{3.34}$$

where λ is the wavelength of the matter wave.

In the one-dimensional case of a matter wave moving along the x-axis, we have $k_y = k_z = 0$ and the amplitude of the wave will be

$$\psi = A\sin(k_x x) = A\sin\left(\sqrt{\frac{2m^*E}{\hbar^2}}x\right) = A\sin\left(\frac{1}{\hbar}\sqrt{2m^*E}x\right)$$

which leads to

$$k_x = \frac{1}{\hbar}\sqrt{2m^*E} \tag{3.35}$$

Solution of the Schrödinger Equation in Three Dimensions

Amplitude of the Matter Wave
The amplitude of the eigenfunction which corresponds to the eigenvalue E in the three-dimensional case is

$$\psi = C\sin\left(\frac{1}{\hbar}\sqrt{2m^*E}x\right)\sin\left(\frac{1}{\hbar}\sqrt{2m^*E}y\right)\sin\left(\frac{1}{\hbar}\sqrt{2m^*E}z\right) \tag{3.36}$$

or

$$\psi = C\sin(k_x x)\sin(k_y y)\sin(k_z z) \tag{3.37}$$

where

$$k_x = k_y = k_z = \frac{1}{\hbar}\sqrt{2m^*E} \tag{3.38}$$

Eigenfunction of the Matter Wave
In the three-dimensional case, the *eigenvalue* is found to be

$$E = \frac{h^2}{8m^*}\left[\left(\frac{n_x}{L_x}\right)^2 + \left(\frac{n_y}{L_y}\right)^2 + \left(\frac{n_z}{L_z}\right)^2\right] \tag{3.39}$$

The squares of the integers, n_x^2, n_y^2 and n_z^2, appear in the eigenvalue expression for the same reason as n^2 did in Example 2.1 on page 57 in Chapter 2. The standing matter wave must have nodes at the crystal surface, i.e. the metal surface. The conditions can be expressed either in terms of the wavelengths or the wave numbers [Equation (3.34)] of the matter wave:

$$\begin{aligned} L_x &= n_x\lambda_x \\ L_y &= n_y\lambda_y \\ L_z &= n_z\lambda_z \end{aligned} \tag{3.40a}$$

or

$$\begin{aligned} k_x &= 2\pi\frac{n_x}{L_x} \\ k_y &= 2\pi\frac{n_y}{L_y} \\ k_z &= 2\pi\frac{n_z}{L_z} \end{aligned} \tag{3.40b}$$

Consequently, an electron which is bound to move within the space (L_x, L_y, L_z) can move only in certain motion modes in agreement with Equation (3.7.2) and with the condition

$$E = \frac{\hbar^2}{2m^*}\left(k_x{}^2 + k_y{}^2 + k_z{}^2\right) \tag{3.41}$$

Two important conclusions are:

- The wavenumbers of the free electrons are quantized.
- The energy states of the free electrons are quantized.

n_x, n_y and n_z are integer quantum numbers. As the number of valence electrons in a macroscopic crystal is always very large, of magnitude $10^{28}\,\mathrm{m}^{-3}$, both E and k are perceived as continuous.

Energy Levels of Free Electrons in a Metal

The free electrons in a metal do not belong to particular metal ions. Consequently, the whole metal represents *one* system with a large number of free electrons and a large number of different energy states. Each electron supplies one eigenfunction and its associated eigenvalue to the pool. Consider, for example, 1 kmol of a metal with valence 1. In this case the number of valence electrons in the metal is equal to Avogadro's number, $N_A = 6.02 \times 10^{26}\,\mathrm{kmol}^{-1}$. In the general case, we have:

- The number of occupied collective electron energy states in a metal crystal is equal to the total number of valence electrons, i.e. the valence number times the number of atoms.

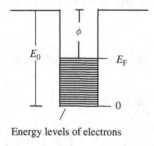

Energy levels of electrons

Figure 3.17 Potential well E_0 and occupied energy levels of free electrons in a metal. E_F = the Fermi level.

Like all other atomic systems, the free electron system must obey the Pauli principle. The Pauli principle, applied to the energy states of the free electron system, gives:

- Each energy state, defined by the quantum numbers n_x, n_y *and* n_z, can only accommodate two electrons with opposite spins.

As the total number of free electrons is equal to the number of energy levels, only half of the available sites are filled. In the absence of thermal excitation, the most low-lying energy levels are filled and the rest are empty (Figure 3.17). Hence the valence electrons are forced to be located not only in each of the lowest energy states, corresponding to the Boltzmann thermal distribution (Chapter 4), but also in higher and higher energy states in agreement with the Pauli exclusion principle. The upper energy limit E_F is called the *Fermi level*, which represents the energy of the most energetic valence electrons in the metal at $T = 0\,\mathrm{K}$.

The Fermi level is of magnitude 5 eV, which is a very high energy. Thermal excitation of classical particles up to such high kinetic energies would require a temperature of magnitude 50 000 K instead of room temperature in combination with the Pauli principle for electrons.

The relationship between the depth of the wall, the Fermi energy and the work function can be obtained from Figure 3.17:

$$E_0 = E_F + \phi \tag{3.42}$$

The quantum-mechanical model for the electron gas, or the 'Fermi sea', is much more likely than the classical model where all the electrons are crowded at the bottom of the potential well.

The next step is to calculate the energy distribution in the electron gas. We still keep the assumption that the mutual electrostatic interaction between the electrons and the electron interaction with the lattice ions are neglected.

Fermi–Dirac Distribution. Fermi Factor

During the discussion of the covalent bond of the H_2 molecule on page 108, it was mentioned that the two valence electrons in H_2 cannot be distinguished. When the Maxwell–Boltzmann statistics are derived, the particles are supposed to be *distinguishable* (Chapter 4, page 174). Hence the Maxwell–Boltzmann statistics cannot be applied on electrons and have to be replaced by some other statistical distribution.

The physicists Fermi and Dirac showed that all particles with half integer spins obey the statistics which named after them. The derivation of Fermi–Dirac statistics is performed in the same way as Maxwell–Boltzmann statistics (Chapter 4, page 174) but with the very important difference that the electrons are *indistinguishable*. The result is that f_{MB} shall be replaced by the so-called *Fermi factor*, f_{FD}:

$$f_{FD} = \frac{1}{e^{\frac{E-E_F}{k_B T}} + 1} \tag{3.43}$$

where E_F is the Fermi level discussed on page 116.

The Fermi factor f_{FD} represents the probability that the energy level E will accommodate an electron at temperature T. The Fermi factor is shown in Figure 3.18 for $T = 0\,K$. The Fermi level is the border between occupied and unoccupied energy levels at $T = 0\,K$.

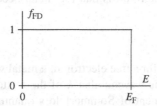

Figure 3.18   The Fermi factor f_{FD} as a function of energy at $T = 0\,K$.

1. The number of free electrons and energy states is very large. Each electron has its own eigenfunction and each energy state can accommodate a maximum of two electrons according to the Pauli principle.
2. At absolute temperature $T = 0\,K$ all energy states $\leq E_F$ will be occupied by two electrons with opposite spins and all energy states $\geq E_F$ will be empty:

$$f_{FD}(E) = \frac{1}{e^{-\infty} + 1} = 1 \quad \text{for } E < E_F$$

$$f_{FD}(E) = \frac{1}{e^{+\infty} + 1} = 0 \quad \text{for } E > E_F$$

3. At temperatures $T > 0\,K$ electrons with kinetic energies of magnitude $E > E_F - k_B T$ may be thermally excited up to available energy levels $> E_F$.

The Fermi factor f_{FD} has the following property. The sharp discontinuity between occupied and empty energy levels at $E = E_F$ is smoothed out at higher temperatures. Some of the most energetic electrons become excited up to empty sites above the Fermi level and leave vacant sites below the Fermi level, as shown in Figure 3.19.

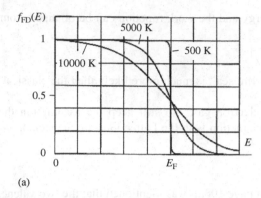
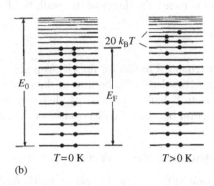

Figure 3.19 (a) The Fermi factor as a function of energy for different values of the temperature T. Reproduced with permission from C. Kittel, *Introduction to Solid State Physics*, 6th edn. © 1986 John Wiley & Sons, Inc.
(b) Distribution of electrons in electron energy states in the vicinity of the Fermi level. Reproduced with permission from M. Alonso and E. Finn, *Fundamental University Physics*. © Addison-Wesley.

The upper temperature limit for solid metals is the melting points of the metals, i.e. $\leq 3500\,\mathrm{K}$, and for most metals considerably lower.

Now it is easy to understand why metals are opaque. The energy states of the valence electrons are so closely spaced that they are practically continuous and form an *energy band* (Figure 3.17 on page 116). Valence electrons in low-lying energy states inside the metal can easily absorb photons of arbitrary energies and be excited to higher empty energy levels in the energy band, above the upper limit of filled energy levels.

If the bound valence electrons with high kinetic energies are excited sufficiently they may escape from the surface of the metal. In this case the energy of the electrons is no longer quantized as arbitrary values of the kinetic energy are allowed outside the metal. Hence photons of all energies can be absorbed. Consequently, all wavelengths of visible light become absorbed and no wavelengths are transmitted. The result is that the metal becomes opaque.

Representation of ψ and E_{kin} in k Space

In Section 3.4.3, where the Schrödinger equation for a free electron in a metal was solved, we found that both the eigenfunction and the eigenvalue were expressed in terms of the wavenumber k of the matter wave of the free electron.

A very fruitful approach for further development of Sommerfeld's quantum mechanical model of the electron gas is to introduce the k space and to discuss the kinetic energy and the energy states of the free electron in this representation. In this way, the Fermi level and the density of energy states per energy unit can be derived. This is the basis of the important band theory of solids which will be discussed in Section 3.5.

The Eigenfunction of the Free Electron

If we use complex functions instead of sine functions and introduce a vector representation of the wavevector, the eigenfunction of the free electron [Equation (3.33) on page 115] can be written in a much more compact way, which simplifies future calculations considerably.

If we make use of Euler's equation:

$$\mathrm{e}^{\mathrm{i}\varphi} = \cos\varphi + \mathrm{i}\sin\varphi \tag{3.44}$$

the eigenfunction can be written as

$$\psi = C\mathrm{e}^{\pm \mathrm{i}k \cdot r} \tag{3.45}$$

The $\pm$ sign in the exponent in Equation (3.45) corresponds to two alternative directions of propagation of the moving wave:

$$\psi = C\sin(\omega t \mp k \cdot r) \tag{3.46}$$

where both k [Equation (3.33) on page 115] and r are vectors:

$$k = k_x\hat{x} + k_y\hat{y} + k_z\hat{z} \tag{3.33}$$

$$r = x\hat{x} + y\hat{y} + z\hat{z} \tag{3.47}$$

Graphical Representation of ψ

Consider a matter wave with eigenfunction ψ, wavelength λ and wavenumber k. We define the *wavevector* k of the matter wave as a vector with the *magnitude* $2\pi/\lambda$ and a *direction* equal to the direction of propagation of the matter wave. In k space, the wavevector k is represented by a *point* (Figure 3.20a), i.e. by a vector from the origin to the point (k_x, k_y, k_z).

The planes in Figure 3.20b are drawn perpendicular to the k vector. It can be seen from Equations (3.45) and (3.46) that the scalar product $k \cdot r$ has the dimension zero and represents a *phase angle*. For the first plane perpendicular to the k vector in Figure 3.20b we have

$$kr_{\mathrm{I}} = \varphi_{\mathrm{I}} \tag{3.48}$$

The value of φ_{I} is constant for all the points in the plane. In addition, the eigenfunction of the free electron has the same value at every point of a plane. If $kr_{\mathrm{I}} = \varphi_{\mathrm{I}}$ is inserted into Equation (3.45), we obtain

$$\psi(\varphi_{\mathrm{I}}) = Ce^{i\varphi_{\mathrm{I}}} \tag{3.49}$$

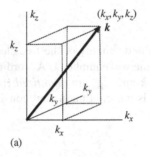
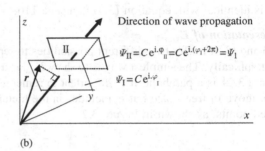

Figure 3.20 (a) In k space the wavevector k is a *point* with the coordinates (k_x, k_y, k_z). (b) The planes I and II in r space are drawn perpendicular to the direction of propagation. The distance between the planes is equal to the wavelength λ.

If the perpendicular plane is displaced in the direction of the normal, the value of the scalar product $k \cdot r$ is increased and the value of eigenfunction varies periodically. In particular, when φ_{I} is increased by 2π ($\varphi_{\mathrm{II}} = \varphi_{\mathrm{I}} + 2\pi$), the value of the eigenfunction will be the same for both the parallel planes:

$$\psi(\varphi_{\mathrm{II}}) = Ce^{i\varphi_{\mathrm{II}}} = Ce^{i(\varphi_{\mathrm{I}}+2\pi)} = Ce^{i\varphi_{\mathrm{I}}} = \psi(\varphi_{\mathrm{I}}) \tag{3.50}$$

The same is true for all parallel and equidistant planes. If 2π in Equation (3.50) is replaced by an integer multiple of 2π, the result will be the same. Hence:

- If the distance between successive planes is λ, the value of the eigenfunction is the same at every point of all parallel equidistant planes, at distances λ from each other.

This set of parallel planes in r space corresponds to the wave vector k in k space.

Kinetic Energy of the Free Electron

In the preceding section, where the solution of the Schrödinger equation is discussed, the wave character of the free electron in a metal was used. The eigenvalue is equal to the kinetic energy of the electron as its potential energy is zero. It is possible to use the particle character of the electron to find its kinetic energy.

The kinetic energy of the electron can be calculated classically in terms of velocities or momentum:

$$E_{\mathrm{kin}} = \frac{m^*v^2}{2} = \frac{m^*\left(v_x^2 + v_y^2 + v_z^2\right)}{2} = \frac{p^2}{2m^*} = \frac{p_x^2 + p_y^2 + p_z^2}{2m^*} \tag{3.51}$$

The momentum vector $\boldsymbol{p}$ of the free electron, written in components, is

$$\boldsymbol{p} = m^*v_x\hat{x} + m^*v_y\hat{y} + m^*v_z\hat{z} \tag{3.52}$$

If we take the duality between waves and particles into consideration, we can express the kinetic energy of the electron in terms of the wavevector $\boldsymbol{k}$. According to Equation (2.8) on page 50, the de Broglie wavelength of the free electron is

$$\lambda_{\text{deB}} = \frac{h}{m^*v} = \frac{h}{p} \quad \Rightarrow \quad p = \frac{h}{\lambda_{\text{deB}}} \tag{3.53}$$

Combining Equation (3.53) with Equation (3.34) on page 115, we obtain

$$p = \frac{h}{\lambda_{\text{deB}}} = \frac{h}{2\pi}\frac{2\pi}{\lambda_{\text{deB}}} = \hbar k \tag{3.54}$$

Combining Equations (3.51) and (3.54), we obtain

$$E_{\text{kin}} = \frac{p^2}{2m^*} = \frac{\hbar^2}{2m^*}k^2 \tag{3.55}$$

or in components

$$E_{\text{kin}} = \frac{p^2}{2m^*} = \frac{\hbar^2}{2m^*}\left(k_x^{\,2} + k_y^{\,2} + k_z^{\,2}\right) \tag{3.56}$$

Equation (3.56) is identical with equation (3.41) on page 116.

Graphical Representation of E_{kin}

Equations (3.55) and (3.56) offer two different possibilities to represent the kinetic energy of the free electron as a function of the wavevector graphically. The simplest way is to show E_{kin} as a function of the wavenumber k. According to Equation (3.55), the curve in Figure 3.21 is a parabola. It is important to remember that both $\boldsymbol{k}$ and E_{kin} are *quantized* (page 116) because the electron does not move in free space but is included in the metal. The curve is *not* continuous but consists of a large number of closely situated points, as shown in Figure 3.21.

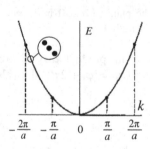

Figure 3.21 Kinetic energy of a free electron in a metal as a function of the wavenumber of the matter wave. $a =$ the lattice constant. Reproduced with permission from M. Alonso and E. Finn, *Fundamental University Physics*. © Addison-Wesley.

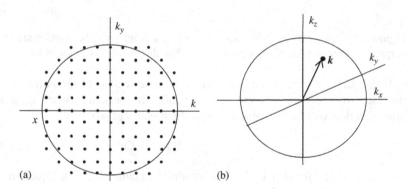

(a) (b)

Figure 3.22 (a) Allowed energy levels of a free electron in a metal are represented by points in $\boldsymbol{k}$ space. Only two dimensions are shown in the figure. (b) Energy levels of a free electron in a metal with constant energy correspond to points in $\boldsymbol{k}$ space on a spherical surface, the so-called *Fermi surface*.

Alternatively, all the allowed values of the $\boldsymbol{k}$ vector can be plotted in $\boldsymbol{k}$ space, i.e. in a three-dimensional k_x, k_y, k_z coordinate system. Two dimensions of such a plot are shown in Figure 3.22a. The circle around the points has a radius that represents the maximum k value which is compatible with Equation (3.55) for a given value of E_{kin}. Each point represents the tip of a $\boldsymbol{k}$ vector, i.e. one of the many possible matter waves inside the metal.

Figure 3.22b is an attempt to show the same thing in three dimensions. Each energy level is represented by a sphere, which contains all the tips of the $\boldsymbol{k}$ vector on its surface. A sphere that represents a given kinetic energy is called a *Fermi surface*. The maximum energy of nonexcited electrons is a sphere with a radius equal to the k value, which corresponds to the Fermi level E_{F}.

Fermi Distribution of Energies in the Electron Gas

Figure 3.19b on page 118 shows that the energy levels are filled with electrons approximately up to the Fermi level E_F. We also know that each energy level can accommodate two electrons.

For full information of the valence electron distribution on different energy levels in a piece of metal, called the crystal below, we have to calculate

- the *density of available electron energy states* $= N(E)$, i.e. the number of available energy states per energy unit and unit volume
- the *density of occupied energy states* $= f_{FD}N(E)$, i.e. the number of occupied energy states per energy unit and unit volume.

Calculation of the Density of Available Electron Energy States

The simplest way to calculate $N(E)$ is to consider the three-dimensional plot of allowed energy levels, represented by points in the k space, in Figure 3.22.

Each k value is represented by a point in k space. The task of finding the number of energy states within the shell is equivalent to finding the number of points within the shell.

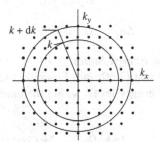

Figure 3.23 Each point in k space represents an energy state.

Each point in the three-dimensional plot (two dimensions of the plot are shown in Figure 3.23) represents an energy state. Hence the number of energy states in the whole crystal within the energy interval $(E + dE)$ and E is equal to the number of points in the volume between the two spherical shells in k space. To obtain the number of points within the shell, we divide the shell volume by the unit volume in k space, i.e. by the volume occupied by one point:

$$\text{Number of points within the shell} = \frac{\text{volume of spherical shell in } k \text{ space}}{\text{volume per point in } k \text{ space}}$$

or

$$\text{Number of points within the shell} = \frac{4\pi k^2 dk}{\Delta k_x \Delta k_y \Delta k_z} \tag{3.57}$$

Calculation of the Shell Volume in k Space in Terms of Energy

The volume of the spherical shell is equal to $4\pi k^2 dk$. We want to express this volume in k space in terms of E. For this purpose, we use Equation (3.35) on page 115. Differentiating Equation (3.35), we obtain

$$dk = \frac{1}{\hbar}\sqrt{2m^*}\frac{dE}{2\sqrt{E}} \tag{3.58}$$

Hence the volume in k space of the spherical shell, expressed in terms of energy, will be

$$4\pi k^2 dk = 4\pi \frac{2m^*E}{\hbar^2}\left(\frac{1}{\hbar}\sqrt{2m^*}\frac{dE}{2\sqrt{E}}\right)$$

or

$$4\pi k^2 dk = 2\pi \left(2m^*\right)^{\frac{3}{2}} \hbar^{-3} E^{\frac{1}{2}} dE \tag{3.59}$$

Calculation of the Volume per Point in k Space

Consider the enlargement in Figure 3.24 of some of the points in k space in Figure 3.23. The distances between adjacent points are Δk_x, Δk_y and Δk_z in the three main directions. Each point in k space disposes a volume $\Delta k_x \Delta k_y \Delta k_z$.

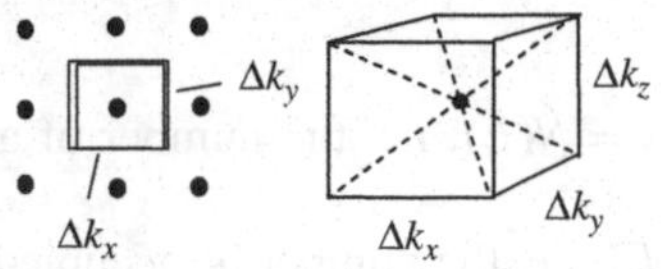

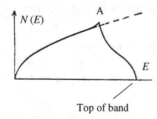

Figure 3.24 Volume per k point.

By differentiating Equations (3.40b) on page 115, we can express this volume in terms of the dimensions of the crystal:

$$\Delta k_x \Delta k_y \Delta k_z = \frac{2\pi}{L_x}\frac{2\pi}{L_y}\frac{2\pi}{L_z} = \frac{(2\pi)^3}{V} \tag{3.60}$$

where V is the crystal volume in r space and $\Delta n_x = \Delta n_y = \Delta n_z = 1$.

Calculation of the Density of Available Electron Energy States

By inserting the expressions in Equations (3.59) and (3.60) into Equation (3.57), we obtain the number of points within the shell in k space. It is also equal to the number of energy levels within the energy interval $(E+dE)$ and E in the whole crystal with volume V. Hence we obtain

$$N(E)\mathrm{d}E \times V = \frac{4\pi k^2 \mathrm{d}k}{\Delta k_x \Delta k_y \Delta k_z} = \frac{2\pi(2m^*)^{\frac{3}{2}}\hbar^{-3}E^{\frac{1}{2}}\mathrm{d}E}{\dfrac{(2\pi)^3}{V}} \tag{3.61}$$

If we divide Equation (3.61) by $V\mathrm{d}E$ we obtain, after reduction, the desired number of available energy states per energy unit and unit volume:

$$N(E) = \frac{(2m^*)^{\frac{3}{2}}}{4\pi^2\hbar^3}E^{\frac{1}{2}} \tag{3.62}$$

The density of electron energy states is plotted as a function of the energy E in Figure 3.25, which shows that deviations from Equation (3.62) occur close to the maximum A of $N(E)$. This phenomenon will be discussed on pages 142–145.

Figure 3.25 Density $N(E)$ of electron energy states as a function of the electron energy E.

As each energy state can accommodate two electrons, one with spin up and the other with spin down, we can also obtain the *electron density* $n(E)$, i.e. the number of electron sites per unit volume with energies between $(E+dE)$ and E:

$$n(E) = 2N(E) \tag{3.63}$$

Equations (3.62) and (3.63) are very useful and will be frequently applied later.

Calculation of the Density of Occupied Electron Energy States

The number of *occupied* electron energy states per unit energy is obtained if we multiply the density of electron energy states by the Fermi factor. Hence the density of occupied electron energy states will be

$$N(E)f_{\text{FD}} = \frac{(2m^*)^{\frac{3}{2}}}{4\pi^2\hbar^3}E^{\frac{1}{2}}f_{\text{FD}} \tag{3.64}$$

The function is derived graphically and shown in Figure 3.26.

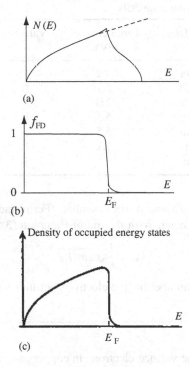

(a)

(b)

(c)

Figure 3.26 (a) Density of electron energy states as a function of the electron energy. (b) The Fermi function. (c) Density of occupied electron energy states as a function of the energy.

Calculation of the Fermi Level

Equation (3.64) can be used for the determination of the value of the Fermi level of the metal. The calculation will be especially simple if the temperature $T = 0\,\text{K}$. In this case $f_{\text{FD}} = 1$ for all energies below E_{F}.

At $T = 0\,\text{K}$ there is no thermal excitation of the electrons and all the valence electrons occupy the lowest possible energy states compatible with the Pauli principle. The highest energy is by definition equal to the Fermi energy E_{F}. The Fermi level is approximately equal to this value even at temperatures $> 0\,\text{K}$ if the temperature is not too high (error of magnitude 0.1 eV). In this case the total number of valence electrons per unit volume n_{total} must be equal to the sum of all occupied energy states per unit volume between $E = 0$ and $E = E_{\text{F}}$.

The total number of valence electrons per unit volume in the metal is obtained by integration of Equation (3.64) from $E = 0$ up to $E = E_{\text{F}}$:

$$n_{\text{total}} = \int\limits_0^{E_{\text{F}}} 2N(E)f_{\text{FD}}\,\text{d}E = \int\limits_0^{E_{\text{F}}} 2 \times \frac{(2m^*)^{\frac{3}{2}}}{4\pi^2\hbar^3}E^{\frac{1}{2}} \times 1 \times \text{d}E \tag{3.65}$$

or

$$n_{\text{total}} = \frac{(2m^*)^{3/2}}{2\pi^2\hbar^3}\frac{2}{3}E_{\text{F}}^{3/2} \tag{3.66}$$

where n_{total} is the number of valence electrons per unit volume.

Equation (3.66) can be used for determination of the Fermi level. If it is solved for E_F we obtain

$$E_F = \frac{\hbar^2}{2m^*}\left(3\pi^2 n_{total}\right)^{2/3} \tag{3.67a}$$

The Fermi level or Fermi energy is a function of the number of electrons per unit volume in the metal. Fermi energies of some common metals are given in Table 3.6.

Table 3.6 Fermi energies of some common metals.

Metal	Fermi energy (eV)	Valence
Na	3.2	1
K	2.1	1
Cu	7.0	1
Ag	5.5	1
Au	5.5	1
Mg	7.1	2
Al	11.6	3

As E_F is a material constant we can calculate a value of the so-called 'Fermi radius' k_F, i.e. the 'radius' in reciprocal space of the Fermi sphere (Fermi surface), for each metal. With the aid of Equation (3.55) on page 120 we obtain

$$k_F = \frac{1}{\hbar}\sqrt{2m^* E_F} \tag{3.67b}$$

If the metal is alloyed with another metal, the number of free electrons per unit volume n total changes and also E_F [Equation (3.67a)] and k_F.

Example 3.1

Calculate the average kinetic energy (eV) of the valence electrons in copper at room temperature. Compare the result with the average thermal energy of a classical particle at $T = 300\,K$. The Fermi level of copper is 7.0 eV.

Solution:

The average kinetic energy can be calculated by use of the density of occupied electron energy states in a metal [Equation (3.65)]:

$$n_{total} = \int_0^{E_F} 2N(E)f_{FD}\,dE = \int_0^{E_F} 2 \times \frac{(2m^*)^{3/2}}{4\pi^2\hbar^3}E^{1/2} \times 1 \times dE$$

The average kinetic energy can be written as

$$\overline{E} = \frac{\int_0^{E_F} E \times N(E)f_{FD}\,dE}{\int_0^{E_F} N(E)f_{FD}\,dE} = \frac{\int_0^{E_F} E\,\dfrac{2\,(2m^*)^{3/2}}{4\pi^2\hbar^3}E^{1/2}\,dE}{\int_0^{E_F} \dfrac{2\,(2m^*)^{3/2}}{4\pi^2\hbar^3}E^{1/2}\,dE}$$

or

$$\overline{E} = \frac{\int_0^{E_F} E^{3/2}\,dE}{\int_0^{E_F} E^{1/2}\,dE} = \frac{\dfrac{E_F^{5/2}}{5/2}}{\dfrac{E_F^{3/2}}{3/2}} = \frac{3}{5}E_F = 0.6 \times 7.0\,\text{eV} = 4.2\,\text{eV}$$

The thermal energy of a classical particle at room temperature is

$$\frac{3}{2}k_{\mathrm{B}}T = 1.5 \times 1.38 \times 10^{-23} \times 300\,\mathrm{J} = \frac{1.5 \times 1.38 \times 10^{-23} \times 300}{1.60 \times 10^{-19}}\,\mathrm{eV} = 0.04\,\mathrm{eV}$$

Answer:

The average kinetic energy of the valence electrons in copper is 4.2 eV, about 100 times higher than the average kinetic energy of a classical particle at room temperature.

3.5 Band Theory of Solids

As we have seen above, Sommerfeld's free electron model of the valence electrons in a metal proved to be most successful compared with the classical model. The quantum mechanical electron model is generally accepted in solid-state physics. However, objections can be raised to his assumption that the potential energy inside a metal is constant. It was natural to improve Sommerfeld's model by replacing the constant potential energy of the electrons with a varying potential energy with the same periodicity as the crystal lattice. This led to the development of *the band theory of solids*.

As an introduction to this theory, we will start with a simple example, which discusses qualitatively the origin of energy bands in solids and illustrates the connection between molecular physics and solid-state physics.

3.5.1 Origin of Energy Bands in Solids

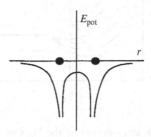

Figure 3.27 Potential energy along the axis between two H nuclei of the H_2 molecule as a function of the distance from the origin.

Figure 3.27 shows the potential energy near the H_2 molecule along the axis through the two H nuclei. The potential energy function is introduced into the Schrödinger equation, which is solved and gives the eigenvalue E and the eigenfunction ψ.

In Section 2.6.1 on pages 74–75 we discussed the formation of an H_2 molecule from two H atoms and the two resulting eigenfunctions which correspond to one stable and one unstable energy state (Figure 3.28). Here we will discuss the formation

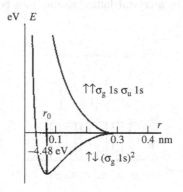

Figure 3.28 Energy of the H_2 molecule as a function of the distance between the two protons. Reproduced with permission from M. Alonso and E. Finn, *Fundamental University Physics*. © Addison-Wesley.

of a 'solid' from six equidistant H atoms in a row. The potential energy in the one-dimensional 'crystal' is shown at the top of Figure 3.29a. The eigenfunction curves (a)–(f) in Figure 3.29a are obtained by combining the six 1s eigenfunctions of the H atoms in various ways.

The six resulting eigenfunctions correspond to six separate eigenvalues, i.e. energy states. The energies of these states vary with the distance R between adjacent H nuclei. As can be seen from Figure 3.29b, the energy levels split when the distance R decreases. At the ground-state distance R_0 the 1s level has widened to a 'band' with six energy levels. The 2s level in the H atom also splits into a band. Between these two bands there is a forbidden region or an *energy gap*.

The width of the band does not increase even if the number of atoms is increased. If the number of atoms is increased, the number of energy levels increases but all the new energy levels lie between the minimum and maximum energies [curves (f) and (a) in Figure 3.29b].

The number of energy levels in the energy band is equal to the number of energy states, i.e. the given number of atoms. Six H atoms have together six 1s states, which give six energy levels in the band. Each energy state can accommodate two electrons owing to the anti-parallel spins of the electrons, which doubles the number of available sites for electrons.

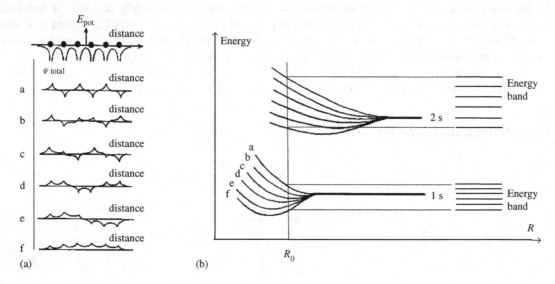

Figure 3.29 (a) Eigenfunctions of the H_6 row. (b) Energy curves of the H_6 row in (a). Reproduced with permission. © E. Lindholm (Deceased).

3.5.2 Energy Bands in Solids. Brillouin Zones

Eigenvalues and Eigenfunctions of Free Electrons in a Crystal Lattice. Brillouin Zones in One Dimension

The electrostatic attraction between the ions in a crystal lattice results in a periodic potential energy (Figure 3.30a) of the same type as in the H_2 molecule (Figure 3.27).

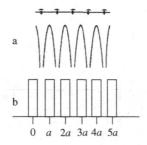

Figure 3.30 (a) Coulomb potential energy due to a row of ions in a crystal lattice.
(b) Periodic potential energy in a linear crystal lattice according to, the so-called Kronig–Penney model.

Three-dimensional Case

The potential energy in the crystal lattice is complicated and depends on the structure of the crystal. It is necessary to use an approximate three-dimensional model such as that in Figure 3.30b rather than that in Figure 3.30a. In a crystal with a simple cubic structure the period of the square well is equal to the lattice constant *a*.

The periodic potential energy $E_{pot}(x, y, z)$ is introduced into the Schröderinger equation, which complicates its solution considerably. In this case, the Schrödinger equation can be written as

$$\frac{\partial \psi^2}{\partial x^2} + \frac{\partial \psi^2}{\partial y^2} + \frac{\partial \psi^2}{\partial z^2} + \frac{8\pi^2 m^*}{h^2}[E - E_{pot}(x, y, z)]\psi = 0 \tag{3.68}$$

The periodic potential energy leads to a periodically varying amplitude $U_k(r)$ of the wave function of the electron instead of a constant amplitude as in the case of a free electron. $U_k(r)$ is one of a numerous group of functions, the so-called *Bloch functions*. They are generally used to describe electrons in lattices.

The wave function is a planar matter wave, which is modulated by the function $U_k(r)$. The amplitude of the wave function of an electron in a three-dimensional lattice can be written as

$$\psi(r) = U_k(r)\,e^{ik \cdot r} \tag{3.69}$$

where

r = vector with components (x, y, z)
k = wavevector in the direction of the matter wave.

The wavevector k was introduced on page 115. The eigenvalues that correspond to the eigenfunctions are complicated functions of k and depend of the geometry of the crystal lattice. The general properties of the eigenfunctions will be described below for the one-dimensional case.

One-dimensional Case. The Kronig–Penney Model. Brillouin Zones in One Dimension

In the one-dimensional case, the Schrödinger equation can be written as

$$\frac{\partial \psi^2}{\partial x^2} + \frac{2m^*}{\hbar^2}[E - E_{pot}(x)]\psi = 0 \tag{3.70}$$

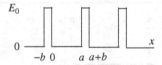

Figure 3.31   One-dimensional Kronig–Penney potential energy model. This function is introduced into the Schrödinger equation. The solutions consist of the eigenvalues and their corresponding eigenfunctions (Chapter 2).

Kronig and Penney suggested in 1930 a simple model of the periodically varying potential energy. They assumed that the potential energy of an electron had the shape of a periodic array of square wells. The distance between successive wells was assumed to be $a + b$, where b is the width of the well (Figure 3.31). The Schrödinger equation for the two regions can be written as

$$\frac{d^2\psi}{dx^2} + \frac{2m^*}{\hbar^2}(E - 0)\,\psi = 0 \quad 0 < x < a \tag{3.71}$$

$$\frac{d^2\psi}{dx^2} + \frac{2m^*}{\hbar^2}(E - E_0)\,\psi = 0 \quad a < x < a + b \tag{3.72}$$

We will omit the further calculations and only describe the solutions. In the case of a linear lattice with the spacing a (b is small), the solution of the Schrödinger equation, i.e. the egenfunction for a given k value, can be written as

$$\psi(x) = e^{ikx}U_k(x) \tag{3.73}$$

where $U_k(x)$ is the *Bloch function*. In addition, $U_k(x)$ must be periodic, i.e. satisfy the condition (Bloch's theorem)

$$U_k(x) = U_k(x+a) \tag{3.74}$$

The corresponding eigenvalue in terms of the wavevector $\boldsymbol{k}$ is complicated and depends on the lattice constant. Both the eigenvalues and the wavevectors are *quantized* just as for a free electron. However, the Bloch function $U_k(x)$, generates a very important difference:

• For the values of k which are given by the conditions

$$k = p\frac{\pi}{a} \quad p = \pm 1, \pm 2, \ldots \tag{3.75}$$

the Schrödinger equation has no unique solutions. For each of these k values there are two eigenvalues, separated by an energy gap.

The condition (3.75) is analogous to Bragg's law, applied to X-ray diffraction in crystals (Chapter 1). The reason for this condition can be understood on basis of physical arguments, which will be discussed on page 135.

Instead of a quasi-continuous curve (consisting of closely spaced points), i.e. the parabola of the free electron (Figure 3.21 on page 120), the Kronig–Penney energy curve (Figure 3.32) shows bands of allowed energy values interrupted by energy gaps, i.e. regions of forbidden energy values. As mentioned above, the eigenvalues and the k numbers are quantized. The broken curve is not continuous but consists of closely spaced points.

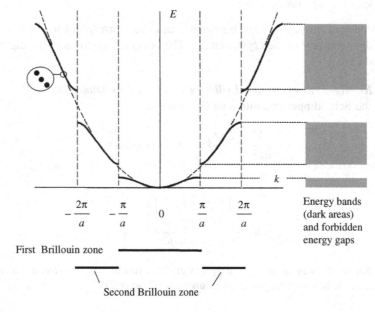

Figure 3.32 Energy levels in a linear periodic lattice as a function of the wavenumber k. a = the lattice constant. Reproduced with permission from M. Alonso and E. Finn, *Fundamental University Physics*. © Addison-Wesley.

Figure 3.32 shows the energy as a function of k. The discontinuities appear at the k values given in Equation (3.75). At k values far from the discontinuity points the energy is nearly the same as for free electrons (dashed curve in Figure 3.32) and the electrons can move freely through the crystal lattice.

For a free electron the potential energy $E_{pot} = 0$. For small values of the potential barriers (Figure 3.31) the discontinuous curve is comparatively close to a parabola. At high barriers the curve has the appearance shown in Figure 3.32.

The k zones of allowed energies are called *Brillouin zones*. Their extensions are described in Figure 3.32. The allowed energy bands are shown on the right in Figure 3.32. The widths of the energy bands increase with increasing energy. The stronger the electron is bound to the lattice ions, the narrower will be the widths of the energy bands.

Inner electron shells around the lattice ions consist of narrow levels in analogy with the conditions in atoms.

The band theory of solids is able to explain properties of solids where the Sommerfeld model fails. It is generally accepted and established in solid-state physics.

Number of Possible Eigenfunctions and Energy States per Band

The concept of *primitive cell* was introduced in Chapter 1 on page 13.

A *space lattice* is a mathematical pattern built of three independent translation base vectors in three dimensions.

A *unit cell* is defined as the volume in a space lattice which by translation movements can fill the whole lattice space without overlapping any other cell or leaving a hollow space inside the lattice.

A *primitive cell* is a special type of unit cell, which contains only one lattice point per unit cell.

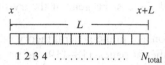

Figure 3.33 Primitive cells in a linear crystal.

Consider a linear crystal of length L which consists of N_{total} primitive cells in a row (Figure 3.33). If the spacing of the crystal lattice is a, the relationship will be

$$L = N_{total}a \qquad (3.76)$$

To solve the Scrödinger equation for the crystal, it is necessary to introduce the boundary condition

$$\psi(x) = \psi(x+L) \qquad (3.77)$$

Using Equation (3.73), we can write Equation (3.77) as

$$e^{ikx}U_k(x) = e^{ik(x+L)}U_k(x+L) \qquad (3.78)$$

If Equation (3.74) is applied to all the primitive cells, we can conclude that

$$U_k(x) = U_k(x+a) = U_k(x+2a) = \ldots = U_k(x+L) \qquad (3.79)$$

Combining Equations (3.78) and (3.79), we realize that the following condition must be valid for a linear crystal:

$$e^{ikx} = e^{ik(x+L)} \qquad (3.80)$$

or

$$kL = p \times 2\pi \quad p = \pm 1, \pm 2, \pm 3, \ldots, \pm N_{total} \qquad (3.81)$$

Hence

$$k = p\frac{2\pi}{L} \qquad (3.82)$$

The upper limit of the k values of the first band is $k = N_{total} \times 2\pi/L$, which corresponds to $k = 2\pi/a$, in agreement with Equation (3.76).

Each point in k space corresponds to an eigenfunction and its corresponding eigenvalue, i.e. an energy state. According to the Pauli exclusion principle, each energy state can be occupied by no more than two electrons with opposite spin vectors. Hence the first band can accommodate a maximum of $2N_{total}$ electrons. The same result is obtained if we examine the second and all the following bands.

Even if we consider a three-dimensional lattice the result will be the same:

- The number of energy states in each Brillouin zone is equal to N_{total}, the total number of atoms of the crystal.
- Each Brillouin zone can accommodate a maximum of $2N_{total}$ electrons.

Brillouin Zones in Two and Three Dimensions

The Brillouin zones in one dimension and the corresponding electron energy bands in a one-dimensional crystal are shown in Figure 3.32. Within the Brillouin zones the energy E is quasi-continuous and forms energy bands. At the Brillouin zone boundaries ($k = p \times \pi/a$ [Equation (3.75)]) the Schrödinger equation has no unique solution but two eigenvalues for each of these k values. At the zone boundaries the energy E is discontinuous, which results in forbidden gaps between the energy bands.

In two- and three-dimensional crystals the energy conditions are much more complicated than in linear crystals. The solution of the Schrödinger equation in the general case and the mathematical calculations to find the eigenvalues are complicated and beyond the scope of this book. Instead, we will discuss the general theory of Brillouin zones in two and three dimensions from a physical point of view.

The reciprocal lattice space is a very useful concept in the general theory of Brillouin zones. The wavevector k and k space were introduced on page 115 and discussed further at pages 118–119. As we shall see later, the k space is identical with the reciprocal space of the ordinary space, which we will call r space.

The general diffraction condition of matter waves in crystals is closely connected with the general theory of Brillouin zones. It will be discussed below.

Reciprocal Lattices of Crystals

Real Space

The real space, or r space for short, is the ordinary three-dimensional space. To describe an arbitrary crystal structure in r space, we choose a coordinate system with the base vectors a, b, c along the axes equal to the translation vectors of the crystal structure. The angles between the coordinate axes are not necessarily $\pi/2$ but depend on the type of crystal structure.

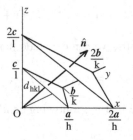

Figure 3.34   Crystal planes in r space.

The atoms in a crystal lattice form several sets of parallel planes. It is well known that radiation of all kinds, for example electrons and X-rays, is diffracted through constructive interference at the atoms which form the crystal planes. This has been discussed in Chapter 1 and will be applied below.

A set of crystal planes has been marked in Figure 3.34. The indices of the multiple set of parallel planes are called (hkl). If we introduce the unit vector of the normal to the set of parallel planes we obtain the distance d_{hkl} from origin to the closest plane by forming the scalar product of the normal unit vector and one of the vectors a/h, b/k or c/l (Figure 3.34):

$$d_{hkl} = \hat{n} \cdot \frac{a}{h} = \hat{n} \cdot \frac{b}{k} = \hat{n} \cdot \frac{c}{l} \tag{3.83}$$

where (hkl) are the Miller indices (Chapter 1, page 16) of the planes.

Reciprocal Space

Next we will define the *reciprocal space* of the r space and use the result to derive an expression of the normal vector to the set of parallel planes in Figure 3.34.

The general definitions of base vectors a^*, b^* and c^* of the reciprocal space as functions of the base vectors a, b and c of the r space are given by Equations (3.84). The word reciprocal also implies the reverse operation: the base vectors of the real space a, b, and c as functions of a^*, b^* and c^* are given by Equations (3.85). The equations are completely symmetrical.

$$a^* = 2\pi \frac{b \times c}{a \cdot (b \times c)}$$

$$b^* = 2\pi \frac{c \times a}{a \cdot (b \times c)} \qquad (3.84)$$

$$c^* = 2\pi \frac{a \times b}{a \cdot (b \times c)}$$

$$a = 2\pi \frac{b^* \times c^*}{a^* \cdot (b^* \times c^*)}$$

$$b = 2\pi \frac{c^* \times a^*}{a^* \cdot (b^* \times c^*)} \qquad (3.85)$$

$$c = 2\pi \frac{a^* \times b^*}{a^* \cdot (b^* \times c^*)}$$

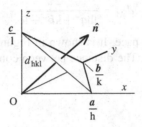

Figure 3.35   Part of Figure 3.34. The distance to the closest plane from the origin.

Figure 3.35 shows the plane closest to the origin. The normal vector to the plane has been drawn. The normal vector to the set of parallel planes can be written as

$$\hat{n} = \frac{\left(\dfrac{a}{h} - \dfrac{b}{k}\right) \times \left(\dfrac{b}{k} - \dfrac{c}{l}\right)}{\left|\left(\dfrac{a}{h} - \dfrac{b}{k}\right) \times \left(\dfrac{b}{k} - \dfrac{c}{l}\right)\right|} \qquad (3.86)$$

After multiplication in the numerator, we obtain

$$\hat{n} = \frac{\left(\dfrac{a \times b}{hk} + \dfrac{c \times a}{lh} + \dfrac{b \times c}{kl}\right)}{\left|\left(\dfrac{a}{h} - \dfrac{b}{k}\right) \times \left(\dfrac{b}{k} - \dfrac{c}{l}\right)\right|} \qquad (3.87)$$

Equation (3.87) is introduced into the first of Equations (3.83):

$$d_{hkl} = \hat{n} \cdot \frac{a}{h} = \frac{\left(\dfrac{a \times b}{hk} + \dfrac{c \times a}{lh} + \dfrac{b \times c}{kl}\right)}{\left|\left(\dfrac{a}{h} - \dfrac{b}{k}\right) \times \left(\dfrac{b}{k} - \dfrac{c}{l}\right)\right|} \cdot \frac{a}{h} = \frac{\dfrac{a \cdot (b \times c)}{hkl}}{\left|\left(\dfrac{a}{h} - \dfrac{b}{k}\right) \times \left(\dfrac{b}{k} - \dfrac{c}{l}\right)\right|}$$

or

$$\frac{1}{\left|\left(\dfrac{a}{h}-\dfrac{b}{k}\right)\times\left(\dfrac{b}{k}-\dfrac{c}{l}\right)\right|}=\frac{d_{hkl}hkl}{a\cdot(b\times c)} \tag{3.88}$$

Equation (3.88) is introduced into Equation (3.87) and we obtain

$$\hat{n}=\frac{d_{hkl}hkl}{a\cdot(b\times c)}\cdot\left(\frac{a\times b}{hk}+\frac{c\times a}{lh}+\frac{b\times c}{kl}\right) \tag{3.89}$$

After transformation of the right-hand side of Equation (3.89), and use of Equations (3.84), we obtain

$$\hat{n}=\frac{d_{hkl}}{2\pi}(ha^*+kb^*+lc^*) \tag{3.90}$$

Next we define *the reciprocal lattice vector* $G(hkl)$:

$$G(hkl)=ha^*+kb^*+lc^* \tag{3.91}$$

$G(hkl)$ is perpendicular to the set of crystal planes (hkl) as it is parallel with the unit vector $\hat{n}$. The magnitude of $G(hkl)$ is

$$|G(hkl)|=\frac{2\pi}{d_{hkl}} \tag{3.92}$$

Equation (3.90) can be written as

$$d_{hkl}=\frac{2\pi}{|ha^*+kb^*+lc^*|} \tag{3.93}$$

The $G(hkl)$ vector is a vector in the reciprocal space. It is shown in Figure 3.36b. The set of parallel planes (hkl) in r space corresponds to the point (hkl) in reciprocal space. The distance between consecutive planes is constant and equal to the value given in Equation (3.93) (Figure 3.36a).

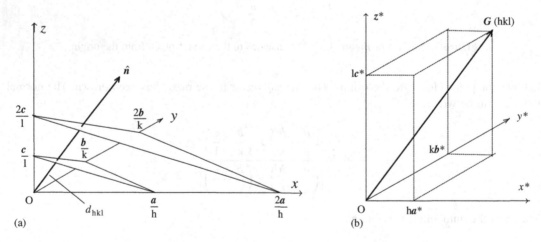

Figure 3.36 (a) Crystal planes in r space. Unit vectors on the x, y and z axes are a, b and c, respectively.
(b) The $G(hkl)$ vector in reciprocal space corresponds to the crystal planes in r space in (a). Unit vectors on the x^*, y^* and z^* axes are a^*, b^* and c^*, respectively.

For the special case of a cubic crystal lattice, we have

$$|a|=|b|=|c|=a \text{ and } |a^*|=|b^*|=|c^*|=\frac{2\pi}{a} \tag{3.94}$$

which gives for a cubic lattice

$$d_{hkl}=\frac{a}{\sqrt{(h^2+k^2+l^2)}} \quad \text{(cubic lattice)} \tag{3.95}$$

von Laue's Diffraction Condition

On page 128, we found that the condition of the Brillouin zone boundaries [Equation (3.75)] in a one-dimensional crystal (Figure 3.32 on page 128) is very reminiscent of a diffraction condition. For future use we will set up the general diffraction condition in a three-dimensional lattice for any kind of wave.

Consider a crystal lattice and a planar wave with the wavevector k. After diffraction at the atoms in a crystal plane (normal vector $\hat{n}$), the wavevector has changed to k'. We want to find an expression for the change $\Delta k = k' - k$ of the wavevector due to the diffraction. The condition for elastic diffraction (no energy loss) can be written as

$$|k| = |k'| = \frac{2\pi}{\lambda} \tag{3.96}$$

where

k = wavevector of the incident wave
k' = wavevector of the diffracted wave
λ = wavelength.

From Figure 3.37a, we obtain the general diffraction condition for constructive interference:

$$|\Delta k| = \frac{r' \cdot k'}{|k'|} - \frac{r' \cdot k}{|k|} = p\lambda \tag{3.97}$$

where p^1 is a positive or negative integer. Equations (3.96) and (3.97) give

$$r' \cdot (k' - k) = p \times 2\pi \tag{3.98}$$

The r' vector is a function of the structure of the crystal lattice:

$$r' = n_1 a + n_2 b + n_3 c \tag{3.99}$$

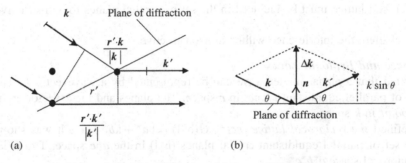

Figure 3.37   (a) Diffraction. (b) Bragg diffraction.

With the aid of Figure 3.37b (Bragg diffraction), we obtain

$$|\Delta k| = |k' - k| = 2k \sin \theta = \frac{2\pi}{\lambda} 2 \sin \theta \tag{3.100}$$

The change Δk of the wavevector can be written as

$$\Delta k = \hat{n} \cdot |\Delta k| = \hat{n} \frac{2\pi}{\lambda} 2 \sin \theta \tag{3.101}$$

If we introduce the expression for the unit vector of the normal to the reflecting crystal plane [Equation (3.90) on page 132] into Equation (3.101), we obtain

---

[1] The usual letter n is avoided in order to eliminate confusion. The letter n is used in connection with the normal of the diffraction plane.

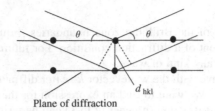

Plane of diffraction

Figure 3.38   The diffraction condition for constructive interference is $2d_{hkl} \sin \theta = p\lambda$.

$$\Delta k = \frac{d_{hkl}}{2\pi}(h\mathbf{a}^* + k\mathbf{b}^* + l\mathbf{c}^*)\frac{2\pi}{\lambda}2\sin\theta \qquad (3.102)$$

or

$$\Delta k = \frac{2d_{hkl}\sin\theta}{\lambda}(h\mathbf{a}^* + k\mathbf{b}^* + l\mathbf{c}^*) = \frac{2d_{hkl}\sin\theta}{\lambda}\mathbf{G}(hkl) \qquad (3.103)$$

If we introduce the condition of constructive interference (Figure 3.39) into Equation (3.103), we obtain

$$\Delta k = p(h\mathbf{a}^* + k\mathbf{b}^* + l\mathbf{c}^*) \qquad (3.104)$$

where p is a positive or negative integer and hkl are the Miller indices.

In crystallography, *Laue indices* are always used instead of Miller indices. Laue indices are equal to the Miller indices except that they may include an integer factor which is equal to the order p.

In the following we will replace the Miller indices by Laue indices (called hkl as before), *which include the order p of diffraction.* In this case, Equation (3.104) can be written as

$$\Delta k = h\mathbf{a}^* + k\mathbf{b}^* + l\mathbf{c}^* = \mathbf{G}(hkl) \qquad (3.105)$$

where hkl now means the *Laue indices*.

Equation (3.105) is the general diffraction condition for a crystal lattice of arbitrary structure. It is important to point out that.

- All the atoms in the crystal lattice must be included in the set of parallel planes for constructive interference.

If this condition is not fulfilled, the interference will be nonconstructive.

Reciprocal Space, k Space and Brillouin Zones

1. On page 119, we found that a planar matter wave can be represented by a wavevector $\mathbf{k}$. The amplitude of the wave is represented by a set of parallel equidistant *planes* in $\mathbf{r}$ space. The planes and the $\mathbf{k}$ vector are perpendicular. The $\mathbf{k}$ vector is represented by a *point* in $\mathbf{k}$ space.
2. On page 132, we defined *the reciprocal lattice vector* $\mathbf{G}(hkl) = h\mathbf{a}^* + k\mathbf{b}^* + l\mathbf{c}^*$. It was shown that the $\mathbf{G}(hkl)$ vector is perpendicular to the set of parallel equidistant crystal planes (hkl) in the $\mathbf{abc}$ space. The $\mathbf{G}(hkl)$ vector corresponds to a *point* (hkl) in the reciprocal space $\mathbf{a}^*\mathbf{b}^*\mathbf{c}^*$.
3. The condition for the Brillouin zone boundaries in the one-dimensional case (page 128) can be interpreted as a diffraction condition. The Laue diffraction condition represents the generalized condition for the Brillouin zone boundaries in the three-dimensional case.
4. The reciprocal lattice vector $\mathbf{G}(hkl)$ appears in von Laue diffraction condition [Equation (3.105)].

Points 1, 2 and 4 lead to the conclusion that the reciprocal space is identical with the $\mathbf{k}$ space while the crystal space is the $\mathbf{r}$ space. In addition,

- The $\mathbf{k}$ space is reciprocal to the $\mathbf{r}$ space and vice versa.

A distance in reciprocal space has the dimension inverse length $(k = 2\pi/\lambda)$. An important property of the reciprocal space is that

- A set of equidistant parallel planes (hkl) in $\mathbf{r}$ space corresponds to a point (hkl) in the reciprocal space ($\mathbf{k}$ space).

This property and the diffraction condition (point 3 above) will be used to find the Brillouin zones boundaries in two and three dimensions. When the crystal structure of a metal ($\mathbf{r}$ space) is known, its Brillouin zone boundaries ($\mathbf{k}$ space) can be constructed.

Brillouin Zones in Two Dimensions

The 'free' electrons in a metal are only free in the sense that they can move within the metal volume but they cannot leave the metal. Each electron in motion corresponds to a matter wave moving in the same direction as the electron. The Brillouin zone boundaries in one dimension are defined by the condition in Equation (3.75) (page 128). As $k = 2\pi/\lambda$, Equation (3.75) can be written as $2\pi/\lambda = p(\pi/a)$. If we replace a by d_{hkl}, which is a more general designation of the distance between consecutive parallel crystal planes, we obtain

$$2d_{hkl} = p\lambda \tag{3.106}$$

Equation (3.106) corresponds to the condition for constructive interference after diffraction at the atoms in a lattice plane (hkl) $2d_{hkl} = p\lambda \sin\theta$ when $\theta = \pi/2$. If λ is the wavelength of the matter wave of free electrons in the metal lattice, d_{hkl} is the distance between consecutive planes, perpendicular to the direction of the motion of the electrons.

- The Brillouin zone boundary condition (3.106) can be interpreted as total reflection of the matter waves of the free electrons and the condition for formation of a standing matter wave.

Brillouin Zones Boundaries of a Simple Square Lattice

Each type of crystal structure gives rise to its own characteristic Brillouin zones. As an introduction, we will study the simplest structure, a square lattice. We want to find the first Brillouin zone which corresponds to the unit cell a^2 in r space.

Consider a two-dimensional, square metal lattice with the lattice constant a. Each set of parallel crystal lines corresponds to a particular point k in k space. The process of finding the k points related to three main directions of lines in r space is shown in Table 3.7.

Table 3.7 Derivation of the reciprocal lattice vectors in k space which correspond to three given sets of parallel crystal lines in r space.

Set of crystal lines in r space	$<$hk$>$ crystal lines in r space	d_{hk} in r space $\|a\| = a$ $\|b\| = a$	Point (hk) in k space	Reciprocal lattice vector G(hk) = point in k space G(hk) = $h a^* + k b^*$ $\|a^*\| = 2\pi/a$ $\|b^*\| = 2\pi/a$	Reciprocal lattice vector in k space $a^* = (2\pi/a)k_x$ $b^* = (2\pi/a)k_y$
	$< 10 >$	$d_{10} = a$	(1,0)	$\left(\dfrac{2\pi}{a}, 0\right)$ Vector OA	
	$< 01 >$	$d_{01} = a$	(0,1)	$\left(0, \dfrac{2\pi}{a}\right)$ Vector OB	
	$< 11 >$	$d_{11} = \dfrac{a}{\sqrt{2}}$	(1,1)	$\left(\dfrac{2\pi}{a}, \dfrac{2\pi}{a}\right)$ Vector OC	

Each reciprocal lattice vector is perpendicular to the corresponding crystal lines and has the same direction as the matter wave of the electrons, which is diffracted by the atoms of the lattice.

A very large number of directions of crystal lines in r space are possible. We let the quasi-continuous direction of the parallel planes vary from 0 to 2π and plot the corresponding G points in k space. The G values form a quasi-continuous square in k space with the lattice points A, B, C, D and the origin in the centre (Figure 3.39a).

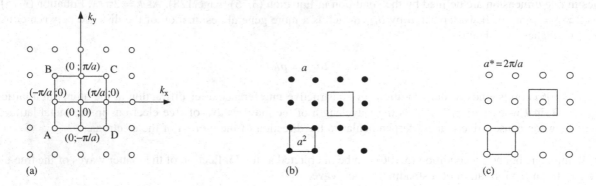

(a) (b) (c)

Figure 3.39   (a) Reciprocal unit cell of a two-dimensional crystal lattice with simple square structure. (b) Unit cell in r space. (c) Unit cells in k space.

The square $(a^*)^2$ in k space (Figure 3.39c) is said to be the reciprocal cell of the unit cell a^2 of the crystal lattice (Figure 3.39b). Each corner in both types of cells is shared between four cells. Hence each cell contains $4 \times (1/4) = 1$ lattice point. This is easier to realize if we change the coordinate system in such a way that an atom in the crystal lattice and a k point in the reciprocal space are placed in the centre of the squares (upper squares in Figures 3.39b and c).

On page 129, we found that each Brillouin zone contains N_{total} energy states if the number of atoms is N_{total}. Hence there are N_{total} allowed values of k (page 129) in the first Brillouin zone. *The unit cell in reciprocal space* (Figure 3.39a) *represents the boundary of the first Brillouin zone in k space.*

The outer boundaries of higher Brillouin zones are derived in a similar way. The general process is described in Figure 3.40a for the boundaries of the second and third Brillouin zones of a two-dimensional square crystal lattice. The outer boundary of

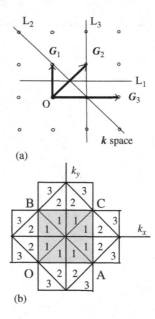

(a)

(b)

Figure 3.40   (a) The three shortest reciprocal lattice vectors G_1, G_2 and G_3 are drawn together with their perpendicular bisectors. By drawing all symmetrical lines of types L_1, L_2 and L_3 around the k point O we obtain enclosed areas, identical with the first, second and third Brillouin zones. Zone 1 depends only on type L_1, the others are bounded by several line types.
(b) First, second and third Brillouin zones of a two-dimensional crystal lattice with simple cubic structure. All the Brillouin zones have equal areas. © 1986 John Wiley & Sons, Inc.

the second Brillouin zone is an inclined square (Figure 3.40b) and the third Brillouin zone is bounded by the symmetrical 'cross' in Figure 3.40b.

Areas of the Brillouin Zone

The areas $A_1, A_2, A_3, \ldots$ of the Brillouin zones in two dimensions can be calculated as follows. The area of the first Brillouin zone, marked in Figure 3.40b, will be

$$A_1 = (OA)^2 = \frac{2\pi}{a}\frac{2\pi}{a} = \left(\frac{2\pi}{a}\right)^2$$

Figure 3.40b shows that the area of the second Brillouin zone is

$$A_2 = (OA\sqrt{2})^2 = \frac{2\pi\sqrt{2}}{a}\frac{2\pi\sqrt{2}}{a} - A_1$$

or

$$A_2 = 2\left(\frac{2\pi}{a}\right)^2 - \left(\frac{2\pi}{a}\right)^2 = \left(\frac{2\pi}{a}\right)^2$$

The first, second and third Brillouin zones in Figure 3.40b are marked by the figures 1, 2 and 3. Each of them consists of eight triangles of equal areas. Consequently, their total areas are equal.

Each Brillouin zone corresponds to an energy band. Each of them contains N_{total} energy levels and can accommodate $2N_{\text{total}}$ electrons. The same is true for all Brillouin zones, as was mentioned on page 129. As we have seen in the one-dimensional case, there is an abrupt and discontinuous change of energy at the Brillouin zone boundaries (page 128).

An important example of two-dimensional crystals is thin films of semiconducting materials.

Brillouin Zones in Three Dimensions

We have seen that the boundaries between Brillouin zones are *points* in one dimension and *lines* in two dimensions. In the general three-dimensional case the boundaries are *planes*, which enclose the three-dimensional Brillouin zones.

We found the *boundary points* in a mathematical way (page 128). We found the *boundary lines* by using the condition for standing matter waves in two dimensions and the theory of reciprocal lattices in the preceding section. In the three-dimensional case we will apply the same principles as in the two-dimensional case to find *boundary planes* of the Brillouin zones.

To find the *planes* which enclose the three-dimensional Brillouin zones in *k* space we will use a so-called *Wigner–Seitz cell*. The construction of a Wigner–Seitz cell in *r* space is described in the next section. The planes which enclose the corresponding Wigner–Seitz cell in *k* space are the Brillouin zone boundaries.

Wigner–Seitz Cell

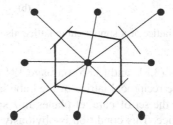

Figure 3.41 Cross-section of a Wigner–Seitz cell in *r* space. Two-dimensional drawing of a Wigner–Seitz cell.

A lattice cell, which contains only one lattice point, is a *primitive unit cell*. A Wigner–Seitz cell is a primitive cell and can be constructed in the following way.

Consider a point in a crystal lattice. Lines are drawn to all nearest neighbours and second nearest neighbours (Figure 3.41). Perpendicular planes are drawn through the midpoint of each of these lines. The nearest and next nearest of these planes (not necessarily all of them) form a polyhedron, that encloses the lattice point. This polyhedron is the Wigner–Seitz cell. Wigner–Seitz cells can be constructed both in crystal space (r space) and in k space.

It can be shown that the Brillouin zone is equivalent with a Wigner–Seitz cell in the reciprocal lattice. When we discussed the Brillouin zones in a two-dimensional reciprocal lattice on page 136 we used in fact a two-dimensional Wigner–Seitz cell (upper square in Figure 3.39c).

General Conditions for Three-dimensional Brillouin Zones

Brillouin zones are closely related to standing matter waves of free electrons in crystals. A point in k space represents the wavevector k of the matter wave with wavelength λ [Equation (3.34) on page 115].

The generalized conditions for Brillouin zones in three dimensions can be summarized as follows:

- The matter wave of the electrons in the crystal lattice of the metal can be reflected at sets of crystal planes under certain conditions [Equation (3.104) on page 134]. The total reflection results in a standing matter wave in the crystal. The reflection condition can be expressed in terms of Δk [Equation (3.105) on page 134].
- All atoms in the crystal must be included in the set of parallel planes.
- The energies of the free electrons within the Brillouin zones form energy bands (Figure 3.32 on page 128). The higher the k value within each zone, the higher will be the energy of the electrons. At the Brillouin zone boundaries the energy of the electrons changes discontinuously.

Other general statements are as follows:

- The volumes of the different Brillouin zones of a given crystal structure are equal.
- Each Brillouin zone can accommodate maximum $2N_{\text{total}}$ electrons.

We will discuss some examples below. As a first example we choose a crystal with a simple cubic structure.

Brillouin Zones of a Simple Cubic Crystal Lattice

The lattice point $(1, 0, 0)$ in crystal space in Figure 3.42a corresponds to the multiple set of planes {100} in the reciprocal space, which are parallel with the $k_y k_z$ plane. For the sake of simplicity, only two planes close to origin are shown in Figure 3.42b. Similarly, the lattice points $(0, 1, 0)$ and $(0, 0, 1)$ in crystal space correspond to sets of planes parallel to the $k_x k_z$ plane and the $k_x k_y$ plane, respectively. Hence the reciprocal lattice of a simple cubic lattice also has a simple cubic structure.

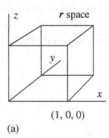
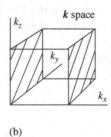

(a) (b)

Figure 3.42 The reciprocal lattice of a simple cubic lattice also has a simple cubic structure.

According to the reverse nature of Equations (3.84) and (3.85) on page 131 and our experiences from the two-dimensional case (pages 135–136), we can conclude that the reciprocal lattice of the cube in k space in Figure 3.43b is the cube in crystal space in Figure 3.43a. As stated on page 134, the set of parallel planes in r space, which corresponds to a reciprocal lattice point, must include all atoms in the crystal lattice. This condition is obviously fulfilled in this case.

Hence the cube in k space in Figure 3.43b, bounded by the six planes:

$$x = \pm\frac{\pi}{a} \quad y = \pm\frac{\pi}{a} \quad z = \pm\frac{\pi}{a} \tag{3.107}$$

represents the boundaries of the first Brillouin zone.

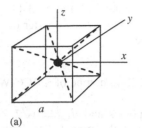 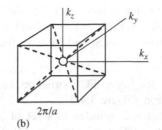 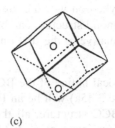

(a)   (b)   (c)

Figure 3.43   (a) Primitive unit cell (Wigner–Seitz cell) of a simple cubic lattice in *r* space. Lattice constant = *a*. • = origin in *r* space.
(b) First Brillouin zone (Wigner–Seitz cell in *k* space) of a simple cubic crystal. ○ = origin in *k* space.
(c) Outer surface of the second Brillouin zone (Wigner–Seitz cell in *k* space) of a simple cubic crystal. ○ = origin in *k* space. Reproduced with permission from W. Hume-Rothery and B. R. Coles, *Atomic Theory for Students of Metallurgy*. (Published by the Institute of Metals, 1969.) now © W. S. Maney & Son Ltd.

The construction of higher Brillouin zone boundaries of a simple lattice in two dimensions was briefly described in Figure 3.40a. Higher Brillouin zones in three dimensions are constructed analogously by use of perpendicular bisector planes to longer *G* vectors than the first ones. The polyhedrons which correspond to the higher Brillouin zone boundaries are generally more complicated than that of the first Brillouin zone.

The outer polyhedron of the second Brillouin zone boundary for a simple cubic crystal is shown in Figure 3.43c.

Brillouin Zones of FCC, BCC and HCP Crystal Lattices

The principles of finding the Brillouin zones of more complicated crystal structures are the same as for the simple cubic structure, but the results are much more complicated. The Brillouin zone polyhedrons are derived by constructing the Wigner–Seitz cells in *k* space. This is equivalent to applying the condition which gives standing matter waves of the moving free electrons in the crystal with all atoms in the crystal lattice included (page 134).

We will restrict the treatment to short descriptions of the three close-packed structures FCC, BCC and HCP, which are common in metals and semiconductors.

FCC

The reciprocal lattice of the FCC crystal lattice has a BCC structure. Hence the first Brillouin zone of an FCC Wigner–Seitz cell (Figure 3.44a) will be a BCC polyhedron (Figure 3.44b).

In the FCC structure $d_{111} = a\sqrt{3}/3$ and $d_{200} = a/2$. Hence the resulting first Brillouin zone is a BCC polyhedron bounded by eight planes of the type[2] {111} at the distance $\pi/d_{111} = \pi\sqrt{3}/a$ from the origin in *k* space and six planes of the type {200} at the distance $\pi/d_{200} = \pi \times 2/a$ from the origin in *k* space.

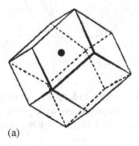 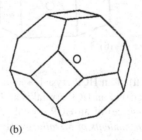 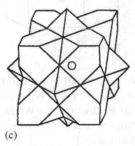

(a)   (b)   (c)

Figure 3.44   (a) Primitive unit cell (Wigner–Seitz cell) of an FCC crystal. • = origin in *r* space. Reproduced with permission from W. Hume-Rothery and B. R. Coles, *Atomic Theory for Students of Metallurgy*. (Published by the Institute of Metals, 1969.) now © W. S. Maney & Son Ltd.
(b) The first Brillouin zone (Wigner–Seitz cell in *k* space) of an FCC crystal is a BCC polyhedron. ○ = origin in *k* space. Reproduced with permission from W. Hume-Rothery and B. R. Coles, *Atomic Theory for Students of Metallurgy*. (Published by the Institute of Metals, 1969.) now © J. S. Maney & Son Ltd.
(c) Outer surface of the second Brillouin zone (Wigner–Seitz cell in *k* space) of an FCC crystal. ○ = origin in *k* space. Adapted with permission from M. J. Sinnott, *The Solid State for Engineers*. © 1958 John Wiley & Sons, Inc.

---

[2] The plane symbol {} means that planes with all possible signs and permutations inside the brackets are included (Chapter 1, page 16).

The polyhedron which corresponds to the outer surface of the second Brillouin zone is shown in Figure 3.44c.

The diamond structure can be described as two FCC lattices where the second one is displaced one-quarter of the principal diameter relative to the first one. The semiconductors Si and Ge have the same structure.

BCC

The reciprocal lattice of the BCC crystal lattice has an FCC structure. Hence the first Brillouin zone of a BCC Wigner–Seitz cell (Figure 3.45a) will be an FCC polyhedron (Figure 3.45b).

In the BCC structure, all the lattice points are included in a set of parallel (110) planes and $d_{110} = a\sqrt{2}/2$. Hence the first Brillouin zone of a BCC crystal is an FCC polyhedron bounded by 12 planes of the type {110} at the 'distance' $\pi/d_{110} = \pi\sqrt{2}/a$ from the origin in k space (Figure 3.45b).

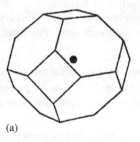

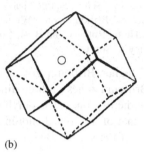

(a)

(b)

Figure 3.45   (a) Primitive unit cell (Wigner–Seitz cell) of a BCC crystal. $\bullet$ = origin in r space. (b) The first Brillouin zone (Wigner–Seitz cell in k space) of a BCC crystal is an FCC polyhedron. $\circ$ = origin in k-space. Reproduced with permission from W. Hume-Rothery and B. R. Coles, *Atomic Theory for Students of Metallurgy*. (Published by the Institute of Metals, 1969.) now © W. S. Maney & Son Ltd.

HCP

The Wigner–Seitz cell of an HCP crystal is shown in Figure 3.46a. In the ideal case the ratio $c/a = \sqrt{8/3} = 1.63$. This ratio is often modified owing to the interaction between the Fermi surface and the faces of the Brilloin zone.

The first Brillouin zone of an HCP crystal, i.e. the Wigner–Seitz cell of HCP in k space, has the same shape as the Wigner–Seitz cell of the crystal.

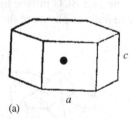

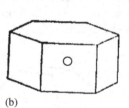

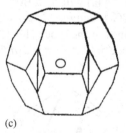

(a)

(b)

(c)

Figure 3.46   (a) Primitive unit cell (Wigner–Seitz cell) of an HCP crystal. $\bullet$ = origin in r space. (b) The first Brillouin zone (Wigner–Seitz cell in k space) of an HCP crystal is an HCP polyhedron. $\circ$ = origin in k space. (c) Outer surface of the second Brillouin zone (Wigner–Seitz cell in k space) of an HCP crystal. $\circ$ = origin in k space. Reproduced with permission from W. Hume-Rothery and B. R. Coles, *Atomic Theory for Students of Metallurgy*. (Published by the Institute of Metals, 1969.) now © W. S. Maney & Son Ltd.

Energy Distribution of Electrons in the Energy States of the Brillouin Zones

Principle of Electron Distribution in Metals

Above we have defined the concept of Brillouin zones and concentrated on finding their shape and size. The reason why we pay such attention to Brillouin zones is that they are closely related to the energies of the free electrons and the energy distribution of the free electrons in crystal lattices. The electron energy distribution controls the lattice structure and is responsible for the properties of metals.

Below we will discuss the relationship between the Brillouin zones and the electron distribution. The outstanding and only principle for the electron distribution and structure of crystals is the *lowest possible total energy*.

Available Energy States Far from the Brillouin Zones

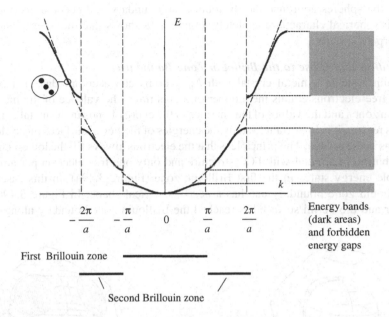

Figure 3.32 Energy levels in a linear periodic lattice as a function of the wavenumber k. $a =$ the lattice constant. Reproduced with permission from M. Alonso and E. Finn, *Fundamental University Physics*. © Addison-Wesley.

The kinetic energy of a free electron, which moves in a linear crystal lattice with a *periodically varying electric potential*, as a function of the wavenumber k is shown in Figure 3.32 on page 128. The abrupt changes of energy at certain k values occur in three dimensions at the planar Brillouin zone boundaries discussed above.

We will use the same representation as in Figure 3.22a on page 120 to show the k points in the three-dimensional reciprocal lattice. Points with the *constant energy* E_F lie on a surface called the *Fermi surface*.

In the three-dimensional case, the energy of a free electron, which moves in a *constant electric potential*, is shown in Figure 3.23 on page 121. This is a good approximation if the Fermi surface is far from the Brillouin zone boundaries. In this case the Fermi surfaces are spheres and the energy can roughly be represented by Equation (3.56) on page 120.

The function is shown in Figure 3.47, which shows a number of Fermi surfaces in k space inside the cube which represents the first Brillouin zone of a simple cubic crystal lattice. The figure shows a cross-section of the spheres.

Available Energy States Close to the Brillouin Zones

Figure 3.48 shows the deformation of the spheres in Figure 3.47 close to the Brillouin zone boundary. The figure can be regarded as a 'topographic map' where the height lines correspond to Fermi surfaces.

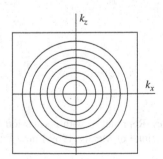

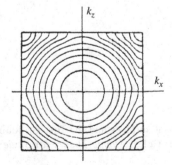

Figure 3.47 Approximately spherical Fermi surfaces in k space within and far from the first Brillouin zone boundary of a simple cubic crystal lattice. It has the shape of a cube.

Figure 3.48 Real Fermi surfaces in k space within the first Brillouin zone of a simple cubic crystal lattice, which has the shape of a cube (in the figure the shape of a square).

In can be concluded from Figure 3.32 that deviations from the spherical shape of the Fermi surfaces are to be expected close to the Brillouin zone boundaries.

With increasing energies the spheres approach the Brillouin zone boundary and become distorted. At still higher energies the Fermi surfaces lose their spherical character completely. Figure 3.48 shows that these Fermi surfaces always intersect the Brillouin zone boundary perpendicularly.

Occupied Energy Levels Within and Close to the Brillouin Zone Boundaries

We know that each Brillouin zone in a metal crystal with N_{total} atoms can accommodate a maximum of $2N_{total}$ electrons (page 129). The number of free electrons equals the number of atoms times the valence of the metal.

The shapes of the Brillouin zones and the values of the energies in the energy bands in the metal control the energy distribution of the free electrons. For this reason, it is important to study the energies of the Fermi surfaces more closely and the ways they are occupied by the free electrons in the crystal. The principle is that the electrons always fill the lowest energy states first.

The normal energy distribution in a metal with FCC structure and only one free electron per atom is that the electrons can only fill half of the available energy states in the first Brillouin zone (Figure 3.49a). In this case the outer occupied Fermi surface is far from the Brillouin zone boundary and has a nearly spherical shape. In Figure 3.49b, the number of occupied energy states is much larger and the Fermi surface has reached the Brillouin zone boundary along the flat circular areas.

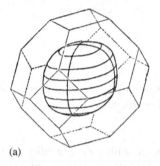
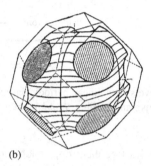

(a)   (b)

Figure 3.49 (a) First Brillouin zone of an FCC metal with one free electron per atom and the Fermi surface of the occupied energy levels far from the Brillouin zone boundaries.
(b) First Brillouin zone of an FCC metal with several electrons per atom and the Fermi surface of the occupied energy levels close to the Brillouin zone boundaries. Reproduced with permission from W. Hume-Rothery and B. R. Coles, *Atomic Theory for Students of Metallurgy*. (Published by the Institute of Metals, 1969.) now © W. S. Maney & Son Ltd.

The exact shapes of the Fermi surfaces are characteristic for each metal. As a real example we choose copper, which has the configuration $1s^2\,2s^2\,2p^6\,3s^2\,3p^6\,3d^{10}\,4s$. Copper has one valence electron per atom. Even at this low electron density, when the Brillouin zone is only half filled, contact is established between the Fermi surface and the Brillouin zone boundaries in the <111> directions (Figure 3.50). These contact areas correspond to the maximum energy at the zone boundary (point A in Figure 3.25 on page 122).

Figure 3.50 Fermi surfaces and the first Brillouin zone boundary of Cu (FCC structure). Reproduced with permission from W. Hume-Rothery and B. R. Coles, *Atomic Theory for Students of Metallurgy*. (Published by the Institute of Metals, 1969.) now © W. S. Maney & Son Ltd.

For most other metals, the Fermi surface is much more complicated. Accurate calculations have been performed only in a few cases.

Occupied Energy Levels Close to the Brillouin Zones

Inside the Brillouin zones the energy increases smoothly with increasing wavenumber k. At each Brillouin zone boundary there is an abrupt and discontinuous increase of the electron energy.

- The magnitudes of the abrupt energy changes at the Brillouin zone boundaries depend on direction.

This is shown for a simple cubic structure in Figure 3.51, which can be used to derive the energy distribution of the free electrons if the energies of consecutive Fermi surfaces are known. The figures inside the Brillouin zone are the k values. Figure 3.51 shows that the abrupt energy change at the zone boundary occurs for a lower k value in the perpendicular directions than in the diagonal direction. The lowest energy states are found by use of the two energy diagrams. The lowest energy states always become filled first. The result is given in Table 3.8.

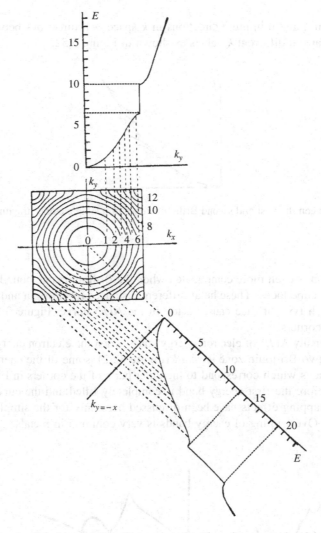

Figure 3.51 First Brillouin zone of a simple cubic structure. Energy as a function of wavenumber k in two different directions. Reproduced with permission from W. Hume-Rothery and B. R. Coles, *Atomic Theory for Students of Metallurgy*. (Published by the Institute of Metals, 1969.) now © W. S. Maney & Son Ltd.

The energy states in the interval 0–6.5 energy units in both the $k_x k_y$ and the diagonal directions become filled first. In the interval 6.5–10 energy units, only energy states in the diagonal directions, corresponding to corner points, are available and become filled next. In the energy interval 10–13 energy units, the remaining k points in the corners of the first Brillouin zone in the diagonal directions and lowest k points of the second zone in the $k_x k_y$ directions become filled. At 13 energy units the first zone is completely filled and additional electrons with energies >13 energy units have to go to the second zone in the $k_x k_y$ directions.

Table 3.8   Lowest energy levels derived from Figure 3.51.

Energy interval Arbitrary units	Direction in k space
0–6.5	$k_x k_y$ 1st zone + diagonal 1st zone
6.5–10	diagonal 1st zone
10–13	diagonal 1st zone + $k_x k_y$ 2nd zone
13–19	$k_x k_y$ 2nd zone

Table 3.8 relates to the maximum and minimum directions in k space. All directions between these two are also possible. This leads to a series of energy gaps at different k values as shown in Figure 3.52.

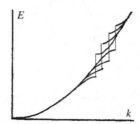

Figure 3.52   Small energy gaps between the first and second Brillouin zones of a simple cubic structure in the directions which correspond to k_x and k_y in Figure 3.51.

For other structures, the situation is even more complicated when the Brillouin zone boundaries consist of different types of faces, for example octahedral and cube faces. These lie at different distances from origin and give another variety of positions and sizes of the energy gaps. Each type of face causes a top in the $N(E)$ curve. Figure 3.53a shows an example with two types of faces and three types of corners.

It is possible to calculate the density $N(E)$ of electron energy states when the electron energy as a function of k is known. If the energy gap between the first two Brillouin zone is *small* (Figure 3.52), some of the energy states of the second Brillouin zone are lower than the energy states which correspond to the outer parts of the corners in Figure 3.51. In this case, electrons are located in the second band before the first energy band is completely filled and the curves of the first and second bands overlap (Figure 3.53b). The overlapping effects have been discussed here only for the simple cubic crystal structure but also occur for other crystal structures. Overlapping of energy bands is very common in metals.

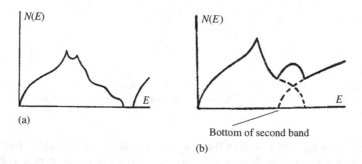

Figure 3.53   (a) Electron distribution (number of electrons per unit volume with energies between E and $E + \mathrm{d}E$) in the first and second Brillouin zones as function of the energy E.
(b) If the energy gaps in Figure 3.52 are small enough, the energy states of the first and second Brillouin zones overlap and there will be no forbidden energy gap.

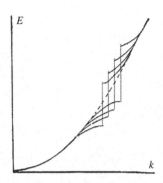

Figure 3.54 Large energy gaps between the first and second Brillouin zones of a simple cubic structure in different directions.

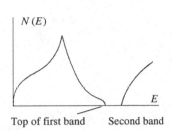

Top of first band   Second band

Figure 3.55 Density of available energy states in the lowest energy band, i.e. in the first Brillouin zone of a simple cubic structure. The energy bands do not overlap. There is only one type of face and one type of corner.

If the energy gaps in different directions are *large* (Figure 3.54), the energy states in the corners, i.e. at the top of the first zone, are lower than all the energy states at the bottom of the second zone, *independent of direction*. Consequently, the corners, i.e. the first Brillouin zone as a whole, will be filled with electrons before any electrons go to the second Brillouin zone. In this case the $N(E)$ curve has a forbidden energy interval such as that in Figure 3.55.

The corner states are rather few and these energy states become filled fairly quickly. Hence the $N(E)$ curve decreases rapidly to zero. This part of the curve corresponds to the top A of the band in Figure 3.25 on page 122, which now has obtained an explanation.

Above we considered the energy distribution in pure metals of a given number of valence electrons. In Section 3.7.2 on page 153 we will consider the valence electron distribution in alloys and particularly the influence of an increasing electron concentration on the positions of the electron bands and the crystal structure.

The Effective Electron Mass as a Function of k

On page 112, it was mentioned that the mass of a valence electron in a metal is not the same as the mass m of an electron in free space. The effective mass of a valence electron in a metal depend on its energy, i.e. m^* must be a function of the wavevector k of the matter wave of the electron. It is beyond of the scope of this book to penetrate this question more closely. On the other hand, it cannot be omitted.

By studying the energy increase of a valence electron when it is accelerated in an electric field, it can be shown theoretically that the relationship in the one-dimensional case between the effective mass, the kinetic energy and the wavenumber k can be written as

$$m^* = \frac{\hbar^2}{\dfrac{d^2 E}{dk^2}} \tag{3.108}$$

The relationship between m^* and k is shown in Figure 3.56. Figure 3.56a shows the kinetic energy E as a function of the wavenumber k of the matter wave of the electron. Equation (3.108) shows that the second derivative of E and especially its values on the curve at the points of inflection are very important. The effective mass becomes infinite at these points ($d^2 E/dk^2 = 0$ at $k = \pm\pi/2a$).

Another startling result is that the effective mass, defined by Equation (3.108), is *negative* at the upper half of the first Brillouin zone. The practical effect of this is that electrons with wavenumbers within the intervals $-\pi/a < k < -\pi/2a$ and $\pi/2a < k < \pi/a$ behave like *positively charged particles* when they are accelerated in an electric field, otherwise like negative particles.

In the three-dimensional case, Brillouin zones in three dimensions, m^* is far more complicated. In the general case $1/m^*$ is a tensor with nine components.

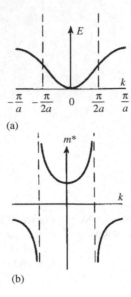

(a)

(b)

Figure 3.56 (a) Kinetic energy of a valence electron within the first Brillouin zone (one-dimensional case) as a function of k. (b) The effective mass m^* of a valence electron as a function of k. Reproduced with permission from A. J. Dekker, *Solid State Physics*. © 1962 Macmillan & Co. Ltd.

3.6 Elastic Vibrations in Solids

So far the theory of solids has dealt entirely with the valence electrons in solids. Owing to the positions of the valence electrons, the bonds in nonmetallic solids are mainly ionic or covalent. The quantum mechanical model of the electron gas, which has been extensively discussed above, explains many of the properties of metals.

In addition, for a more complete and general theory of solids, we have to consider the *elastic vibrations* in the crystal lattice. These elastic vibrations are of great importance when *thermal properties of solids*, e.g. heat capacity and heat conduction, are treated.

In Chapter 2, Section 2.7.3 we discussed harmonic vibrations in diatomic molecules. We found that the vibrations are quantized and that the molecules always have a zero point vibrational energy.

A crystal is a regular lattice of atoms which are held in their equilibrium average positions by strong cohesive forces. The atoms are in continuous motion and perform vibrations with small amplitudes around their equilibrium positions. Owing to the strong cohesive forces between neighbouring atoms, there is strong coupling between the atoms in the lattice. It is impossible to set one atom into vibration without disturbing nearby atoms and indirectly the whole crystal. Collective vibrations arise which travel through the whole lattice.

The collective vibrations move back and forth through the crystal and form travelling waves. Vibrations of many different frequencies and energies are present. The frequencies depend on the shape and size of the solid and are in some respects analogous to electromagnetic waves in a cavity.

3.6.1 Phonons

Properties of Phonons

We remember from Chapter 2 that the study of the radiation inside a cavity led to the first assumption of quantization of energy, the introduction of the concept of photon $E = h\nu$ and Planck's radiation law.

In the case of elastic waves, the duality between waves and particles has led to the introduction of the concept of phonon. Phonons are quasi-particles associated with the elastic waves in solids. The vibrations inside a solid may be regarded as manifold phonons with different frequencies and energies.

- Phonons have quantized energies just like photons.

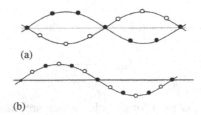

Figure 3.57 Two perpendicular transverse modes of vibration in a linear crystal lattice with two kinds of atoms: (a) transverse optical mode; (b) transverse acoustic mode.

Elastic waves have three different modes of vibration: two perpendicular transverse modes (Figure 3.57) and one longitudinal mode. The *energy* of a phonon can not have an arbitrary value but is quantized according to quantum mechanics. In analogy with the vibrations in diatomic molecules, the vibrational energy of each elastic mode can be written as

$$G(n) = \hbar\omega(n + \tfrac{1}{2}) \qquad (3.114)$$

where
  $G(n)$ = energy of the elastic wave in vibrational mode n
  ω = angular frequency of the elastic wave
  n = integer quantum number of the mode of vibration.

Two or more phonons may interact, which results in permanent changes of both the wavelength and direction of propagation of the phonons. The process can be regarded as a collision, where the total momentum and energy are conserved. The interaction between phonons increases with increasing amplitude of vibration.

Phonons may interfere constructively in a localized region of the lattice and cause a large displacement of one or two atoms. Phonons carry momentum and energy and interact with other types of primary imperfections in crystals. In all such processes the total momentum and energy are conserved.

The phonon energy distribution will be discussed in Section 5.2.2 and applied in Section 5.

A striking difference between photons and phonons concerns their speed of propagation. *Photons* move with the velocity of light, $c_0/n_{crystal}$, where $c_0 = 3 \times 10^8$ m/s and $n_{crystal}$ is the refractive index of the crystal.

Phonons propagate with velocities which vary with their frequencies. Their velocities are much lower than that of electromagnetic waves. Phonons are elastic waves of the same kind as sound waves and have the same velocity as sound waves within the frequency range of sound. For sound waves we have

$$v = \sqrt{\frac{E}{\rho}} \qquad (3.115)$$

where
  v = velocity of the wave
  E = modulus of elasticity of the crystal
  ρ = density of the crystal.

The velocity of sound and phonons of sound frequencies in metals is of the magnitude $(2-5) \times 10^3$ m/s.

Phonon Statistics

We remember that the discrepancy between the classical and the quantum mechanical models of the electron gas is caused by the difference between classical Maxwell–Boltzmann statistics and Fermi–Dirac statistics, which is valid for electrons. The phonons and photons obey none of these statistics. Instead, they obey the so-called Bose–Einstein statistics.

- The classical Maxwell–Boltzmann statistics is valid for particles which are *identical* and *distinguishable*.

$$f_{MB} = \frac{1}{e^{\frac{E-E_0}{k_B T}}}$$

- The Bose–Einstein statistical distribution is valid for particles which are *identical* but *indistinguishable* and belong to such a system where the Pauli exclusion principle is *not* valid.

$$f_{BE} = \frac{1}{e^{\frac{E-E_0}{k_B T}} - 1}$$

- The Fermi–Dirac statistics is valid for particles which are *identical* and *indistinguishable* and belong to such a system where *the Pauli exclusion principle is valid*.

$$f_{FD} = \frac{1}{e^{\frac{E-E_0}{k_B T}} + 1}$$

The Fermi–Dirac statistics was used in Section 3.4.3 on page 117 in connection with the quantum mechanical model of the electron gas in a metal. Figure 3.58 shows a comparison between the three statistics.

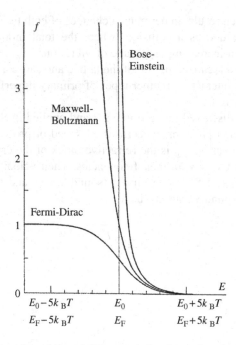

Figure 3.58 Maxwell–Boltzmann, Bose–Einstein and Fermi–Dirac statistics. Reproduced with permission from I. Lindgren *et al.*, © 1971, Almqist & Wiksell.

In the Fermi–Dirac statistics, $E_0 = E_F$. In Bose–Einstein statistics we have the restriction $E > E_0$ otherwise f_{BE} becomes negative.

The differences between the three statistics are very large for small energy values. At high energy values they coincide and both the Bose–Einstein and Fermi–Dirac statistics can be approximated by the mathematically simpler Maxwell–Boltzmann statistics.

A crystal at equilibrium at a given temperature contains a variety of phonons of different frequencies. The energy distribution of the phonons is identical with Planck's distribution law because both phonons and photons obey Bose–Einstein statistics. We will discuss this topic in Chapter 6 in connection with the heat capacity of solids.

Angular Frequency of Phonons as a Function of Wavenumber

Just as for matter waves of electrons, it is very convenient to introduce the *wavenumber* k of the phonon and express its total energy E and its momentum p as functions of the wavenumber k and the wavevector $\mathbf{k}$, respectively:

$$E = \hbar\omega(k) \tag{3.116}$$

$$\mathbf{p} = \hbar\mathbf{k} \tag{3.117}$$

where the angular frequency $\omega = 2\pi\nu$.

For simplicity, we will first study the deviation of an atom or ion from its equilibrium position in a *one-dimensional crystal lattice* which consists of only *one type of atom* with mass M in terms of change of distance to the two nearest neighbours (Figure 3.59).

Figure 3.59   One-dimensional crystal lattice.

If we solve the differential wave equation and disregard all end effects, we obtain the following relationship between the angular frequency ω and the wavenumber k of the phonon (Figure 3.60):

$$\omega = 2\sqrt{\frac{\beta}{M}}\sin\frac{ka}{2} \tag{3.118}$$

where
ω = angular frequency of the phonon motion
β = elastic constant of the adjacent atoms in the lattice
M = mass of the atom
k = wavenumber of the phonon
a = lattice constant, i.e. distance between consecutive atoms in the lattice.

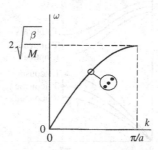

Figure 3.60   Angular frequency of lattice vibrations of equal atoms as a function of the wavenumber. Reproduced with permission from M. Alonso and E. Finn, *Fundamental University Physics*. © Addison-Wesley.

Obviously there is an upper limit of the angular frequency of the phonon [Equation (3.118)]. This cut-off frequency is of magnitude 10^{15} Hz for most substances, which is beyond the ultrasonic frequencies.

Figure 3.61 One-dimensional crystal lattice with two kinds of atoms.

It should be pointed out that k is quantized because the phonon energy has discrete values. The energy levels are very close and the curve in Figure 3.60 appears to be continuous. It is said to be quasi-continuous.

For a *one-dimensional crystal lattice* consisting of *two kinds of atoms* (Figure 3.61) with masses M_1 and M_2, for example an ionic crystal, the angular frequency will be

$$\omega^2 = \beta\left(\frac{1}{M_1} + \frac{1}{M_2}\right) \pm \beta\left[\left(\frac{1}{M_1} + \frac{1}{M_2}\right)^2 - \frac{4\sin^2 ka}{M_1 M_2}\right]^{\frac{1}{2}} \tag{3.119}$$

If the third term is small compared with the second term, the angular frequency becomes approximately constant:

$$\omega_0^2 = 2\beta\left(\frac{1}{M_1} + \frac{1}{M_2}\right) \tag{3.120}$$

The angular frequency ω_0 belongs to the infrared region (Figure 3.62). Hence ionic crystals have absorption maxima in this region.

Equation (3.119) expresses the relationship between ω and k. Elastic lattice vibrations have several different modes of vibration: two perpendicular transverse vibration modes (page 147) and one longitudinal mode of vibration. Two opposite directions are possible in each case. Hence for each value of k there are *six* different values of $\omega(k)$, three for each sign in Equation (3.119).

At wavelengths that correspond to acoustic waves or sound waves, the angular frequency is inversely proportional to $1/\lambda$ and the velocity of the elastic wave is constant.

The average energy and the total number of phonons increase with temperature. The number of phonons can be increased in several ways. The crystal can, for example, be attached to a piezoelectric oscillator or put into contact with a heat source. The phonons will flow from the mechanical or thermal source into the crystal.

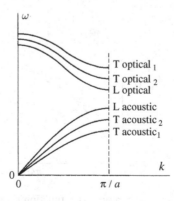

Figure 3.62 Angular frequency of phonons as a function of the wavenumber k in a linear lattice of an ionic crystal. Reproduced with permission from M. Alonso and E. Finn, *Fundamental University Physics*. © Addison-Wesley.

3.7 Influence of Lattice Defects on Electronic Structures in Crystals

Lattice defects have been introduced and briefly discussed in Chapter 1. It is easy to realize that if the regular order of the atoms or ions in a crystal lattice is disturbed in one way or an other, this must necessarily affect the energy levels of the valence electrons in the solid.

In this section, we will study the influence of the most important lattice defects, vacancies, interstitials and foreign atoms, on the electron structure of both nonmetallic and metallic solids.

3.7.1 Influence of Lattice Defects on Electronic Structures in Nonmetallic Crystals

Vacancies and Interstitials

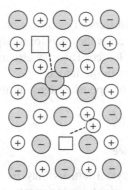

Figure 3.63   Vacancies and interstitials in a crystal lattice.

A missing ion or atom in a crystal lattice is called a *vacancy*. It is customary to illustrate a vacancy as a *square* in a crystal lattice (Figure 3.63). An *interstitial* is an excess atom or ion outside the normal sites in the lattice.

A certain number of vacancies and interstitials are always present in a crystal lattice at a given temperature. They are generated by phonons. The density of the vacancies depends strongly on the temperature and increases rapidly with increase in temperature. The upper limit is of magnitude ≥ 0.1 at-% close to the melting point temperature for most solids. The vacancies cause an expansion of the crystal lattice. This is one of the reasons for thermal expansion of solids.

Both vacancies and interstitials can move within the crystal lattice. This process, which implies mass transport and is called *diffusion*, plays a very important role in the solidification of metals. The mobility of interstitials is a matter of space and energy. The smaller the interstitials are, the more easily they can move if the necessary energy for the process is available.

Normally vacancies and interstitials are generated in equal numbers. The migration of these imperfections explains the electrolytic conductivity in pure salts. In ion crystals there are two different types of defects, named Schottky defects and Frenkel defects.

A *Schottky defect* is a vacancy in a crystal lattice where the ion has been removed from its site to the surface of the lattice (Figure 3.65). The vacancy is not coupled to any interstitial.

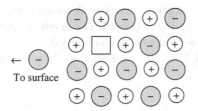

Figure 3.64   A Schottky defect is a vacancy which is not coupled to any interstitial in the crystal lattice.

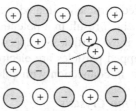

Figure 3.65   A Frenkel defect consists of a vacancy and a nearby interstitial in a crystal lattice.

A *Frenkel defect* is formed by excitation and migration of an ion from its normal site in the lattice, which is left empty. The ion becomes an interstitial. Each process results in a vacancy–interstitial pair (Figure 3.65).

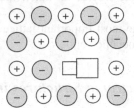

Figure 3.66 Coupled pair of vacancies of opposite signs.

Interstitial atoms may interact and form stable clusters of two or more interstitials. Similarly, vacancies can combine to form clusters of two or more vacancies. In extreme cases they can form large voids in the crystal lattice. Pairs of vacancies of opposite signs may form a unit and move within the lattice. The may also annihilate, i.e. both types of imperfections disappear and the energy released is transferred to the phonons in the lattice.

Figure 3.66 shows a coupled pair of vacancies of opposite signs in alkali metal halides. The coupling energy has been estimated to be of magnitude 1 eV.

Trapping of Charged Particles in a Crystal Lattice

All charged particles in asymmetric positions in crystal lattices may be *trapped*. The concept of trapping implies that the electric field caused by the charged particle induces slight displacements of neighbouring ions (Figure 3.67). Ions of the same type of charge are repelled and their distances to the charged particle are increased. Similarly, particles with unlike charges are attracted by each other and decrease their distances to the charged particle. The displacements of the ions produce a *local polarized region*.

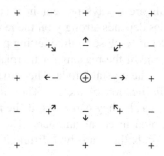

Figure 3.67 A trapped electron (circle near a positive ion) in the vicinity of a positive ion in the crystal lattice causes small displacements of both the positive and negative neighbouring ions.

Hence the charged particle is situated at the centre of the polarized region. At larger distances weak forces from the polarized centre are acting, which result in a potential well with some discrete energy levels (see below). The depth of such a potential well, where the charged particle is trapped, is normally of the magnitude a few tenths of an electronvolt.

Examples of trapped particles are electrons, holes and also interstitial ions. Examples and effects of trapped particles will be discussed below in connection with colour centres.

Colour Centres in Ion Crystals

A *colour centre* is a *lattice defect that is able to absorb visible light*. Crystals absorb electromagnetic radiation and vacancies may be formed, but the required energy quanta have energies which correspond to ultraviolet radiation.

Experience shows that there are several methods for colouring crystals, for example by introduction of chemical impurities, by X-ray or γ-radiation or by heating them in an atmosphere of metal vapour.

An example of the first method is aluminium oxide. Pure Al_2O_3 crystals are transparent. If small amounts of chromium are added, the crystals acquire an intense red colour. If pure alkali metal halide crystals, which are uncoloured, are heated in an atmosphere of the metal vapour in question they become coloured. In both cases crystal defects are introduced.

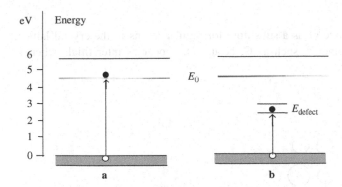

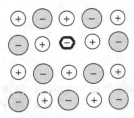

Figure 3.68   Absorption process in (a) a perfect crystal and (b) a crystal with lattice defects.

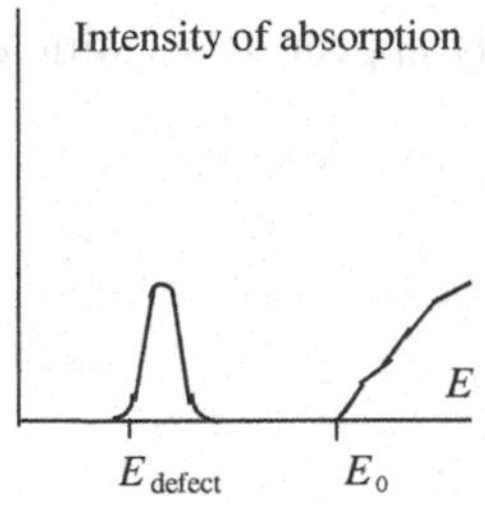

Figure 3.69   Absorption spectrum of a crystal with energy levels due to crystal defects.

Photons in the visible region have energies roughly between 1.6 and 3.2 eV. Transparent crystals do not absorb light quanta in the visible region of spectrum if the energy gap is >3.2 eV (Figure 3.68a). In presence of crystal defects there are additional energy levels above the valence band (Figure 3.68b). If the crystals are exposed to white light and the absorption spectrum is recorded, its appearance will in principle be that in Figure 3.69. There is a strong absorption band in the visible region.

When such crystals are exposed to white light, they absorb the wavelengths which correspond to the absorption band and the reflected and/or transmitted light has the complementary colour.

Aluminium oxide with small amounts of chromium has an absorption band in the green part of the spectrum, resulting in a red colour of the solid when exposed to white light. Similarly, NaCl has a blue absorption band and emits yellow reflected light when it is illuminated with white light.

Figure 3.70   F centre.

The simplest type of colour centre is the so-called F centre (F stands for Farbe, the German word for colour). An F centre consists of a negative vacancy, which is equivalent to a positive ion, and an excess electron bound to the vacancy.

The symbol of an F centre is a hexagon (instead of a square for the vacant negative ion) with a minus sign in the middle for the bound electron (Figure 3.70).

An M centre consists of two adjacent F centres. A so-called V centre consists of a trapped hole, bound to a pair of negative ions. Such centres have been observed in metal oxides.

3.7.2 Influence of Lattice Defects on Electronic Structures in Metals and Semiconductors

A defect, which is of particular importance in metals, is *foreign atoms* in crystal lattices. Foreign atoms may be present in crystals, either as impurities or added on purpose. When large amounts of foreign metal atoms are added to a base metal on purpose, an *alloy* is formed. Alloys are of great technical and practical importance.

When small amounts of foreign atoms are added to a semiconductor, the material is said to be doped. This is of great importance for the electrical conductivity of the material. This topic will be treated in Chapter 7.

Foreign Atoms as Interstitials and Substitutionals

Foreign atoms can either appear as *interstitials* in the crystal lattice or as a substitute for regular atoms in the crystal lattice, so-called *substitutionals* (Figure 3.71). In metals only a few elements, such as C, N and O, appear as interstitials whereas metal atoms normally appear as substitutionals for space reasons.

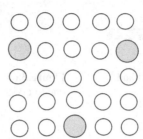

Figure 3.71 Substitutionals in a metal lattice.

Even small amounts of foreign atoms change the properties of the solid dramatically. They may introduce new electronic energy levels in an insulator. Such levels may permit absorption of light in a previously transparent part of the spectrum.

In metals, the foreign atoms alter the density of free electrons in the crystal lattice and cause extensive changes to the electron distribution and the energy levels. New energy levels, such as acceptor and donor levels in semiconductors, may appear.

Electron Structures in Alloys

When foreign atoms are introduced in a metal, the number of free electrons in the crystal changes and the *electron concentration,* i.e. the average number of free electrons per atom, changes. When alloying atoms with more or fewer valence electrons than the host atoms are added, the average number of free electrons per atom changes. The crystal as a whole is, of course, electrically neutral as the positive charges of the foreign nuclei in the crystal lattice balance the increased number of free electrons. There is clear experimental evidence that

- The electron density distribution and the energy levels of the Brillouin zones, i.e. the positions of the energy bands, in metals change when alloying elements are introduced.

These changes may cause a total change in the structure of the crystal lattice even if they are extremely small.

A change in the electron concentration of a metal also leads to a change in its properties. One example is the solubility of hydrogen in solid copper, which is considerably reduced by alloying copper with zinc. Copper has one and zinc has two valence electrons per atom. The use of zinc as an alloying element increases the average electron concentration. The same effect has been observed in copper for other alloying elements. The solubility of hydrogen in liquid copper is significantly reduced by addition of tin or aluminium.

Example 3.2

An alloy consist of 70 at-% Cu and 30 at-% Zn. Calculate the electron concentration, i.e. the average number of valence electron per atom in the alloy.

Solution:

Consider 100 alloy molecules: 70 of them are Cu atoms with one valence electron per atom and 30 of them are Zn atoms with two valence electrons per atom. Hence we obtain the average number of valence electrons per atom:

$$\frac{(70 \times 1) + (30 \times 2)}{100} = 1.3 \text{ electrons/atom}$$

Answer:

The average electron concentration in the alloy is 1.3 valence electrons per atom.

Generally, the energy of an alloy crystal depends on many factors, such as the sizes and chemical nature of the atoms and the number of electrons per atom. In some cases the dominant factor seems to be the electron concentration. This fact was first discovered empirically and was later treated theoretically by Jones. He assumed that the specific characteristics of the atoms can be neglected and considered the alloy as a simple mixture of atoms and free electrons. He stated the principle that

- An alloy adopts the structure which accommodates the valence electrons with the lowest possible energy.

This principle is illustrated in Figure 3.72. The steeper the $N(E)$ curve is, the lower will be the energy with which a given number of electrons can be accommodated. Obviously the structure in Figure 3.72b gives the lowest total energy of the crystal.

A necessary condition for the conclusion above is the assumption that *the excess electrons introduced into the lattice with the alloying element do not change the N(E) function of the solvent.* This simple theory represents the *rigid band approximation.* It holds in some cases but not in the general case. For this so-called *flexible band approximation,* a more sophisticated theory has been developed that is built on the Brillouin zone theory.

Jones's principle is valid provided that the number of free electrons can be accommodated within the first Brillouin zone and the $N(E)$ curve is as simple as Figure 3.72 shows, i.e. until the Fermi surface approaches the first Brillouin zone boundary and ceases to be spherical (Figure 3.48 on page 141). As an example, we will discuss the FCC and BCC structures by comparing their $N(E)$ curves.

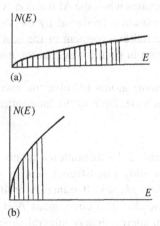

Figure 3.72   Energy distribution of valence electrons. Reproduced with permission from W. Hume-Rothery and B. R. Coles, *Atomic Theory for Students of Metallurgy.* (Published by the Institute of Metals, 1969.) now © W. S. Maney & Son Ltd.

The shapes of the $N(E)$ curves of FCC and BCC structures are drawn in Figure 3.73. Peak A, which corresponds to the situation when the Fermi surface approaches the closest face of the first Brillouin zone, is reached at an electron concentration of 1.36 electrons per atom for the FCC structure and 1.48 electrons per atom for the BCC structure. The $N(E)$ curve of the BCC structure continues to rise within the interval 1.36–1.48 electrons per atom whereas the $N(E)$ curve of the FCC structure decreases rapidly and the Fermi surface becomes strongly distorted.

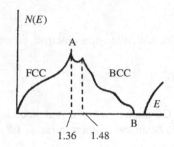

Figure 3.73   Change of structure of Cu–Zn alloys due to change in electron density caused by modification of composition.

The conclusion is that electrons can be accommodated most favourably if the crystal structure is FCC for electron concentrations ≤ 1.36 electrons per atom. At electron concentrations in the interval 1.36–1.48 electrons per atom a BCC structure corresponds to the lowest possible total energy of the crystal. In excellent agreement with this prediction, it has been found experimentally that Cu–Zn alloys change structure from FCC (α brass) to BCC (β brass) at an electron concentration of 1.4 electrons per atom.

When the electron concentration has reached the upper limit which corresponds to point B in Figure 3.73, the first Brillouin zone is completely filled and additional electrons would have to be accommodated in the second Brillouin zone at a high energy because of the energy gap at the zone boundary. Instead, the alloy changes structure as this process requires less energy than excitation of valence electrons up to the second Brillouin zone.

However, further investigations show deviations from the developed theory which accounts for the distortion of the Fermi surface. These discrepancies can be understood with the aid of the Brillouin zone theory. The Brillouin zones in k space result from the Bragg condition of the electronic matter waves (page 135). The same general Brillouin zone relationships will hold even if a few atoms drop out of the lattice and leave a defect structure with some vacant sites. The conclusion is that in the general case *the number of valence electrons per unit cell is a more significant quantity than the number of electrons per atom* when the crystal structures of metals are discussed.

In most normal structures, all lattice points are occupied and the number of valence electrons per unit cell is simply a multiple of the number of electrons per atom. In a structure with vacancies this is no longer the case. Even if atoms drop out, the number of electrons per unit cell may be constant. When the electron concentration increases in such an alloy, the structure does not necessarily change as indicated above. An example of this is given below.

In Cu–Al alloys, the electron concentration increases when the Al fraction is increased. It is found that the predicted 'γ brass' type of structure is formed at the electron concentration predicted by the theory. When the fraction of Al is increased, the structure of the γ phase remains stable up to the full complement of the unit cell. A further increase in the proportion of Al results in atoms dropping out by vacancy formation in such a way that the number of electrons per unit cell remains constant and the structure of the lattice remains intact.

In this case, the crystal 'prefers' to give up some atoms to solve the energy problem rather than change its structure or place additional electrons in the second Brillouin zone. Both of the latter alternatives require more energy than the first one.

Intermediate Phases

So far we have implicitly assumed that interstitial and substitutional atoms in alloys have random distributions. Normally it is possible to change the composition of an alloy almost continuously by varying the addition of alloying elements. This is not the case in the so-called *intermediate phases*. If either A or B atoms in a binary alloy AB attract electrons more strongly than each other, then characteristic chemical compounds A_xB_y with bonds of ionic character are formed. An intermediate phase of two or more metals has an approximately integral stoichiometric composition and shows a low ability to solve additional atoms of either component. Compounds of such intermediate phases still exist to a certain extent in a melt.

Examples of intermediate phases are Mg_2Pb, Cd_3Sb_2 and Mg_3Bi_2. The simplest and most reasonable explanation of the observations is that the valence electrons are distributed in such a way in intermediate phases that ions are formed. In the cases mentioned above, the ions would be $(Mg^{2+})_2(Pb^{4-})$, $(Cd^{2+})_3(Sb^{3-})_2$ and $(Mg^{2+})_3(Bi^{3-})_2$. Ion formation leads to a strongly reduced electrical conductivity as there are no free valence electrons in ionic crystals. This reduces the number of free electrons and the electrical conductivity considerably. This topic will be discussed further in Chapter 7.

Vacancies in Metals and Alloys

Vacancies of the Schottky defect and Frenkel defect types (page 151) also appear in metals and alloys. A vacancy can be regarded and treated as a special case of foreign atoms, i.e. absence of a foreign atom.

When a vacancy is formed in a perfect crystal lattice, two energy changes appear. The potential energy of the valence electrons *increases* as the negative contribution of the missing positive ion in the empty site disappears. In addition, the average kinetic energy of the valence electrons *decreases* slightly because the enlargement of the crystal volume from N_{total} sites to $N_{total} + 1$ sites requires energy, which is taken from the kinetic energy of the electron gas.

Fumi estimated the total energy change, i.e. the energy required to form a vacancy in a monovalent metal, to be $E_F/6$, where E_F is the Fermi energy of the metal. The agreement between the theoretical and experimental values is good for most metals.

Vacancies become trapped when pure metals and alloys solidify from metal vapours and metal melts.

Summary

■ *Definitions*

Ionization Energy

The energy required to move an electron from its orbit in the atom to infinity.

Electron Affinity

The energy which is released when an electron is moved from infinity to lowest possible orbit in an atom and a stable negative ion is formed.

Sublimation Energy

The energy required to move an atom from its position in the solid to infinity.

Cohesion Energy

The energy which has to be added to one unit of a crystal to separate its components into neutral free atoms at rest at infinite distance from each other.

Lattice Energy

The energy which has to be added to one unit of a crystal to separate its component ions into free ions at rest at infinite distance from each other.

■ *Bonds in Solids*

Crystalline solids can be classified according to the predominant type of bonds between the atoms in the crystal lattice. The main types of bonds are

- molecular bonds
- ionic bonds
- covalent bonds
- metallic bonds.

The three first types of bonds occur in nonmetallic solids.

Molecular Bonds in Nonmetallic Solids

Solids with molecular bonds are soft and fragile. The forces between the molecules are van der Waals forces between permanent, fluctuating or induced dipoles in the solid. The weak, attractive van der Waals forces are inversely proportional to the *seventh* power of the distance between the molecules:

Ionic Bonds in Nonmetallic Solids

Ionic crystals are hard and brittle and have high melting points. The ionic bonds are very strong.

The dominant contribution to the potential energy is due to the strong attractive electrostatic forces between the ions but four additional energy terms are also involved:

$$E_{total} = E_{attr} + E_{rep} + E_{covalent} + E_{vib} + E_{pol}(E_{ion} - E_{aff})$$

$$E_{attr} = \frac{A}{4\pi\varepsilon_0} \frac{-e}{R_0} \text{ (eV)} \tag{3.9b}$$

A is the Madelung constant.

The second term is caused by the short-range repulsive forces between the outer filled electron shells. It balances the attraction forces between the ions. The covalent contribution can often be neglected.

Covalent bonds in Nonmetallic Solids

Covalent bonds occur in nonionic solids. The bonds are very strong. Covalent solids are characterized by high melting points and high mechanical strength.

This type of bond is called covalent bonds or electron pair bonds or homopolar bonds. Two atoms share an electron pair, which gives the lowest possible energy of the system.

Covalent bonds between equal atoms are strengthened by the presence of exchange energy, which is a quantum mechanical effect.

Hybridization in Carbon

The C atom has the electron configuration $1s^2 2s^2 2p^2$ in its ground state. If energy is added, a 2s electron can be excited to a 2p orbital. The wave functions of the 2s electron and the three 2p electrons are combined to give four wave functions which are symmetrical in space. This results in four tetrahedral bonds and the lowest possible energy of the system.

Such a process is called hybridization. Examples of sp^3 hybridization in carbon are the CH_4 molecule and diamond. Examples of sp^2 hybridization are benzene and other organic aromatic compounds and graphite. The consecutive layers of carbon rings in graphite act as macromolecules which are held together by molecular bonds (van der Waals forces).

■ *Metallic Bonds*

Most metallic solids have good mechanical strengths, high melting points and excellent thermal and electrical conductivities. They have relatively low ionization energies and are opaque.

The explanation of the strong bonds in metals and their thermal and electrical properties is the free electron model.

Metals have high coordination numbers and not enough valence electrons to form electron pair bonds. Instead, all the valence electrons belong to the whole crystal lattice. The valence electrons can move within the lattice like the molecules in a gas. In ionic and covalent crystals the valence electrons are bound and cannot move within the crystal.

The mobility of the electron varies with its energy, which is considered by replacing the electron mass m by the effective mass m^*, which is a function of the kinetic energy of the electron.

Simple Quantum Mechanical Model of the Electron Gas

The Schrödinger equation for a valence electron:

$$\frac{\partial^2 \psi}{\partial x^2} + \frac{\partial^2 \psi}{\partial y^2} + \frac{\partial^2 \psi}{\partial z^2} + \frac{2m^*}{\hbar^2}[E - 0]\psi = 0$$

where the potential function is approximately constant and is set to *zero*.

Solution in one dimension
Eigenvalue:

$$E = \frac{\hbar^2}{2m^*}k_x^2$$

Amplitude of standing matter wave:

$$\psi = A\sin(k_x x) = A\sin\left(\sqrt{\frac{2m^*E}{\hbar^2}}\,x\right) \quad 0 < x < L$$

Wave vector and wave numbers

$$\mathbf{k} = k_x\hat{x} + k_y\hat{y} + k_z\hat{z}$$

$$|\mathbf{k}| = \frac{2\pi}{\lambda}$$

Solution in three dimensions
Eigenvalue:

$$E = \frac{\hbar^2}{2m^*}\left(k_x^2 + k_y^2 + k_z^2\right)$$

Amplitude of the standing matter wave:

$$\psi = C\sin\left(\frac{1}{\hbar}\sqrt{2m^*E}\,x\right)\sin\left(\frac{1}{\hbar}\sqrt{2m^*E}\,y\right)\sin\left(\frac{1}{\hbar}\sqrt{2m^*E}\,z\right)$$

or

$$\psi = C\sin(k_x x)\sin(k_y y)\sin(k_z z)$$

where

$$k_x = k_y = k_z = \frac{1}{\hbar}\sqrt{2m^*E}$$

- The wavenumbers of the valence electron are quantized.
- The energy levels of the valence electron are quantized.

■ *Energy States of Free Electrons in a Metal*

The whole metal represents *one* system with a large number of free electrons and a large number of different energy states. Each electron supplies one eigenfunction and its associated eigenvalue to the pool.

- The number of collective electron energy states in a metal crystal is equal to the total number of valence electrons, i.e. the valence number times the number of atoms.

The Pauli principle is valid for the valence electrons:

- Each energy state, defined by the quantum numbers n_x, n_y and n_z, can only accommodate two electrons with opposite spins.

If the total number of free electrons is equal to the number of energy states, only half of the available sites are occupied. In the absence of thermal excitation, the lowest-lying energy states are filled and the rest are empty.

The Fermi level E_F represents the energy of the most energetic valence electrons in the metal at $T = 0\,\mathrm{K}$. At higher temperatures some electrons become excited above the Fermi level and leave an equal number of empty sites below the Fermi level.

Fermi–Dirac Distribution

The Fermi factor f_{FD} represents the probability at temperature T that the energy state E will accommodate an electron:

$$f_{FD} = \frac{1}{e^{\frac{E-E_F}{k_B T}} + 1}$$

Representation of Electron Energies in k Space

Kinetic energy of an electron:

$$E_{kin} = \frac{\hbar^2}{2m}\left(k_x^{\,2} + k_y^{\,2} + k_z^{\,2}\right)$$

Allowed energy states are represented by points in k space. Electron states with a given energy correspond to points on a spherical surface, the so-called *Fermi surface*, in k space.

Density of Electron Energy States as a Function of Energy

Density of *available* electron energy states:

$$N(E) = \frac{(2m^*)^{\frac{3}{2}}}{4\pi^2 \hbar^3} E^{\frac{1}{2}}$$

Fermi factor:

$$f_{FD} = \frac{1}{e^{\frac{E-E_F}{k_B T}} + 1}$$

Density of *occupied* electron energy states:

$$N(E)f_{FD} = \frac{(2m^*)^{\frac{3}{2}}}{4\pi^2 \hbar^3} E^{\frac{1}{2}} \frac{1}{e^{\frac{E-E_F}{k_B T}} + 1}$$

Each electron state can accommodate two electrons (spin up and spin down). Electron density:

$$n(E) = 2N(E)$$

Fermi Level

Total number of free electrons per unit volume:

$$n_{total} = \frac{(2m^*)^{3/2}}{2\pi^2 \hbar^3} \frac{2}{3} E_F^{\,3/2}$$

Fermi level:

$$E_F = \text{the maximum energy of the free electrons at } T = 0\,\text{K}.$$

$$E_F = \frac{\hbar^2}{2m^*}\left(3\pi^2 n_{total}\right)^{2/3}$$

'Radius' of the Fermi sphere (Fermi level) in reciprocal space:

$$k_F = \frac{1}{\hbar}\sqrt{2m^* E_F}$$

■ *Energy Bands in Solids*

The quantum mechanical free electron model was very successful compared with the classical theory of the electron gas in a metal. The theory has been developed further, as it obviously is a rough approximation to assume a constant potential energy of the electron inside the metal.

The positive ions in the lattice give a periodic potential energy of the same type as in atoms and molecules. Kronig and Penney suggested a three-dimensional square function as a reasonable approximation.

The Kronig–Penney Model. Brillouin Zones in One Dimension

In the case of a linear lattice with spacing a, for the solution of the Schrödinger equation the wavefunction can be written as

$$\psi(x) = e^{ikx}U_k(x)$$

where $U_k(x)$ is a so-called *Bloch function*. In addition, $U_k(x)$ must be periodic, i.e. satisfy the condition

$$U_k(x) = U_k(x + a)$$

Both the eigenfunction and the wave vector are *quantized* just as for a free electron. A very important difference, caused by the Bloch function $U_k(x)$, is that
For the values of k which are given by the conditions

$$k = p\frac{\pi}{a} \quad p = \pm 1, \pm 2, \ldots$$

the Schrödinger equation has no unique solution. For each of these k values there are two eigenvalues, separated by an energy gap.

The Kronig–Penney energy curve shows discontinuities. Bands of allowed energy values are interrupted by energy gaps, i.e. regions of forbidden energy values. The eigenvalues and the k vectors are quantized. The broken curve is not continuous but consists of closely spaced points.

The k zones of allowed energies are called *Brillouin zones*. The construction and shape of Brillouin zones in two and three dimensions are discussed extensively in the text. The Brillouin zones correspond to the k values which are multiples of π/a.

The Brillouin zones represent electron energy bands, interrupted by forbidden energy gaps. The widths of the energy bands, which correspond to the k zones, increase with increasing energy.

Number of Possible Eigenfunctions and Energy States per Band

The number of energy states in each Brillouin zone is equal to N_{total}, the total number of atoms. Each Brillouin zone (energy band) can accommodate a maximum of $2N_{total}$ electrons.

Energy Distribution of Electrons in the Energy States in the Brillouin Zones

The theory of Brillouin zones in three dimensions in the so-called k space or reciprocal space is closely coupled to the energy distribution in the energy bands of the valence electrons in metals. It can be used to understand how the energy bands in metals are successively filled. The simplest Brillouin zones belong to crystals with a cubic structure. Initially the electrons fill the bottom of the first Brillouin zone and are represented by spheres in the reciprocal space. When the number of electrons increases, the radius (a k value) of the so-called Fermi sphere increases, i.e. the energy of the most energetic electrons increases. As long as the spheres are undisturbed, i.e. are far from the upper Brillouin zone boundary, the energy density in the energy band follows the relationship

$$N(E) = \frac{(2m^*)^{\frac{3}{2}}}{4\pi^2 \hbar^3} E^{\frac{1}{2}}$$

When the Fermi surface approaches the first Brillouin zone, the maximum of $N(E)$ is reached and the relationship is no longer valid for higher E values.

Each point in k space represents an energy state. When the Fermi surface approaches the Brillouin zone there are still empty energy levels or points outside the maximum sphere which represents the Fermi level E_F. If the energy gap up to the second Brillouin zone is large enough, the empty levels become filled before electrons occupy energy levels in the second Brillouin zone. The points in the Brillouin zone corners are rather few and the band soon becomes filled. This is the explanation of the electron distribution of the upper part of the first energy band.

Electrons always go to the energy states with the lowest possible energy. If the energy gap between the first and second zones is small, electrons may go to the second Brilloin zone before the corners in the first zone become filled. This results in a complicated energy distribution where the two energy bands overlap.

■ Effective Mass of Electron in a Metal

The effective mass of a valence electron in a metal depends on its energy, i.e. m^* must be a function of the wavevector k of the matter wave of the electron.

By studying the energy increase of a valence electron when it is accelerated in an electric field, it can be shown theoretically that the relationship in the one-dimensional case between the effective mass, the kinetic energy and the wavenumber k can be written as

$$m^* = \frac{\hbar^2}{\dfrac{d^2 E}{dk^2}}$$

■ Elastic Vibrations in Solids

Elastic vibrations in solids are of great importance for the thermal properties of solids. Collective vibrations, which move back and forth through the crystal, form travelling waves. Vibrations of many different frequencies and energies are present. Three modes of vibration appear: one longitudinal mode and two perpendicular modes of transverse waves. Elastic energy is quantisized:

$$G(n) = constant \times (n + \tfrac{1}{2})$$

Phonons

Phonons can be regarded as particles associated with elastic waves in solids. Phonons and photons both obey Bose–Einstein statistics. Phonons have quantized energies just like photons:

$$E = \hbar \omega (k) \quad p = \hbar k$$

■ Influence of Lattice Defects on Electronic Structures in Nonmetallic Crystals

Vacancies and Interstitials

A vacancy is a missing atom or ion in a crystal lattice. An interstitial is an excess atom or ion in a site between the lattice atoms. Vacancies and interstitial occur in all sorts of crystals.

A *Schottky defect* is a vacancy in a crystal lattice where the ion has been removed from its site to the surface of the lattice. The vacancy is not coupled to any interstitial.

A *Frenkel defect* is formed by excitation and migration of an ion from its normal site in the lattice, which is left empty. The ion becomes an interstitial. Each process results in a vacancy–interstitial pair.

Trapping of Charged Particles in a Crystal Lattice

All charged particles in asymmetric positions in crystal lattices may be *trapped*. The electrostatic interaction between the ions cause displacements of neighbouring atoms, which produces a local polarized region and a shallow potential well, i.e. energy states in the forbidden zone.

Colour Centres in Ion Crystals

A colour centre is a lattice defect which is able to absorb visible light.

In presence of crystal defects, there are additional energy states above the valence band. If the crystals are exposed to white light and the absorption spectrum is recorded, there is a strong absorption band in the visible region. When such crystals are exposed to white light, they absorb the wavelengths which correspond to the absorption band and the reflected light has the complementary colour.

An F centre is a negative vacancy, which is equivalent to a positive ion, and a trapped excess electron bound to the vacancy.

■ Influence of Lattice Defects on Electronic Structures in Metals and Semiconductors

Foreign Atoms

Foreign or impurity atoms can either appear as *interstitials* in the crystal lattice or as substitute regular atoms in the crystal lattice, so-called *substitutionals*. In metals only a few elements, such as C, N and O, appear as interstitials whereas metal atoms normally appear as substitutionals for space reasons.

Small amounts of foreign atoms are used to dope semiconductors. Large amounts of foreign atoms added to a base metal give an alloy.

Electron Structures in Alloys

The electron density distribution and the positions of the energy bands in metals change when alloying elements are introduced. Addition of foreign atoms changes the electron concentration, i.e. the number of valence electrons per atom.

An alloy adopts the structure which accommodates the valence electrons with the lowest possible energy.

Exercises

3.1   What is the difference between cohesive energy and lattice energy?

3.2   The distance r_0 between two neighbouring atoms in NaCl is 0.281 nm. Each ion is influenced by electrostatic attractive forces from ions with opposite charge and repulsive forces from ions of equal charge. The potential energy of an ion in the crystal, due to Coulomb interaction with the rest of the crystal, is

$$E_{attr} = -\frac{A}{4\pi\varepsilon_0}\frac{e^2}{r_0}$$

The Madelung constant A is 1.75 for NaCl.

   (a)   Calculate the energy (eV) required to release an ion from the interior of the crystal, due to the Coulomb interaction. Owing to a certain overlap of filled electron shells between neighbouring ions there are also repulsive forces between the ions. Assume that the repulsive energy term can be written as $E_{rep} = Br^n$, where B and n are positive constants. The experimental value of the binding energy of an ion is $E_B = 7.93\,\text{eV}$.
   (b)   Calculate the value of n, provided that the difference between E_B and the Coulomb energy, given above, depends entirely on E_{rep}.

3.3 There are three types of strong bonds in solids.

(a) Discuss briefly the theoretical basis and nature of the forces in each case. Give examples.
(b) Explain the concept of hybrid formation and give an example.

3.4 Carbon appears in several crystallographic shapes, the most common being diamond and graphite. The lattice constant of diamond is $a = 0.3567$ nm. The graphite structure can be described as a hexagonal lattice with four atoms in the base. The lattice constants of graphite are $a = 0.2461$ nm and $c = 0.6709$ nm. Calculate the density of

(a) diamond
(b) graphite.

3.5 (a) Describe the main features of the free electron model or quantum mechanical model of the electron gas in a metal. Define the work function and the Fermi level. Explain why the energy levels of the free electrons in a metal are much higher than expected from the classical point of view.
(b) Define the wavevector and give its relationship to the eigenfunction ψ of the free electron in one dimension.
(c) Give the kinetic energy of a free electron as a function of the wavevector. Illustrate the function graphically. Is the curve continuous?

3.6 A hot metal filament emits electrons. The number of electrons emitted per second and unit area of the metal surface depends on the temperature of the filament and the work function of the metal. The electrons can be absorbed by an anode and measured as an emission current. The relationship between the saturation current I_s and the absolute temperature T is given by Richardson–Dushman's law:

$$I_s = constant \times T^2 e^{-\frac{\phi}{k_B T}}$$

where ϕ is the work function of the metal.
  A series of measurements on a molybdenum cathode resulted in the following measurements:

$I_s (\mu A)$	1.31	4.11	12.00	32.90	85.18
T (K):	1500	1550	1600	1650	1700

Determine the work function of Mo graphically and expressed in eV.

3.7 (a) Define the Fermi factor and show graphically how it depends on temperature.
(b) Give the expressions of the Fermi distribution of energies in the electron gas, i.e. the densities of available and occupied energy states as a function of the electron energy.
(c) Why is the mass of the electron designated m^* instead of m?
(d) Calculate the total number of valence electrons per unit volume in the metal when the answer in (a) is known.
(e) Give the Fermi energy as a function of n_{total} (the total number of valence electrons per unit volume).

3.8 The Fermi energy of copper is known to be 7.04 eV.

(a) What is the maximum velocity of the conduction electrons in copper?
(b) Calculate the average velocity of the conduction electrons.

Assume that the temperature is 0 K and that the free electron model is valid.

3.9 Calculate the change in Fermi energy of sodium when the temperature increases from −30 to 70 °C. Assume that the free electron model can be applied and that the energy change depends only on the expansion of the crystal.

3.10 Calculate the Fermi energies of Na, Li and Al with the aid of convenient information from a standard table. The valences of the metals are normal.

3.11 The band theory of solids is based on and represents an improvement of the free electron model.

(a) Describe this improvement.
(b) Is the Kronig–Penney curve continuous? Give a motivation to your answer.
(c) Draw the Kronig–Penney curve, i.e. the electron energies as a function of the wavevector k. Which k values give energy gaps, i.e. discontinuities in the curve?

3.12 (a) Define the concepts of unit cell and primitive cell.

    (b) Consider a linear crystal of length L, which consists of N_{total} primitive cells. Calculate the lattice constant a. Which k values are allowed?

    (c) What is a Wigner–Seitz cell? How is it constructed? Wigner–Seitz cells are used when Brillouin zones are constructed. Explain.

    (d) Consider a three-dimensional crystal which contains N_{total} atoms. What is the total number of energy states in each Brillouin zone?

    (e) What is the maximum number of electrons that can be accommodated in each Brillouin zone? Motivate your answer.

3.13 (a) The Brillouin zones in a one-dimensional crystal are shown in Figure 3.32 on page 128 in Chapter 3. The band theory can be extended to two- and three-dimensional lattices. Why is the reciprocal space such a useful concept in the general theory of Brillouin zones?

    (b) It can be shown that the normal vector of a set of parallel planes (hkl) in r space is

$$\hat{n} = \frac{d_{hkl}}{2\pi} (h\boldsymbol{a}^* + k\boldsymbol{b}^* + l\boldsymbol{c}^*) \tag{1}$$

The set of parallel planes (hkl) in real space corresponds to a *point* in reciprocal space, defined by the vector $\boldsymbol{G}[hkl]$ in reciprocal space:

$$\boldsymbol{G}[hkl] = h\boldsymbol{a}^* + k\boldsymbol{b}^* + l\boldsymbol{c}^* \tag{2}$$

Give the size and direction of the lattice vector $\boldsymbol{G}[hkl]$ in reciprocal space in the case of Cartesian coordinates.

    (c) Calculate the distance d_{hkl} from the origin in r space to the first parallel plane (hkl) outside the origin.

    (d) Consider a cubic crystal lattice $|\boldsymbol{a}| = |\boldsymbol{b}| = |\boldsymbol{c}| = a$. Determine the values of $|\boldsymbol{a}^*| = |\boldsymbol{b}^*| = |\boldsymbol{c}^*|$ and d_{hkl} in terms of the lattice constant a.

3.14 The well-known Bragg diffraction condition for constructive interference is

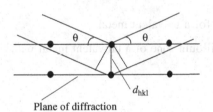

$$2d_{hkl} \sin \theta = p\lambda \tag{1}$$

or, applied to matter waves of free electrons:

$$\Delta k = p(h\boldsymbol{a}^* + k\boldsymbol{b}^* + l\boldsymbol{c}^*) \tag{2}$$

    (a) Define Laue indices.

    (b) Write Equation (2) in terms of Laue indices instead of Miller indices.

    (c) In addition to Equation (2), another condition must be fulfilled for constructive interference. Which one?

3.15 Find the plane in the FCC structure that has the maximum atomic density.

3.16 (a) What is the condition for Brillouin zone boundaries?

    (b) Introduce the wavelength λ of the matter wave instead of the wavenumber k in the condition in (a) and give the physical interpretation of the result.

    (c) Describe how the Brillouin zone boundaries can be constructed in k space when the structure of a two- dimensional (flat) crystal in r space is known.

3.17 Construct the reciprocal lattice, the first and second Brillouin zones of a two-dimensional crystal of a primitive, regular lattice with the axes a and $2a$, respectively. Calculate the 'areas' (in k space) of the first and second Brillouin zones.

3.18 (a) List the two conditions which are used to find the Brillouin zones of a three-dimensional crystal with a given structure.

(b) Describe the process of finding the first and second Brillouin zones that correspond to a crystal lattice of a simple cubic structure.

3.19 (a) Determine the first Brillouin zone of a simple cubic lattice.

(b) A simple cubic crystal has N_{total}^3 primitive unit cells. Prove that the number of independent values of the wavevector k can have within the first Brillouin zone is exactly N_{total}^3.

3.20 (a) Discuss the number and types of planes which enclose the first Brillouin zone of a BCC crystal.

   Hint: Consider the condition for X-ray reflections (page 19 in Chapter 1).

(b) Verify that $k_{max}[100] = 2\pi/a$, $k_{max}[110] = \pi\sqrt{2}/a$ and $k_{max}[111] = \pi\sqrt{3}/a$ for the first Brillouin zone of a BCC crystal lattice.

3.21 (a) Discuss the number and types of planes which enclose the first Brillouin zone of an FCC crystal.

   Hint: Consider the conditions for X-ray reflections (page 19 in Chapter 1).

(b) Verify that $k_{max}[100] = 2\pi/a$, $k_{max}[110] = 3\pi/(a\sqrt{2})$ and $k_{max}[111] = \pi\sqrt{3}/a$ for the first Brillouin zone of an FCC crystal lattice.

3.22 (a) Define the concepts of Fermi sphere and Fermi radius.

(b) Consider a metal. Give the relationship between the Fermi radius, the number of electrons per unit volume and the number of electrons per metal atom.

   Calculate the Fermi radius for a univalent metal with

(c) BCC structure

(d) FCC structure

as a function of the lattice constant a.

3.23 Calculate the shortest distance from the origin in k space to the surface of the first Brillouin zone in

(a) a BCC crystal lattice

(b) an FCC crystal lattice

as a function of the lattice constant a for a univalent metal.

3.24 Calculate the 'volume' of the first Brillouin zone of a univalent metal with

(a) SC structure

(b) BCC structure

(c) FCC structure

as a function of the lattice constant a of the crystal.

   Hint: Compare the volumes of the primitive cells and not those of the unit cells, which do not correspond to each other.

3.25 (a) Find the minimum 'distance' from the origin to a first Brillouin zone surface in a potassium crystal (BCC structure). Its lattice constant is 0.5225 nm.

(b) What fraction of this 'distance' in reciprocal space is located inside the Fermi sphere?

3.26 Lattice defects in a general sense, such as vacancies, interstitials and substitutionals, influence the electronic structure in both non-metallic and metallic solids.

(a) Two examples are Schottky defects and Frenkel defects. Define these two types of defects.

(b) What is a colour centre? In which type of crystals do colour centres appear? Explain the mechanism behind the phenomenon.

3.27 Generally, the energy of an alloy crystal depends on many factors, such as size, chemical nature of the atoms and number of electrons per atom. In some cases the dominant factor seems to be the concentration of valence electrons. Addition of alloying elements causes a change of the electron concentration.

(a) What is the general rule concerning the structure of the alloy?

(b) What is an intermediate phase? What are the conditions for formation of an intermediate phase?

3.28 Show that the whole Fermi sphere is located inside the first Brillouin zone of copper (FCC). Explain the physical signification of this statement.

3.29 Suppose that a piece of pure copper metal is successively alloyed with zinc, i.e. Cu atoms are gradually replaced by Zn atoms and the number of free electrons increases. When the electron concentration increases, the Fermi sphere expands.

    (a) Show that this sphere just touches the first Brillouin zone when the free electron concentration (number of electrons per atom) equals 1.36 and calculate the concentration of Zn atoms (at-%) in the corresponding alloy.

    (b) Show that the Fermi sphere just touches the first Brillouin zone of a BCC lattice when the free electron concentration reaches the value 1.48.

3.30 Assume that a univalent metal crystallizes into a simple cubic lattice.

    (a) Calculate the atom percent of a divalent metal that has to be added to bring the Fermi sphere and the first Brillouin zone in touch with each other. The free electron model is assumed to be valid.

    (b) What fraction of the lowest band of the alloy is occupied by free electrons in this case?

3.31 Characterize a phonon (energy, momentum, statistics, formation and annihilation).

3.32 A photon with the wavelength 500 nm is diffracted in a crystal with refractive index 1.5. The diffraction angle is 29°. On diffraction, a phonon with the frequency $\nu = 3.0 \times 10^9$ Hz is created. Calculate the wavevector of the phonon. The diffraction is supposed to be regular.

4

Properties of Gases

4.1	Introduction	170
4.2	Kinetic Theory of Gases	170
	4.2.1 Pressure	170
	4.2.2 Thermal Velocity Distribution in a Gas	172
4.3	Energy Distribution in Particle Systems: Maxwell–Boltzmann Distribution Law	174
	4.3.1 Maxwell–Boltzmann Distribution	174
	4.3.2 Thermal Energy Distribution in a Gas	175
4.4	Gas Laws	178
	4.4.1 General Law of Ideal Gases	178
	4.4.2 Interaction Between Molecules in Gases. Lennard-Jones Potential Energy	180
	4.4.3 Laws of Real Gases. Van der Waals Equation	183
4.5	Heat Capacity	185
	4.5.1 Heat Capacities of Ideal Monoatomic Gases	186
	4.5.2 Heat Capacities of Diatomic Gases	187
	4.5.3 Heat Capacities of Polyatomic Gases	192
4.6	Mean Free Path	192
4.7	Viscosity	196
	4.7.1 Kinetic Theory of Viscosity in Gases	196
4.8	Thermal Conduction	198
	4.8.1 Kinetic Theory of Thermal Conduction in Gases	198
4.9	Diffusion	199
	4.9.1 Kinetic Theory of Diffusion in Gases	200
	4.9.2 Fick's Laws	201
4.10	Molecular Sizes	204
4.11	Properties of Gas Mixtures	205
	4.11.1 Nonmolecular-specific Properties	205
	4.11.2 Dalton's Law	205
	4.11.3 Molecular-specific Properties	206
	4.11.4 Diffusion in Gas Mixtures	206
4.12	Plasma – The Fourth State of Matter	208
	4.12.1 Plasmas in Nature	208
	4.12.2 Laboratory Plasmas	209
	4.12.3 Applications	211
Summary		213
Exercises		216

Physics of Functional Materials   Hasse Fredriksson and Ulla Åkerlind
© 2008 John Wiley & Sons, Ltd

4.1 Introduction

Democritos in ancient Greece claimed that matter consists of atoms. In the absence of experimental verifications, it took a long time before this incredible idea was generally accepted in science. At the end of the 17th century, the idea that a gas consists of atoms or molecules in random motion was presented for the first time. This hypothesis was further developed by Clausius and especially by Maxwell and Boltzmann about 200 years later, when they presented their kinetic theory of gases, one of the cornerstones of classical physics.

4.2 Kinetic Theory of Gases

The fundamental postulates of the kinetic theory of gases are:

1. Gas molecules behave like hard elastic spheres.
2. Gas molecules do not attract each other.
3. Gas molecules are in permanent random motion. The higher the temperature, the more violent will be the motion.
4. The space occupied by the molecules is very small compared with the available volume of the gas.

These postulates will be used below in our brief discussion of the kinetic theory of gases and the properties of gases which can be derived on this basis.

4.2.1 Pressure

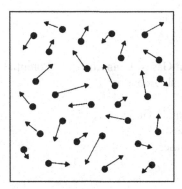

Figure 4.1 Model of molecules in a gas.

Consider a gas containing N_0 molecules, included in a container of volume V. The gas consists of a great number of molecules, which are in incessant random motion with high speeds in all directions (Figure 4.1). The molecules collide with each other and the wall of the container and change directions and sizes of their velocities very frequently.

We assume that the gas is *ideal*, i.e. the extension of the gas molecules is negligible and the forces between them are zero. This is a good model if the temperature is far above the condensation temperature of the gas.

As a first rough approximation, we assume that all the molecules have the same velocity. By studying the total momentum of the gas molecules we can derive a relationship between the pressure of the gas and the velocity of the molecules, independent of direction.

Consider a particular molecule with the velocity components v_x, v_y and v_z in a cubic container. As the velocity v is constant for all the molecules independent of direction, we have

$$v^2 = v_x^2 + v_y^2 + v_z^2 \qquad (4.1)$$

All directions are equivalent, which can be expressed as

$$v_x^2 = v_y^2 = v_z^2 \qquad (4.2)$$

and we obtain

$$v^2 = 3v_x^2 \qquad (4.3)$$

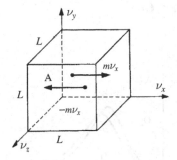

Figure 4.2   Momentum of a molecule before and after collision with the yz plane.

When a molecule with mass m collides with the yz plane (Figure 4.2), the net change in momentum before and after the impact will be $mv_x - (-mv_x) = 2mv_x$. If we assume that the container is cubic and has a length L, the molecule can traverse its length in a time L/v_x. The time interval between two successive collisions with the yz plane will be $2L/v_x$ and the number of collisions per unit time will be $v_x/2L$.

According to Newton's second law, the force f exerted by the molecule on the yz plane equals the net change in momentum per unit time:

$$f = 2mv_x \frac{v_x}{2L} = \frac{mv_x^2}{L} \tag{4.4}$$

The total force F of all the N_0 molecules in the cubic container on the surface with area $A = L^2$ is

$$F = N_0 f = \frac{N_0 m}{L} v_x^2 \tag{4.5}$$

However, as we will see in Section 4.2.2, the molecules in the gas do *not* have the same velocity. On the contrary, they are distributed from zero up to very high velocities. If we take this into consideration, the proper force exerted by all the molecules on the surface A is obtained if we replace the square of the velocity by its mean value:

$$F = \frac{N_0 m}{L} \overline{v_x^2}$$

The pressure which the molecular collisions exert on the surface A is equal to the force per unit area F/L^2, which gives

$$p = \frac{N_0 m}{L^3} \overline{v_x^2} \quad \text{or} \quad p = \frac{N_0 m}{V} \overline{v_x^2}$$

Next we introduce mean values into Equation (4.3), i.e. $\overline{v^2} = 3\overline{v_x^2}$, and obtain

$$p = \frac{N_0 m}{3V} \overline{v^2} \tag{4.6}$$

where
  p = pressure of the gas
  N_0 = number of molecules in the gas
  m = mass of a molecule
  V = volume of the gas
  $\overline{v^2}$ = mean value of the squared molecular velocities.

Equation (4.6) is generally valid as the pressure does not depend on the shape of the container.

The concept of *mean free path* is closely related to pressure and collisions between molecules. It is defined as the average distance between two random collisions of a molecule with other molecules in a gas. Mean free path will be treated in Section 4.6 in connection with viscosity, thermal conduction and diffusion.

4.2.2 Thermal Velocity Distribution in a Gas

Maxwell's Velocity Distribution Law

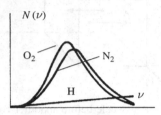

Figure 4.3 Velocity distribution curves of some gases at $T =$ constant. Reproduced from P. W. Atkin, *Atkin's Physical Chemistry*. © Oxford University Press.

The velocity of an individual molecule changes continuously owing to collisions with other molecules and with the wall. Because of the great number of molecules there is, however, a constant statistical distribution of velocities in the gas. The velocity distribution, independent of velocity direction, is a result of the Maxwell–Boltzmann kinetic theory of gases, derived at the end of the 19th century. The well-known *Maxwell distribution law* for the velocities of the molecules in a gas can be written as

$$dN = N(v)dv = N_0 \times 4\pi \left(\frac{m}{2\pi k_B T} \right)^{3/2} e^{-\frac{mv^2}{2}/k_B T} v^2 dv \qquad (4.7)$$

where

$v =$ velocity of a molecule
$dN =$ number of molecules per unit volume that have velocities in the interval v to $v + dv$, independent of direction
$N_0 =$ total number of molecules per unit volume
$m =$ mass of a molecule
$k_B =$ Boltzmann's constant
$T =$ absolute temperature.

Maxwell's velocity distribution law has been confirmed by experiments of the type described on page 174. The agreement between experiment and theory is excellent.

The function is shown in Figures 4.3–4.5. The influence of *mass* is shown in Figure 4.3. Figure 4.5 shows that the higher the *temperature*, the wider will be the distribution curve and the more it will be displaced towards higher velocities.

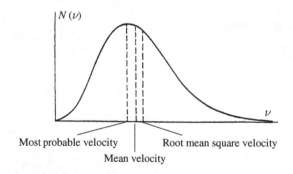

Figure 4.4 The distribution function $N(v)$ as a function of the molecular velocity v. The calculated mean velocities are marked.

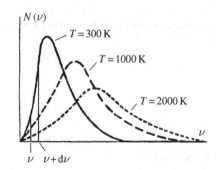

Figure 4.5 Distribution curves at various temperatures. The number of molecules within the velocity interval marked increases strongly with decrease in temperature.

The area under part of the curve in Figure 4.5 represents the number of molecules per unit volume which have velocities within the marked interval v to $v + dv$, independent of direction. Figure 4.5 shows that it varies strongly with temperature.

In Figure 4.4, three different velocities are marked: the most probable velocity, the mean velocity and the root mean square velocity.

The *most probable velocity* v_{mp} is the velocity which corresponds to the maximum value of $N(v)$. It is obtained by taking the derivative of Equation (4.7) with respect to v. The derivative equal to zero gives

$$v_{mp} = \sqrt{\frac{2k_B T}{m}} = \sqrt{\frac{2RT}{M}} \tag{4.8}$$

where M is the mass of 1 kmol of the gas.

The *mean velocity* can be calculated from the Maxwell distribution of velocities in a gas:

$$\bar{v} = \frac{\int\limits_0^\infty v N(v)\, dv}{\int\limits_0^\infty N(v)\, dv} \tag{4.9}$$

The result is

$$\bar{v} = \sqrt{\frac{8k_B T}{\pi m}} = \sqrt{\frac{8RT}{\pi M}} \tag{4.10}$$

The *root mean square velocity* $v_{rms} = \sqrt{\bar{v^2}}$ can be calculated analogously if we replace the factor v by v^2 in the numerator of Equation (4.9). The calculation gives

$$v_{rms} = \bar{v^2} = \frac{\int\limits_0^\infty v^2 N(v)\, dv}{\int\limits_0^\infty N(v)\, dv} = \frac{3k_B T}{m} = \frac{3RT}{M} \tag{4.11}$$

and we obtain

$$v_{rms} = \sqrt{\frac{3k_B T}{m}} = \sqrt{\frac{3RT}{M}} \tag{4.12}$$

Equation (4.11) can be used to calculate the *average kinetic energy of the molecules*. The average kinetic energy per molecule is

$$\bar{u} = \frac{m\bar{v^2}}{2} = \frac{3}{2}k_B T \tag{4.13}$$

where $\bar{u}$ is the average kinetic energy per molecule.

The kinetic energy U per kmol is

$$U = N_A \frac{m\bar{v^2}}{2} = \frac{3}{2}RT \tag{4.14}$$

where N_A is Avogadro's number = the number of molecules per kmol = $6.022 \times 10^{26}\, \text{kmol}^{-1}$.

Experimental Determination of Molecular Velocities in a Gas

The velocity distribution in the gas discussed above can be verified by direct 'counting' of the number of molecules in a gas which have velocities between v and $v + dv$. One method is illustrated in Figure 4.6.

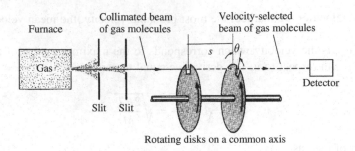

Figure 4.6 Equipment for determination of molecular velocities in a gas. Adapted with permission from O. Beckman, B. Kjollerstrom and T. Sundstrom, *Energi Lara.*

The gas is kept at a constant temperature in a furnace. Some gas molecules escape through a small hole and pass through two slits and then a velocity selector consisting of two disks which rotate with known angular frequency ω. The disks are equipped with two slots, displaced relative each other by an angle θ. Molecules can pass through both slots only if their velocity is

$$v = s\omega/\theta \qquad (4.15)$$

As the slots have a finite width, the number of transmitted molecules with velocities between v and $v + \mathrm{d}v$ is registered by the detector. By changing either θ or ω the velocity of the transmitted molecules can be varied and the whole velocity distribution can be obtained. In complete agreement with the theory, it was found that

- Molecules in a gas have velocities which vary in principle between zero and infinity with a maximum number at a certain velocity.
- The maximum number and the corresponding velocity are functions of temperature.

4.3 Energy Distribution in Particle Systems: Maxwell–Boltzmann Distribution Law

4.3.1 Maxwell–Boltzmann Distribution

In Section 4.2.2 we used Maxwell's velocity distribution law for molecules in a gas. Boltzmann treated the more general problem of a large number of particles distributed among a number of different available energy states. This topic will be discussed in this section for future use in this and later chapters.

Consider a particle system of a fixed volume and temperature T containing N_0 *identical*, but *distinguishable*, and non-interacting particles. Their total energy is U. The energy can be distributed among the particles in many different ways. The energy distribution will be described by specification of the number N_i of the particles which have the energy u_i. The problem is to find the most likely distribution, i.e. to derive N_i as a function of u_i.

Suppose that N_1 particles have the energy u_1, N_2 particles have the energy u_2, $\ldots$, N_i particles have the energy u_i. This partition N_1, N_2, $\ldots$, N_i represents a so-called *macrostate* of the system. For the macrostate the following subconditions are valid:

$$\sum_i N_i = N_0 \qquad (4.16)$$

$$\sum_i N_i u_i = U \qquad (4.17)$$

Boltzmann assumed that the probability of each partition is proportional to the number of different ways in which the particular partition can be obtained. Permutations of particles within each energy level produce no new distribution and consequently no new macrostate. Only if the *number* of particles within two or more energy levels is changed is a new macrostate obtained.

Boltzmann deduced the probability for an arbitrary partition, which is proportional to the total number of distinguishable different ways to obtain the partition. The equilibrium of the particle system corresponds to the most probable partition, i.e. the partition which can be obtained in the maximum number of distinguishable different ways. By use of this maximum condition, he found the equilibrium distribution:

$$N_i = \frac{N_0}{e^{-\frac{u_1}{k_B T}} + e^{-\frac{u_2}{k_B T}} + \ldots + e^{-\frac{u_i}{k_B T}}} \, e^{-\frac{u_i}{k_B T}} \tag{4.18}$$

where

u_i = energy of particle i
N_i = number of particles which have the energy u_i
N_0 = total number of particles
k_B = Boltzmann's constant
T = absolute temperature of the system.

We have assumed that all energy levels are equally probable. If this is not the case, we have to introduce the *statistical weight* g_i of energy level u_i and the *Maxwell–Boltzmann distribution law* in its general form becomes

$$N_i = \frac{N_0}{Z} \, g_i e^{-\frac{u_i}{k_B T}} \tag{4.19}$$

where Z is called the *partition function*, given by

$$Z = g_1 e^{-\frac{u_1}{k_B T}} + g_2 e^{-\frac{u_2}{k_B T}} + \ldots + g_i e^{-\frac{u_i}{k_B T}} = \sum_i g_i e^{-\frac{u_i}{k_B T}} \tag{4.20}$$

The fraction f_i is defined as *the fraction of the N_0 particles which has the energy u_i*:

$$f_i = \frac{N_i}{N_0} = \frac{g_i e^{-\frac{u_i}{k_B T}}}{g_1 e^{-\frac{u_1}{k_B T}} + g_2 e^{-\frac{u_2}{k_B T}} + \ldots + g_i e^{-\frac{u_i}{k_B T}}} \tag{4.21}$$

The Maxwell–Boltzmann distribution function can be applied to all sorts of thermal equilibrium distributions. Depending on the circumstances, u_i may be potential energy, kinetic energy or other types of energy.

The distribution is also very useful for calculating average values. In analogy with Equation (4.9) on page 173, we obtain the average value of the quantity X:

$$\overline{X} = \frac{\sum_i X_i g_i e^{-\frac{u_i}{k_B T}}}{\sum_i g_i e^{-\frac{u_i}{k_B T}}} \tag{4.22}$$

Some examples are the velocity distribution in a gas, energy distribution in a gas, distribution of vacancies in a solid and distribution of atoms and molecules in different energy levels. Here we will restrict the discussion to the thermal equilibrium distribution of velocities and energies in gases.

4.3.2 Thermal Energy Distribution in a Gas

The number of molecules in a gas is so large that the energy distribution can be regarded as continuous rather than discrete. In this case, integrals instead of sums are involved in the Maxwell–Boltzmann distribution. By replacing N_i by $N(u)du$ and g_i by $g(u)du$ we obtain

$$dN = N(u)du = \frac{N_0}{Z}g(u)e^{-\frac{u}{k_B T}}du = \frac{N_0 g(u)e^{-\frac{u}{k_B T}}du}{\int_0^\infty g(u)e^{-\frac{u}{k_B T}}du} \qquad (4.23)$$

where

dN = the number of molecules which have energies between u and $u+du$

$g(u)du$ = statistical weight within the energy interval u to $u+du$.

In the present case, u is the kinetic energy of a molecule. By calculations of $g(u)$ and Z, which will be omitted here, it can be shown that the Maxwell–Boltzmann energy distribution will be

$$dN = N(u)\,du = \frac{2\pi N_0}{(\pi k_B T)^{3/2}}u^{1/2}\,e^{-\frac{u}{k_B T}}\,du \qquad (4.24)$$

The distribution $N(u)$ is shown in Figure 4.7 for two different temperatures. The distribution is independent of molecular mass and consequently the same for all ideal gases.

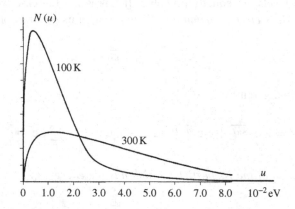

Figure 4.7 Maxwell–Boltzmann energy distribution in a gas. Reproduced with permission from M. Alonso and E. Finn, *Fundamental University Physics*. © Addison-Wesley.

The velocity distribution in an ideal gas can also be derived by use of the Maxwell–Boltzmann distribution function. This is done in a simple way in Example 4.1 below. The energy distribution and the velocity distributions are related but *not* identical.

Example 4.1

Show that Maxwell's distribution law for the velocities of the molecules in a gas is a special case of the general Maxwell–Boltzmann distribution law.

The Maxwell–Boltzmann energy distribution is assumed to be known.

Solution:

The energy distribution function can be written as [Equation (4.24)]

$$N(u) = \frac{dN}{du} = \frac{2\pi N_0}{(\pi k_B T)^{3/2}}u^{1/2}\,e^{-\frac{u}{k_B T}} \qquad (1')$$

We want to derive $N(v) = dN/dv$. The relationship between the two distributions is

$$\frac{dN}{dv} = \frac{dN}{du}\frac{du}{dv}$$ (2')

By taking the derivative of $u = mv^2/2$ with respect to v, we obtain

$$\frac{du}{dv} = mv$$ (3')

We insert dN/du [Equation (1')] and du/dv [Equation (3')] into Equation (2'):

$$\frac{dN}{dv} = \frac{2\pi N_0}{(\pi k_B T)^{3/2}} u^{1/2} e^{-\frac{u}{k_B T}} mv$$ (4')

We use the relationship $u = mv^2/2$ and obtain after reduction of Equation (4')

$$N(v) = \frac{dN}{dv} = N_0 \times 4\pi \left(\frac{m}{2\pi k_B T}\right)^{3/2} e^{-\frac{mv^2}{2}{k_B T}} v^2$$ (5')

Answer:

Equation (5') is the same as the Maxwell velocity distribution law [Equation (4.7) on page 172 divided by dv], which had been derived earlier by Maxwell on the basis of the kinetic theory of gases. Obviously this equation can also be derived from the Maxwell–Boltzmann distribution law.

Example 4.2

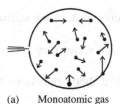

(a)   Monoatomic gas

The Maxwell–Boltzmann statistical thermodynamics including the distribution law, derived at the end of the 19th century, was regarded as one of the most successful and fundamental parts of classical physics. No more important discoveries were to be expected. Over 100 years later we know that this was completely wrong. The beginning of the 20th century was the most fruitful and exciting period in the history of physics.

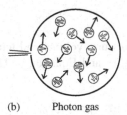

(b)   Photon gas

The key to modern physics was *quantization*, introduced by Max Planck and applied for the first time to the theory of blackbody radiation (Chapter 2, Section 2.3.1 on page 48).

Consider the Maxwell–Boltzmann distribution law, applied to the energy distribution of the molecules in a gas, and Planck's blackbody radiation law as a link between classical and modern physics. Compare the two laws and describe similarities and dissimilarities between the two energy distributions.

Solution and Answer:

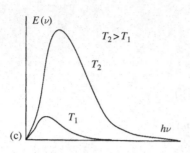

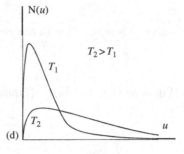

(c) Intensity distribution of blackbody radiation from the interior of a sphere of temperature T:

(d) Distribution of the kinetic energies of the molecules in a gas of temperature T:

$$E(\nu)\,d\nu = \frac{8\pi\nu^2}{c^3}\,\frac{h\nu\,d\nu}{e^{\frac{h\nu}{k_BT}}-1}$$

$$N(u)\,du = \frac{2\pi N_0}{(\pi k_B T)^{3/2}}\,u^{1/2}\,e^{-\frac{u}{k_BT}}\,du$$

Similarities
- If the electromagnetic radiation is considered as photons, both distributions deal with energy distribution of particles included in a given volume.
- Both functions are zero at the energies zero and infinity and have an intermediate maximum.
- The maxima of both curves are displaced towards higher energy values when the temperature T increases.

Dissimilarities
- The gas molecules are *classical* particles, which obey Boltzmann statistics. The photons are *quantum* particles which obey the so-called Bose–Einstein statistics.
- The energy of the molecules is *kinetic* energy, which varies *continuously*. The energy of the photons is the *total* energy. It is *quantized*.
- The area under the curve increases with T^4 for blackbody radiation. The area under the gas molecules curve is constant ($=$ number of molecules).

At *small* values of the photon energy ($h\nu \ll k_B T$), the quantization of the photon energy makes no sense because the number of energy levels within an energy range of the magnitude $k_B T$ is very large or approximately an energy continuum.

4.4 Gas Laws

4.4.1 General Law of Ideal Gases

A gas phase is characterized by three quantities: pressure, volume and temperature. We want to find a relationship between them and use Equations (4.6) on page 171 $p = \frac{N_0 m}{3V}\overline{v^2}$ and Equation (4.11) on page 173 $\overline{v^2} = \frac{3RT}{M}$ in combination with the relationship

$$M = N_A m \tag{4.25}$$

where

   M = mass of 1 kmol of the gas
   N_A = Avogadro's number (page 173)
   m = mass of a single gas molecule.

We eliminate $\overline{v^2}$ with the aid of Equations (4.6), (4.11) and (4.25):

$$pV = \frac{N_0 m}{3} \overline{v^2} = \frac{N_0 m}{3} \frac{3RT}{M} = \frac{N_0 m}{3} \frac{3RT}{N_A m} = \frac{N_0}{N_A} RT$$

The ratio N_0/N_A equals n, the number of kilomoles of the gas, and we obtain the well-known general law of ideal gases:

$$pV = nRT \qquad\qquad (4.26)$$

where R is the *general gas constant*. Its value in SI units is 8.314 kJ/kmol K.

Experimental Relationships. Absolute Temperature

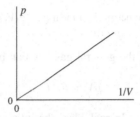

Figure 4.8   Pressure of a gas as a function of its volume.

Experiments on gases were performed as early as at the end of the 17th century. A gas always fills the whole available volume. It is also easy to measure the pressure and temperature of a gas. The English physicist Boyle varied the volume of the gas and measured carefully the corresponding pressures while the temperature of the gas was kept constant (Figure 4.8). He found the relationship

$$pV = constant \qquad \text{Boyle's law} \qquad\qquad (4.27)$$

Boyle's law is valid if the temperature of the gas is constant and far from the condensation temperature.

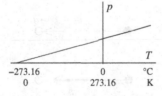

Figure 4.9   Pressure of a gas as a function of its temperature. Both the Celsius scale (°C) and the absolute temperature scale (K) are used in the diagram.

At the end of the 18th century, the French physicist Charles kept the volume of the gas constant, varied its temperature and measured the corresponding pressures (Figure 4.9). He found that there is a linear relationship between the pressure and temperature of the gas. In a graphical representation the line seemed to intersect the temperature axis at the same value, roughly −273°C, for different gases. This has been confirmed by careful and numerous experiments on many gases, for example hydrogen and nitrogen. The temperature −273 °C was called the *absolute zero point* and was used as the basis of the *absolute temperature scale*.[1]

Today the absolute temperature scale is defined by the *triple-point of water*, 273.16 K, i.e. the only temperature when ice, water and water vapour are in equilibrium with each other, and the absolute zero point = 0 K. The temperature unit of the

---

[1] Usually Celsius temperatures are denoted by t and Kelvin temperatures by T. Unfortunately, this distinction has not been possible everywhere in this book as time may also be involved; t is then used for time and T for temperature even if it is expressed in degrees Celsius. Hopefully no severe misunderstandings will appear for this reason.

absolute temperature scale is called the kelvin; $1\,K = 1\,°C$ and $273.16\,K$ is the fixed point of the absolute temperature scale.

If the absolute temperature scale is used, Charles's results can be written in a simple and compact form:

$$p = constant \times T \qquad \text{Charles's law} \tag{4.28}$$

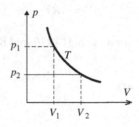

Figure 4.10 Isotherm of an ideal gas. $p_1 V_1 = p_2 V_2 = nRT$.

By combination of Equations (4.27) and (4.28), the general gas law can be derived:

$$pV = nRT \tag{4.26}$$

It was found to be strictly valid only for ideal gases. In real gases the interaction between the molecules has to be considered.

4.4.2 Interaction Between Molecules in Gases. Lennard-Jones Potential Energy

Dipole–Dipole Interaction

In Chapter 2, Sections 2.6 and 2.7, we have seen in a pure mathematical and abstract way that two atoms may form a stable molecule. The potential energy of the molecule as a function of the interatomic distance was derived and the dissociation energy for such a diatomic molecule was found to be of the magnitude of a few electronvolts.

In this section, we will discuss the nature of the acting forces and describe qualitatively and in more physical terms how the potential energy changes when two molecules in a gas approach each other.

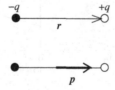

Figure 4.11 The electric dipole moment of a dipole is a vector defined as $p = qr$, where q is the electric charge and r a vector directed from the negative towards the positive charge.

The interaction between the molecules in a real gas consists of an *electric dipole–dipole interaction*, which is called the *van der Waals interaction*. This topic has been discussed in Chapter 3 on page 101. A molecule is a permanent dipole if the centre of its orbital electrons does not coincide with the centre of its positive charges (Figure 4.11). The dipole moment is often zero for symmetry reasons, i.e. the centres of the positive and negative charges coincide.

Hence the electric dipole moments of the molecules are zero in many gases. This is the case in gases such as H_2, N_2 and O_2 and the inert gases. The inert gases consist of monoatomic molecules with filled shells resulting in zero dipole moment.

Consider two molecules in a gas at an infinite distance r from each other. The forces between the molecules are zero and we use this state to define the zero level of the potential energy, $E_p = 0$.

When the two molecules at an infinite distance are brought closer to each other, they will be influenced by electrostatic forces. Even if a molecule has an average dipole moment equal to zero, it may become a temporary dipole if the vibrations in the molecule result in fluctuations of the positions of the positive and negative electrical centres. In the vicinity of a permanent or temporary dipole the orbitals of other neighbouring molecules become slightly displaced and the molecules become *induced dipoles*.

The mutual interaction between different types of dipoles is the origin of the van der Waals forces between the molecules in a real gas. As has been mentioned in Chapter 3 on page 101, the *attractive van der Waals forces* are assumed to be inversely proportional to the *seventh power* of the distance between the molecules.

When the molecules approach still further, a new type of force appears. When the molecules come so close to each other that their filled electron shells begin to overlap, very strong short-range repulsive forces appear. In the Lennard-Jones potential model (page 182), the *repulsive forces* are supposed to be inversely proportional to the *thirteenth power* of the distance between the molecules. When the forces are known it is possible to calculate the total potential energy of the van der Waals forces as a function of the intermolecular distance.

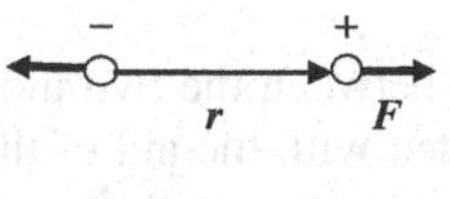

Figure 4.12 The force F marked in the figure is repulsive and positive. r is also a vector. It is directed from the negative charge towards the positive charge of the dipole.

The general relationship between force and potential energy is

$$F = -\frac{dE_{pot}}{dr} \qquad \text{or} \qquad dE_{pot} = -Fdr \tag{4.29}$$

which gives

$$E_{pot} = \int\limits_{0}^{E_{pot}} dE_{pot} = \int\limits_{\infty}^{r} -Fdr \tag{4.30}$$

where
  F = force acting on one of the molecules
  r = distance between the molecules.

Figure 4.12 shows that a repulsive force is positive and an attractive force is negative. The potential energy corresponding to each of two forces is calculated separately in the box below by applying Equation (4.30).

Forces Between Dipole Molecules and Their Corresponding Potential Energies

Repulsive force: Attractive force:

$$F^r = \frac{C_1}{r^{13}} \qquad\qquad F^a = -\frac{C_2}{r^7}$$

Repulsive potential energy: Attractive potential energy:

$$E^r_{pot} = \int\limits_{\infty}^{r} -\frac{C_1}{r^{13}}dr = \frac{C_r}{r^{12}} \qquad E^a_{pot} = \int\limits_{\infty}^{r} -\left(-\frac{C_2}{r^7}\right)dr = -\frac{C_a}{r^6}$$

The total potential energy of the molecules is equal to the algebraic sum of the repulsive and attractive potential energies. This relationship can be expressed as

$$E_{pot} = 4E_e \left[\left(\frac{r_0}{r} \right)^{12} - \left(\frac{r_0}{r} \right)^6 \right] \tag{4.31}$$

where

E_{pot} = total potential energy
E_e = depth of potential well
r = intermolecular distance
r_0 = the r value when $E_{pot} = 0$.

Equation (4.31) is the *Lennard-Jones potential energy*.

According to Equation (4.29), the force is the negative derivative of the potential energy. This will be kept in mind when we consider and compare Figure 4.14a and b.

At small r values, the strong repulsive, positive force dominates over the attractive, negative force and results in a strong repulsive net force (Figure 4.14a) and a very steep potential curve (Figure 4.14b). Much energy is required to reduce the distance between the molecules.

At the equilibrium distance r_e, the resulting force between the two molecules is zero, which corresponds to a minimum of the potential energy. The net force F can be calculated with the aid of the potential energy in Equation (4.31):

$$F = -\frac{dE_{pot}}{dr} = -4E_e \left(\frac{12r_0^{12}}{r^{13}} - \frac{6r_0^6}{r^7} \right) \tag{4.32}$$

The minimum condition $F = 0$ gives the relationship $r_e = r_0 \sqrt[6]{2}$ between r_e and r_0.

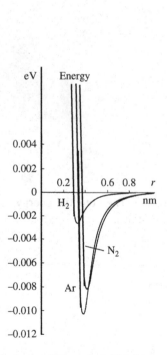

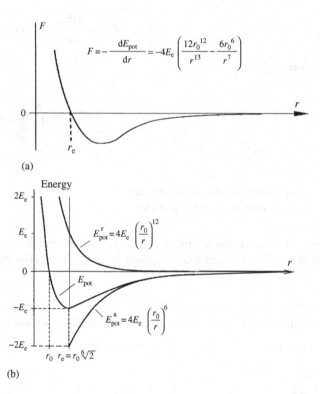

Figure 4.13 Lennard-Jones potential for some common molecules without permanent electric dipole moments. Reproduced with permission from W. J. Moore, *Physical Chemistry*, 5th edn. © Pearson Education.

Figure 4.14 (a) The force between two molecules as a function of the intermolecular distance.
(b) Potential energy of dipole–dipole interaction between molecules as a function of the intermolecular distance. Reproduced with permission from W. J. Moore, *Physical Chemistry*, 5th edn. © Pearson Education.

When r increases beyond r_e, the net force becomes attractive and negative, as the attractive force dominates over the repulsive force. The attractive force passes through a minimum (maximum attraction force) at a certain r value and approaches zero at infinity (Figure 4.14a). The corresponding potential energy increases slowly and becomes zero at infinity (Figure 4.14b).

The interaction between molecular dipoles is very reminiscent of the interaction between atoms and molecule formation, which has been treated in Chapter 2, Section 2.6. A comparison between the energy scales in Figure 2.35 in Chapter 2 on page 74 and Figure 4.13 shows that the potential energy is much smaller for the van der Waals interaction than for the interaction between atoms in molecules. The van der Waals forces are very weak.

4.4.3 Laws of Real Gases. Van der Waals Equation

Law of Real Gases Based on Lennard-Jones Potential Energy

The Lennard-Jones potential can be used to derive a gas law for real gases:

$$p = \frac{nRT}{V}\left[1 + B(T)\frac{n}{V} + \ldots\right] \tag{4.33a}$$

where $B(T)$ can be calculated with the aid of statistical mechanics. The result is

$$B(T) = 2\pi N_0 \int\limits_0^\infty \left[1 - e^{-\frac{E_{pot}(r)}{k_B T}}\right] r^2 dr \tag{4.33b}$$

where

N_0 = number of gas molecules in volume V
E_{pot} = Lennard-Jones potential energy [Equation (4.31)].

Equation (4.33a) has the advantage of being based on an analysis of the nature of the forces between the molecules. However, it is not so easy to handle and it is less used than the simpler and well-established empirical relationship suggested by van der Waals on the basis of qualitative arguments.

Van der Waals Equation for Real Gases

The general gas law is not valid for nonideal gases. It must be replaced by some other relationship between p, V and T. Many attempts have been made to find an adequate model. The simplest and most frequently used equation was suggested by van der Waals in 1873:

$$\left(p + \frac{n^2 a}{V^2}\right)(V - nb) = nRT \tag{4.34}$$

where

a, b = two constants, specific for each type of gas
n = number of kilomoles of the gas.

The term $n^2 a/V^2$ is introduced in order to take the influence of the attraction forces between the molecules into consideration. The pressure p has been derived by discussing the collisions between a wall and the molecules in the absence of intermolecular forces (Figure 4.15). If this interaction is taken into consideration, the pressure at the wall will no longer be the same as that in the interior of the gas. Owing to attractive forces between the molecules there is a resulting force acting on molecules close to the wall which reduces the pressure. The true pressure inside the gas is higher and a positive correction term has to be added.

The smaller the volume V is, the closer the molecules will be to each other and the stronger the intermolecular forces will be. The term nb is a correction due to the volume of the molecules, which no longer can be neglected. Two molecules cannot come closer to each other than the distance $2r$ if we assume that the molecules are spheres with radius r. Each pair of molecules results in an excluded volume equal to $4\pi(2r)^3/3$. Hence the excluded volume per molecule is $4 \times 4\pi r^3/3$ (Figure 4.16). The total excluded volume will be four times the sum of the molecular volumes.

The van der Waals equation can be visualized by a p–V diagram with the absolute temperature as a parameter. This is done in Figure 4.17.

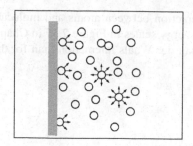

Figure 4.15 Intermolecular forces on some molecules inside a gas and close to the wall. Reproduced with permission from W. J. Moore, *Physical Chemistry*, 5th edn. © Pearson Education.

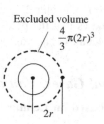

Figure 4.16 Excluded volume due to molecule size.

At high temperatures the isotherm approaches the appearance of a branch of a hyperbola and the general gas law may be approximately valid. The lower the temperature, the more extensive will be the deviations from Boyle's law.

At lower temperatures, for example T_4, the isotherm contains a horizontal part, which corresponds to the condensation process. Within this region gas and condensed liquid exist in equilibrium with each other and the van der Waals equation has no physical relevance. The higher the temperature, the shorter will be the horizontal line. Finally, the critical point (p^*, V^*, T^*) is reached (Figure 4.17). The critical temperature T^* is the highest temperature at which it is possible to condense the gas into a liquid. At the critical point there is no difference between gas and liquid.

Table 4.1 shows the values of a and b for some common gases. The constants are based on experimental data.

Table 4.1 Experimental values of constants a and b in the van der Waals equation.

Type of gas	a $(\mathrm{N\,m^4/kmol^2})$	b $(\mathrm{m^2/kmol})$
He	3.4×10^3	24×10^{-3}
H_2	24.7×10^3	26×10^{-3}
N_2	140×10^3	39×10^{-3}
O_2	137×10^3	32×10^{-3}
H_2O	552×10^3	30×10^{-3}

The van der Waals equation [Equation (4.34)] describes the conditions of the gas very well in most cases.

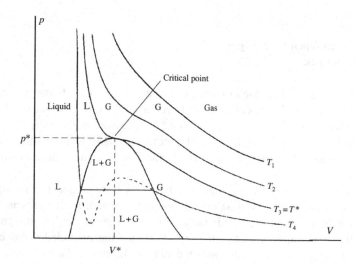

Figure 4.17 Isotherms of a van der Waals gas. Reproduced with permission from O. Beckman.

Example 4.3

If the forces between the molecules in a real gas are unknown, the most general equation of state can be expressed as a power series development of the pressure p as a function of n/V:

$$p = \frac{nRT}{V}\left[1 + B(T)\frac{n}{V} + C(T)\left(\frac{n}{V}\right)^2 \cdots\right]$$

where the so-called virial coefficients B and C are specific for each gas and functions of temperature only. $B(T)$ and $C(T)$ are determined from experimental data.

(a) Transform the van der Waals empirical equation into such a power series and identify the virial coefficients $B(T)$ and $C(T)$ as functions of a, b and T.

(b) Find the condition for the minimum deviation of the gas from the ideal gas law.

Solution:

The van der Waals equation [Equation (4.34)] can be written as

$$p = \frac{nRT}{V\left(1 - \dfrac{nb}{V}\right)} - a\left(\frac{n}{V}\right)^2 \tag{1'}$$

Series development of the first term on the right-hand side is allowed because $nb \ll V$. Rearrangement of the second term and series development of the first term gives

$$p = \frac{nRT}{V}\left[1 + \frac{nb}{V} + \left(\frac{nb}{V}\right)^2 + \cdots\right] - \frac{an}{RTV}\frac{nRT}{V} \tag{2'}$$

Equation (2') is compared with the equation given in the text:

$$p = \frac{nRT}{V}\left[1 + B(T)\frac{n}{V} + C(T)\left(\frac{n}{V}\right)^2 \cdots\right] \tag{3'}$$

Identification of terms gives

$$B(T) = b - \frac{a}{RT} \quad \text{and} \quad C(T) = b^2$$

Answer:

(a) The virial coefficients are $B(T) = b - \dfrac{a}{RT}$ and $C(T) = b^2$.

(b) Minimum deviation from an ideal gas behavior is obtained if $B(T) = 0$, i.e. at the so-called Boyle temperature, $T_{Boyle} = a/bR$.

4.5 Heat Capacity

The *internal energy* U of a system of molecules is defined as the sum of the kinetic energy, vibrational energy, rotational energy and potential energy of all the molecules:

$$U = E_{kin} + E_{vibr} + E_{rot} + E_{pot} \tag{4.35}$$

This concept of internal energy, which can be applied on gas molecules, is very useful when the heat capacity of a gas is to be derived.

4.5.1 Heat Capacities of Ideal Monoatomic Gases

Internal Energy

Consider 1 kmol of a monoatomic gas, which contains N_A molecules, where N_A is Avogadro's number $= 6.022 \times 10^{26}\,\mathrm{kmol}^{-1}$. In the case of an ideal monoatomic gas, the potential energy is constant in the absence of forces between the atoms. We choose the value of the potential energy equal to zero. As atoms have neither vibrational nor rotational energies, the only internal energy is the kinetic energy:

$$U = N_A \frac{m\overline{v^2}}{2} \tag{4.36}$$

Equation (4.6) on page 171 can be applied:

$$pV = \frac{N_A m}{3}\,\overline{v^2} = \frac{2}{3}\left(N_A \frac{m\overline{v^2}}{2}\right) \tag{4.37}$$

If we combine Equations (4.36) and (4.37) with the general gas law, applied to 1 kmol ($n = 1$), we obtain

$$pV = \frac{2}{3}\left(N_A \frac{m\overline{v^2}}{2}\right) = \frac{2}{3}U = 1 \times RT$$

which gives

$$U = 3\frac{RT}{2} \tag{4.38}$$

in agreement with Equation (4.14) on page 173. Equation (4.13) on page 173 gives the mean kinetic energy per atom:

$$\overline{u} = \frac{m\overline{v^2}}{2} = 3\frac{k_B T}{2} \tag{4.39}$$

The same relationship can be obtained by dividing Equation (4.38) by Avogadro's number N_A and use of the relationship $k_B = R/N_A$.

The internal energy U of a gas depends on its number of *degrees of freedom*. A monoatomic gas has three degrees of freedom, i.e. the molecules can move in the x, y and z directions. For symmetry reasons, each of them has the internal energy $RT/2$. From Equation (4.38) we can conclude that

- The internal energy per kmol and degree of freedom $= \frac{1}{2}RT$.

Heat Capacity

The first law of thermodynamics can be written as

$$Q = U + W \tag{4.40}$$

where
  Q = heat absorbed by the system
  U = increase in the internal energy of the system
  W = external work done by the system.

The heat capacity per kilomol C_V of a gas at constant volume is defined as

$$C_V = \left(\frac{\mathrm{d}Q}{\mathrm{d}T}\right)_V \tag{4.41}$$

If the volume is constant, no external work is done and we obtain by combining Equations (4.38), (4.40) and (4.41).

$$C_V = \left(\frac{dQ}{dT}\right)_V = \frac{dU}{dT} = 3\frac{R}{2} \tag{4.42}$$

The heat capacity per kilomol of a gas at constant pressure C_p is defined analogously as

$$C_p = \left(\frac{dQ}{dT}\right)_p \tag{4.43}$$

In this case external work is done because the volume changes. It is equal to pdV. For an ideal gas, this work can be obtained by differentiating the general gas law. As $dp = 0$ we obtain

$$dW = d(pV) = pdV + Vdp = pdV = RdT \tag{4.44}$$

By combining Equations (4.38), (4.40), (4.43) and (4.44), we obtain

$$C_p = \left(\frac{dQ}{dT}\right)_p = \frac{dU}{dT} + \frac{dW}{dT} = 3\frac{R}{2} + R = 5\frac{R}{2} \tag{4.45}$$

or, in general terms,

$$C_p = C_V + R \tag{4.46}$$

Measurements of heat capacities of inert gases show excellent agreement between experiment and theory (Table 4.2).

Table 4.2 Heat capacity of He as a function of temperature.

T (K)	No. of degrees of freedom	C_V
20	3	1.5R
300	3	1.5R
600	3	1.5R
4000	3	1.5R

4.5.2 Heat Capacities of Diatomic Gases

In addition to kinetic energy, i.e. *translational motion* in the x, y and z directions, diatomic molecules also have other modes of motion.

Rotational motions around three perpendicular axes are possible in principle. However, for diatomic molecules the moment of inertia I is zero in the axial direction and the energy associated with rotation around this axis is zero. Hence only two rotational directions remain.

The *vibrational energy* of a diatomic molecule consists of kinetic energy and potential energy of the nuclei (Chapter 2, pages 80–81).

The relationship $C_p = C_V + R$ holds strictly for both ideal monoatomic and diatomic gases. Hence the ratio C_p/C_V can be written as

$$\frac{C_p}{C_V} = \frac{C_V + R}{C_V} \tag{4.47}$$

The heat capacity at constant volume equals the number of degrees of freedom times $R/2$. The maximum number of degrees of freedom is 3 for translation, 2 for rotation and 2 for vibration, or maximum 7 in all. However, as we shall see below, not all the degrees of freedom are normally developed.

Heat Capacities of Diatomic Gases as a Function of Temperature

The variation of the heat capacities of diatomic molecules with temperature can be explained in terms of the kinetic theory of gases and quantum mechanics. The internal energy of a diatomic molecule can be written as

$$U = E_{kin} + E_{rot} + E_{vibr} \qquad (4.48)$$

Each of these quantities is a function of temperature. As before, we consider 1 kmol of the gas, i.e. N_A molecules.

Kinetic Energy and Its Contribution to the Heat Capacity

The kinetic motion of the diatomic molcules in the gas follows exactly the same law as monoatomic molecules [Equation (4.38)].

• The *kinetic energy* of the N_A molecules is a continuous function of temperature. The energy distribution of the molecules follows the Maxwell–Boltzmann distribution law and the average energy has been calculated as [Equation (4.38) on page 186]

$$U = 3\frac{RT}{2} \qquad (4.38)$$

which gives $C_V = 3R/2$. This value is valid at all temperatures except in the vicinity of the absolute zero point. At $T = 0$ there is no translation motion at all in a diatomic gas.

Rotational Energy and Its Contribution to the Heat Capacity

In order to calculate the internal rotational energy, we have to take into consideration that the rotational energy is quantized. The population of molecules in the different energy levels is a function of temperature and follows the Maxwell–Boltzmann distribution law (page 175):

$$N_i = \frac{N_0}{Z} g_i e^{-\frac{u_i}{k_B T}} \qquad (4.19)$$

where
  u_i = energy of particle i
  N_i = number of particles which have the energy u_i
  N_0 = total number of particles
  Z = the partition function (Equation (4.20), page 175)
  k_B = Boltzmann's constant.

If we apply Equation (4.19) to the rotational energy levels of N_A molecules and assume that the levels are equally probable ($g_i = 1$), we obtain

$$N_J = \frac{N_A}{Z} e^{-\frac{E_{rot}}{k_B T}} \qquad (4.49)$$

where N_A/Z is a constant. N_J represents the number of molecules which have the rotational energy E_{rot} and correspond to a particular rotational quantum number J according to Equation (2.64) in Chapter 2 on page 78.

$$E_{rot} = hc \times BJ(J+1) \qquad (4.50)$$

Hence Equation (4.49) can be written as

$$N_J = \frac{N_A}{Z} e^{-\frac{hc \times BJ(J+1)}{k_B T}} \qquad (4.51)$$

The energy levels are shown in Figure 2.41 on page 77 in Chapter 2 and here in Figure 4.18.

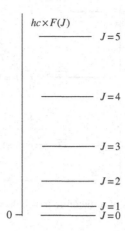

Figure 4.18   Rotational energy levels of a diatomic molecule.

● At very low temperatures, i.e. when $k_B T \ll hc \times BJ(J+1)$, very few molecules are excited to higher rotational energy levels, even for the lowest J values. The reason for this is that all N_J values in Equation (4.51) are approximately zero because the exponential factor is approximately zero ($e^{-\infty} = 0$). In this case, practically all molecules are in the lowest energy level $J = 0$ and do not rotate at all. Their contributions to the heat capacity are close to zero.

● When the temperature increases, the number of excited molecules increases gradually and these molecules begin to contribute to the heat capacity. At a certain temperature T_1 the contribution increases rapidly until practically all molecules are excited to higher rotational energy states The rotation is fully developed for all temperatures $> T_1$ and contribute to the molar heat capacity by R (2 times $R/2$).

Figure 4.21 on page 191 shows the contribution of rotation to the heat capacity at various temperatures.

Vibrational Energy and Its Contribution to the Heat Capacity

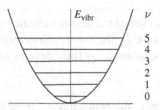

Figure 4.19   Vibrational energy levels in a potential well of a diatomic molecule (compare Chapter 2).

Just like the rotational energy levels, the vibrational energy levels of diatomic molecules are quantized and the distribution of molecules on different energy levels is found by the Maxwell–Boltzmann distribution law. In analogy with Equation (4.49), we obtain

$$N_v = \frac{N_A}{Z} e^{-\frac{E_{vibr}}{k_B T}} \tag{4.52}$$

where N_A/Z is a constant and N_v represents the number of molecules which have the vibrational energy E_{vibr} and correspond to a particular vibrational quantum number v.

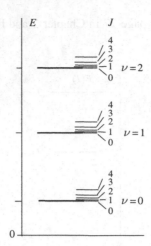

Figure 4.20 Sketch of the energy levels of a diatomic molecule with simultaneous rotation and vibration.

According to Equation (2.76) on page 81 we have

$$E_{\text{vibr}} = hc \times \omega_e \left(v + \tfrac{1}{2}\right) \tag{4.53}$$

Hence Equation (4.52) can be written as

$$N_v = \frac{N_A}{Z} e^{-\frac{hc \times \omega_e (v + 1/2)}{k_B T}} \tag{4.54}$$

The vibrational energy levels, are shown by Figure 2.43 on page 80 and by Figure 4.19 here. Figure 4.20 shows schematically the added energy levels of simultaneous rotational and vibrational motion.

- At low and medium or even rather high temperatures, i.e. $k_B T << hc \times \omega_e (v + \tfrac{1}{2})$, very few molecules are excited to higher vibrational energy levels, even for the lowest v values. The reason for this is that all N_v values in Equation (4.52) are approximately zero because the exponential factor is approximately zero ($e^{-\infty} = 0$). In this case, practically all molecules are in the lowest energy level $v = 0$ and their vibrational energies are not zero but constant. Their contributions to the heat capacitivity are close to zero.
- When the temperature increases, the number of excited molecules increases gradually and these molecules begin to contribute to the heat capacitivity. At a certain temperature T_2 the contribution increases rapidly until practically all molecules are excited to higher vibrational energy states. The vibration is fully developed for all temperatures $> T_2$ and contribute to the molar heat capacity by R (2 times $R/2$).

Figure 4.21 shows the contribution of vibration to the heat capacity at various temperatures.

Temperature Depencence of Heat Capacities of Diatomic Gases

The number of degrees of freedom of diatomic gases depends on the temperature of the gas. As an example, the number of degrees of freedom of H_2 as a function of temperature, obtained from measurement of C_V, are listed in Table 4.3.

- At extremely low temperatures, the rotational and vibrational motions are not developed and only three degrees of freedom are active.
- At room temperature and temperatures of several hundred degrees Celsius, the rotational motion is developed in hydrogen while the vibrational motion is still 'frozen'. The number of active degrees of freedom is then $3 + 2 = 5$.
- At very high temperatures, the vibrational motion is also developed and the number of active degrees of freedom becomes 7.

Table 4.3 Degrees of freedom of H_2 as a function of temperature.

T (K)	Active Number of degrees of freedom	C_V
20	3	1.5R
300	5	2.5R
600	5	2.5R
4000	7	3.5R

The same is true for other gases with the restriction that the temperatures at which rotational and vibrational motion of the molecules develop are specific for each gas. The most common gases, such as N_2 and O_2, have five degrees of freedom at room temperature. Hence we normally have for diatomic gases

$$C_V = 5\frac{R}{2} \tag{4.55}$$

C_p and C_V can be determined experimentally. The results agree well with the theory given above. The C_p/C_V ratios for monoatomic and diatomic molecules are given in Table 4.4.

Table 4.4 C_p/C_V for monoatomic and diatomic gases.

Type of gas	C_V	C_p/C_V	Remarks
Monoatomic	1.5R	$5/3 = 1.67$	All temperatures
Diatomic	1.5R	$5/3 = 1.67$	At very low temperatures
Diatomic	2.5R	$7/5 = 1.40$	At medium temperatures (50 K $< T <$ 1000 K for most gases)
Diatomic	3.5R	$9/7 = 1.29$	At very high temperatures

Figure 4.21 shows the contributions to the heat capacity of the gas of simultaneous translation, rotation and vibration of the diatomic molecules at various temperatures.

The reason why the vibrational degrees of freedom are 'frozen' at much higher temperatures than the rotational ones is that E_{vibr} is much larger than E_{rot}. This is shown in Figure 4.20. Consequently, $T_2 \gg T_1$ as shown in Figure 4.21.

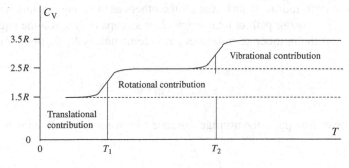

Figure 4.21 Sketch of C_V of a diatomic molecule as a function of temperature T.

4.5.3 Heat Capacities of Polyatomic Gases

Molecules with more than two atoms have many more degrees of freedom due to several different rotational and vibrational modes. Consequently, the C_p/C_V ratios of polyatomic molecules have even lower values than those given in Table 4.4. It is difficult to predict the values of C_p/C_V or use the model for conclusions concerning the pattern of motion of the molecules.

Even if the pattern of motion is complicated and varies with temperature, the relationship $C_p = C_V + R$ is always strictly valid.

4.6 Mean Free Path

The random motion of the molecules in a gas can indirectly be observed by the study of the so-called *Brownian motion* under a microscope, i.e. the random motion of small, visible particles in liquids or gases. In analogy with the discussion of the van der Waals equation, we will assume that the gas molecules have a finite extension.

If smoke is enclosed in a transparent box and illuminated from the side with laser light, the irregular motion of the smoke particles can be observed using a microscope. Their motion reveals indirectly the random motion of the gas molecules.

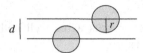

Figure 4.22 Collisions between molecules.

The gas molecules travel in a zigzag manner owing to collisions with each other and the walls of the container. The average distance between two successive collisions is called the *mean free path*, *l*. To find the mean free path we will examine the collisions between the molecules. Let the radius of each molecule be *r*. Collisions occur when the distance *d* between two molecules is $\leq 2r$ (Figure 4.22).

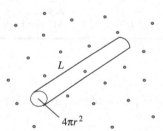

Figure 4.23 The zigzag paths have been added, independent of direction, to a straight tube. It makes no difference for the resultant value of the mean free path.

Consider one selected molecule with radius $2r$ and treat all the others as stationary points with no extension (Figure 4.23). When the selected molecule travels a zigzag path of total length L, it sweeps over a volume equal to $L \times 4\pi r^2$ and is exposed to x collisions within this volume. If the number of molecules per volume unit is N, we obtain two expressions for the number of molecules within the volume

$$x = L \times 4\pi r^2 N$$

According to the definition of mean free path, the average distance l between two collisions is

$$l = \frac{L}{x} = \frac{1}{4\pi r^2 N} \tag{4.56}$$

If the motion of all the molecules is taken into account, a more adequate calculation shows that the number of collisions within the volume $4\pi r^2 L$ becomes $x\sqrt{2}$ instead of x. Hence the average path between two collisions will decrease by a factor of $\sqrt{2}$ and the expression for the mean free path will be

$$l = \frac{1}{4\sqrt{2}\pi r^2 N} \tag{4.57}$$

where

l = mean free path
N = number of molecules per unit volume
r = radius of the atom or molecule.

Some examples of the mean free paths of gases are given in Table 4.5.

Table 4.5 Mean free path at STP of some common gases.

Gas	l (m)
H_2	112×10^{-9}
N_2	60×10^{-9}
O_2	65×10^{-9}
Ar	64×10^{-9}

The mean free path can alternatively be expressed as a function of pressure p instead of molecule density N. The general gas law $pV = nRT$ gives the number of kilomoles per unit volume. The number of molecules per unit volume is obtained by multiplying n/V (the number of kilomoles per unit volume) by Avogadro's number, $N_A = 6.022 \times 10^{26}\,\text{kmol}^{-1}$:

$$N = \frac{n}{V}N_A = \frac{p}{RT}N_A \tag{4.58}$$

This expression for N is introduced into Equation (4.57), which gives

$$l = \frac{RT}{4\sqrt{2}\pi r^2 p N_A} \tag{4.59}$$

The mean free path in a gas is an important quantity when transport phenomena in gases are studied. It will be used later when we discuss viscosity, thermal conduction and diffusion in gases.

Example 4.4

The densities of all gases decrease with increasing altitude above sea level. With decreasing density the collisions in a gas become more and more rare and the mean free path becomes longer and longer. The critical height z_{cr} is defined as the height above sea level where the density of the gas is so low that the fraction $1/e$ of the molecules moving upwards experience no collisions with other molecules. Hence these molecules are able to leave Earth if they have sufficient kinetic energy. The region $z > z_{cr}$ is called the *exosphere*.

At high altitudes, gases dissociate into atoms owing to the ultraviolet radiation emitted by the Sun. The critical level z_{cr} for atomic oxygen is $\sim$54 km and that for atomic hydrogen is $\sim$850 km.

The average temperature of the upper atmosphere as a function of altitude is shown in the figure (a). It can be seen that the temperature in the upper atmosphere is fairly constant at a given solar activity. The figure (b) shows the temperature variation over the equator at the exospheric level as a function of local time.

Hydrogen appear in the upper atmosphere as hydrogen atoms. They originate from molecules of water vapour and methane which are dissociated by the ultraviolet radiation from the Sun.

Discuss whether the Earth's atmosphere may contain hydrogen permanently or not, on the basis of reasonable approximations, simple calculations, known facts and the information given above.

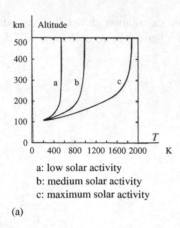

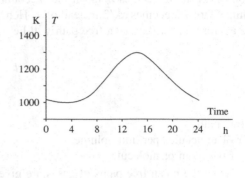

a: low solar activity
b: medium solar activity
c: maximum solar activity

(a)

(b)

Reproduced with permission from J. Houghton, *The Physics of Atmospheres*. © Cambridge University Press.

Solution:

The atmosphere of the Earth at sea level consists of approximately 21% O_2, 78% N_2 and 1% other gases. In the upper atmosphere at the critical level z_{cr}, O atoms are present in equilibrium with O_2 molecules and indirectly with the O_2 molecules at sea level.

We will initially calculate the root mean square (rms) velocities of O and H atoms at their z_{cr} levels at two temperatures, 1000 K and an upper temperature, which corresponds to very high solar activity and a maximum of the exospheric temperature during the day, which is roughly at least 2000 K.

A comparison between the rms velocities of the two types of molecules and the escape velocity may result in an answer to the above question.

Escape Velocity
In the absence of collisions with other atoms, the escape velocity from the Earth of an atom can be calculated.

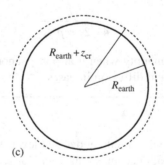

(c)

The kinetic energy of the atom must exceed or at least be equal to the potential graviational energy of the atom with mass m:

$$\frac{mv_{esc}^2}{2} = \frac{GmM_{Earth}}{R_{Earth} + z_{cr}} \tag{1'}$$

where G is the general gravitation constant and M_{Earth} and R_{Earth} are the mass and radius of the Earth, respectively.

At sea level, we have the relationship

$$mg = \frac{GmM_{Earth}}{R_{Earth}^2} \tag{2'}$$

Equations (1') and (2') gives the escape velocity, which is independent of the atomic mass:

$$v_{esc} = \sqrt{\frac{2gR_{Earth}^2}{R_{Earth} + z_{cr}}}$$ (3')

Root Mean Square (RMS) Velocity

The rms velocity of the atoms in a gas can be calculated from Equation (4.12) on page 173:

$$\sqrt{\overline{v^2}} = \sqrt{\frac{3k_B T}{m}}$$ (4')

where the Bolzmann constant $k_B = 1.38 \times 10^{-23}$ J/K.

Results of Calculations and Discussion

The lower and the upper atmosphere are in equilibrium with each other. Consequently, oxygen atoms must be present in the upper atmosphere as the lower atmosphere contains 21% O_2 at sea level. Oxygen is used to check the results of our calculations. They are given in the table below.

Atom	M (kg)	z_{cr} (km)	v_{esc} (km/s) at z_{cr}	RMS velocity at $T = 1000$ K (km/s)	RMS velocity at $T = 2000$ K (km/s)
O	16	54	11.1	1.3	1.8
H	1	850	10.5	5.0	7.0

For O atoms at $T = 1000$ K, the escape velocity is $11.1/1.3 \approx 8.5$ times larger than the rms velocity. The Maxwell velocity distribution curve in the figure (a) shows that the fraction of atoms with such velocities is practically zero. At $T = 2000$ K, roughly six times the rms velocity is required for escape. The figure shows that the fraction of atoms with this velocity is small. Hence very few oxygen atoms escape from the Earth and oxygen is a part of its atmosphere.

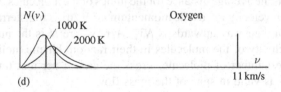

(d)

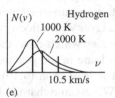

(e)

For H atoms at $T = 1000$ K the escape velocity is roughly twice the rms velocity. At $T = 2000$ K only $10.5/7 = 1.5$ times the rms velocity is required for escape. The Maxwell velocity distribution curve [figure (b)] shows that in both cases there is a considerable fraction of H atoms which fulfils this requirement. When the high-speed H atoms escape, a new velocity distribution is established and new high-speed H atoms escape and so on.

Hence hydrogen atoms escape from Earth and hydrogen is not a part of its atmosphere.

Answer:

The Earth's atmosphere does not contain hydrogen because H atoms in the upper atmosphere escape.

4.7 Viscosity

Viscosity, which also is called internal friction, is a measure of the frictional resistance which appears between layers of fluid, which move with different velocities relative to each other. The theory given below is valid provided that the flow is *laminar* and not turbulent.

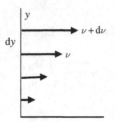

Figure 4.24 Laminar fluid flow with a constant velocity gradient.

Consider a fluid which flows along a stationary plane surface at rest. The fluid layer close to the plate is stagnant while higher layers move with linearly increasing velocities parallel to the plate (Figure 4.24).

Some of the molecules in a layer, which move with velocity v, diffuse into the next layer, which moves with velocity $v + dv$, and retard the fluid flow there. The retardation is equivalent to a frictional force, acting in the opposite direction to the velocity on the upper layer at the interface between the two layers. The frictional force is proportional to the interface area A and the velocity gradient dv/dy:

$$F = -\eta A \frac{dv}{dy} \tag{4.60}$$

where η is the *dynamic coefficient of viscosity*. The higher the value of η, the more viscous the fluid will be.

The aim of this section is to derive an expression for η as a function of atomistic quantities and to find the temperature dependence of η. For this purpose we will use the concept of mean free path in a gas.

4.7.1 Kinetic Theory of Viscosity in Gases

In order to find the influence of temperature on the viscosity coefficient, we will use one of the concepts of the kinetic theory of gases, i.e. the mean free path.

We consider two parallel layers in a stream line gas flow (Figure 4.25). The flow is laminar but the random kinetic motion in the gas is always present. We assume that the mean free path l represents the average distance for the jumps of the molecules. When a molecule jumps from a layer with velocity v to the adjacent layer with velocity $v + dv$ the momentum $ml\,dv/dy$ is transferred.

The number of molecules which cross the area A per unit time on their way upwards is $N\overline{v_{kin}}A/6$, where N is the number of molecules per unit volume in the gas and $\overline{v_{kin}}$ [1] is the average velocity of the molecules in their random kinetic motion at the cross-sectional area A [Equation (4.10) on page 173]. The same number of molecules cross the area A per unit time on their way downwards. We assume that the Maxwell distribution law is valid in spite of the mass flow.

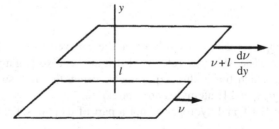

Figure 4.25 Laminar flow.

---

[1] The designation may seem odd – all velocities are kinetic. The subscript 'kin' refers to the disordered thermal motion of particles (or molecules or phonons) and is introduced to distinguish this velocity from v, the velocity of the gas layer.

The momentum transported across the area A per unit time can be written as

$$F = \underbrace{ml\frac{dv}{dy}}_{\substack{\text{Momentum transfer} \\ \text{per jump of molecule}}} \times \underbrace{\left[\frac{N\overline{v_{kin}}A}{6} - \frac{N(-\overline{v_{kin}})A}{6}\right]}_{\substack{\text{Number of jumping molecules} \\ \text{across the area } A \text{ per unit time}}} = \frac{Nml\overline{v_{kin}}A}{3}\frac{dv}{dy}$$

where m is the mass of a gas molecule. The momentum transfer across the area A per unit time is equal to the frictional force on the area A. The expression above in combination with Equation (4.60) gives

$$\frac{F}{A} = -\eta\frac{dv}{dy} = -\frac{1}{3}Nm\overline{v_{kin}}l\frac{dv}{dy} \tag{4.61}$$

where v is the velocity of the gas at position y.

Equation (4.61) gives

$$\eta = \frac{\rho\overline{v_{kin}}l}{3} \tag{4.62}$$

where $\rho = Nm$ is the density of the gas. If we introduce the value of l [Equation (4.57) on page 193] into Equation (4.62), we obtain

$$\eta = \frac{m\overline{v_{kin}}}{12\sqrt{2}\,\pi r^2} \tag{4.63}$$

Temperature Dependence of Viscosity of Gases

Next we introduce the expression for $\overline{v_{kin}}$ [Equation (4.10) on page 173] into Equation (4.63). The result is

$$\eta = \frac{m\sqrt{RT}}{6\pi r^2\sqrt{\pi M}} \tag{4.64}$$

where

η = viscosity coefficient of the gas
m = mass of a gas molecule
T = absolute temperature
r = radius of a gas molecule
M = mass of 1 kmol of the gas.

If we introduce the relationship $m = M/N_A$, the viscosity coefficient can be written as

$$\eta = constant \times \frac{1}{r^2}\sqrt{TM} \tag{4.65}$$

The viscosity coefficients at STP of some common gases are given in Table 4.6.

Table 4.6 Viscosity coefficients of some gases at STP.

Gas	η (kg/ms)
H_2	8.4×10^{-6}
N_2	16.7×10^{-6}
O_2	18.1×10^{-6}
CO_2	13.8×10^{-6}
Ar	21.0×10^{-6}

4.8 Thermal Conduction

We have seen that gas viscosity can be explained in terms of transport of momentum under the influence of a momentum (velocity) gradient. Analogously, thermal conduction depends on transport of kinetic energy under the influence of a temperature gradient.

The amount of heat transported across an area A per unit time is proportional to the area and to the temperature gradient, perpendicular to the area:

$$\frac{\mathrm{d}Q}{\mathrm{d}t} = -\lambda A \frac{\mathrm{d}T}{\mathrm{d}y} \qquad (4.66)$$

where
$\mathrm{d}Q/\mathrm{d}t$ = heat flow
λ = thermal conductivity
A = cross-sectional area
$\mathrm{d}T/\mathrm{d}y$ = temperature gradient.

4.8.1 Kinetic Theory of Thermal Conduction in Gases

The mechanism of transport of heat is analogous to that of momentum. The difference is that kinetic energy is transported in the case of thermal conduction instead of momentum in the case of viscosity.

The kinetic energy transported across the area A per unit time can be written as

$$\frac{\mathrm{d}Q}{\mathrm{d}t} = \underbrace{mc_{\mathrm{v}}l\frac{\mathrm{d}T}{\mathrm{d}y}}_{\substack{\text{Transfer of kinetic}\\\text{energy per jump}\\\text{of molecule}}} \times \underbrace{\left[\frac{N\overline{v_{\mathrm{kin}}}A}{6} - \frac{N(-\overline{v_{\mathrm{kin}}})A}{6}\right]}_{\substack{\text{Number of jumping molecules}\\\text{across the area } A \text{ per unit time}}} = \frac{Nmc_{\mathrm{v}}l\overline{v_{\mathrm{kin}}}A}{3}\frac{\mathrm{d}T}{\mathrm{d}y}$$

where
m = mass of a gas molecule
c_{V} = heat capacitivity (J/kg K)
T = temperature at cross-sectional area A.
l = mean free path of the gas molecules at temperature T
N = number of gas molecules per unit volume
$\overline{v_{\mathrm{kin}}}$ = average velocity of the molecules in their random kinetic motion at the cross-sectional area A (pages 196 and 173).

The transfer of kinetic energy across the area A per unit time is equal to the heat flow through the area A. The expression above in combination with Equation (4.66) gives the heat flux:

$$\frac{1}{A}\frac{\mathrm{d}Q}{\mathrm{d}t} = -\lambda\frac{\mathrm{d}T}{\mathrm{d}y} = -\frac{1}{3}Nmc_{\mathrm{V}}\overline{v_{\mathrm{kin}}}l\frac{\mathrm{d}T}{\mathrm{d}y} \qquad (4.67)$$

As Nm is equal to the density ρ, Equation (4.67) gives

$$\lambda = \frac{\rho c_{\mathrm{v}} \overline{v_{\mathrm{kin}}}l}{3} \qquad (4.68)$$

Equation (4.68) in combination with Equation (4.62) gives

$$\lambda = \eta c_{\mathrm{V}} \qquad (4.69)$$

Instead of c_{V}, we want to introduce the molar heat capacity $C_{\mathrm{V}} = 3R/2$ into Equation (4.69). We use the relationship $C_{\mathrm{V}} = Mc_{\mathrm{V}}$ and obtain

$$c_{\mathrm{V}} = \frac{C_{\mathrm{V}}}{M} = \frac{3R}{2M} \qquad (4.70)$$

Hence Equation (4.69) can alternatively be written as

$$\lambda = \eta \frac{3R}{2M} \qquad (4.71)$$

where

η = viscosity coefficient of the gas
M = mass of 1 kmol of the gas.

Temperature Dependence of Thermal Conductivity in Gases

From Equations (4.65) and (4.71), we can conclude that the rate of thermal conduction varies with T, M and r according to the relationship

$$\lambda = constant \times \frac{1}{r^2} \sqrt{\frac{T}{M}} \qquad (4.72)$$

The thermal conductivity at STP of some common gases are given in Table 4.7.

Table 4.7 Coefficients of thermal conductivity of some gases at STP.

Gas	λ (J/ms K)
H_2	170×10^{-3}
N_2	24.3×10^{-3}
O_2	24.6×10^{-3}
CO_2	14.4×10^{-3}
Ar	16.2×10^{-3}

4.9 Diffusion

If a gas volume with a uniform pressure is included in a space capsule, external influences on the gas molecules, i.e. pressure differences and gravitation, are completely eliminated. In this case the random kinetic motion of the gas molecules results in no net motion.

In other cases, the random kinetic motion of the molecules may be overlapped by a systematic motion in a particular direction. This phenomenon, which is called *diffusion*, results in *mass transport*. The diffusion flux of molecules is in most cases caused by a *concentration difference* in the direction of the flux.

As an example we will study a pure gas when radioactive tracer molecules of the same kind are added (page 208). The tracer molecules will diffuse into the gas. The diffusion process does not stop until the concentration gradient is zero and the tracer molecules are uniformly distributed.

The mass transport of *tracer* molecules per unit time across a surface is proportional to the concentration gradient and the cross-sectional area. The molecules always move from higher to lower concentration, which gives the minus sign in Equation (4.73).

$$\frac{dm}{dt} = -DA \frac{dc}{dy} \qquad \text{Fick's first law} \qquad (4.73)$$

where

dm/dt = mass transported per unit time (kg/s) (See also page 254 in Chapter 5 for other units in Fick's law)
D = diffusivity or diffusion coefficient of the diffusing gas component (m^2/s)
A = cross-sectional area (m^2)
c = concentration of the diffusing gas component expressed as mass per unit volume (kg/m^3)
dc/dy = concentration gradient of the diffusing gas component (kg/m^4).

Equation (4.73), which is known as *Fick's first law*, is the basic law of diffusion. It is generally valid for diffusion in gases, liquids and solids.

The concept of diffusion is more relevant in systems with two or more components. This topic will be discussed in Section 4.11.4 on page 206, where diffusion in gas mixtures and mixing of two separate gases are treated.

4.9.1 Kinetic Theory of Diffusion in Gases

Figure 4.26 Diffusing atoms within the volume $A \times 2\pi\overline{\nu_{\text{kin}}}\mathrm{d}t$.   **Figure 4.27** Concentration gradient of gas molecules.

Consider the dark volume element in Figure 4.26 with N^*_{total} tracer molecules of a gas with a *tracer* concentration gradient $\mathrm{d}N^*/\mathrm{d}y$ in the y direction (Figure 4.27). N^* is the number of molecules per unit volume. The tracer molecules, which cross the area A at $y = 0$ in both vertical directions, will on average have travelled one mean free path l between two successive collisions (page 193).

One-sixth of the tracer molecules in the upper half of the volume element move in the downwards direction for symmetry reasons. Only the molecules within the y interval $\overline{\nu_{\text{kin}}}\mathrm{d}t$ to 0 will pass the plane $y = 0$, where $\mathrm{d}t$ is the time between two successive collisions. $\overline{\nu_{\text{kin}}}$ is defined on page 196.

Analogously, one-sixth of the tracer molecules within the y interval $-\overline{\nu_{\text{kin}}}\mathrm{d}t$ to 0 move upwards and pass the plane $y = 0$. Only these molecules have to be taken into account during the time $\mathrm{d}t$.

The net number $\mathrm{d}N^*_{\text{total}}$ of tracer molecules which pass the area A in the plane $y = 0$ in the upwards direction during the time $\mathrm{d}t$ can be written as

$$\mathrm{d}N^*_{\text{total}} = \frac{N^*(-l)}{6} A \overline{\nu_{\text{kin}}}\mathrm{d}t - \frac{N^*(l)}{6} A \overline{\nu_{\text{kin}}}\mathrm{d}t$$

or

$$\frac{\mathrm{d}N^*_{\text{total}}}{\mathrm{d}t} = -\frac{\overline{\nu_{\text{kin}}}A}{6}\left[N^*(l) - N^*(-l)\right] \tag{4.74}$$

where N^* is the number of tracer atoms per unit volume.

The mean free path l is small and with the aid of series development we can write

$$N^*(l) = N^*_0 + \frac{\mathrm{d}N^*}{\mathrm{d}y}l \quad \text{and} \quad N^*(-l) = N^*_0 + \frac{\mathrm{d}N^*}{\mathrm{d}y}(-l)$$

These expressions are introduced into Equation (4.74):

$$\frac{\mathrm{d}N^*_{\text{total}}}{\mathrm{d}t} = -\frac{\overline{\nu_{\text{kin}}}A}{6}\frac{\mathrm{d}N^*}{\mathrm{d}y} \times 2l$$

and we obtain

$$\frac{\mathrm{d}N^*_{\text{total}}}{\mathrm{d}t} = -\frac{\overline{\nu_{\text{kin}}}Al}{3}\frac{\mathrm{d}N^*}{\mathrm{d}y} \tag{4.75}$$

If Fick's first law [Equation (4.73)] is applied to the tracer molecule diffusion above and divided by the mass per molecule, m_0 ($dm = m_0 dN^*_{total}$ [kg] and $dc = m_0 dN^*$[kg/m^3]), it will be transformed into

$$\frac{dN^*_{total}}{dt} = -DA\frac{dN^*}{dy} \tag{4.76}$$

Identification of Equations (4.75) and (4.76) gives

$$D = \frac{\overline{v_{kin}}l}{3} \tag{4.77}$$

A comparison between Equation (4.77) and Equation (4.62) on page 197 gives the simple relationship

$$D = \frac{\eta}{\rho} \tag{4.78}$$

   Light molecules diffuse more rapidly than heavy molecules. The diffusion constants at STP of some common gases are given in Table 4.8.

Table 4.8   Coefficients of diffusion of some gases at STP, calculated from Table 4.6 and Equation (4.78).

Gas	D(m^2/s)
H$_2$	93×10^{-6}
N$_2$	13.4×10^{-6}
O$_2$	12.7×10^{-6}
CO$_2$	7.0×10^{-6}
Ar	11.7×10^{-6}

Temperature Dependence of Diffusion in Gases

The expressions for mean free path [Equation (4.59) on page 193] and the mean velocity (Equation (4.10) on page 173) can be introduced into Equation (4.77). The result is

$$D = \frac{RT}{6\pi r^2 p N_A}\sqrt{\frac{RT}{\pi M}} \tag{4.79}$$

The diffusion rate depends on the pressure of the gas, the temperature, the molar weight and the radius of the molecules:

$$D = constant \times \frac{1}{pr^2}\sqrt{\frac{T^3}{M}} \tag{4.80}$$

where M is the mass of 1 kmol and p is the gas pressure.

4.9.2 Fick's Laws

Fick's first law, introduced on page 199, can be used to derive the second fundamental equation of diffusion.

   Consider a gas with a tracer concentration gradient dc/dy, directed upwards in Figure 4.28. The concentration of gas molecules increases upwards and the molecules diffuse downwards. The concentration c of gas molecules is a function of both position y and time t.

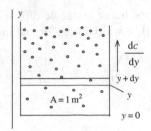

Figure 4.28 Concentration gradient of tracer atoms in a gas.

If we consider a small volume element with a cross-sectional area A equal to 1 unit area and thickness dy (Figure 4.28), the diffusive flux J can be written as

$$J = -D\frac{\partial c}{\partial y} \tag{4.81}$$

The concentration c inside the volume element changes with time. The change can be written in two ways:

$$\frac{\partial c}{\partial t} \times (1 \times dy) = J(y) - J(y+dy) \tag{4.82}$$

Increase in the number of tracer molecules per unit time within the volume $(1 \times dy)$ Net number of molecules which enter and leave the two end cross-sections during the time dt

The right-hand side of Equation (4.82) can be written with the aid of series development of the last term:

$$J(y) - J(y+dy) = J(y) - \left[J(y) + \frac{\partial J}{\partial y}dy\right] = -\frac{\partial J}{\partial y}dy \tag{4.83}$$

Taking the partial derivatives of Equation (4.81) with respect to y and inserting the result into Equation (4.83) gives, in combination with Equation (4.82), after division with dy

$$\frac{\partial c}{\partial t} = D\frac{\partial^2 c}{\partial y^2} \qquad\qquad \text{Fick's second law} \tag{4.84}$$

This equation will be used in several of the following chapters. Fick's first and second laws are the tools which are used to study diffusion processes and calculate diffusion rates and concentration distributions.

In most cases, but not always, a concentration gradient leads to diffusion. In Example 4.5 below the gravitational field is the cause of a concentration gradient, which is maintained at equilibrium. The concentration gradient is balanced by the gravitational field and no net diffusion occurs in the system.

Example 4.5

Consider a mixture of two gases, a heavy one with molar weight M and a light one with molar weight M'. The gas mixture is included in a container of considerable height z.

Show that the gravitational field roughly separates the two gases, i.e. at equilibrium the heavier gas (black circles) is mainly found in the lower part of the container whereas the lighter gas (open circles) is concentrated to the upper part of the container if the condition $(M - M') gz \gg RT$ is valid.

Solution:

Consider the volume element Adz in the figure. The gravitational force is balanced by the net force on the volume element of the pressure from outside:

$$-pA + (p - dp) A = Adz\rho g \tag{1'}$$

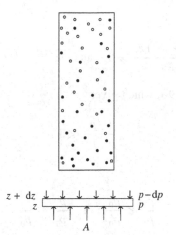

or

$$-\mathrm{d}p = \rho g \mathrm{d}z \qquad (2')$$

The density of the gas can be calculated with the aid of the general gas law:

$$pV = \frac{m}{M}RT \quad \Rightarrow \quad \rho = \frac{m}{V} = \frac{pM}{RT} \qquad (3')$$

This expression for ρ is introduced into Equation (2'):

$$-\mathrm{d}p = \frac{pM}{RT}g\mathrm{d}z \qquad (4')$$

The variables are separated and the equation is integrated:

$$\int_{p_0}^{p} \frac{-\mathrm{d}p}{p} = \int_{0}^{z} \frac{Mg}{RT}\mathrm{d}z \qquad (5')$$

which gives the pressure as a function of the height:

$$p = p_0 e^{-\frac{Mgz}{RT}} \qquad \text{the hydrostatic equation} \qquad (6')$$

Equation (6') is applied on the two gases. We use the following nomenclature:

1. subscript 0 refers to the lower part of the container
2. subscript 1 refers to the upper part of the container
3. no superscript refers to the heavy gas
4. a prime (') refers to the lighter gas.

$$p_1 = p_0 e^{-\frac{Mgz}{RT}} \qquad \text{heavy gas} \qquad (7')$$

$$p_1' = p_0' e^{-\frac{M'gz}{RT}} \qquad \text{light gas} \qquad (8')$$

The two equations are divided, which gives

$$\frac{p_1/p_0}{p_1'/p_0'} = \frac{e^{-\frac{Mgz}{RT}}}{e^{-\frac{M'gz}{RT}}} = e^{-(M-M')\frac{gz}{RT}} \qquad (9')$$

Because $(M - M') \, gz \gg RT$, we can conclude that

$$\frac{(M - M') \, gz}{RT} \gg 1 \quad \text{and} \quad e^{-(M - M')\frac{gz}{RT}} \ll 1 \tag{10'}$$

The latter relationship is applied to Equation (9'), which gives

$$\frac{p_1/p_0}{p_1'/p_0'} \ll 1 \tag{11'}$$

which can be written as

$$\frac{p_1}{p_1'} \ll \frac{p_0}{p_0'} \quad \text{or} \quad \frac{p_1'}{p_1} \gg \frac{p_0'}{p_0} \tag{12'}$$

Equations (12') can be transformed mathematically into

$$\frac{p_1}{p_1' + p_1} \ll \frac{p_0}{p_0' + p_0} \quad \text{or} \quad \frac{p_1'}{p_1' + p_1} \gg \frac{p_0'}{p_0' + p_0} \tag{13'}$$

The ratios in Equations (13') represent the partial pressures of the two gases.

Answer:

The partial pressure of the heavier gas is much lower at height z than at the bottom of the container. The partial pressure of the lighter gas is much higher at height z than at the bottom of the container. This is equivalent with the description of the gas distribution in the text.

The process in Example 4.5 leads to a certain degree of gas separation. A better method to separate gases is diffusion in several steps. Unequal diffusion coefficients lead to a difference in the diffusion rates of the gases. The degree of separation of the gases, i.e. the composition of the gas mixture, varies with the number of diffusion steps.

4.10 Molecular Sizes

The molar weights and the radii of some common gas molecules are listed in Table 4.9. The data show that measurements of the radii of the gas molecules differ, depending on the method of measurement. Obviously, the theoretical models that are the basis of the calculations are not accurate enough for total agreement.

Table 4.9 Molar weights and radii of some common gas molecules.

Molecule	Molar weight	Radius measured from gas viscosity (nm)	Radius measured from van der Waals constant b (nm)	Radius measured from closest packing (nm)
H_2	2	0.218	0.276	
N_2	28	0.316	0.314	0.400
O_2	32	0.296	0.290	0.375
Ar	40	0.286	0.286	0.383
CO_2	44	0.460	0.324	

4.11 Properties of Gas Mixtures

4.11.1 Nonmolecular-specific Properties

Some properties of ideal gases are independent of the mass and radius of the molecules. In these cases, pure gases and gas mixtures show no observable differences. Examples of such properties, which we call nonspecific, are energy distribution, heat capacity, kinetic energy and pressure. None of them depend on the type of gas. As an example we will discuss Dalton's law.

4.11.2 Dalton's Law

Ideal Gases

The pressure, for example, depends only on the volume and the temperature of the gas and the number of kilomoles. The total pressure of a gas mixture is equal to the sum of the partial pressures. This statement is called Dalton's law. It can be derived by introduction of the mole fractions x_i and use of the gas law:

$$\sum_i p_i = \sum_i n_i \frac{RT}{V} = \sum_i x_i n \frac{RT}{V} = n \frac{RT}{V} \sum_i x_i = n \frac{RT}{V} = p_{total} \tag{4.85}$$

where

p_{total} = total pressure of the gas mixture
p_i = partial pressure of gas component I
n = total number of kilomoles in gas volume V
n_i = total number of kilomoles of gas component i in volume V
x_i = mole fraction of gas component i
V = volume of the gas mixture.

Dalton's law can be verified experimentally with the aid of semipermeable membranes, which are composed of substances which transmit some gases much more readily than others. Iron, for example, is more permeable to H_2 than to other gases. At high temperature iron is quite permeable to N_2 but impermeable to inert gases. Silver transmits O_2 at high temperatures. Gases are generally transmitted through metals in which they dissolve readily. Rubber is very permeable to CO_2 and more permeable to O_2 than to N_2.

Deviations from Dalton's law occur if a chemical reaction between different components in the gas results in a change in the total number of kilomoles before and after the reaction.

Deviation from the gas laws may occur both in pure gases and gas mixtures, owing to dissociation of one or more of the gas components, if the total number of kilomoles in the pure gas or gas mixture changes during the dissociation process.

Relationship Between Pressure and Kinetic Energy

The partial pressures can also be related to the kinetic energy of the gases in a gas mixture. For each gas we have, according to Equation (4.6) on page 171,

$$p_i = \frac{N_i m_i}{3V} \overline{v_{kin\ i}^2}$$

where

N_i = total number of component molecules i
m_i = mass of component molecule i
$\overline{v_{kin\ i}^2}$ = mean value of the squared molecular velocities of component molecules i.

This form of Equation (4.6) can be combined with an expression for the total kinetic energy of the component molecules i:

$$U_i = N_i \frac{m_i \overline{v_{kin\ i}^2}}{2} \tag{4.86}$$

Elimination of $N_i m_i \overline{v_{\text{kin } i}^2}$ between Equations (4.6) and (4.86) gives

$$p_i = \frac{2}{3} \frac{U_i}{V} \qquad (4.87)$$

This relationship is generally valid. For a pure gas or a gas mixture we have

$$p_{\text{total}} = \frac{2}{3} \frac{U}{V} \qquad (4.88)$$

The energy law requires that the total kinetic energy of the gas mixture is the sum of the kinetic energy of the component molecules:

$$U = \sum_i U_i \qquad (4.89)$$

which gives Dalton's law as

$$p_{\text{total}} = \sum_i p_i \qquad (4.85)$$

Real Gases

Dalton's law is strictly valid only for ideal gases. The forces between the molecules in real gases lead to interaction and energy exchange between the molecules and deviations from Dalton's law.

4.11.3 Molecular-specific Properties

Velocity distribution, mean free path, viscosity, thermal conduction and diffusion are examples of properties which differ in pure gases and gas mixtures. These quantities depend on the molar weight and/or the radius of the gas molecules.

No general equation, valid for all properties, can be offered to take all molecular differences between the components in a gas mixture into consideration. In each case equations have to be set up for each gas component and for the gas mixture. Additional equations of the type representing Daltons's law are also introduced.

Different types of mean values may help to simplify a problem. An example of a constructive mean value which simplifies the treatment of the harmonic oscillator in molecular physics is the introduction of the *reduced mass* μ, which is the harmonic mean value of two masses m_1 and m_2 (Chapter 2, page 80):

$$\frac{1}{\mu} = \frac{1}{m_1} + \frac{1}{m_2} \quad \Rightarrow \quad \mu = \frac{m_1 m_2}{m_1 + m_2} \qquad (4.90)$$

As an example of special interest we will discuss diffusion in a gas mixture.

4.11.4 Diffusion in Gas Mixtures

Mean Free Path of Molecules in a Gas Mixture

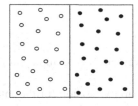

Figure 4.29 When the wall that separates two gases B and C is removed, the two gases diffuse into each other. A steady state is achieved when the gas mixture becomes homogeneous.

Consider a binary gas mixture of two gases B and C (Figure 4.29). The mean free path of component B can be found by a derivation analogous to that given in Section 4.6 on page 192. A mobile molecule with radius $r_B + r_C$ moves among other stationary molecules with no extension and the number of collisions per unit time is studied. If the motion of the other molecules and their different molecular weights are taken into consideration, the mean free path of gas B can be described by the relationship

$$l_B = \frac{1}{\pi (r_B + r_C)^2 N_C} \sqrt{\frac{M_C}{M_B + M_C}} \qquad (4.91)$$

where

$l_{B, C}$ = mean free path of molecules B and C in the gas mixture
$r_{B, C}$ = radius of molecules B and C
$M_{B, C}$ = molar weight of molecules B and C
$N_{B, C}$ = number of molecules B and C per unit volume in the gas mixture.

The mean free path of gas C can be calculated by exchanging indices B and C in Equation (4.91):

$$l_C = \frac{1}{\pi (r_C + r_B)^2 N_B} \sqrt{\frac{M_B}{M_C + M_B}} \qquad (4.92)$$

If the molecules B and C are of the same kind, Equations (4.91) and (4.92) become identical with the expression of the mean free path [Equation (4.57)] given on page 193.

Diffusion Coefficient of a Two-component Gas Mixture

According to the kinetic theory of diffusion in gases, the diffusion coefficient was calculated on page 201 as

$$D = \frac{\overline{v_{kin}} l}{3} \qquad (4.77)$$

The same equation can be used to calculate the diffusion coefficient of each gas in the gas mixture if l is replaced by the weighted average value of the mean free paths of the molecules B and C in proportion to their concentrations in the gas mixture. The diffusion coefficient of gas B becomes

$$D_B = \frac{1}{3} \frac{N_B l_C \overline{v_{kin\ C}} + N_C l_B \overline{v_{kin\ B}}}{N_B + N_C} \qquad (4.93)$$

D_C can be calculated by exchanging indices B and C. The symmetry of Equation (4.93) shows that the diffusion coefficients of the two gases are equal:

$$D_B = D_C \qquad (4.94)$$

which gives

$$D = D_B = D_C = \frac{x_B l_C \overline{v_{kin\ C}} + x_C l_B \overline{v_{kin\ B}}}{3} \qquad (4.95)$$

where x_B and x_C are the mole fractions of the two components.

- The diffusion coefficients of the two gases in a two component gas mixture are *equal* for any composition of the gas mixture.
- The *value* of the diffusion coefficient depends on the composition of the gas mixture.

Tracer Diffusion in a Two-component Gas Mixture

A smart and frequently used approach to study diffusion processes in gases is to use radioactive tracer methods. A small amount of an isotope B* of component B is added to a mixture of gases B and C and its diffusion is studied by measuring the activity of the isotope as a function of time and position by detectors. The diffusion coefficient D_B^* can be calculated from the measurements. D_B^* is defined by a relationship analogous to Equation (4.81) on page 202:

$$J_B^* = -D_B^* \frac{dc_B^*}{dy} \tag{4.96}$$

where J_B^* is the diffusive flux of B*.

The diffusion coefficient D_C^* can be determined in the same way by adding a small amount of C* molecules to the gas mixture:

$$J_C^* = -D_C^* \frac{dc_C^*}{dy} \tag{4.97}$$

We have seen above that in a mixture of two gas components the diffusion coefficients of gas B and gas C are *equal* for any composition of the gas. This is valid only in binary mixtures and *not* in mixtures with more than two components. Equations (4.93) and (4.94) show that $D_B^* = D_C^*$. If the number of components is >2, $D_B^* \neq D_C^*$.

In the ideal case, valid for gases at moderate pressures, the diffusion coefficient D of the main components B and C in the gas mixture is a weighted average of D_B^* and D_C^*:

$$D = x_B D_B^{0*} + x_C D_C^{0*} \tag{4.98}$$

where D_B^{0*} and D_C^{0*} are the diffusion coefficients of the pure gases B and C, respectively. They can easily be derived from Equation (4.93). When D_B^* and D_C^* have been determined experimentally, the diffusion coefficient D can be calculated.

4.12 Plasma – The Fourth State of Matter

So far we have discussed the properties of ordinary gases at temperatures within the temperature interval 0–4000 K.

At the end of the 1920s, Langmuir and co-workers studied ionized gases. Such gases proved to have very divergent properties, compared with ordinary gases, which justifies the designation 'fourth state of matter'. Langmuir gave the new state of matter the name *plasma*, which is the Greek word for jelly.

Plasma is a completely or partially ionized gas, which consists of electrons and highly ionized atoms or even naked nuclei. There is a strong electrical interaction between the adjacent charged particles in dense plasmas, which means that strong electrical fields are present. Particle motion in a plasma gives rise to currents and results in magnetic fields. The presence of electrical and magnetic interactions is responsible for the extraordinary properties of plasmas.

4.12.1 Plasmas in Nature

Figure 4.30 The corona around the Sun is the source of the solar wind which sweeps over the planetary system with a velocity of 300–900 km/s.

Plasmas appear in Nature. The most striking and nearby example is *the Sun*, which consists of several layers of plasmas, the corona with a temperature of roughly 10^6 K (Figure 4.30), the 'surface' with a temperature of approximately 6000 K and the interior with extremely high temperatures and pressures. Other cosmic plasmas are all the different types of *stars* and the thin *interstellar gas*.

Examples on Earth are the *van Allen belts*, which are permanent layers of trapped charged particles around the Earth (Figure 4.31). The so-called *solar wind* is a thin plasma stream, emitted by the Sun, which fills the solar system and continuously strikes the Earth.

Figure 4.31   Part of the van Allen belts. The belts and the Earth are not drawn to the same scale.

At heights of 80 km above sea level and upwards the air is strongly ionized owing to the ultraviolet radiation from the Sun. This layer is called the *ionosphere*, which among other levels contains the so-called E layer (Heaviside layer, 100 km above sea level) and F layer (Appleton layer, 240 km above sea level), which reflect radio waves back to Earth. They consist mainly of protons and electrons, which move in screwed orbits back and forth, trapped by the magnetic field of the Earth.

4.12.2 Laboratory Plasmas

Production of Plasmas

The transformation from one state of matter to another requires a supply of energy. When a solid is heated enough, the thermal motion of the atoms becomes so violent that the crystal structure breaks and the solid becomes a liquid. Additional energy supply increases the thermal motion of the atoms and leads to vaporization, i.e. the liquid becomes a gas (Figure 4.32).

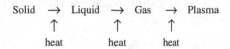

Figure 4.32   Survey of aggregation state changes.

Two main methods are used to produce plasmas in the laboratory:

- thermal excitation
- electrical excitation.

At increasing temperature the molecules of the gas dissociate into atoms. The kinetic motion of the atoms becomes more and more violent as the temperature increases, leading to frequent collisions and ionization of the atoms. However, it is hard to reach the high temperatures required to form a plasma with the aid of thermal energy. Other sources of excitation are used. A common way is to use an electrical discharge in a tube or an electrical arc discharge or induction methods.

Electrical excitation is achieved by running an electric current through the gas, exposing it to radio waves or using induction primarily to excite the electrons in the gas. By collisions with the ions, energy will be transferred from the electrons to the ions and the plasma obtains a reasonably homogeneous temperature.

The easiest way to produce a plasma is to use an electrical arc discharge. Such a device definitely does not behave like an ordinary resistor. With increasing voltage the current may even decrease, which corresponds to a negative resistance.

Electron Temperature in Plasmas

The properties of the plasma are determined preferably by two variables, the *particle density* and the *temperature* of the plasma. The particle density is the *number of particles per unit volume*.

Electron Temperature

The temperature is *not* measured in kelvin, the SI unit of temperature. Electrons play a very important role in a plasma and the 'temperature of the electrons', T_e, is expressed in *electronvolts* (eV).

It is convenient to define a phase-space particle density f of the particles in the plasma:

f = number of particles per unit of a six-dimensional phase-space
    (volume element = $\mathrm{d}x\,\mathrm{d}y\,\mathrm{d}z\,\mathrm{d}v_x\,\mathrm{d}v_y\,\mathrm{d}v_z$,)

where v is the velocity of the particles.

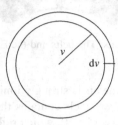

Figure 4.33 The 'volume element' in velocity space is a spherical shell with an area $4\pi v^2$ and a thickness $\mathrm{d}v$.

If we integrate f over a 'volume element' in velocity space, independent of direction, we obtain the fraction $\mathrm{d}N$ of the normal particle density which has velocities between v and $v+\mathrm{d}v$, expressed as the number of particles per unit volume (Figure 4.33):

$$\mathrm{d}N = \int f\mathrm{d}v_x\mathrm{d}v_y\mathrm{d}v_z = f\int \mathrm{d}v_x\mathrm{d}v_y\mathrm{d}v_z = f\times 4\pi v^2\mathrm{d}v \tag{4.99}$$

The Maxwell-Boltzmann statistics is certainly not valid for relativistic electrons ($mv^2/2$ is not correct and electrons obey the so-called Fermi-Dirac statistics). However, it is a 'short cut' to introduce the concepts 'thermal velocity' and 'electron temperature'. If $\mathrm{d}N$ were given by Maxwell's velocity distribution law [Equation (4.7) on page 172]. Elimination of $\mathrm{d}N$ by combining Equations (4.7) and (4.99) gives

$$f = N_0\left(\frac{m}{2\pi k_B T}\right)^{3/2}\mathrm{e}^{-\frac{mv^2}{2}{k_B T}} = \frac{N_0}{\left(\sqrt{2\pi}v_{\text{thermal}}\right)^3}\mathrm{e}^{-\frac{v^2}{2v_{\text{thermal}}^2}} \tag{4.100}$$

where v_{thermal} is called the *thermal velocity*, defined as

$$v_{\text{thermal}} = \sqrt{k_B T/m} \tag{4.101}$$

where
    k_B = Boltzmann's constant
    T = temperature (K)
    m = electron mass.

In plasma physics, it is convenient to express temperature in energy units. To do this we take the bold step of dropping the factor k_B in Equation (4.101) and introduce the *electron temperature*:

$$T_e = k_B T \tag{4.102}$$

which gives

$$v_{\text{thermal}} = \sqrt{T_e/m} \tag{4.103}$$

Equation (4.103) shows that the dimension of the electron temperature T_e has the dimension of energy.

Example 4.6

Consider an electron in a plasma which has a kinetic energy of 1 eV. The temperature of the plasma is said to have the electron temperature $T_e = 1\,\text{eV}$. What temperature in kelvin corresponds to an electron temperature of 1 eV?

Solution:

The temperature 1 eV, expressed in kelvin, is obtained from the relationship $k_B T = 1\,\text{eV} = 1.60 \times 10^{-19}\,\text{J}$:

$$T = \frac{T_e}{k_B} = \frac{1.60 \times 10^{-19}\,\text{J}}{1.38 \times 10^{-19}\,\text{J/K}} = 11\,600\,\text{K}$$

Answer:

The electron temperature 1 eV corresponds to $\sim 11\,600\,\text{K}$.

4.12.3 Applications

Figure 4.34 and Table 4.10. present a rough survey of natural and laboratory-produced plasmas as a function of their particle densities and electron temperatures. They show that the electron temperatures and particularly the particle densities of plasmas vary within wide limits. Some examples of applications of plasmas are illustrated in Figure 4.34.

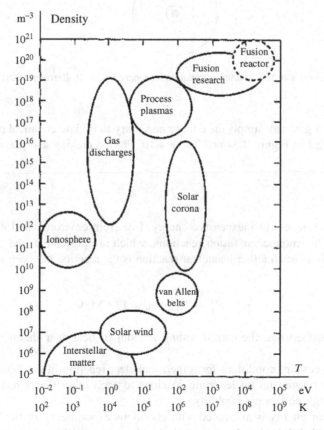

Figure 4.34 Schematic survey of natural and laboratory-produced plasmas as a function of their densities and electron temperatures. Reproduced with permission from R. J. Goldston and P. H. Rutherford, *Introduction to Plasma Physics*. © 1995 IOP Publishing Ltd (now Taylor & Francis Group).

Table 4.10 Characteristic properties of some natural and laboratory–produced plasmas.

Type of plasma	Magnitude of particle density in the plasma (m^{-3})	Magnitude of electron temperapture of the plasma plasma (eV)	Magnitude of corresponding plasma temperature (K)
Solar wind	10^7	10	10^5
Van Allen belts	10^9	10^2	10^6
Ionosphere of earth	10^{11}	0.1	10^3
Solar corona	10^{13}	10^2	10^6
Gas discharges	10^{18}	2	2×10^4
Process Plasmas	10^{18}	2	2×10^4
Plasmas used in fusion experiments	10^{19}–10^{20}	10^3–10^4	10^7–10^8
Calculated plasma in a tentative future fusion reactor	10^{20}	10^4	10^8

The method for generating the population inversion in highly excited atomic states necessary for laser action in a gas laser (Figure 4.35) is to run a discharge through the gas, which becomes a plasma. The excited electrons collide with atoms in the gas and excite them to highly excited energy states.

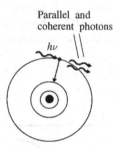

Figure 4.35 A great number of excited atoms (inverted population compared with Boltzmann distribution) is a necessary condition for stimulated emission.

Highly excited atomic states in a gas may supply the energy necessary to initiate chemical processes. Plasmas used for these purposes are called process plasmas in Figure 4.34 and Table 4.10. Plasma etching and plasma deposition in semiconductors are important applications.

Thermonuclear Reactions

The dream of humans is unlimited access to inexpensive energy, free from environmental disadvantages. The energy of the Sun and the stars originate from thermonuclear fusion reactions, which are possible at the extreme density and temperature conditions in the stars. An example of such a thermonuclear reaction is the reaction between a deuterium nucleus and a tritium nucleus:

$$H_1^2 + H_1^3 \rightarrow He_2^4 + n_0^1 + 17.6\,\text{MeV}$$

As a comparison, it can be mentioned that the corresponding amount of heat at a chemical combustion reaction is of the magnitude of eV instead of MeV.

The problem is to create the necessary conditions for a thermonuclear reaction in the laboratory. A temperature of at least $10^8\,\text{K}$ is necessary. At this temperature, the nuclei come so close to each other that a reaction can occur. A high particle density is necessary to obtain a high collision probability.

The production of thermonuclear energy is associated with enormous experimental difficulties. The major problems are

- to create the necessary conditions for a thermonuclear reaction and keep them during a sufficiently long time
- to insulate the extremely hot and dense plasma thermally form the walls of the container.

If the latter condition is not fulfilled, the plasma will never reach the necessary high temperature. The plasma is usually kept away from the walls by magnetic fields which force the plasma to move in helical orbits. So far no breakthrough in fusion research has occurred even though promising advances have been made. Unlimited and inexpensive energy is still a dream.

Summary

■ *Kinetic Theory of Gases*

$$\text{Pressure: } p = \frac{N_0 m}{3V}\, \overline{v^2}$$

Thermal Velocity Distribution

Maxwell Velocity Distribution Law

$$dN = N(v)\, dv = N_0 \times 4\pi \left(\frac{m}{2\pi k_B T}\right)^{3/2} e^{-\frac{mv^2}{2 k_B T}} v^2 dv$$

Most probable velocity:

$$v_{\text{mp}} = \sqrt{\frac{2k_B T}{m}} = \sqrt{\frac{2RT}{M}}$$

Mean velocity:

$$\overline{v} = \sqrt{\frac{8k_B T}{\pi m}} = \sqrt{\frac{8RT}{\pi M}}$$

RMS velocity:

$$\sqrt{\overline{v^2}} = \sqrt{\frac{3k_B T}{m}} = \sqrt{\frac{3RT}{M}}$$

Average kinetic energy per molecule:

$$\overline{u} = \frac{m\overline{v^2}}{2} = \frac{3}{2k_B T}$$

Kinetic energy per kilomol:

$$U = N_A \frac{m\overline{v^2}}{2} = \frac{3}{2}RT$$

■ *Energy Distribution in a Gas*

Maxwell–Boltzmann Energy Distribution Law

$$dN = N(u)\, du = \frac{2\pi N_0}{(\pi k_B T)^{3/2}} u^{1/2}\, e^{-\frac{u}{k_B T}}\, du$$

■ *Gas Laws*

Ideal gas

$$pV = nRT$$

Real gases

The interaction between the molecules in a real gas consists of dipole–dipole interaction or van der Waals interaction.

The repulsive force due to overlap of electron clouds of diatomic molecules is proportional to r^{-13}.

The attractive force due to dipole–dipole interaction is proportional to r^{-7}.

Lennard-Jones potential: $\quad E_{\text{pot}} = 4E_e \left[\left(\frac{r_0}{r}\right)^{12} - \left(\frac{r_0}{r}\right)^{6}\right]$

van der Waals law: $\quad \left(p + \frac{n^2 a}{V^2}\right)(V - nb) = nRT$

■ *Heat Capacity*

General Relationship

$$C_p = C_V + R$$

Internal energy per kilomol and degree of freedom $= \dfrac{RT}{2}$

Heat Capacities of Ideal Monoatomic Gases

$$C_V = 3\frac{R}{2} \quad \text{(three degrees of freedom)} \quad C_p = 5\frac{R}{2}$$

Heat Capacities of Ideal Diatomic Gases

The heat capacities depend on temperature. The energy distribution of the molecules follows the Maxwell–Boltzmann distribution law. At low temperature some degrees of freedom are not developed, owing to a lack of population of vibrational or both rotational and vibrational energy states.

$$N_v = \frac{N_A}{Z}e^{-\frac{hc \times \omega_e\left(v+\frac{1}{2}\right)}{k_B T}}$$

$$N_J = \frac{N_A}{Z}e^{-\frac{hc \times BJ(J+1)}{k_B T}}$$

Diatomic Molecules

Very low temperature: $C_V = 1.5R$ (translation)
Medium temperatures: $C_V = 2.5R$ (translation + rotation)
High temperatures: $C_V = 3.5R$ (translation + rotation + vibration)

■ *Mean Free Path*

$$l = \frac{1}{4\sqrt{2}N\pi r^2} = \frac{RT}{4\sqrt{2}\pi r^2 p N_A}$$

■ *Viscosity. Transport of Momentum*

$$F = -\eta A \frac{dv}{dy}$$

$$\eta = \frac{\rho \overline{v_{\text{kin}}}\, l}{3}$$

$$\eta = constant \times \frac{\sqrt{TM}}{r^2}$$

■ Thermal Conductivity. Transport of Kinetic Energy

$$\frac{dQ}{dt} = -\lambda A \frac{dT}{dy}$$

$$\lambda = \frac{\rho c_v \overline{v_{kin}} l}{3}$$

$$\lambda = \eta c_v$$

$$\lambda = constant \times \frac{1}{r^2} \sqrt{\frac{T}{M}}$$

■ Diffusion. Transport of Mass

$$\frac{dm}{dt} = -DA \frac{dc}{dy}$$

$$D = \frac{\overline{v_{kin}} l}{3}$$

$$D = \frac{\eta}{\rho}$$

$$D = constant \times \frac{1}{pr^2} \sqrt{\frac{T^3}{M}}$$

■ Fick's Laws

$$\frac{dm}{dt} = -DA \frac{dc}{dy} \qquad \text{Fick's first law}$$

$$\frac{\partial c}{\partial t} = D \frac{\partial^2 c}{\partial y^2} \qquad \text{Fick's second law}$$

■ Properties of Gas Mixtures

Dalton's law : $\quad p_{total} = \sum_i p_i$

Diffusion in Gas Mixtures

$$D = D_B = D_C = \frac{x_B l_C \overline{v_{kin\ C}} + x_C l_B \overline{v_{kin\ B}}}{3}$$

The diffusion coefficients of the two gases in a two-component gas mixture are equal for any composition of the gas mixture.

Tracer Diffusion in a Two-component Gas Mixture
The value of the diffusion coefficient depends on the composition of the gas mixture:

$$D = x_B D_B^{0*} + x_C D_C^{0*}$$

■ *The Four States of Matter*

The four states of matter are

1. solid
2. liquid
3. gas
4. plasma.

Plasma is a completely or partly ionized gas, which consists of a mixture of electrons and often highly ionized atoms.

Exercises

4.1 Use Boltzmann's distribution law to calculate the temperature at which 1% of the molecules in a container with hydrogen gas are excited up to the first rotational level above the ground state. The remaining 99% of the molecules are assumed to be in the ground state. The distance between the protons in the H_2 molecule is 0.074 nm.

4.2

(a) Consider Equation (4.24) and Figure 4.7 on page 176. Calculate the particular kinetic energy of the molecules which corresponds to the maximum of the Maxwell–Boltzmann energy distribution curve. Use $k_B T$ as the energy unit.

(b) Use the result in (a) to calculate the u value (energy unit eV) for the maximum point of the Maxwell–Boltzmann energy distribution curve when $T = 100 \, \text{K}$. Compare your result with Figure 4.7.

(c) What fraction of N_2 molecules has kinetic energies within the interval $0.95 \, k_B T \le u \le 1.05 \, k_B T$ at 300 and 1500 K, respectively?

Hint: Integration is not necessary. Choose $\Delta u = 0.10 \, k_B T$.

4.3 What fraction of the molecules in an ideal gas have velocities between $\bar{v}$ and $\sqrt{\overline{v^2}}$?

Hint: Choose $v_{\text{ave}} = \dfrac{\bar{v} + \sqrt{\overline{v^2}}}{2}$ and $\Delta v = \sqrt{\overline{v^2}} - \bar{v}$.

4.4 A gas mixture consists of two gases A and B. The mole fractions of the gas mixture are x_A and x_B.

(a) Calculate the weight percent c_A and c_B of the two gases in terms of x_A and x_B and the molar weights of the two gases. Use air as an example ($x_A = 0.80$ and $x_B = 0.20$).

(b) Calculate x_A and x_B if c_A and c_B are known. Use air to check the result.

(c) What are the partial pressures of A and B in terms of x_A and x_B and the total pressure p of the gas?

(d) What is the average molar weight of a gas mixture when x_A, x_B, M_A and M_B are known? Use air as an example.

(e) Calculate the density of helium at STP.

(f) Calculate the number of He molecules per cubic metre at STP.

4.5

An air balloon consists of a large silicone rubber balloon with a vent and a cover, which limits its maximum diameter to 15 m. The total weight of balloon, ropes, equipment and basket with passengers is 900 kg.

To keep the balloon and its passengers floating in the air, a combustion aggregate is used (dark box in the figure). When the air inside the balloon is heated, it rises.

On one occasion the height of the balloon above the ground was 100 m and the temperature outside the balloon was 0 °C. The balloon was kept at an excess pressure of 10% compared with the STP pressure in order to balance the surface tension and reach its maximum volume.

What minimum temperature of the air inside the balloon is required to keep the height of the balloon unchanged?

4.6 The general gas law does not hold for gases close to their condensation temperature. For nonideal gases the van der Waals equation often is a good model:

$$\left(p + \frac{n^2 a}{V^2}\right)(V - nb) = nRT$$

where n is the number of kilomoles of the gas. It is illustrated in a p–V diagram in Figure 4.17 on page 184.

(a) What is the physical significance of the critical point?
(b) An amount of gas at temperature T_4 is compressed at constant temperature. Follow the isotherm in Figure 4.17 and describe the physical process and the ratio of the liquid and gas amounts during the compression.

4.7 A nonideal gas at STP is compressed isothermally at 0 °C to one-hundredth of its original volume. Calculate the final pressure of the gas provided that the van der Waals equation is valid, and compare the result with that for an ideal gas. T_{cr} of the gas is 130 K and $p_{cr} = 40$ atm.

Hint: The relations between the constants a and b and the coordinates of the critical point can be found by taking the derivative of the pressure p in the van der Waals equation twice with respect to the volume V. The derivatives dp/dV and d^2p/dV^2 at the critical point are both zero, as it is an inflection point with a horizontal tangent to the curve (Figure 4.17 on page 184). The constants a and b can be calculated from these equations. The result is

$$b = RT_{cr}/8p_{cr} \text{ and } a = p_{cr} \times 27b^2.$$

4.8 The Lennard-Jones potential can be written as

$$E_{pot} = 4E_e\left[\left(\frac{r_0}{r}\right)^{12} - \left(\frac{r_0}{r}\right)^{6}\right]$$

where E_e is the depth of the potential well (Figure 4.14 b on page 182).

(a) The equation describes the potential energy of a system. What type of system? Give examples of such systems.
(b) List the expressions for the forces associated with this potential energy, discuss their nature and origin and explain the significance of all the quantities in the equations.
(c) Compare the energy required to separate two H_2 molecules from each other, shown in Figure 4.13 on page 182, with the dissociation energy of the H_2 molecule found in Figure 2.37 on page 75. Comment on the result.

4.9 (a) Prove that $C_p - C_V = R$ for an ideal gas.
(b) Analyse and explain Figure 4.21 on page 191, which concerns C_p/C_V for diatomic molecules as a function of temperature.

4.10 The ratio C_p/C_V has been determined for three pure transparent gases at room temperature. The result was (a) 1.42, (b) 1.32 and (c) 1.70. What types of molecules do the three gases consist of?

4.11 (a) The mean free path at STP of Ar is 64 nm (Table 4.5 on page 193). Use this information to calculate a value of the radius of the Ar atom.
(b) The mean free path at STP of hydrogen is 112 nm. Calculate the mean free path of hydrogen gas at standard temperature and pressures of 10^6 and 10^{-6} atm.

4.12 In the determination of the coefficient of thermal conduction of a gas, the equipment consisted of a straight vertical resistance filament (1.5 ohm/m) along the axis of a narrow tube, which was kept at constant temperature. When the tube was evacuated a current of 90 mA through the filament was required to keep maintain a temperature difference of 1.0 °C between the filament and the tube constant. When the gas was present in the tube, the current had to be increased to 270 mA to maintain the same temperature difference as before. Calculate the thermal conductivity of the gas at the given pressure. The inner diameter of the tube was 10 mm and the thickness of the filament was 0.50 mm.

4.13 A methane molecule, CH_4, can be regarded as spherical with a volume five times the volume of an argon atom. What are the (CH_4/Ar) ratios of

(a) the viscosity coefficients
(b) the thermal conductivities

of the two gases, when measured at the same temperature?

4.14 (a) Define the self-diffusion constant of a pure gas. How can it be measured?
(b) Prove the relationship $D = \dfrac{\eta}{\rho}$.
(c) Which are the two basic equations of diffusion?

4.15 Diffusion of isotope gases has proved to be a possible way to separate them or at least concentrate one of the components in a mixture of several isotope gases. Calculate the ratio of the diffusion constants of

(a) H_2/D_2 and $U^{235}F_6/U^{238}F_6$. Careful motivations of equations are required.
(b) Compare the results in (a).

5

Transformation Kinetics: Diffusion in Solids

5.1	Introduction	220
5.2	Thermodynamics	220
	5.2.1 First Law of Thermodynamics	220
	5.2.2 Enthalpy	221
	5.2.3 Second Law of Thermodynamics	221
	5.2.4 Entropy. Third Law of Thermodynamics	224
	5.2.5 Entropy and Probability	228
	5.2.6 Thermodynamics of Ideal and Nonideal Solutions	231
	5.2.7 Thermodynamics of Phase Transformations	232
	5.2.8 Gibbs Free Energy. Chemical Potential	235
5.3	Transformation Kinetics	236
	5.3.1 Reactions and Transformations	237
	5.3.2 Thermodynamic Condition for Equilibrium	237
	5.3.3 Stable and Metastable States	237
	5.3.4 Activation Energy	238
	5.3.5 Driving Force	239
	5.3.6 Endothermic and Exothermic Reactions and Transformations	239
	5.3.7 Thermal Energy Distribution in Particle Systems	240
5.4	Reaction Rates	242
	5.4.1 Reaction Rates of Thermally Activated Reactions and Transformations	242
	5.4.2 Definition of Reaction Rate	242
	5.4.3 Reaction Rates of Simple Reactions and Transformations	242
	5.4.4 Determination of Reaction Rates	243
5.5	Kinetics of Homogeneous Reactions in Gases	245
	5.5.1 Collision Theory of Homogeneous Chemical Reactions	245
	5.5.2 Temperature Dependence of the Rate Constant	247
	5.5.3 Rate Constant as a Function of Thermodynamic Quantities	247
	5.5.4 Driving Force and Reaction Rate of Homogeneous Chemical Reactions	248
	5.5.5 Activated Complex Theory	250
5.6	Diffusion in Solids	253
	5.6.1 Basic Theory of Diffusion. Diffusion Coefficient	254
	5.6.2 Diffusion Mechanisms	254

5.6.3 Theory of Diffusion   255
5.6.4 Diffusion in Alloys   265
Summary   276
Exercises   283

5.1 Introduction

In Chapter 4 the properties of gases were discussed. The properties of solids, which represent an extensive subject, will be discussed in Chapters 5, 6 and 7.

This chapter starts with some essential thermodynamic concepts and relationships and a general section on transformation kinetics. These sections form the basis for the subsequent treatment of diffusion in solids. This topic can be regarded as an important application of transformation kinetics.

Diffusion is one of the three transport properties of solids. The other two transport properties of solids will be discussed in Chapter 7.

Diffusion during and after solidification processes is extremely important for the material properties of alloys and semiconductors.

5.2 Thermodynamics

5.2.1 First Law of Thermodynamics

The most fundamental law of physics is the law of energy conservation or the principle of conservation of energy.

- The total energy of a closed system is conserved.

This is one of the most fundamental laws of physics. No exceptions of this law have ever been found so far.

When the law of energy conservation is applied in thermodynamics, it is called the *first law of thermodynamics* and deals with thermal energy.

If an amount of heat Q is added to a a closed system and no heat leaves the system, the added energy is used to increase the internal energy of the system and to do work, done by the system. Internal energy U^1 of a system is defined as the sum of its kinetic and potential energy:

$$U = U_{\text{kin}} + U_{\text{pot}} \tag{5.1}$$

The principle of conservation of energy, applied to a closed system, can be written as

$$Q = U + W \tag{5.2}$$

where
  Q = energy added to a closed system
  U = internal energy of the system
  W = work done by the system.

If the volume V of the system is extended by an amount $\mathrm{d}V$, the work done by the system against the external pressure p is $p\mathrm{d}V$.

Taking the derivative of Equation (5.2) gives the relationship

$$\mathrm{d}Q = \mathrm{d}U + \mathrm{d}W = \mathrm{d}U + p\mathrm{d}V \tag{5.3}$$

---

[1] Terminology: 1 kmol of all the thermodynamic quantities are usually designated by capital letters. In addition, capital letters with the subscript 'a' are used to designate thermodynamic quantities of a single atom.

5.2.2 Enthalpy

The enthalpy of a system is defined as

$$H = U + pV \tag{5.4}$$

Enthalpy is closely related to heat and is frequently used in chemistry, for example in connection with chemical reactions.
Taking the derivative of Equation (5.4) gives

$$dH = dU + pdV + Vdp \tag{5.5}$$

Using Equation (5.3), we obtain

$$dH = dQ + Vdp$$

- At constant pressure the enthalpy increase of a system is equal to the heat absorbed by the system:

$$(dH)_p = (dQ)_p \tag{5.6}$$

As an example of enthalpy change we will discuss exothermic and endothermic chemical reactions. The *molar enthalpy* of a chemical reaction is the *heat per kilomol developed in the reaction*.
If the reaction is *exothermic*, heat is *emitted* by the system and the enthalpy increase is *negative*.
If the enthalpy increase is *positive*, heat is *absorbed* by the system and the chemical reaction is said to be *endothermic*.

5.2.3 Second Law of Thermodynamics

Heat Engines

As the name thermodynamics indicates, this field of physics deals with the *dynamics of heat*. Examples are transfer of heat and transformation of heat into work and vice versa. As an introduction to the second law of thermodynamics, we will briefly discuss heat engines of two types.

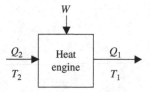

Figure 5.1 Energy balance.

In the *first* type of heat engine pumps heat is transferred from a lower to a higher temperature. It is described in Figures 5.1 and 5.2a. Work is done by a compressor on an enclosed gas at A. Its pressure and temperature increase. At B the medium is condensed. The energy Q_1 is emitted to the surroundings at the high temperature T_1. The pressure of the liquid is reduced by the expansion valve. On the low-pressure side the liquid boils and evaporates to a gas at D. The required heat is taken from the cold surroundings and the heat Q_2 is absorbed from the heat source at the low temperature T_2 (Figure 5.1).
The first law of thermodynamics gives the relationship

$$W + Q_2 = Q_1 \tag{5.7}$$

if heat losses are disregarded. The machine does *not* work unless the compressor work is supplied.
When the engine is a freezer or a refrigerator, the latter is located at the evaporator D. The interior of the refrigerator is the heat source. T_2 is the interior temperature of the refrigerator and T_1 is the surrounding room temperature (Figure 5.2a).
In the case of a heat pump D corresponds to the evaporator and a heat exchanger, coupled to a secondary loop including the evaporator and the heat source, from which heat is withdrawn. It can be the air, a lake, the ground or a 100–200 m deep

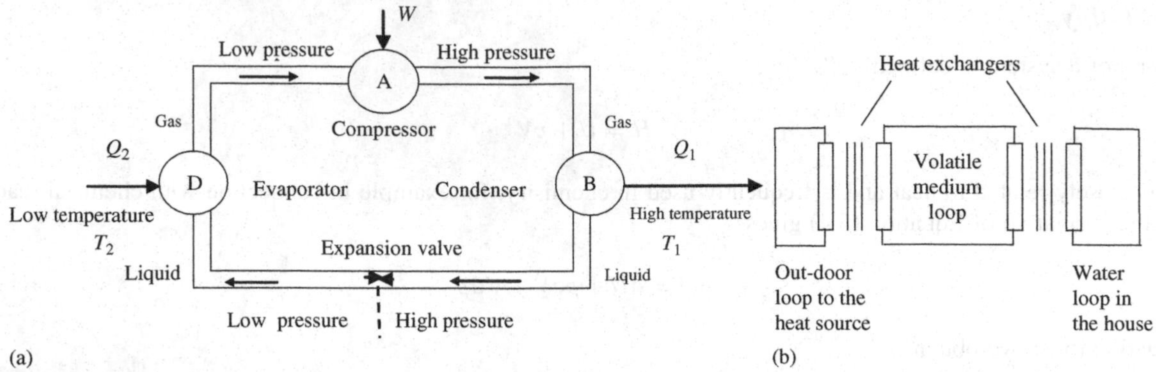

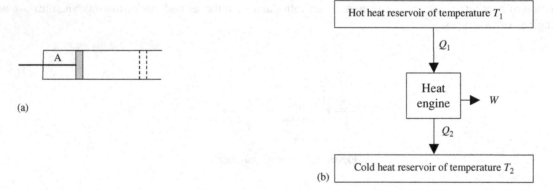

(a)   (b)

Figure 5.2 (a) The essential part of a heat pump or a refrigerator consists of a volatile medium in a closed loop. (1) At A the medium is compressed from a cold, low-pressure gas to a warm, high-pressure gas. Both the pressure and temperature of the gas increase. (2) At B heat is withdrawn from the gas by the colder surroundings. The warm gas is condensed. Its temperature decreases and the medium becomes a liquid. (3) The expansion valve reduces the pressure of the liquid. (4) Owing to the low pressure, the boiling point of the medium is low and the liquid boils in the evaporator and becomes gaseous. The heat required for the evaporation is taken from the heat source, which is cold but warmer than the boiling liquid.
(b) The medium loop and two heat exchangers are included in the white heat pump box in the house. The volatile medium may be propane or other freon- or chlorine-free substance for environmental reasons. The three loops are not drawn to scale.

hole drilled in rocky ground. The compressor A, the condenser B, the expansion valve and the evaporator D and the closed medium loop are located inside the heat pump box in the house. The medium loop is coupled to another heat exchanger at B. This secondary loop is a separate water loop, which transports heat to different parts of the house.

The *second* type of heat engine transforms heat to work. Steam engines and combustion engines, for example a car engine, belong to this type. The principle of a car engine is shown and described in Figure 5.3a and b.

Figure 5.3 (a) Combustion phase in a car engine. (b) A mixture of gaseous fuel and air is ignited by an electric spark in section A (a). Q_1 is the developed combustion heat. A minor part of Q_1 is transformed to the work W done by the combustion gases when they expand and move the piston to the dotted position. The rest Q_2 of the energy Q_1 is transferred to the surroundings.

The first law of thermodynamics, applied to the second type of heat engines, gives

$$Q_1 = W + Q_2 \tag{5.8}$$

This is identical with Equation (5.7).

Second Law of Thermodynamics

The second law of thermodynamics deals with the direction of spontaneous thermodynamic processes. There are many ways to express this law, based on experience from heat engines and everyday evidence. Some of them are listed below.

- Heat is transferred spontaneously from a warmer to a colder body, never the contrary.
- Heat can be transferred from a colder to a warmer body only if work is supplied. This is not a spontaneous process.
- Heat can be transformed spontaneously to work but never 100%.

The last statement induces the question of the efficiency of thermodynamic processes. For this reason, we will discuss the ideal Carnot cycle (Figure 5.4):

1. A gas absorbs the amount of heat Q_1 at temperature T_1 and expands isothermally, i.e. at constant temperature.
2. The gas expands adiabatically, which means that dQ is zero. The energy required for the expansion is taken from the internal energy of the gas. Its temperature decreases from T_1 to T_2.
3. The gas is compressed isothermally at temperature T_2 and emits an amount of heat Q_2 to the surroundings.
4. The gas is compressed adiabatically. Its temperature increases from T_2 back to T_1.

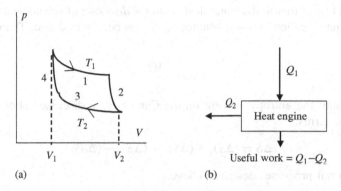

(a) (b)

Figure 5.4 (a) The Carnot cycle. The enclosed area represents the work done by the heat engine. (b) Efficiency of the Carnot cycle.

Q_1 is calculated with the aid of the ideal gas law (n kmol):

$$Q_1 = \int (dU + p\,dV) = 0 + \int_{V_1}^{V_2} \frac{nRT_1}{V} dV = nRT_1 \ln \frac{V_2}{V_1} \qquad (5.9)$$

Similarly we obtain

$$Q_2 = \int (dU + p\,dV) = \int_{V_2}^{V_1} \frac{nRT_2}{V} dV = -nRT_2 \ln \frac{V_2}{V_1} \qquad (5.10)$$

The minus sign in Equation (5.10) indicates that Q_2 is emitted and not absorbed. Equations (5.9) and (5.10) are divided, which gives

$$\left| \frac{Q_1}{Q_2} \right| = \frac{T_1}{T_2} \qquad (5.11)$$

The efficiency η of a thermodynamic process is defined as

$$\eta = \frac{\text{useful energy or work done by the system}}{\text{delivered energy to the system}} \qquad (5.12)$$

The efficiency η of the Carnot cycle will be

$$\eta = \frac{Q_1 - Q_2}{Q_1} = \frac{T_1 - T_2}{T_1} \qquad (5.13)$$

where
  η = efficiency of the Carnot engine
  Q_1 = combustion heat
  T_1 = combustion temperature
  Q_2 = heat absorbed by the surroundings
  T_2 = temperature of the surroundings.

The definition of η and Equation (5.11) can also be applied to other thermodynamic processes.

Owing to heat losses to the surroundings, friction and other reasons, real engines always have lower efficiencies than the Carnot engine.

5.2.4 Entropy. Third Law of Thermodynamics

Entropy Change in Reversible Processes

As mentioned above, the second law of thermodynamics deals with *the direction of spontaneous processes*. In order to describe the direction of a process, a quantity called *entropy*, denoted by S, has been introduced. The change of entropy of a process is defined by the relationship

$$dS = \frac{dQ}{T} \tag{5.14}$$

As an application we will apply the entropy concept on the Carnot cycle discussed above. The total change of entropy during a cycle is the sum of four terms:

$$\Delta S = (\Delta S)_1 + (\Delta S)_2 + (\Delta S)_3 + (\Delta S)_4$$

each of them referring to the partial processes described above.

$$\Delta S = \frac{Q_1}{T_1} + \int \frac{dQ}{T} + \frac{-Q_2}{T_2} + \int \frac{dQ}{T} = \frac{Q_1}{T_1} + 0 + \frac{-Q_2}{T_2} + 0$$

The second and fourth partial processes are adiabatic, which means that $dQ = 0$ and the value of the integrals is zero. Hence the total entropy change of each cycle is

$$\Delta S = \frac{Q_1}{T_1} - \frac{Q_2}{T_2} \tag{5.15}$$

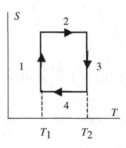

Figure 5.5 The Carnot cycle in a T–S diagram.

The first term in Equation (5.15) represents the entropy change at the isothermal expansion at temperature T_1. The entropy change is positive, which means that the process occurs spontaneously. The second term in Equation (5.15) is negative. This part of the cycle is not spontaneous: work is done on the gas to compress it.

The Carnot cycle can be represented in a T–S diagram. The isothermal expansion 1 and compression 3 are represented by vertical lines. The horizontal lines 2 and 4 illustrate the adiabatic steps of the cycle.

The Carnot cycle is reversible, which means that the gas is in equilibrium at every stage and the process can proceed in either direction. In all reversible processes the entropy will be the same when the system returns to the starting point. Hence the total entropy change will be zero and Equation (5.15) becomes be identical with Equation (5.11).

Entropy Change in Irreversible Processes

Below we will consider two examples of isothermal, irreversible processes and calculate the entropy change. In both cases the process in question is expansion of ideal gases.

Example 5.1

Calculate the change in entropy when n_A kmol of an ideal gas A with pressure p and volume V_1 expands irreversibly to the volume $V_1 + V_2$ in the way shown in the figure. The final pressure is p_A.

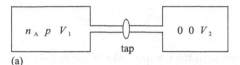

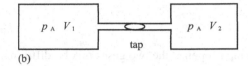

Solution:

There are no forces between the molecules in an ideal gas and therefore there is no change in internal energy when the gas expands.

When the tap shown in figure (a) is opened, the gas expands from volume V_1 to $V_1 + V_2$ and the pressure p becomes the same in the whole volume.

The first law of thermodynamics and the definition of entropy give the entropy change of the gas:

$$\Delta S = \int \frac{dQ}{T} = \int \frac{0 + p\,dV}{T} = \int_{V_1}^{V_1+V_2} \frac{p\,dV}{T} \tag{1'}$$

Using the ideal gas law $pV = nRT$ to eliminate p/T, we obtain

$$\Delta S = \int_{V_1}^{V_1+V_2} \frac{nR\,dV}{V} = nR\ln\frac{V_1 + V_2}{V_1} \tag{2'}$$

Boyles's law gives

$$pV_1 = p_A(V_1 + V_2) \qquad (or) \qquad \frac{V_1 + V_2}{V_1} = \frac{p}{p_A} \tag{3'}$$

Answer:

The entropy increase is $n_A R \ln \dfrac{p}{p_A}$.

As a second example of deriving entropy changes, we will calculate the entropy change when two gases mix.

Example 5.2

Two ideal gases A and B of equal pressures, each in a separate closed container, are connected by a short tube and a closed tap. When the tap is opened the two gases mix irreversibly. No changes in pressure and temperature are observed.

Calculate the total change of entropy as a function of the initial pressure p and the final partial pressures when the two gases mix. The data for the gases are given in the figure. n_A and n_B are the number of kmol of the gases A and B.

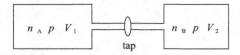

Solution:

When the tap is opened the two gases mix by diffusion. The diffusion goes on until the composition of the gas is homogeneous. It is far more likely that the gase will mix by diffusion than remain separate. Therefore, the total entropy change is expected to be positive.

In a gas, the distances between the molecules are large and the interaction between them can be neglected. Consequently, the diffusion of each gas is independent of the other. The total entropy change can be regarded as the sum of the entropy change of each gas after its separate diffusion from one container into the other:

$$\Delta S_{mix} = \Delta S_A + \Delta S_B \tag{1'}$$

Gas A changes its pressure from p to p_A, where p_A is its final partial pressure. In the same way, gas B changes its pressure from p to p_B.

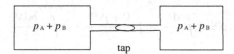

The initial pressure and the final total pressure are equal as no pressure change is observed:

$$p = p_A + p_B \tag{2'}$$

Using the result in Example 5.1, we obtain

$$\Delta S_A = n_A R \ln \frac{p}{p_A} \tag{3'}$$

and

$$\Delta S_B = n_B R \ln \frac{p}{p_B} \tag{4'}$$

The total entropy change is

$$\Delta S_{mix} = \Delta S_A + \Delta S_B = n_A R \ln \frac{p}{p_A} + n_B R \ln \frac{p}{p_B} \tag{5'}$$

The ratio of the pressures is >1 and the entropy change is therefore positive, as predicted above.

Answer:

The total entropy change when the gases mix is $\qquad n_A R \ln \dfrac{p_A + p_B}{p_A} + n_B R \ln \dfrac{p_A + p_B}{p_B}$.

Comparison of the Entropy Changes in Reversible and Irreversible Processes

On page 224, we found that the total entropy change of the ideal reversible Carnot cycle is zero. The entropy changes in the irreversible processes in Examples 5.1 and 5.2 were found to be positive in both cases. Experience shows that the results concerning reversible and irreversible processes are true in all other cases also. The following statements are generally valid:

- If a process is reversible, the entropy change is zero: $\Delta S = 0$.
- The entropy increases in all irreversible spontaneous processes: $\Delta S > 0$.

The final states in Examples 5.1 and 5.2 are far more likely than the initial states. When the tap in Example 5.1 is opened, the molecules move into the empty container until the pressures in the two containers are equal rather than that no change at all occurs. In Example 5.2, the molecules of each gas spontaneously distribute in such a way that the final partial pressure becomes the same in both containers.

In the case of reversible processes, both directions of the process are equally probable at equilibrium.

In both cases, the system changes spontaneously from one state to another more likely state and the entropy increases. Entropy seems to be connected with probability in one way or an other. This topic will be discussed on page 228.

Third Law of Thermodynamics

So far we have only dealt with entropy changes ΔS and not with the absolute value of entropy S. To obtain an absolute entropy scale we start with a fixed value S_0 of the entropy of a pure crystalline substance at absolute zero temperature.

To find the entropy of the substance at an arbitrary temperature T we will calculate the entropy change by starting with S_0 and integrating the definition equation

$$dS = \frac{dQ}{T} \tag{5.14}$$

We consider a reversible process of the substance at constant pressure and integrate Equation (5.14) from temperature $0\,K$ to T. The corresponding integration limits are S_0 and S.

$$\int_{S_0}^{S} dS = \int_{0}^{T} \frac{dQ}{T} = \int_{0}^{T} \frac{n_0 C_p dT}{T} \tag{5.16}$$

where

n_0 = number of kilomol of pure crystalline substance
S = entropy of the system (n_0 kmol of the substance)
S_0 = entropy of the system at absolute zero temperature
C_p = molar heat capacity of the element at constant pressure
T = absolute temperature.

C_p varies strongly with temperature at low temperatures (Chapter 6) and approaches zero at $T = 0\,K$ and the integral can be written as

$$S = S_0 + \int_{0}^{T} \frac{n_0 C_p dT}{T} \tag{5.17}$$

The constant S_0 is the zero point of the entropy scale. The German chemist Nernst made extensive studies to determine S_0 for pure elements. His result is called *Nernst's theorem*.

- The entropy at absolute zero temperature of a pure crystalline substance is zero.

Nernst's theorem was regarded as so important that it is called the *third law of thermodynamics*.

As it is impossible to reach the absolute zero point, it is not possible to verify Nernst's theorem by direct measurements. However, entropy values calculated from measurements of heat capacities at constant pressure at very low temperatures agree well with entropy values based on theoretical calculations (statistical mechanics, Section 5.2.5) if the value of S_0 is zero.

The entropy of any closed system can be built up by starting with pure crystalline substances and Equation (5.17) and adding entropy contributions due to mixing of substances.

5.2.5 Entropy and Probability

The discussion on page 227 indicates that the entropy change of a process is related in some way to its probability. The probability function can be found by the following arguments.

Consider N molecules in a container of volume V. The molecules do not interact at all; each molecule is free to move within the volume V and the probability of finding it within a unit volume is the same everywhere. Hence the probability of finding a molecule within a volume V_1 is V_1/V. The probability of finding two molecules within the same volume V_1 equals the product of their probabilities $(V_1/V)^2$. The probability of finding N molecules within a particular volume V_1 is $(V_1/V)^N$.

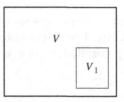

Figure 5.6 Volume element.

Equation (2') in Example 5.1 on page 225 and Equation (5') in Example 5.2 on page 226 give us the clue to relating entropy and probability. We have seen above that the overall probability is the *product* of the probabilities of independent events (here particle distributions). We also know that the total entropy change of partial systems equals the *sum* of their entropy changes. It is striking that the logarithmic function converts the multiplicative property of probability to the additive property of entropy.

These arguments led the Austrian physicist Ludwig Boltzmann in 1877 to suggest a relationship between entropy and probability. The entropy S was assumed to be a function of the probability of the system:

$$S = F(P) \tag{5.18}$$

The function $F(P)$ can be derived in the following way. Consider two partial systems A and B at the same temperature and pressure. The two systems are joined together to a single system AB. As temperature and pressure are constant, the entropy of the united system equals the sum of the entropies of the partial systems:

$$S_A + S_B = S_{AB} \tag{5.19}$$

which can be written with the aid of Equation (5.18) as

$$F(P_A) + F(P_B) = F(P_{AB}) \tag{5.20}$$

Assume that the most probable distributions of two partial systems A and B are P_A and P_B. The most probable distribution of the united system must be the product of P_A and P_B:

$$P_{AB} = P_A P_B \tag{5.21}$$

This expression is introduced into Equation (5.20):

$$F(P_A) + F(P_B) = F(P_A P_B) \tag{5.22}$$

By taking the derivative of Equation (5.22) with respect to each of the two independent variables P_A and P_B, we obtain two relationships which after multiplication with P_A and P_B, respectively, are found to be equal:

$$P_A \frac{dF(P_A)}{dP_A} = P_B \frac{dF(P_B)}{dP_B} \left(= P_A P_B \frac{dF(P_A P_B)}{dP_A P_B} \right) \tag{5.23}$$

Because P_A and P_B are independent of each other, the expressions in Equation (5.23) must be constant:

$$P_A \frac{dF(P_A)}{dP_A} = constant \quad \text{or} \quad dF(P_A) = constant \times \frac{dP_A}{P_A} \tag{5.24}$$

Hence, we have for an arbitrary system

$$dF(P) = constant \times \frac{dP}{P} \tag{5.25}$$

The function $F(P)$ can be replaced by S in Equation (5.25), which gives

$$dS = constant \times \frac{dP}{P} \tag{5.26}$$

Equation (5.26) is integrated and the solution can be written as

$$S = S_0 + k_B \ln P \tag{5.27}$$

where
  S = entropy of the system
  S_0 = integration constant
  k_B = Boltzmann's constant
  P = probability of the system.

Equation (5.27) is the fundamental relationship between entropy and probability.

The integration constant S_0 can be determined with the aid of Nernst's theorem. The entropy S of a pure crystalline substance is zero at absolute zero temperature. At this temperature the system is in its ground state and the probability $P = 1$. If we insert these values of S and P into Equation (5.27), we obtain $S_0 = 0$. Hence the entropy of a pure crystalline substance can be written as

$$S = k_B \ln P \tag{5.28}$$

Boltzmann interpreted entropy as a measure of the order, or rather disorder, of a system. The more probable a state of a system is and the greater its disorder, the higher will be its entropy.

Entropy Change on Mixing Two Components

As a test of the probability expression of entropy, we will use Equation (5.28) for calculation of the entropy change when two components are mixed.

Consider a binary system, a liquid or a solid, of two components B and C. N_B atoms of B are arranged at random among $N = N_B + N_C$ sites. This can be done in many different ways and is equivalent to mixing the two components. The probability P is defined as the number of independent alternative ways of arranging the B and C atoms among the N sites which gives the same energy of the system. Statistical considerations give the result

$$P = \frac{N!}{N_B! N_C!} = \frac{N!}{N_B!(N - N_B)!} \tag{5.29}$$

We use Stirling's equation:

$$\lim_{N \to \infty} N! \to \sqrt{2\pi} N^{N+\frac{1}{2}} e^{-N}$$

for the very large numbers N, N_B and N_C and obtain

$$P = \frac{N^{N+\frac{1}{2}}}{\sqrt{2\pi} N_B^{N_B+\frac{1}{2}} (N - N_B)^{N - N_B + \frac{1}{2}}} \quad \frac{e^{-N}}{e^{-N_B} e^{-(N - N_B)}} \tag{5.30}$$

The last factor in Equation (5.30) is equal to 1. The term $\frac{1}{2}$ can be neglected in comparison with N and N_B in the exponents of the first factor. Taking the logarithm of both sides of Equation (5.30), we obtain

$$\ln P = N \ln N - N_B \ln N_B - (N - N_B) \ln (N - N_B) - \ln \sqrt{2\pi} \tag{5.31}$$

The last term can be neglected in comparison with the others. If we introduce the mole fractions

$$x_B = \frac{N_B}{N} \text{ and } x_C = \frac{N - N_B}{N}$$

and the relationship $x_B + x_C = 1$, Equation (5.31) can, after some calculation, be transformed into

$$\ln P = N \left(-x_B \ln x_B - x_C \ln x_C \right) \tag{5.32}$$

Instead of N we introduce $n =$ the number of kilomol in the container:

$$N = n N_A \tag{5.33}$$

where N_A is Avogadro's number, i.e. the number of atoms or molecules in 1 kmol. Hence we obtain

$$\ln P = n N_A \left(-x_B \ln x_B - x_C \ln x_C \right) \tag{5.34}$$

Instead of N_A we introduce R/k_B, where k_B is Boltzmann's constant:

$$k_B \ln P = n R \left(-x_B \ln x_B - x_C \ln x_C \right) \tag{5.35}$$

According to Equation (5.28), $k_B \ln P$ equals the entropy change ΔS_{mix} when the two components B and C mix:

$$\Delta S_{\text{mix}} = -n R \left(x_B \ln x_B + x_C \ln x_C \right) \tag{5.36}$$

Equation (5.36) is valid for homogeneously mixed solids and liquids. It becomes identical with the answer in Example 5.2 on page 226 if the partial pressures p_B and p_C are replaced by the mole fractions x_B and x_C (this has been done on page 232). The derivation of Equation (5.36) is based on $S = k_B \ln P$ whereas Example 5.2 is deduced from the thermodynamic definition of entropy, $dS = dQ/T$.

Hence ΔS_{mix} has been calculated in two entirely different and independent ways, which give identical results. The conclusion must be that the concept of entropy as a function of probability is in complete agreement with the classical definition of entropy and does not lead to any contradictions.

In fact, Equation (5.28) can alternatively be used as the definition of entropy and the old definition of entropy can be derived from the new definition with the aid of the Maxwell–Boltzmann distribution law.

5.2.6 Thermodynamics of Ideal and Nonideal Solutions

Ideal and Nonideal Solutions

An *ideal solution* is defined as a solution in which

- The homogeneous attractive forces (A–A and B–B) and the heterogeneous attractive forces (A–B) are equal.
- The heat of mixing is zero. The solubility is complete.

Solutions which do not fulfil the conditions above are *nonideal*. Nonideal solutions can deviate from ideal solutions in two different ways.

When the forces between the A and B atoms are *stronger* than the forces between A–A atoms and B–B atoms, the heat of mixing is *negative*. Heat is *released* and given to the surroundings when the two components mix or the temperature of the system *increases*.

When the forces between the A and B atoms are *weaker* than the forces between A–A atoms and B–B atoms, the heat of mixing is *positive*. The solution process to break strong A–A and B–B bonds and replace them with weaker A–B bonds requires additional energy. Heat is *consumed* and taken from the surroundings when the two components mix or the temperature of the system decreases.

Enthalpy Change on Mixing of Two Components. Heat of Mixing of Ideal and Nonideal Solutions

Consider a system of two components, for example a binary alloy with randomly distributed atoms. If the attractive forces between the atoms are known, one can easily calculate the change in enthalpy ($-\Delta H_{mix}$) or the *molar heat of mixing* ($-\Delta H_{mix}$) released when the components A and B mix:

$$-\Delta H_{mix} = N_{AB} \left[E_{AB} - \frac{1}{2} \left(E_{AA} + E_{BB} \right) \right] \tag{5.37}$$

where
  N_{AB} = the number of atom pairs AB
  E_{ij} = bonding energy between specified types of atoms.

The number of mixed atom pairs N_{AB} is proportional to the presence of the two types of atoms, i.e. the concentrations of A and B. The heat of mixing $-\Delta H_{mix}$, based on the forces between neighbouring atoms, can then be written as

$$-\Delta H_{mix} = L_{mix} x_A x_B \tag{5.38}$$

where L_{mix} is a constant, which can be derived by identification of Equations (5.37) and (5.38). L_{mix} includes both the proportionality constant and the brackets factor, as the binding energies of the three molecule types are constant.

For *ideal solutions*, L_{mix} and the heat of mixing are zero. The components A and B mix at all proportions:

$$-\Delta H_{mix}^{ideal} = 0 \tag{5.39}$$

For *nonideal solutions* the constant L_{mix} and hence the heat of mixing $-\Delta H_{mix}$ can be either positive or negative. The sign of L_{mix} depends on the interatomic forces and the sign of the brackets factor in Equation (5.37).

The enthalpy of mixing of a nonideal solution can be defined as an excess quantity, i.e. the excess enthalpy of mixing:

$$-\Delta H_{mix}^{nonideal} = H_{mix}^{Ex} \tag{5.40}$$

The excess enthalpy is simply the heat of mixing $-\Delta H_{mix}$ of the nonideal solution. The excess enthalpy of mixing is zero for an ideal solution.

Entropy Change on Mixing Two Components. Entropy Change of Ideal and Nonideal Solutions

In Example 5.2 on page 226 we found an expression for the entropy change when two different gases mix by diffusion. Diffusion occurs not only in gases but also in liquids and solids. The entropy change ΔS_{mix} due to mixing of two compounds in a melt or a solid to give an *ideal solution* can be calculated if we make a minor modification of Equation (5′) in Example 5.2 on page 226. Instead of the partial pressures of the two gases, we introduce the mole fractions x_A and x_B:

$$x_A = \frac{p_A}{p_A + p_B} \text{ and } x_B = \frac{p_A}{p_A + p_B}$$

If we introduce the mole fractions, Equation (5′) on page 226 can be written as

$$\Delta S_{mix} = -n_A R \ln x_A - n_B R \ln x_B \tag{5.41}$$

or using the relationship $n = n_A + n_B$, where n is the total number of kilomol:

$$\Delta S_{mix} = -nR(x_A \ln x_A + x_B \ln x_B) \tag{5.42}$$

In agreement with Equation (5.36), Equation (5.42) is directly applicable to mixtures of gases but also to liquids and solids in general, which form ideal solutions. Such applications will be discussed later in this chapter.

In analogy with the enthalpy above, we define the *excess entropy* of a nonideal solution as

$$S_{mix}^{Ex} = S_{mix}^{nonideal} - S_{mix}^{ideal} \tag{5.43}$$

S_{mix}^{ideal} is equal to the entropy change when two pure components, both with $S = 0$, are mixed. According to Equation (5.42), the entropy change equals ΔS_{mix} and $S_{mix}^{ideal} = \Delta S_{mix}$. For an ideal solution, the excess entropy S_{mix}^{Ex} is zero.

The *entropy of an ideal solid solution* consists of two terms, one due to lattice vibrations and the other originating from the binding energy of the component atoms.

The vibrational entropy of an ideal solution is approximately the same as that of pure components and does not contribute to the excess entropy.

For *nonideal solutions*, the entropy can be regarded as the sum of two terms, the entropy of the ideal solution plus the excess entropy:

$$S_{mix}^{nonideal} = S_{mix}^{ideal} + S_{mix}^{Ex} = \Delta S_{mix} + S_{mix}^{Ex} \tag{5.44}$$

where ΔS_{mix} is defined by Equation (5.42).

5.2.7 Thermodynamics of Phase Transformations

Phase Transformations

Figure 5.7 shows a survey of the phase transformations in a pure substance.

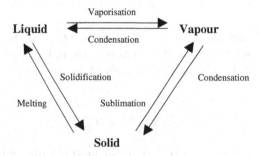

Figure 5.7 Survey of phase transformations in a pure substance.

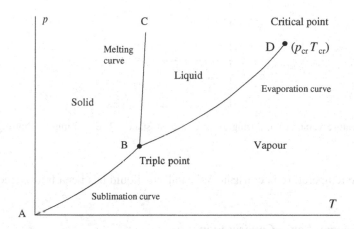

Figure 5.8 Phase diagram of a pure substance.

Figure 5.8 shows the phase diagram of a pure substance. The three areas represent solid phase, liquid phase and vapour phase. The curve AB is called the sublimation curve. At the pressures and temperatures along the curve, solid phase and vapour phase exist in equilibrium with each other, otherwise either as a solid or a vapour. Along the sublimation curve, solid phase can be transformed into vapour or vice versa.

The point B is called the triple point. It represents the only temperature and pressure where solid phase, liquid phase and vapour phase can exist at equilibrium with each other. It is an important fixed point of a substance. The triple point of water (273.16 K) is used for definition of the thermodynamic temperature scale.

The steep curve BC is the melting curve. It represents the temperatures and pressures where the solid and liquid phases are in equilibrium with each other. Very high pressure changes are required for changing the melting point of the substance. Normally the slope of the curve is positive but some substances, for example the melting curve of water, have a negative derivative. In the latter cases the melting point decreases with increasing pressure.

The curve BD is the evaporation curve. At the pressures and temperatures along the curve, liquid phase and vapour phase exist in equilibrium with each other. The slope of the curve is not as steep as that of the melting curve. For this reason, the boiling point of the liquid varies strongly with the external pressure. At low pressures, for example at high altitudes, water boils at temperatures considerably below 100 °C.

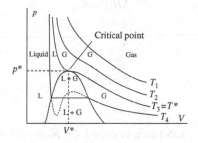

Figure 5.9 Isotherms of a van der Waals gas. Reproduced with permission from *Fysik I*, © O. Beckman.

At point D in Figure 5.8, the evaporation curve stops abruptly. Point D corresponds to the critical point, where there is no difference between the liquid and the vapour. Above the critical temperature, the vapour cannot be compressed to a liquid, even at extremely high pressures (Figure 5.9). This topic has been discussed in connection with real gases on pages 183–184 in Chapter 4.

Enthalpy and Entropy Changes in Phase Transformations

When heat is added continuously to a solid body, its temperature increases until the melting point is reached (Figure 5.10). During the melting time the temperature remains constant. The temperature of the liquid rises until the boiling point is

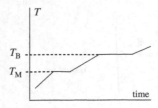

Figure 5.10 Temperature versus time during heating of a substance. The melting and boiling points are marked.

reached. During the boiling, the temperature is constant. When all the liquid has been transferred into vapour, the temperature increases again.

Melting – Solidification and Evaporation – Condensation

At phase transformations the system is in equilibrium and the temperature and pressure are constant. Heat is transferred reversibly and isothermally between the system and the surroundings.

At constant pressure, the heat *absorbed* by a system equals its enthalpy *increase* [Equation (5.6) on page 221]. Phase changes cause changes in the molecular order of the substance and are therefore expected to lead to changes in entropy. As the temperature is constant during the transformation, the entropy change will be

$$\Delta S_{tr} = \frac{\Delta H_{tr}}{T_{tr}} \tag{5.45}$$

On *melting* and *evaporation* the system *absorbs* heat and the phase transition is endothermic. In this case the enthalpy change is *positive*. The system changes from an ordered state to a more disordered state (solid to liquid and liquid to vapour, respectively). Heat has to be added to the system and its disorder increases. The entropy change is *positive*.

In the case of *solidification* and *condensation*, the opposite is true. The phase transition is exothermic. Heat is *emitted* by the system to the surroundings and the enthalpy change is *negative*. The system changes from a disordered to a more ordered state (liquid to solid and vapour to liquid, respectively). The entropy change is *negative*.

The heat required to melt 1 kmol of a substance is called the *molar enthalpy of melting*, L_M (M = melting). The same amount of heat is released when the substance solidifies:

$$\Delta H_M = L_M \tag{5.46}$$

The molar entropy of melting is

$$\Delta S_M = \frac{L_M}{T_M} \tag{5.47}$$

where T_M is the melting point.

The heat required to evaporate 1 kmol of a substance is called the *molar enthalpy of evaporation*, L_B (B = boiling). The same amount of heat is released when the substance condenses:

$$\Delta H_B = L_B \tag{5.48}$$

The molar entropy of evaporation is

$$\Delta S_B = \frac{L_B}{T_B} \tag{5.49}$$

where T_B is the boiling point.

A survey of the transformations is given in Table 5.1. The phase transformations are accompanied by discontinuous changes of volume and entropy (Figure 5.11).

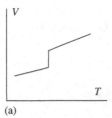 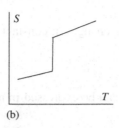

(a)   (b)

Figure 5.11   (a) Volume as a function of temperature. (b) Entropy as a function of temperature.

Table 5.1   Survey of phase transformations.

Process	Molar enthalpy change ΔH	Molar entropy change
Melting	$\Delta H_M = L_M$	$\Delta S_M = L_M/T_M$
Solidification	$\Delta H_s = -L_M$	$\Delta S_s = -L_M/T_M$
Evaporation	$\Delta H_v = L_B$	$\Delta S_v = L_B/T_B$
Condensation	$\Delta H_c = -L_B$	$\Delta S_c = -L_B/T_B$

5.2.8 Gibbs Free Energy. Chemical Potential

Consider a system in thermal equilibrium with its surroundings at temperature T. An irreversible spontaneous change of state of the system always leads to an increase in the entropy of the system. This can be expressed mathematically by Clausius's inequality:

$$dS - \frac{dQ_{\text{irreversible}}}{T} > 0 \tag{5.50}$$

If the process is *reversible*, the inequality is replaced by an equality:

$$dS - \frac{dQ_{\text{reversible}}}{T} = 0 \tag{5.51}$$

Heat transfer at constant pressure is of special interest in chemistry and metallurgy. If there is no work other than expansion work, $dQ = dH$ (page 221) and Equations (5.50) and (5.51) can be summarized as

$$dS - \frac{dQ}{T} \geq 0$$

which can be written as

$$TdS - dH \geq 0 \tag{5.52}$$

Equation (5.52) is valid if either of the following two conditions is fulfilled:

1. The enthapy of the system is constant:

$$dS_{H,p} \geq 0 \tag{5.53}$$

2. The entropy of the system is constant:

$$dH_{S,p} \leq 0 \tag{5.54}$$

Conditions 1 and 2 can be expressed in a simpler and more understandable way by introduction of another thermodynamic function, the *molar Gibbs free energy*, defined as

$$G = H - TS \tag{5.55}$$

When the state of the system changes at constant temperature, the change in G is given by

$$dG = dH - TdS \tag{5.56}$$

Equations (5.53) and (5.54) can be expressed in terms of the Gibbs free energy as

$$dG_{T,p} \leq 0 \tag{5.57}$$

At equilibrium G has a minimum and $dG_{T,p} = 0$. The change in the thermodynamic function G can be regarded as driving force (page 239) in chemical and metallurgical reactions. The general rule is

• At constant temperature and pressure processes always occur spontaneously in the direction of decreasing G.

The Gibbs free energy is a most useful device for studying the driving forces of various processes, for example chemical reactions, solution processes, melting/solidification and evaporation/condensation processes.

Chemical Potential

Chemical potential is a very useful and important concept, widely used in chemistry. It is closely related to the Gibbs free energy. It will be defined below.

Consider a binary solution of two elements A and B. The solution is characterized by its composition, given in mole fractions x_A and x_B. The chemical potential of a pure element A is defined as the Gibbs free energy of the element:

$$\mu_A^0 = G_A^0 \tag{5.58}$$

The chemical potential of A varies with the concentration of A in the solution as

$$\mu_A = \mu_A^0 + RT \ln(x_A \gamma_A) \tag{5.59}$$

where γ_A is the activity coefficient. If the mole fraction x_A is small, $\gamma_A \approx 1$, otherwise <1. The last term in Equation (5.59) can be split into two terms. One of them, $RT \ln\gamma_A$, describes the deviation of the solution from an ideal solution.

Hildebrand proposed in 1929 the so-called *regular solution model*, which means that

$$RT \ln \gamma = L_{mix} x_B^2 \tag{5.60}$$

where L_{mix} is given by Equation (5.38) on page 231.

Chemical potential will be used in connection with diffusion in binary alloys later in this chapter.

5.3 Transformation Kinetics

One of the most important aims of material science is to develop materials with suitable properties for various purposes. By using various types of production methods, alloying and heat treatment are applied to cause changes in the structure of the materials to promote the desired properties. Changes in the microstructure may, for example, concern composition, crystal structure and grain size.

The structure changes occur as a result of rearrangements of atoms in the material, for example phase changes, diffusion and chemical reactions. However, very few products are thermodynamically stable. It is necessary to examine the stability of the new materials to make sure that they retain their optimal properties.

Hence it is of great interest to study the thermodynamic conditions of transformations, the transformation rates and the stability of the desired structure of the material. These topics will be discussed in Sections 5.3, 5.4 and 5.5, mainly in terms of chemical reactions. The derived equations and other statements are valid also for other kinds of transformations. The introductory general parts are followed by some specific applications to gases.

5.3.1 Reactions and Transformations

Chemical reactions can be either homogeneous or heterogeneous. *Homogeneous reactions* occur, as the name indicates, at the same time over a volume of a single phase. This is the case when two gas components react with each other, giving a third gas component or several gas components.

Heterogeneous reactions occur when two phases are involved. The reaction occurs at the interface between the two phases. The reaction rate is influenced by the transport rate of the reactants to the interface.

In metallurgy, most reactions are heterogeneous. However, the theory of heterogeneous reactions is developed from the theory of homogeneous reactions. Hence it is natural to start with a short review of homogeneous reactions. The sections include the theory of homogeneous reactions and applications on chemical reactions in gases.

5.3.2 Thermodynamic Condition for Equilibrium

Most transformations occur at constant temperature and pressure. On page 236 we found that:

- At constant temperature and pressure processes always occur spontaneously in the direction of decreasing Gibbs free energy.

Or in other words:

- A system is in equilibrium when the Gibbs free energy of the system is a minimum.

The equilibrium condition can also be written as

$$dG_{T,p} = 0 \qquad (5.61)$$

where dG is the change in Gibbs free energy as a result of an infinitesimal change in the system. If we differentiate the definition of Gibbs free energy $G = H - TS$ we obtain

$$dG = dH - TdS \qquad (5.62)$$

where

$$dH = dU + pdV \qquad (5.63)$$

In liquids and solids the change in volume associated with a change in the system is normally very small and can be neglected. Then $dH \approx dU$ and Equation (5.62) can be replaced by

$$dG \approx dU - TdS \qquad (5.64)$$

5.3.3 Stable and Metastable States

There are two types of equilibrium. They are best understood by a mechanical analogy.

Consider the block in Figure 5.12. Figure 5.12a represents an *unstable* state of the system. The slightest deviation means that the system does not return to its former position.

In Figure 5.12b, the centre of gravity of the block has its lowest possible position. The system is in a *stable* state. After a slight deviation in position the system goes back to its original position spontaneously.

In Figure 5.12c, the centre of gravity is higher than in Figure 5.12b. An infinitesimal deviation makes the system return to its former position. If the centre of mass is moved to its highest position, it may turn over into the stable state in Figure 5.12b. Figure 5.12c does not represent a true stable state but is locally stable. Such a state is called *metastable*. It is necessary to add energy to a metastable state to transform it into a stable state.

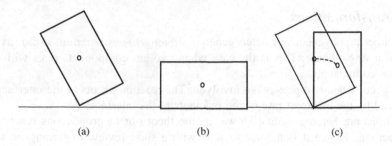

Figure 5.12 Mechanical equilibria. (a) Unstable state; (b) stable state; (c) metastable state. Reproduced from J. W. Martin and R. D. Doherty, *Stability of Microstructures in Metallic Systems*, 1st edn. © 1976 Cambridge University Press.

Example 5.3

A block with the dimensions $10 \times 10 \times 20$ cm is transferred from position c in Figure 5.12 to position b. Its mass is 16 kg. The gravity constant $g \approx 10 \, \text{m/s}^2$. Calculate

(a) the difference in potential energy between the initial state and the final state
(b) the energy which has to be added to transfer the block from position c to position b.

Solution:

(a) The energy difference equals the change in potential energy of the centre of mass of the block:

$$U_{\text{pot i}} - U_{\text{pot f}} = mg(h_i - h_f) = 16 \times 10 \times \left(\frac{0.20}{2} - \frac{0.10}{2} \right) = 8.0 \, \text{J}$$

(b) It is necessary to add enough energy to move the block into its highest position (Figure 5.12c) before it can spontaneously move into its stable state.

$$\Delta U = mg(h_{\text{max}} - h_i) = 16 \times 10 \times \left(\frac{0.10\sqrt{5}}{2} - 0.10 \right) = 1.9 \, \text{J}$$

Answer:

(a) The difference in energy is 8 J.
(b) The energy barrier that has to be overcome is ~ 2 J.

The additional energy which has to be added to a metastable state to transform it into a stable state is called *activation energy*.

5.3.4 Activation Energy

Metastable states are very common in material science. Energy is required to bring an atom from one site to another in the case of diffusion, for instance. In chemical reactions the bonds existing between the atoms must be broken before new ones can be formed.

In general, there exists an energy barrier which has to be overcome before a transformation of a system can occur, in analogy with Figure 5.12 and Example 5.3. This is shown in Figure 5.13.

Initially the system is in a metastable state and has the Gibbs free energy G_i. In the final stable state the free energy is G_f. A transformation of the system from the metastable to the stable state requires addition of energy to overcome the energy barrier and 'lift' the system into the *activated state* or *transition state*.

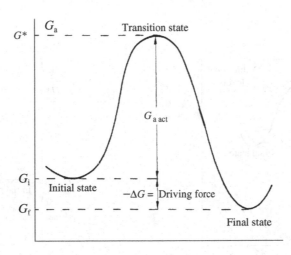

Figure 5.13 Variation of Gibbs free energy $G_{a\ act}$ during a transformation of a system. The reaction coordinate can be anything which defines the progress of the transformation, i.e. time or concentration of a component. Reproduced with permission from J. Burke, *The Kinetics of Phase Transformations in Metals.* © 1965 Pergamon Press (now with Elsevier).

The necessary additional energy is called the *activation energy*. It is defined as the height of the energy barrier and is denoted $G_{a\ act}$,[2] where the subscript 'a' refers to 'atom'.

The activation energy represents an energy difference and could be denoted $\Delta G_{a\ act}$. However, we restrict the use of Δ to energy differences between final and initial states in order to avoid confusion.

5.3.5 Driving Force

The total change in free energy associated with the transformation is

$$\Delta G = \int_{\text{initial}}^{\text{final}} \mathrm{d}G = G_f - G_i \qquad (5.65)$$

ΔG is always *negative* for a spontaneous process.

As a measure of the probability of the transformation, the concept of *driving force* is used. It is defined as

$$\text{Driving force} = -\Delta G = -\int_{\text{initial}}^{\text{final}} \mathrm{d}G = -(G_f - G_i) \qquad (5.66)$$

This term is not very suitable as the driving force has the dimension of energy, but it is so established and generally accepted that it is difficult to alter.

The driving force of a spontaneous process is always a *positive* quantity. The larger the driving force is, the more likely will be the transformation.

5.3.6 Endothermic and Exothermic Reactions and Transformations

Figures 5.14 and 5.15 show the variation of the internal energy U during a reaction or transformation. They appear very similar to Figure 5.13, but there is a very important difference:

---

[2] If the activation energy refers to 1 kmol it is designated G_{act} without the subscript 'a'. Compare page 220.

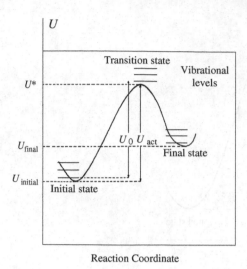

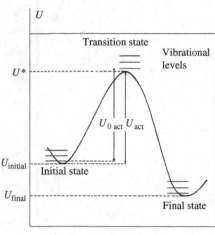

Figure 5.14 Endothermic reaction. The variation of the internal energy during an *endothermic* transformation. Vibrational energy (indicated schematically) should be included in U. $\Delta U = U_f - U_i > 0$. Energy is *absorbed* during the transformation. Reproduced with permission from J. Burke, *The Kinetics of Phase Transformations in Metals*. © 1965 Pergamon Press (now with Elsevier).

Figure 5.15 Exothermic reaction. The variation of the internal energy during an *exothermic* transformation. Vibrational energy (indicated schematically) should be included in U. $\Delta U = U_f - U_i < 0$. Energy is *released* during the transformation. Reproduced with permission from J. Burke, *The Kinetics of Phase Transformations in Metals*. © 1965 Pergamon Press (now with Elsevier).

- ΔG is always <0 whereas ΔU can be positive *or* negative.
- If $U_f > U_i$ the transformation is *endothermic*.
- If $U_f < U_i$ the transformation is *exothermic*.

Vibrational energy has to be included in the total energy. It depends on the vibrational quantum number v. It can, as a first approximation, be written as (Chapter 2)

$$\text{Vibrational energy} = constant \times (v + \tfrac{1}{2}) \tag{5.67}$$

Equation (5.67) shows that vibrational energy is present even at the lowest vibrational quantum number $v = 0$. This is in agreement with quantum mechanics but a deviation from classical physics.

5.3.7 Thermal Energy Distribution in Particle Systems

In connection with the kinetic theory of gases, the energy distribution of noninteracting particles in a gas as a function of temperature was discussed in Chapter 4.

The distribution of particles in different available energy states influences the reaction rates of thermally activated reactions and transformations. The fraction of particles with energies above a given energy is of particular interest and will be considered below.

Maxwell–Boltzmann Distribution Law

The Maxwell–Boltzmann distribution law in its general form can be written as [Equations (4.19) and (4.20) on page 175]

$$N_i = \frac{N_0}{Z} g_i e^{-\frac{u_i}{k_B T}} \tag{5.68}$$

Z is called the *partition function* and is given by

$$Z = g_1 e^{-\frac{u_1}{k_B T}} + g_2 e^{-\frac{u_2}{k_B T}} + \ldots + g_i e^{-\frac{u_i}{k_B T}} = \sum_i g_i e^{-\frac{u_i}{k_B T}} \tag{5.69}$$

where

  u_i = energy of particle i
  N_i = number of particles which have the energy u_i
  N_0 = total number of particles
  g_i = statistical weight of energy level u_i
  k_B = Boltzmann's constant
  T = absolute temperature of the system.

If all the energy levels have equal statistical weight, the distribution function will be simplified to

$$N_i = \frac{N_0}{e^{-\frac{u_1}{k_B T}} + e^{-\frac{u_2}{k_B T}} + \ldots + e^{-\frac{u_i}{k_B T}}} e^{-\frac{u_i}{k_B T}} \tag{5.70}$$

Thermal Energy Distribution in a Gas

In a gas, the mean kinetic energy of the particles is a function of the temperature T:

$$\overline{u_i} = \frac{3}{2} k_B T \tag{5.71}$$

The Maxwell–Boltzmann distribution law applied to a gas at temperature T can be written as

$$N_i = constant \times e^{-\frac{u_i}{k_B T}} \tag{5.72}$$

The constant can be determined by use of the relationship

$$N_0 = \sum N_i = constant \times \sum_i e^{-\frac{u_i}{k_B T}} \tag{5.73}$$

Fraction of Particles with Energies Equal to or Greater than a Given Energy
The fraction f_i of the N_0 particles which have the thermal energy u_i per particle is then

$$f_i = \frac{N_i}{N_0} = \frac{e^{-\frac{u_i}{k_B T}}}{\sum e^{-\frac{u_i}{k_B T}}} \tag{5.74}$$

where

  f_i = the fraction of the N_0 particles which has the kinetic energy u_i per particle.

The high-energy part of the Maxwell–Boltzmann distribution function is shown in Figure 5.16. The larger u_i is, the smaller will be the fraction f_i.

The denominator of the expression in Equation (5.74) is the *partition function Z*:

$$Z = \sum e^{-\frac{u_i}{k_B T}} \tag{5.75}$$

Application to Reactions and Transformations

Equation (5.74) can be applied to reactions and transformations. For better agreement with the terminology in this chapter, we will replace u_i from the statistical derivation in Chapter 4 by $U_{a\,i}$. Both designations mean the internal energy per atom.

The fraction f^* of atoms which have enough thermal energy to overcome the energy barrier, i.e. the activation energy $G_{a\,act}$, discussed in Section 5.3.4 (Figure 5.13 page 239), is found if we insert $u_i = U_{a\,act}$ into Equation (5.74). $U_{a\,act}$ is the internal energy related to $G_{a\,act}$ by the relationship ($H_{a\,act} \approx U_{a\,act}$)

$$G_{a\,act} = U_{a\,act} - TS_{a\,act} \tag{5.76}$$

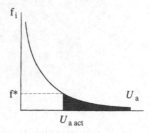

Figure 5.16 The fraction f_i as a function of the energy u. The dark area under the curve (extrapolated to infinity) is proportional to the total number of atoms with a thermal energy $U_a > U_{a\ act}$.

The total number of atoms which have a thermal energy $\geq U_{a\ act}$ is represented by the black area under the curve in Figure 5.16. The critical fraction f^* can be written as

$$f^* = \frac{\mathrm{e}^{-\frac{U_{a\ act}}{k_B T}}}{Z} \qquad (5.77)$$

where Z is the partition function [Equation (5.75)]. If $U_{a\ act} >> k_B T$ the fraction is very small and the transformation rate will be very low. This case will be discussed in the next section.

5.4 Reaction Rates

5.4.1 Reaction Rates of Thermally Activated Reactions and Transformations

A necessary but not sufficient condition for a transformation is a positive driving force. Another condition is that energy corresponding to the activation energy must be added.

The available energy for this is the random thermal energies of the atoms. If the activation energy is comparatively low and the temperature is high, many atoms have kinetic energies high enough to overcome the energy barrier and the transformation occurs readily. The reaction rate is high. The definition of reaction rate will be given below.

If the activation energy is high compared with the thermal energies of the atoms, very few of them have energies high enough to overcome the barrier. The reaction rate becomes very low and the transformation or reaction will be prohibited in practice.

Obviously the thermal energy distribution among the atoms is essential for the reaction rate. This will be discussed below. The result will be used to derive a mathematical model for the reaction rate.

5.4.2 Definition of Reaction Rate

The reaction rate at the time t of a transformation is defined as

$$k = \frac{\mathrm{d}f_{trans}(t)}{\mathrm{d}t} \qquad (5.78)$$

where the fractional transformation $f_{trans}(t)$ is defined as

$$f_{trans}(t) = \frac{\text{number of atoms per unit volume in the final state at the time } t}{\text{total number of atoms per unit volume available for transformation at } t = 0}$$

The reaction rate k can also be expressed as *the fraction of the total number of particles which reach the final state per unit time*.

The reaction rate is normally a function of time.

5.4.3 Reaction Rates of Simple Reactions and Transformations

Consider a transformation which involves only one basic atomic process associated with the internal activation energy $U_{a\ act}$. The reaction rate is proportional to three factors:

- the frequency ν with which the atoms have the opportunity of reaction or transformation
- the fraction f^* of the total number of atoms in the initial state, which has enough thermal energy to overcome the energy barrier
- a probability factor associated with the activation entropy term in the relationship ($H_{a\ act} \approx U_{a\ act}$):

$$G_{a\ act} = U_{a\ act} - TS_{a\ act} \tag{5.76}$$

For a gas, the *frequency factor* ν equals the collision frequency between the atoms. In a solid it equals the vibration frequency ν_{vibr} of the atoms in the crystal lattice.

The *fraction factor* f^* is that given in Equation (5.77) on page 242.

The *probability factor* $P_{a\ act}$ is given by the relationship

$$S_{a\ act} = k_B \ln P_{a\ act}$$

or

$$P_{a\ act} = e^{\frac{S_{a\ act}}{k_B}} \tag{5.79}$$

where $S_{a\ act}$ is the entropy and k_B is Boltzmann's constant. The probability can be related to concrete matters depending of the type of transformation.

Hence the reaction rate can be written as

$$k = \frac{\nu P_{a\ act}}{Z} e^{-\frac{U_{a\ act}}{k_B T}} \tag{5.80}$$

or

$$k = \frac{\nu}{Z} e^{-\frac{S_{a\ act}}{k_B}} e^{-\frac{U_{a\ act}}{k_B T}} = \frac{\nu}{Z} e^{-\frac{U_{a\ act} - S_{a\ act} T}{k_B T}} = \frac{\nu}{Z} e^{-\frac{G_{a\ act}}{k_B T}} \tag{5.81}$$

where

k = reaction rate
ν = the frequency with which the atoms have the opportunity of reaction or transformation
Z = partition function (page 175 in Chapter 4)
k_B = Boltzmann's constant
$U_{a\ act}$ = activation energy of the reaction
$G_{a\ act}$ = free activation energy of the reaction.

The entropy factor and ν/Z are often combined into one factor A and Equation (5.81) can be written as

$$k = A e^{-\frac{U_{a\ act}}{k_B T}} \tag{5.82}$$

Equation (5.82) is known as the *Arrhenius equation*.

5.4.4 Determination of Reaction Rates

Model Restrictions

The theory of reaction rates given above is simplified in the following respects:

1. No attention has been paid to the thermal activation of the atom in the initial and activated states. This changes the activation free energy $G_{a\ act}$ and the internal activation energy $U_{a\ act}$.
2. The transformation process may not occur in a single well-defined way. There may be alternative ways with different activation energies.
3. The transformation process may consist of several consecutive steps with several atoms involved. Each step has its own activation energy.

The first objection is not serious. In most cases it is most reasonable to neglect the thermal excitation as we have done above.

Concerning the second objection, it is true that a transformation process can occur in many different ways. One example is diffusion where the atoms move from one site to another. Two common diffusion mechanisms (pages 255–256) are

- interstitial motion of the atom in the crystal lattice from one site to another
- exchange of an atom and a vacant site (substitutional motion).

Each alternative transformation process has its own activation energy, which does not affect the final result. One example is the use of catalysts which promote chemical reactions by lowering the activation energy without changing the final products at all.

Each alternative transformation is associated with a free energy curve like that in Figure 5.13 on page 239. All the curves form a free energy surface. The alternative with the lowest possible activation energy is that which is the most probable. It corresponds to the curve in Figure 5.13, which can be described as a path on the free energy surface leading from the initial state over the pass down to the final state. The activated state corresponds to a saddle point on the free energy surface as shown in Figure 5.17.

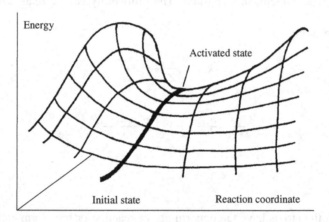

Figure 5.17 Free energy surface with a saddle point.

The third objection against the simple theory is that a transformation may be a multiple-step process with several atoms involved. This objection is serious. Equations (5.79)–(5.82) in Section 5.4.3 are *not* valid in such cases.

Determination of Reaction Rates

Calculation of theoretical values of reaction rates requires estimation of the internal activation energy $U_{a\ act}$ and the frequency factor A.

The calculation of $U_{a\ act}$ is no easy mathematical task. It is necessary to know the binding forces of the atom or group of atoms in the initial state and the interaction forces when the atoms move from their initial sites to the saddle point position, corresponding to the activated state. Very approximate calculations have been performed only in a few cases referring to relatively simple atomic processes.

By studying the reaction rate as a function of temperature, it is possible and easy to derive experimental values of $U_{a\ act}$ and A. The Arrhenius equation (page 243) can be written as

$$\ln k = \ln A - \frac{U_{a\ act}}{k_B T} \tag{5.83}$$

The reaction rate is measured as a function of temperature. If both A and $U_{a\ act}$ are independent of temperature, Equation (5.83) gives a straight line when the experimental values of $\ln k$ are plotted versus $1/k_B T$. The graph can be used for calculation of the constants A and $U_{a\ act}$, as shown in Figure 5.18.

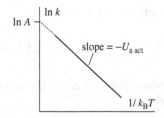

Figure 5.18   Graphical derivation of the reaction rate of a transformation or a chemical reaction.

Numerous chemical reactions and other transformation processes follow the Arrhenius equation.

A discrepancy between theoretical and experimental values is likely if the reaction or transformation involves several consecutive steps. In this case, the simple theory is *not* valid and the corresponding curve in Figure 5.18 will be bent.

The free activation energy [Equation (5.76) on page 241] can be calculated if the entropy $S_{a\ act}$ can be estimated. Such calculations require a detailed knowledge of the atomic configuration in the initial and activated states. Only semiempirical methods are available as $S_{a\ act}$ cannot be measured separately by experiments.

5.5 Kinetics of Homogeneous Reactions in Gases

The basic theories and general results in the preceding sections will be applied to the concrete case of reactions in gases. Reactions in gases are always homogeneous.

5.5.1 Collision Theory of Homogeneous Chemical Reactions

Consider, for example, two gas components A and B, which react with each other and form a third component AB. A normal assumption is that the rate of formation of AB is proportional to the concentrations of A and B:

$$A + B \rightarrow AB$$

$$-\frac{dx'_A}{dt} = -\frac{dx'_B}{dt} = \frac{dx'_{AB}}{dt} = k_1 x_A x_B \tag{5.84}$$

where
  x_i = mole fraction of component i (A, B, AB)
  t = time
  k_1 = reaction rate in the forward direction.

However, a chemical reaction is always proceeding in both directions as soon as some AB molecules have been formed. The reaction is reversible.

$$A + B \leftarrow AB$$

A and B are formed at a rate proportional to the concentration of AB. This can be written in the following way:

$$\frac{dx''_A}{dt} = \frac{dx''_B}{dt} = -\frac{dx''_{AB}}{dt} = k_2 x_{AB} \tag{5.85}$$

where k_2 is the reaction rate in the backward direction.

The total change of each component (i = A, B, AB) can be written as

$$\frac{dx_i}{dt} = \frac{dx'_i}{dt} + \frac{dx''_i}{dt} \tag{5.86}$$

By combining Equations (5.84)–(5.86), the total change of components can be described by

$$\frac{dx_A}{dt} = \frac{dx_B}{dt} = -\frac{dx_{AB}}{dt} = -k_1 x_A x_B + k_2 x_{AB} \tag{5.87}$$

Equation (5.87) contains two different rate constants. It can be written as

$$\frac{dx_A}{dt} = \frac{dx_B}{dt} = -\frac{dx_{AB}}{dt} = -k_1 x_A x_B \left(1 - \frac{k_2}{k_1} \frac{x_{AB}}{x_A x_B} \right) \qquad (5.88)$$

By applying the basic thermodynamic relationship between free energy and concentration ($G = G^0 + RT \ln x$) we can derive an expression for the driving force of the reaction:

$$-\Delta G = G_i - G_f = G_{AB} - G_A - G_B$$

or

$$-\Delta G = G^0_{AB} - G^0_A - G^0_B + RT \ln x_{AB} - RT \ln x_A - RT \ln x_B$$

which gives the driving force of the reaction:

$$-\Delta G = -\Delta G^0 + RT \ln \frac{x_{AB}}{x_A x_B} \qquad (5.89)$$

where
 $-\Delta G$ = driving force of the reaction
 ΔG^0 = free energy difference of the standard state.

After transformation of Equation (5.89), we have at any time during the reaction

$$\frac{x_{AB}}{x_A x_B} = e^{\frac{-\Delta G + \Delta G^0}{RT}} \qquad (5.90)$$

At equilibrium $-\Delta G = 0$ and the time derivatives in Equation (5.88) are zero, which gives

$$\frac{k_2}{k_1} = \frac{x_A x_B}{x_{AB}} = e^{\frac{-\Delta G + \Delta G^0}{RT}} = e^{-\frac{\Delta G^0}{RT}} \qquad (5.91)$$

Inserting the expression for k_2/k_1 from Equation (5.91) and that for $x_{AB}/x_A x_B$ from Equation (5.90) into Equation (5.88), we obtain after reduction

$$\frac{dx_A}{dt} = \frac{dx_B}{dt} = -\frac{dx_{AB}}{dt} = -k_1 x_A x_B \left(1 - e^{-\frac{\Delta G}{RT}} \right) \qquad (5.92)$$

Hence it is sufficient to determine only one rate constant to describe the dynamics of the chemical reaction.

If the exponent $-\Delta G/RT$ is small, series expansion of the exponential function gives a simplified expression for the reaction rate:

$$\frac{dx_A}{dt} = \frac{dx_B}{dt} = -\frac{dx_{AB}}{dt} = -k_1 x_A x_B \frac{-\Delta G}{RT} \qquad (5.93)$$

The study of reaction rates is performed far from equilibrium, often in the initial stage. *Real time analysis* means that the composition of the gas mixture is followed as a function of time during the reaction, for example by registering the absorption spectra of each component. Another method is *quenching*, when the reaction is stopped for example by sudden cooling and the composition is analysed.

There are many types of homogeneous chemical reaction. For each component the reaction rate can generally be written as

$$\text{Reaction rate} = k x_A^p x_B^q$$

where p is the order in component A, q the order of component B and p + q the *overall order of the reaction*.

It is not the aim of this section to discuss the overall reaction rates of the various types of reactions. Instead, we concentrate on studying the basic rate constant k, its temperature dependence and the possibility of expressing k as a function of thermodynamic quantities.

5.5.2 Temperature Dependence of the Rate Constant

It has been known since the end of the 19th century that there is a very strong temperature dependence of the reaction rate of a chemical reaction. Arrhenius showed that the temperature dependence of the rate constant can be described empirically by Equation (5.83) on page 244, where A and $U_{a\,act}$ are two constants. They can easily be derived from experimental values by plotting $\ln k$ against $1/k_B T$ (Figure 5.18 on page 245). The slope of the straight line is $-U_{a\,act}$. The constant A can be derived from the intersection of the line and the $\ln k$ axis.

Arrhenius claimed that

1. A is proportional to the total number of atomic collisions in the gas per unit time and unit volume.
2. $U_{a\,act}$ is the minimum kinetic energy of the atoms necessary for a reaction to occur.

Collisions between two atoms result in a chemical reaction only if the available energy exceeds the activation energy $U_{a\,act}$.

- The reaction rate k is proportional to the number of collisions per unit time and unit volume which result in a chemical reaction.

The reaction can be described by the empirical equation

$$I = \nu_{coll}\, e^{-\frac{U_{a\,act}}{k_B T}} \tag{5.94}$$

where
$U_{a\,act}$ = activation energy of the reaction
I = number of collisions per unit time and unit volume which lead to a chemical reaction
ν_{coll} = total number of collisions per unit time and unit volume
T = absolute temperature.

5.5.3 Rate Constant as a Function of Thermodynamic Quantities

The basic rate constant of a reaction can be expressed in terms of thermodynamic quantities. It is also possible to evaluate the overall reaction rate in terms of concentrations and thermodynamic quantities.

According to the kinetic theory of gases, the collision frequency equals the mean velocity divided by the mean free path. Using Equation (4.10) on page 173 and Equation (4.57) on page 193, we obtain

$$\nu_{coll} = \frac{\overline{\nu_{kin}}}{l} = 16Nr^2\sqrt{\frac{\pi RT}{M}} \tag{5.95}$$

where
ν_{coll} = total number of collisions per unit time and unit volume
ν_{kin} = mean velocity of the molecules
l = mean free path
N = number of molecules per unit volume
r = radius of the molecules
R = gas constant
T = absolute temperature
M = molar weight of the molecules.

If Equation (5.95) is applied to a homogenous chemical reaction of the first order, it will be identical with empirical Equation (5.94) if $k = constant \times I$ and $A = constant \times \nu_{coll}$ and we obtain

$$k = 16Nr^2\sqrt{\frac{\pi RT}{M}}\, e^{-\frac{U_{a\,act}}{k_B T}} \tag{5.96}$$

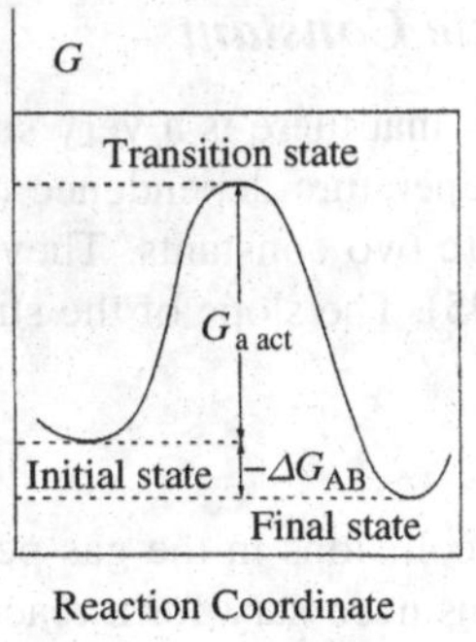

Figure 5.19 According to Eyring the activation energy $= G_{a\ act}$. Reproduced with permission from J. Burke, *The Kinetics of Phase Transformations in Metals*. © 1965 Pergamon Press (now with Elsevier).

where the right-hand side represents the number of collisions per unit time and unit volume times the fraction of intermediate activated complex $(AB)^*$.

The calculated pre-exponential factor ν_{coll} in Equation (5.94) turns out to be too large compared with experimental values. In order to obtain a better description of experiments, Eyring among others suggested that the activation energy $U_{a\ act}$ could be replaced by and identical with the change in Gibbs free activation energy, which is shown in Figure 5.19:

$$G_{a\ act} = H_{a\ act} - TS_{a\ act} \tag{5.76}$$

If this expression for $G_{a\ act}$ is introduced into Equation (5.96) instead of $U_{a\ act}$, we obtain

$$k = 16Nr^2 \sqrt{\frac{\pi RT}{M}} e^{-\frac{G_{a\ act}}{k_B T}} \tag{5.97}$$

or

$$k = 16Nr^2 \sqrt{\frac{\pi RT}{M}} e^{\frac{S_{a\ act}}{k_B}} e^{-\frac{H_{a\ act}}{k_B T}} \tag{5.98}$$

As before (Figure 5.18 on page 245), the pre-exponential factor

$$k = 16Nr^2 \sqrt{\frac{\pi RT}{M}} e^{\frac{S_{a\ act}}{k_B}}$$

and also the activation enthalpy $(-H_{a\ act})$ can easily be determined experimentally. The calculated values agree better with reality than the pure value of ν_{coll} in Equation (5.96) as the activation entropy $S_{a\ act}$ is negative. We will come back to the important quantities $H_{a\ act}$ and $S_{a\ act}$ in Section 5.5.5.

$-H_{a\ act}$ is often designated Q and is identical with the activation energy of the reaction. $H_{a\ act} = U_{a\ act}$ if the volume is constant.

5.5.4 Driving Force and Reaction Rate of Homogeneous Chemical Reactions

A chemical reaction always occurs under the influence of a driving force.

Reaction $A + B \rightarrow AB$

First we consider *the reaction of formation of component* AB by collision of A and B:

$$A + B \rightarrow AB$$

The driving force is the difference between the free energy in the initial state and the final state:

$$\text{Driving force} = -\Delta G_{AB} = -\int_{initial}^{final} dG = -(G_f - G_i) = G_A + G_B - G_{AB} \tag{5.99}$$

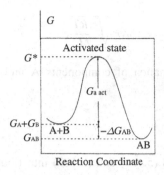

Figure 5.20 Driving force $-\Delta G_{AB}$ of a chemical reaction. Reproduced with permission from J. Burke, *The Kinetics of Phase Transformations in Metals*. © 1965 Pergamon Press (now with Elsevier).

where $-\Delta G_{AB}$ is the driving force for formation of component AB. The driving force is always >0 for spontaneous processes. The driving force is a thermodynamic quantity. It is only a function of the final and initial states and is independent of the activation energy $G_{a\ act}$ (Figure 5.20).

As we have seen in Section 5.4.3, the rate of the reaction is strongly influenced by the activation energy. The rate constant can be written as

$$k_1 = 16Nr^2 \sqrt{\frac{\pi RT}{M}} e^{-\frac{G_{a\ act}}{k_B T}} \qquad (5.100)$$

Reaction AB $\rightarrow$ A + B

The driving force of the *reverse reaction, the reaction of formation of components A and B*:

$$AB \rightarrow A + B$$

is

$$\text{Driving force} = -\Delta G_{A+B} = -\int_{\text{initial}}^{\text{final}} dG = -(G_f - G_i) = G_{AB} - (G_A + G_B) \qquad (5.101)$$

The driving force has the same size in the two cases but opposite sign, depending on the direction of the reaction.

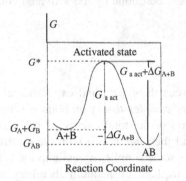

Figure 5.21 Driving force $-\Delta G_{A+B}$ of the chemical reaction AB $\rightarrow$ A + B. Reproduced with permission from J. Burke, *The Kinetics of Phase Transformations in Metals*. © 1965 Pergamon Press (now with Elsevier).

Figure 5.21 shows that the activation energy of formation of A + B by decay of AB can be written as $G_{a\ act} + \Delta G_{A+B}$. The energy is supplied either by collisions or by loss of vibrational energy of the AB molecule. Hence the rate constant can be written as

$$k_2 = 16Nr^2 \sqrt{\frac{\pi RT}{M}} e^{-\frac{G_{a\,act}+\Delta G_{A+B}}{k_B T}} \tag{5.102}$$

where $-\Delta G_{A+B}$ is the driving force for formation of components A and B. It is independent of the activation energy $G_{a\,act}+\Delta G_{A+B}$.

Reaction AB ⇌ A + B

If we introduce the rate constants in Equations (5.100) and (5.102) into Equation (5.87) on page 245, the reaction rate can finally be written as

$$\frac{dx_A}{dt} = \frac{dx_B}{dt} = -\frac{dx_{AB}}{dt} = 16Nr^2 \sqrt{\frac{\pi RT}{M}} e^{\frac{S_{a\,act}}{k_B}} e^{-\frac{H_{a\,act}}{k_B T}} \left(-x_A x_B + x_{AB} e^{-\frac{\Delta G_{A+B}}{k_B T}} \right) \tag{5.103}$$

where
x_i = mole fraction of component i (i = A, B, AB)
t = time
N = number of molecules per unit volume
M = molecular weight of component
r = radius of the molecules
$S_{a\,act}$ = activation entropy related to the activation free energy $G_{a\,act}$
$-H_{a\,act}$ = activation enthalpy related to the activation free energy $G_{a\,act}$
$-\Delta G_{A+B}$ = driving force of the reaction AB → A + B.

5.5.5 Activated Complex Theory

The simple hard-sphere collision theory (Section 5.5.1) on page 245 as a model for homogeneous gas reactions often gives poor agreement between theoretical and experimental values of the frequency factor A of the reaction rate [Equation (5.82) on page 243]. The theoretically calculated values are in most cases considerably higher than the experimental values. The calculated values for different reactions become essentially the same, independent of the structure of the reactants.

As we have mentioned on page 248, better agreement is obtained if the internal activation energy $U_{a\,act}$ is replaced by the free energy difference $G_{a\,act}$.

A still better method is to apply the so-called *activated complex theory*. The problem is mainly to calculate the quantities $S_{a\,act}$ and $H_{a\,act}$ with sufficient accuracy. This is best done by the activated complex theory, which is alternatively called the *transition state theory*.

Activated Complex Theory

An approach to improve the agreement between theoretical and experimental values of the reaction rate is to abandon the simple collision theory and apply the *activated complex theory* on homogeneous gas reactions.

The theory is based on accurate calculation of the total energy of the system. It includes the interactions between the nuclei, their potential and kinetic energies and their interaction with the electrons and the interaction between the electrons. The calculations cannot be performed without some simplifying assumptions. The motions of the nuclei and of the electrons are treated independently of each other. The methods of quantum chemistry and high-capacity computers are necessary for the calculations.

An energy surface can be constructed which gives a detailed description of the reaction from beginning to end. Initially the reacting atoms A and B are far from each other. The reaction process is described by a path along the energy surface. The reaction follows the path of lowest possible energy as shown in Figure 5.22.

The reactants A and B initially are in the reactant valley. During the reaction they have to move across a 'mountain pass' or a col on their way to the product valley.

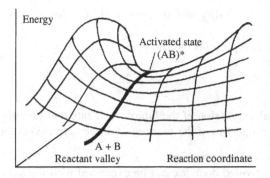

Figure 5.22 Energy surface illustrating a homogeneous gas reaction. The reactant valley is on the nearest side of the 'mountain chain'. The product valley is on the other side of the ridge.

At the saddle point, a so-called activated complex (AB)* is formed. The activated complex is similar to a molecule except that it has no stable state. It can either form a product, i.e. move on into the product valley, or decay, i.e. go back into the reactant valley.

The lowest point of the mountain pass represents the transition state. At the same time it is the highest point on the 'minimum energy path' $(A+B) - (AB)^* - (AB)$.

The energy of the activated complex is a function of the structures of the reactants and the product or products.

Calculation of the Rate Constant

Rate Constant in Terms of Statistical Mechanics Functions

According to the activated complex theory, the reaction rate can be written as

$$-\frac{dA}{dt} = k_1 [A][B] = v^* \left[(AB)^*\right] \tag{5.104}$$

where

$[\ ]$ = concentration

k_1 = rate constant of formation of AB* from A and B

v^* = frequency of passage of the activated complex over the 'mountain' pass.

The reaction rate equals the frequency v^* times the number of activated complex. v^* also represents the frequency with which a complex splits apart into products or forms a stable product.

k_1 can be expressed as a function of the partition functions Z of the particles in their energy levels at the bottom of the energy curves (Figures 5.14 and 5.15 on page 240):

$$k_1 = \kappa \frac{k_B T}{h} \frac{Z_{(AB)^*}}{Z_A Z_B} e^{-\frac{U^0}{k_B T}} \tag{5.105}$$

where

k_1 = rate constant of formation of (AB)* from A and B

κ = transmission coefficient

k_B = Boltzmann's constant

$Z_{(AB)^*}$ = partition function of the activated complex in its lowest vibrational level

Z_A = partition function of particles A in their lowest vibrational level

Z_B = partition function of particles B in their lowest vibrational level

$U_0 = U_{0(AB)^*} - (U_{0A} + U_{0B})$ = the difference between the lowest energy level of the activated complex and the sum of the lowest energy levels of particles $A + B$

h = Planck's constant

The transmission coefficient is the probability that the activated complex will form products instead of going back to reactants.

The second factor in Equation (5.105) represents the frequency v^*:

$$v^* = \frac{k_B T}{h} \tag{5.106}$$

Equation (5.105) is the final theoretical expression of the bimolecular rate constant given by the activated complex theory. It includes factors which depend on the properties of the reactants and the activated complex.

Rate Constant in Terms of Thermodynamic Functions

The rate constant of formation of an activated complex can be expressed with the aid of thermodynamic functions instead of partition functions:

$$A + B \rightarrow (AB)^* \rightarrow \text{products}$$

The chemical equilibrium constant K^* can be written as

$$K^* = \frac{[(AB)^*]}{[A][B]} \tag{5.107}$$

Combining equations (5.104), (5.106) and (5.107), we obtain

$$k_1 = \frac{k_B T}{h} K^* \tag{5.108}$$

Using thermodynamic relationships we obtain for the transition state

$$G^* = -RT \ln K^* \tag{5.109}$$

and

$$G^*_{a\ act} = H^*_{a\ act} - TS^*_{a\ act} \tag{5.110}$$

where $G^* = G^*_{a\ act} N_A$ (N_A = Avogadro's number).

The final expression will be

$$k_1 = \frac{k_B T}{h} e^{-\frac{G^*_{a\ act}}{k_B T}} = \frac{k_B T}{h} e^{\frac{S^*_{a\ act}}{k_B}} e^{-\frac{H^*_{a\ act}}{k_B T}} \tag{5.111}$$

where
  k_1 = rate constant
  $G^*_{a\ act}$ = free energy of activation
  k_B = Boltzmann's constant
  $S^*_{a\ act}$ = activation entropy
  $H^*_{a\ act}$ = activation enthalpy.

$S^*_{a\ act}$ and $H^*_{a\ act}$, which depend on the structure, shape and other properties of the reactants and the activated complex, can be calculated theoretically.

Comparison Between the Collision Model and the Activated Complex Model of Reaction Rates

The hard-sphere model [Equation (5.103) on page 250] will be compared with the activated complex model [Equation (5.105) on page 251 and Equation (5.111) above].

The activated complex theory gives much better agreement with experimental values of the frequency factor A than the simple hard-sphere collision theory. In the latter case, the model gives values which are practically constant, independent of reaction.

The reason is that the activated complex theory involves the properties of the reactants and products. Its disadvantage is the complex and extensive calculations.

Sections 5.3–5.5 may give the impression that reaction kinetics concern only chemical reactions. This is not the case. The laws of reaction kinetics are valid for many other types of time-dependent processes, for example solidification and diffusion. The rest of this chapter will be devoted to diffusion in solids, which is of great importance in many phase transformation processes in alloys.

5.6 Diffusion in Solids

In most crystal growth and solidifications processes, mass transport by diffusion is very important. Detailed descriptions of various crystallization processes require knowledge of the mechanisms of diffusion in gases, liquids and solids.

The origin of diffusion is the *random motion* of atoms, ions or molecules in a medium. An example of a pure random motion is the motion of the molecules in a gas, where the net transfer of molecules is zero. Even if the mobilities of atoms and ions in liquids and solids are much lower than those in gases, diffusion occurs in the latter cases also. Diffusion in liquids will be treated briefly in Chapter 8. The mechanism of diffusion in solids will be discussed below.

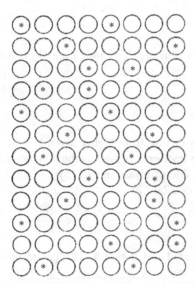

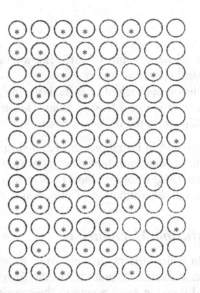

Figure 5.23   In the absence of a concentration gradient the tracer atoms are uniformly distributed on average. The number of radioactive atoms is greatly exaggerated in Figures 5.23 and 5.24 compared with the number of ordinary atoms. Reproduced with permission from A. G. Guy, *Elements of Physical Metallurgy*, 2nd edn. © Addison-Wesley Publishing Company, Inc. (now under Pearson Education).

Figure 5.24   The diffusion of the radioactive atoms occurs in both directions but is larger from left to right than in the opposite direction owing to the horizontal concentration gradient. Reproduced with permission from A. G. Guy, *Elements of Physical Metallurgy*, 2nd edn. © Addison-Wesley Publishing Company, Inc. (now under Pearson Education).

In a pure solid, *self-diffusion*, i.e. random motion of the atoms, which is always present, is hard to observe as all atoms are identical. As mentioned in Chapter 4 on page 208, diffusion in a pure medium can be studied by the use of radioactive atoms of the same kind, so-called *tracer atoms*. At equilibrium the tracer atoms are uniformly distributed (Figure 5.23) and the numbers of jumps per unit time of atoms across an arbitrary plane in opposite directions are equal.

If the distribution of tracer atoms in a pure solid is made uneven (Figure 5.24) or the alloying atoms in a binary alloy are *not* uniformly distributed, the random motion of the atoms is overlapped by a *systematic* net motion of the atoms, in a special direction. Such a combined motion is called *diffusion* and results in *mass transport*.

The systematic motion is always caused by a driving force, which in most cases is a concentration difference. Mass transport caused by concentration gradients is very common in crystal growth and solidification processes. Diffusion occurs from higher towards lower concentration.

The diffusion phenomena in solids have been studied from atomic and statistical points of view. The results of such analyses can be related to the macroscopic diffusion equations and quantities. This topic will be discussed in this chapter as a basis for applications to crystallization and other processes.

5.6.1 Basic Theory of Diffusion. Diffusion Coefficient

The basic law of diffusion relates the net flux of diffusing atoms to the concentration gradient of the atoms in the medium, which can be a gas, a solid or a liquid.

The *diffusion coefficient* or *diffusivity* is defined with the aid of *Fick's first law* (Chapter 4 on page 199), provided that we use numbers/unit volume or kmol/unit volume as the concentration unit. If mole fraction x is used as the concentration unit, a factor $1/V_m$ (molar volume) has to be inserted on the right-hand side of the equation.

Fick's law can alternatively be written as

$$J = -D \text{ grad } c \quad \text{Fick's first law} \tag{5.112}$$

In one dimension (y axis), we obtain

$$J = -D \frac{dc}{dy} \tag{5.113}$$

where

J = flux or net amount of diffusing atoms passing a cross-section per unit area and unit time (kg/m^2 s or number of atoms/m^2 s or kmol/m^2 s)

D = diffusion coefficient or diffusivity (m^2/s)

c = concentration of diffusing atoms (kg/m^3 or number of atoms/m^3 or kmol/m^3)

y = coordinate in the diffusion direction.

The minus sign in Equations (5.112) and (5.113) indicates that the atoms diffuse from higher towards lower concentration. The flux and the concentration gradient always have opposite signs. If mole fraction x is used as the concentration unit instead of c, a factor $1/V_m$ (molar volume) has to be introduced on the right-hand side of Equation (5.113).

Experimental determination of diffusion coefficients is described on pages 273–274.

5.6.2 Diffusion Mechanisms

Diffusion in solids differs strongly from diffusion in gases, which was treated in Chapter 4. The gas molecules are free to move in any direction at random and the forces between them can be neglected.

In a solid the atoms are bound to their positions in the crystal lattice. The forces between the atoms are strong in a solid. However, the system is not rigid as the atoms vibrate with high frequencies around their equilibrium positions. This motion, which enables the atoms to move if they have enough energy, makes diffusion within the solid possible.

When an atom moves from one site to another in a crystal lattice it has to overcome an energy barrier of the kind which has been sketched in the preceding sections of transformation kinetics. Diffusion is a typical example of a process to which the theory of transformation kinetics can be applied.

The Diffusion Process

Each atom in a crystal lattice is incessantly exposed to vibrations of various, very high frequencies. In most cases the atoms return to their original sites. Occasionally, but seldom, there will be an opportunity for an atom to change its site from one position to another. Lucky circumstances, which all must coincide in time, are the following:

- The atom must gain enough energy due to random interference of phonons to overcome the energy barrier.
- Simultaneously the neighbouring atoms must be in such vibrational positions that there is enough space for the jumping atom to pass.

The probability of a jump is very low for each single atom but the number of atoms in a crystal is very large. Hence diffusion does occur in solids (compare pages 261–262 for a concrete example).

Diffusion Mechanisms

Diffusion of atoms in the interior of a crystalline solid can occur in many different ways. The eight mechanisms, presented in three subgroups below, are explained briefly in Figure 5.25a–h:

- exchange mechanism
- ring mechanism
- interstitial mechanism
- indirect interstitial mechanism
- crowdion mechanism

- single vacancy mechanism
- divacancy mechanism
- relaxation mechanism

The most common diffusion mechanisms in solids are the *interstitial* mechanism and the *vacancy* mechanism.

In addition to the 'volume' diffusion mechanisms described above, there are other mechanisms such as *dislocation diffusion pipe mechanisms, grain boundary mechanisms* and *surface diffusion mechanisms*. These mechanisms appear in open regions where the regular lattice structure no longer exists and where diffusion occur more easily than in the interior of a crystal.

These diffusion mechanisms are difficult to analyse because the detailed atomic paths at grain boundaries, surfaces and dislocations are difficult to calculate. However, the numbers of dislocations, grain boundaries and surfaces are relatively independent of temperature while the volume effects increase rapidly with temperature. Hence the volume mechanisms dominate at high temperatures and the others are in most cases of minor importance in solids. They may be of importance in liquids (Chapter 8).

Below we will restrict the discussion to the interstitial and vacancy mechanisms, which are the most common. The former is important for interstitially dissolved alloying elements in metals. The latter is applied in connection with substitutionally dissolved alloying elements in metals.

5.6.3 Theory of Diffusion

All diffusion mechanisms are accompanied by displacements of neighbouring atoms. In this section we will study the diffusion process from an atomistic point of view and relate the diffusion coefficient D to thermodynamic and atomic quantities.

Concentration of Point Defects in a Crystalline Solid at Equilibrium

All crystals contain defects to a smaller or greater extent. At a given temperature there is an equilibrium number of defects per unit volume. The equilibrium concentration of defects depends on the temperature.

Addition of a point defect, for example a vacancy or an interstitial, always increases the total energy of a crystal. We will calculate the equilibrium concentration of defects as a function of temperature. For this purpose we will use the condition given on page 236:

- At constant temperature and pressure processes always occur spontaneously in the direction of decreasing G.

At equilibrium the Gibbs free energy G has a minimum, which corresponds to the condition

$$dG_{\text{T,p}} = 0 \tag{5.57}$$

Equilibrium Vacancy Concentration as a Function of Temperature
We consider a system consisting of 1 kmol of a pure element. The free energy of the system (page 235) is

$$G = H - TS \tag{5.55}$$

where H is the molar enthalpy (page 221) and S the molar entropy.

Group 1:

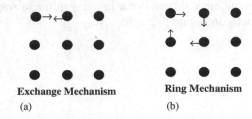

Exchange Mechanism

(a)

Ring Mechanism

(b)

Figure 5.25 (a) Direct exchange of two adjacent atoms. Unlikely process in close-packed crystal structures. (b) Direct exchange of three or more adjacent atoms. Complex and unlikely mechanism in solids but likely in liquids.

Group 2:

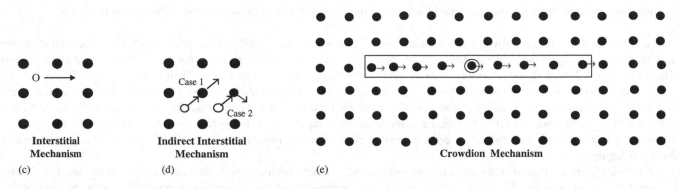

Interstitial Mechanism

(c)

Indirect Interstitial Mechanism

(d)

Crowdion Mechanism

(e)

Figure 5.25 (c) An atom moves from one interstitial site to an adjacent interstitial site. Very frequent mechanism in imperfect crystal lattices especially for *small* interstitial atoms. (d) An interstitial atom of nearly the size of a lattice atom replaces a lattice atom which moves to an interstitial site, either by a collinear impact (case 1) or by a non-collinear impact (case 2). Two successive indirect interstitial mechanisms are required to move the interstitial atom from one interstitial site to another. (e) Crowdion mechanism is a third type of interstitial diffusion. The additional atom is included in the *crowdion configuration*, a row of up to 10 atoms. Each atom in the configuration is slightly displaced from its equilibrium lattice position. The entire crowdion configuration can move along the row. After passing one configuration length, each atom is displaced one atomic distance. The centre of the configuration is marked by a circle.

Group 3:

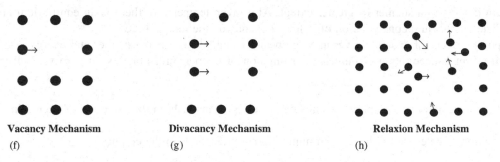

Vacancy Mechanism

(f)

Divacancy Mechanism

(g)

Relaxion Mechanism

(h)

Figure 5.25 (f) A lattice atom moves from its position to a vacant lattice site. The vacancy moves in the opposite direction. The vacancy mechanism is very frequent in imperfect crystal lattices, especially at high temperatures. (g) If bound vacancy pairs are present in a crystal, lattice diffusion by divacancies may be appreciable, especially at high temperatures. The mechanisms of single and double vacancies are similar but there are differences owing to the vacancy binding and the lack of symmetry. (h) If the atoms in the region of a vacancy relax inwards into a vacant site the lattice structure in the region may disappear. The region behaves like a liquid. The atoms in the region of 'localized melting' can diffuse relatively freely. Figures 5.25 a–h are reproduced from J. R. Manning, *Diffusion Kinetics for Atoms in Crystals.* © D. Van Nostrand Company, Inc., Princeton, NJ.

The condition (5.57) on page 236 will be used to find the equilibrium vacancy concentration as a function of temperature.

When a small amount, n_{vac} kmol, of vacancies are formed at constant temperature, the Gibbs free energy of the system increases by the amount

$$\Delta G = \Delta H - T dS = \Delta H_{vac\ form} n_{vac} - T\Delta S \tag{5.114}$$

where $\Delta H_{vac\ form}$ is the molar formation enthalpy of vacancies.

Provided that the number of vacancies is small, the vacancies do not interact, which means that the vacancy enthalpy dH increases linearly with n_{vac}. This fact has been used in Equation (5.114).

The entropy change dS is the sum of two terms:

$$dS = \Delta S_{vac\ form} n_{vac} + \Delta S_{mix} \tag{5.115}$$

where $\Delta S_{vac\ form}$ is the molar formation entropy of vacancies.

The first term corresponds to the entropy increase, related to the molar formation energy, when n_{vac} kmol of vacancies are formed.

The second entropy term, ΔS_{mix}, is due to the mixing of 1 kmol of the pure element and n_{vac} kmol of vacancies. This term can be found by application of Equation (5.42) on page 232, $n = 1 + n_{vac}$, $x_A = n_{vac}/(1 + n_{vac})$ and $x_B = 1/(1 + n_{vac})$ and we obtain

$$\Delta S_{mix} = -R\left[\frac{n_{vac}}{1+n_{vac}}\ln\left(\frac{n_{vac}}{1+n_{vac}}\right) + \frac{1}{1+n_{vac}}\ln\left(\frac{1}{1+n_{vac}}\right)\right] \tag{5.116}$$

As $n_{vac} << 1$ we may use $(1 + n_{vac}) \approx 1$ and the second term in Equation (5.116) can be neglected. Hence we obtain

$$\Delta S_{mix} = -Rn_{vac}\ln n_{vac} \tag{5.117}$$

Combinination of Equations (5.114), (5.115) and (5.117) gives

$$\Delta G = (\Delta H_{vac\ form} n_{vac} - T\Delta S_{vac\ form} n_{vac} + RTn_{vac}\ln n_{vac}) \tag{5.118}$$

As $n_{vac} \neq 0$, the equilibrium condition (5.57) above gives

$$\Delta H_{vac\ form} - T\Delta S_{vac\ form} + RT\ln n_{vac} = 0 \tag{5.119}$$

We solve for n_{vac} in Equation (5.119):

$$n_{vac} = e^{-\frac{\Delta H_{vac\ form} - T\Delta S_{vac\ form}}{RT}} = e^{-\frac{\Delta H_{a\ vac\ form} - T\Delta S_{a\ vac\ form}}{k_B T}} \tag{5.120}$$

or

$$n_{vac} = e^{-\frac{\Delta G_{vac\ form}}{RT}} = e^{-\frac{\Delta G_{a\ vac\ form}}{k_B T}} \tag{5.121}$$

where $\Delta G_{vac\ form}$ is the Gibbs free energy of formation of 1 kmol of vacancies and $\Delta G_{a\ vac\ form}$ the free energy of forming one vacancy.

The concentration of vacancies can be expressed as the fraction $x_{vac\ form}$ of the total number of atoms in the crystal. It is equal to n_{vac}:

$$x_{vac} = e^{-\frac{\Delta G_{vac\ form}}{RT}} = e^{-\frac{\Delta G_{a\ vac\ form}}{k_B T}} \tag{5.122}$$

Equation (5.122) is the equilibrium fraction of vacancies as a function of temperature.

Equilibrium Concentration of Large Interstitials as a Function of Temperature
Large Interstitials

The same type of equation can be derived in the same way for large interstitials as Equations (5.120)–(5.122) for vacancies:

$$x_{i\ form} = n_i = e^{-\frac{\Delta H_{i\ form} - T\Delta S_{i\ form}}{RT}} = e^{-\frac{\Delta G_{i\ form}}{RT}} \tag{5.123}$$

or in terms of one interstitial atom instead of 1 kmol of interstitial atoms:

$$x_{i\ form} = n_i = e^{-\frac{\Delta H_{a\ i\ form} - T\Delta S_{a\ i\ form}}{k_B T}} = e^{-\frac{\Delta G_{a\ i\ form}}{k_B T}} \tag{5.124}$$

where $x_{i\ form}$ is the fraction of interstitials of the total number of atoms of the same kind in the crystal. The subscript 'i' stands for interstitial.

If we insert reasonable values of ΔH_{vac}, ΔS_{vac}, ΔH_i and ΔS_i into Equations (5.120) and (5.123) or (5.124) for metals, we find that the heat of formation of a large interstitial is much higher than that of a vacancy. Calculations show that the fraction of large interstitials is very much lower than the fraction of vacancies and can be neglected in comparison with the vacancy fraction, which normally is of the magnitude 10^{-3}–10^{-4} at the melting point of the metal.

Energy Barrier of a Jumping Atom

A necessary condition for a jump of an atom is that the atom must have enough energy to overcome the energy barrier between the initial and final sites.

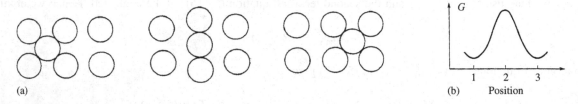

(a)   (b)   Position

Figure 5.26 (a) Sequence of configurations involved when an interstitial atom moves from one normal site to an adjacent one. At a normal site, which represents an equilibrium position, the free energy has a minimum value. If the interstitial atoms are smaller than the solvent atoms the diffusion is greatly facilitated.
(b) Variation of the free energy of the entire crystal lattice when a diffusion atom is reversibly moved from position 1 (the initial state) to 2 (the intermediate state) to 3 [the final state in (a)]. Reproduced from P. Shewmon, *Diffusion in Solids*. © 1963 McGraw-Hill Book Company, Inc.

In Figure 5.26b, the free energy of a jumping interstitial atom is plotted as a function of its position. The curve has a maximum corresponding to the saddle point in Figure 5.22 on page 251. The free energy difference $\Delta G_{i\ barrier}$ between the saddle point and the normal sites of the interstitial atom is the energy barrier which has to be overcome by the jumping atom.

The fraction of the available interstitials which occasionally have sufficient energy from phonon collisions to pass the energy barrier can be written as

$$f_{i\ barrier} = e^{-\frac{\Delta G_{a\ i\ barrier}}{k_B T}} \tag{5.125}$$

where $\Delta G_{a\ i\ barrier}$ is the energy barrier for an interstitial atom.

Analogously, we obtain the fraction of atoms nearby a vacancy which have sufficient energy to jump:

$$x_{vac\ barrier} = e^{-\frac{\Delta G_{a\ vac\ barrier}}{k_B T}} \tag{5.126}$$

where $G_{a\ vac\ barrier}$ is the energy barrier for a lattice atom jump.

The lower the energy barrier is, the more frequent will be the atom jumps. A high-energy barrier corresponds to infrequent atom jumps.

Interstitials and vacancies must be treated separately during calculations as their energy barriers differ. As the fraction of large interstitials is very small compared with the fraction of vacancies, we will concentrate on jumps of lattice atoms to vacancies below.

Jump Frequency of Lattice Atoms to Vacancies. Activation Energy

The influence of an energy barrier, i.e. activation energy $U_{a\ act}$, on reaction rates has been treated in Section 5.4.3 on page 242. It resulted in the Arrhenius equation [Equation (5.82) on page 243]. Eyring has shown that if the internal energy $U_{a\ act}$ is replaced by the Gibbs free energy $G_{a\ act}$, better agreement with experiments is obtained (Section 5.5.3 on page 247).

The transformation kinetics of homogeneous chemical reactions and diffusion in a solid are analogous. The collision frequency of reacting atoms corresponds to the vibrational frequency of the atoms in the crystal lattice. Activation energy is present in both cases. The reaction rate corresponds to the jump frequency of the diffusing atoms. Below we will derive an expression for the jump frequency and analyse the origin of the activation energy.

The number of lattice atom jumps per unit volume and unit time is a function of three factors:

1. the number of tentative jumping atoms per unit volume, i.e. atoms which have a vacancy as nearest neighbour
2. the fraction of the number of atoms in the initial state which have enough energy to overcome the energy barrier
3. the frequency with which the atoms have the opportunity to jump.

Considering the conditions above, we can write the number of atom jumps per unit volume and unit time as a product of five factors:

$$F_{jump} = N_0 x_{vac} Z_{coord} f_{barrier} \nu_{vibr} \qquad (5.127)$$

where

F_{jump} = number of atom jumps per unit volume and unit time
N_0 = number of atoms per unit volume
x_{vac} = fraction of vacancies
Z_{coord} = number of adjacent equivalent sites per atom, i.e. number of nearest neighbours per atom
$f_{barrier}$ = fraction of atoms with a vacancy as nearest neighbour which have enough energy to overcome the energy barrier
ν_{vibr} = vibration frequency of the lattice atoms.

The number of vacancies per unit volume is $N_0 x_{vac}$. The number of tentative jumping atoms per unit volume which surround the vacancies is $N_0 x_{vac} Z_{coord}$. Of these, only $N_0 x_{vac} Z_{coord} f_{barrier}$ have enough energy to pass the energy barrier between the initial and final sites. This is the number of atoms which fulfil conditions 1–3 above. Each of them has ν_{vibr} chances per second to jump as the number of vibrations per second is equal to the vibration frequency ν_{vibr}.

The number of atom jumps per unit volume and unit time equals the number of atoms which fulfil conditions 1–3 times the vibration frequency of the atoms.

Using Equations (5.122) and (5.126), we can rewrite Equation (5.127) as

$$F_{jump} = N_0 e^{-\frac{\Delta G_{a\ form}}{k_B T}} Z_{coord} e^{-\frac{\Delta G_{a\ barrier}}{k_B T}} \nu_{vibr} \qquad (5.128)$$

If we divide Equation (5.128) by N_0 and introduce the jump frequency of an atom or for short the *jump frequency* f_{jump}:

$$f_{jump} = \frac{F_{jump}}{N_0} \qquad (5.129)$$

we obtain

$$f_{jump} = Z_{coord} \nu_{vibr} e^{-\frac{G_{a\ form} + G_{a\ barrier}}{k_B T}} \qquad (5.130)$$

If we apply the relationship $G = H - TS$ to both exponential terms, we obtain

$$f_{jump} = Z_{coord} v_{vibr} e^{\frac{S_{a\ form}}{k_B}} e^{-\frac{H_{a\ form}}{k_B T}} e^{\frac{S_{a\ barrier}}{k_B}} e^{-\frac{H_{a\ barrier}}{k_B T}} \tag{5.131}$$

In Equation (5.131), the two entropy factors can be included in a constant.

The sum of the formation energy of a vacancy and the energy barrier is called the *activation energy of diffusion*. Using an analogous designation of the activation energy as in earlier sections of transformation kinetics, we can write

$$H_{a\ act} = H_{a\ form} + H_{a\ barrier} \tag{5.132}$$

In solids the product pV is small compared with H and U. Hence the enthalpy H can be replaced by the internal energy U and the activation energy is also given by the expression

$$U_{a\ act} = U_{a\ form} + U_{a\ barrier} \tag{5.133}$$

The jump frequency can be written as

$$f_{jump} = Z_{coord} v_{vibr} e^{-\frac{U_{a\ act}}{k_B T}} \tag{5.134}$$

where $U_{a\ act}$ = activation energy of diffusion.

Equation (5.134) has the same form as Equation (5.82) on page 243 but should rather be compared with Equation (5.97) on page 248.

Jump Frequency of Small Interstitial Atoms

In the case of small interstitials, the conditions are different compared with vacancies. Vacancies are formed in the crystal lattice and their concentration varies strongly with temperature. Small interstitials such as H and C are solved foreign atoms with a given concentration independent of temperature and there is no formation energy. The interstitial sites exist in excess and the majority of them are empty.

The same equation as for lattice atom jumps to vacancies:

$$f_{jump} = Z_{coord} v_{vibr} e^{-\frac{U_{a\ act}}{k_B T}} \tag{5.134}$$

is valid for small interstitials, but with an important modification: the jump frequency depends on the activation energy but $U_{a\ act} = U_{a\ barrier}$ as $U_{a\ form} = 0$.

Small interstitals jump if they have energy enough to overcome the energy barrier between adjacent sites. The interstitials become activated by phonon collisions.

The Diffusion Coefficient of Lattice Atoms as a Function of Atomic Quantities and Temperature

The atoms in all crystal lattices vibrate around their equilibrium positions. Owing to the Maxwell–Boltzmann distribution of their thermal energies (Section 5.3.7 on page 240), some atoms have energies large enough to leave their lattice site and jump to an adjacent site. Diffusion in crystals occurs by atom jumps by means of some of the mechanisms mentioned on pages 255–258, preferably the vacancy and interstitial mechanisms.

Therefore, the diffusion coefficient is related to atomic quantities and to the atom motion in the lattice. In this section we will derive such relationships. We start with the simplest case, a one-dimensional random walk in a crystal with a single jump distance.

One-dimensional Random Walk in a Crystal with a Single Jump Distance
Consider a bar of a binary crystalline material with a cubic structure. It has a concentration gradient of the solute along the y axis. The solute atoms are assumed to jump only upwards and downwards. When they change position, they jump a distance $\pm d_j$ along the y axis.

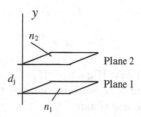

Figure 5.27   Parallel planes in a cubic crystal structure.

Consider two adjacent (010) lattice planes 1 and 2 in a crystal with cubic structure (Figure 5.27). The following assumptions are made:

n_1 = number of diffusing atoms per unit area and unit time which jump from plane 1 to plane 2
f_{12} = jump frequency of an atom from plane 1 to plane 2
n_2 = number of diffusing atoms per unit area and unit time which jump from plane 2 to plane 1
f_{21} = jump frequency of an atom from plane 2 to plane 1.

The net flux J in the diffusion direction, i.e. from plane 1 to plane 2, will be

$$J = n_1 f_{12} - n_2 f_{21} \qquad (5.135)$$

To obtain a relationship between the diffusion coefficient D and atomic quantities we must find a relationship between the surface concentration n of atoms and the volume concentration c of atoms in the crystal lattice. Consider the slice of the crystal in Figure 5.28.

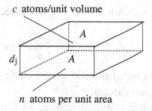

Figure 5.28   Part of a crystal with cubic structure.

The number of diffusing atoms included in the slice with a thickness d_j, equal to the distance between planes 1 and 2, and a cross-section of A can be written in two ways:

$$nA = d_j A c \qquad (5.136)$$

where d_j is the jump distance. n depends on position. For small changes of y we have

$$n_2 = n_1 + d_j \frac{dn}{dy}$$

or

$$n_1 - n_2 = -d_j \frac{dn}{dy} = -d_j \frac{dn}{dc} \frac{dc}{dy} = -d_j d_j \frac{dc}{dy} \qquad (5.137)$$

where the relationship $dn/dc = d_j$ is obtained if Equation (5.136) is divided by A and we take the derivative of Equation (5.136) with respect to c.
   The expression

$$n_1 - n_2 = -d_j^2 \frac{dc}{dy} \qquad (5.137)$$

is introduced into Equation (5.135) and we obtain for $f_{12} = f_{21} = f_j$ because of the symmetry of a cubic structure

$$J = -d_j^2 f_j \frac{dc}{dy} \tag{5.138}$$

This equation can be identified with Fick's first law [Equation (5.113) on page 254], which gives an expression of the coefficient of diffusion in one direction. In Figure 5.27 we have chosen the y direction. In the general case we choose an arbitrary direction [uvw] (Chapter 1, pages 15–16) and obtain

$$D = d_j^2 f_j \tag{5.139}$$

where
  $d_j = $ jump distance in the diffusion direction
  $f_j = $ jump frequency.

Equation (5.139) has been derived for a single atom jump only in a diffusion direction along one of the x, y or z axes.

Cubic Crystals
In the three-dimensional cubic crystal the jump distances are equal and the atoms can jump to any vacant nearest neighbour site (Figure 5.29 on page 263). Hence there are six equivalent nearest neighbour directions: [100], [$\bar{1}$00], [010], [0$\bar{1}$0], [001] and [00$\bar{1}$]. The total jump frequency (compare pages 259–260) will be

$$f_{j \text{ total}} = Z_{\text{coord}} f_j = 6f_j = \sum_{uvw} f_{uvw} \tag{5.140}$$

where Z_{coord} is the number of adjacent sites (nearest neighbours) and the subscript uvw indicates the six jump directions.

   The relationship $f_j = f_{j \text{ total}}/6$ is introduced into Equation (5.139). Equation (5.139) corresponds to two jump directions, i.e. jumps along *one* axis (x, y or z). The diffusion coefficient for diffusion in a cubic crystal along *three* axes and six jump directions will be three times higher. As the effective jump distance in the direction of the diffusion direction varies for the six jump directions, D must be written as a sum, extended over the six directions:

$$D = \frac{3}{6} \sum_{uvw} d_{uvw}{}^2 f_j = \frac{1}{2} \sum_{uvw} d_{uvw}{}^2 f_j \tag{5.141}$$

The effective jump distance d_{uvw} is equal to d_j in the direction of the diffusion and zero for perpendicular jumps. The jump frequency is assumed to be the same in all the jump directions.

   Equation (5.141) is valid for crystals with simple cubic structure, FCC structure and BCC structure. An example of the calculation of D with the aid of Equation (5.141) will be given in the next section.

   To obtain an idea of the magnitudes of the quantities involved, we realize that d_j must be of the same magnitude as the inter-atomic distances, 10^{-10} m. Near their melting point most FCC metals have diffusion coefficients of magnitude 10^{-12} m²/s. Using Equation (5.140) we obtain a magnitude of f_{jump} equal to 10^8 s^{-1}.

   Hence the atoms jump from one site to another in the lattice 100 million times per second. This seems to be incredibly high, but the figure should be compared with the vibration frequencies of the atoms in the lattice. From measurements of the vibration energies one knows that the vibration frequency equals 10^{12}–10^{13} s^{-1}. That means that only 0.01–0.001% of the vibrations lead to a jump.

General Case
Most crystals have more complicated structures than the cubic structure. The diffusion coefficient is not necessarily the same in different crystallographic directions as both f_{uvw} and d_{uvw} may vary. In the general case, the diffusion coefficient can be calculated from the relationship

$$D = \frac{1}{2} \sum_{uvw} f_{uvw} d_{uvw}^2 \tag{5.142}$$

where
  $f_{uvw} = $ jump frequency of possible atom jumps to neighbouring sites in the jump direction [uvw]
  $d_{uvw} = $ effective jump distance, i.e. its projection on the direction of diffusion.

   Equation (5.142) is very useful for the calculation of the diffusion coefficient in various directions of crystals. To show its practical use we will apply the general equation to a cubic structure and a hexagonal close-packed structure.

Simple Cubic Structure

Figure 5.29   Simple cubic crystal structure.

Consider a simple cubic crystal where the diffusion is directed along the *x* axis. Equation (5.142) can be checked by applying it to a simple cubic structure (Figure 5.29). Consider an atom in position (0, 0, 0) and let the frequency subscripts refer to the nearest neighbouring sites of possible jumps. Equation (5.142) gives

$$D = \frac{1}{2}\left[f_{100}\, d_j^2 + f_{\bar{1}00}\left(-d_j\right)^2 + \left(f_{010} + f_{0\bar{1}0} + f_{001} + f_{00\bar{1}}\right) \times 0^2\right] \tag{5.143}$$

It may seem confusing that the jump distances are set to zero for jumps perpendicularly to the diffusion direction, but it is important to observe that it is only the *projections* on these directions that count and *contribute to the diffusion*. The frequency is the same.

For symmetry reasons, all the frequencies are equal and Equation (5.143) can be reduced to

$$D = \frac{1}{2}\left[f_j d_j^2 + f_j\left(-d_j\right)^2\right] = f_j d_j^2 = \frac{1}{6}f_{j\text{ total}}\, d_j^2 \tag{5.144}$$

in agreement with Equation (5.141) and Equation (5.140), which give

$$D = \frac{1}{2}\sum_{uvw} d_{uvw}{}^2 f_j = \frac{1}{2} \times 2d_j^2 \times \frac{f_{j\text{ total}}}{6} = \frac{1}{6}f_{j\text{ total}}\, d_j^2$$

Hexagonal Close-packed Structure

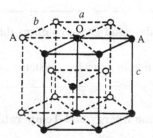

Figure 5.30   HCP crystal structure. See pages 23–24 in Chapter 1. Reproduced with permission from B. D. Cullity, *Elements of X-Ray Diffraction*, © Addison-Wesley Publishing Company, Inc. (now under Pearson Education).

Consider the central atom O in the upper horizontal plane in Figure 5.30. The diffusion occurs along the vertical *c* axis and we want to find the coefficient of diffusion in this direction. Jumps to the six symmetrical adjacent sites in the horizontal plane at distances *a* and the three sites in horizontal planes above and below the central atom are the nearest neighbour sites which have to be taken into consideration.

Equation (5.142) is applied to the 12 possible jumps from O to each of the 12 sites:

$$D_c = \frac{1}{2}\left[6f_c\left(\frac{c}{2}\right)^2 + 6f_a \times 0^2\right] = \frac{3}{4}f_c c^2 \tag{5.145}$$

where

f_c = jump frequency from O to the six 'middle' sites
$f_a = f_b$ = jump frequency from O to the six symmetrical sites.

The jumps from O to the symmetrical sites in the upper horizontal plane are perpendicular to the diffusion direction and do not contribute to the diffusion. The jumps to the middle sites have jump distance projections equal to $|d_j| = c/2$.

Example 5.4

Find the diffusion coefficient in an HCP crystal in terms of jump frequencies and jump distances provided that the diffusion has the direction AA′ shown in Figure 5.30.

Solution:

The diffusion coefficient along the axis AA′ in the horizontal plane will be

$$D_{AA'} = \frac{1}{2}f_a \left[a^2 + (-a)^2 + 2\left(\frac{a}{2}\right)^2 + 2\left(-\frac{a}{2}\right)^2 \right] + \frac{1}{2}f_c \left[2\left(\frac{a}{2}\right)^2 + 2 \times 0 + 2\left(-\frac{a}{2}\right)^2 \right]$$

The 6 atoms in the horizontal plane   The 2×3 'middle' atoms

where the projections of the jump distances from O to the middle atoms on the axis AA′ are $-a/2$, 0 and $a/2$, respectively. After reduction, we obtain

$$D_{AA'} = \frac{3}{2}f_a a^2 + \frac{1}{2}f_c a^2$$

Answer:

The diffusion coefficient in the direction AA′ is

$$D_{AA'} = \frac{3}{2}f_a a^2 + \frac{1}{2}f_c a^2$$

A comparison between Equation (5.145) and the result above confirms that

- The diffusion coefficient varies with the direction in noncubic crystals.

Diffusion Coefficient as a Function of Activation Energy and Temperature

In the preceding sections we found that the diffusion coefficient is proportional to the jump frequency and a function of the square of the jump distance. The latter is substantially independent of temperature whereas the jump frequency varies strongly with temperature. According to Equation (5.134) on page 260, we have

$$f_{jump} = f_0 e^{-\frac{U_{a\ act}}{k_B T}} \tag{5.134}$$

Hence the diffusion coefficient can be written as

$$D = D_0 e^{-\frac{U_{a\ act}}{k_B T}} \tag{5.146}$$

where D_0 is a material constant, which includes f_0 and the influence of the jump distance.

Equation (5.146) is the important relationship between the diffusion coefficient and the activation energy of diffusion. The higher the activation energy is, the lower will be the diffusion coefficient. The diffusion coefficient increases with increase in temperature.

Self-diffusion

The theory of diffusion, which we have discussed so far in Sections 5.6.1–5.6.3, includes only one type of diffusing atoms. It is strictly valid only for diffusion in pure solids and valid to a good approximation for dilute binary alloys, where only the solute atoms diffuse. In these cases the diffusion coefficient is defined by

$$J_A = -D_A \frac{dc_A}{dy} \tag{5.147}$$

where

J_A = flux of diffusing atoms A relative to the crystal lattice
D_A = diffusion coefficient of atoms A
c_A = concentration of atoms A in terms of number of atoms per unit volume
y = coordinate.

In the case of diffusion of radioactive tracer atoms A* in a metal of pure A atoms, the diffusion coefficient in Equation (5.147) is called the *self-diffusion* coefficient or *intrinsic* diffusion coefficient.

As we shall see in Section 5.6.4, this quantity is of minor interest in systems with more than one diffusing component.

5.6.4 Diffusion in Alloys

So far we have only discussed diffusion in solids with a single diffusing component, for example self-diffusion by the vacancy mechanism. In these cases the theory given in the preceding sections is valid.

In alloys with two or more components, diffusion is in most cases more complicated, mainly for two reasons:

- There are several diffusing components.
- Atoms of different kinds interact, which influences the diffusion and cannot be neglected.

The treatment of diffusion with several diffusing components requires an extension of the theory of diffusion and a new interpretation of the concept of diffusion coefficient. We shall mainly discuss binary alloys.

To extend the theory of diffusion we need both thermodynamics and the theory of transformation kinetics. A more general condition for the driving force of diffusion will be introduced. The diffusion rates of the components in an alloy are different. Each component has its own driving force. This leads to an equation system, one equation for each component, where the interactions between atoms of different kinds are considered.

Driving Force of Diffusion

Fick's first law on page 254 is the common law of diffusion. It gives the impression that the driving force of diffusion is the concentration gradient of the diffusing element. This is true in most cases, but there are rare cases of diffusion when the diffusing element diffuses from lower to higher concentrations or diffuses with no concentration difference at all. An example is given on page 274.

This is a contradiction of Fick's first law but not to the general physical laws. Diffusion is a spontaneous process and at constant pressure and temperature the general condition for spontaneous processes, given on page 236, must be valid.

- At constant temperature and pressure, processes always occur spontaneously in the direction of decreasing G:

$$dG_{T,p} < 0 \tag{5.57}$$

To be able to use this condition to derive an expression for the driving force of diffusion, we must relate the Gibbs free energy and the concentration of the diffusing element. This can be done by using the concept of *chemical potential*, which we introduced on page 236:

The chemical potential of a pure element A is defined as the Gibbs free energy of the element:

$$\mu_A^0 = G_A^0 \tag{5.58}$$

The chemical potential of A varies with the concentration of A in the solution as

$$\mu_A = \mu_A^0 + RT \ln x_A \gamma_A \tag{5.59}$$

The activity coefficient γ_A is a function of concentration. In solutions with a composition close to the pure element A, the activity coefficient approaches the value 1.

We assume that the chemical potential is a function of no other quantities than the concentration distribution of the solution. In this case the driving force of diffusion can be written as

$$-\Delta G^{\text{diff}} = \text{grad } G = \text{grad } \mu \tag{5.148}$$

The rigorous condition for the driving force of diffusion is obtained by replacing the concentration gradient by grad μ in Equation (5.112) on page 254. Instead of Equation (5.113) on page 254, we obtain the proper basic diffusion condition in one dimension:

$$J = -L \frac{d\mu}{dy} \tag{5.149}$$

where L is the diffusion coefficient related to the gradient of the chemical potential.

In the absence of external influences, Equations (5.149) and (5.113) never contradict each other in cases with a single diffusion component or in binary alloys. The diffusing atoms move from higher towards lower concentration.

If the solid is exposed to electrical fields or other influences and in ternary alloys there are cases when diffusing atoms move from lower towards higher concentration. In these cases, Equations (5.149) and (5.113) contradict each other and Equation (5.149) determines the direction of the diffusion flow.

Calculation of the Diffusion Coefficient D as a Function of the Diffusion Coefficient L in Solid Binary Alloys
Equation (5.149) can be applied on the diffusing A atoms:

$$J_A = -L_A \frac{d\mu_A}{dy} \tag{5.150}$$

We take the derivative of Equation (5.59) on page 236 is with respect to y, which gives

$$\frac{d\mu_A}{dy} = RT \left(\frac{1}{x_A} \frac{dx_A}{dy} + \frac{d \ln \gamma_A}{dy} \right) \tag{5.151}$$

Combination of Equations (5.149) and (5.151) gives

$$J_A = -L_A \frac{d\mu_A}{dy} = -L_A RT \left(\frac{1}{x_A} \frac{dx_A}{dy} + \frac{d \ln \gamma_A}{dy} \right) \tag{5.152}$$

Alternatively, the diffusion flux can be written, based on Equation (5.113) on page 254 as

$$J_A = -D_A \frac{dc_A}{dy} \tag{5.113}$$

We want to identify Equations (5.113) and (5.152). Hence we must introduce c_A instead of x_A into Equation (5.152). The relationship between the x_A and c_A can be written as

$$x_A = \frac{c_A}{c_0} \tag{5.153}$$

where
x_A = mole fraction of A atoms in the alloy
c_A = number of A atoms per unit volume of the alloy
c_0 = total number of atoms per unit volume of the alloy.

Taking the derivative of Equation (5.153) with respect to y gives

$$\frac{dx_A}{dy} = \frac{dc_A}{c_0 dy} \tag{5.154}$$

The expressions in Equations (5.153) and (5.154) are introduced into Equation (5.152):

$$J_A = -L_A RT \left(\frac{c_0}{c_A} \frac{dc_A}{c_0 dy} + \frac{d \ln \gamma_A}{dy} \right) = -L_A \frac{RT}{c_A} \frac{dc_A}{dy} \left(1 + \frac{c_A d \ln \gamma_A}{dc_A} \right)$$

or, as $d \ln c_A = 1/c_A$

$$J_A = -L_A \frac{RT}{c_A} \frac{dc_A}{dy} \left(1 + \frac{d \ln \gamma_A}{d \ln c_A} \right) \tag{5.155}$$

Identification of Equations (5.113) and (5.155) gives

$$D_A = L_A \frac{RT}{c_A} \left(1 + \frac{d \ln \gamma_A}{d \ln c_A} \right) \tag{5.156}$$

Equation (5.156) is the desired relationship, which holds for all values of c_A. In a *dilute* solution $\gamma_A = 1$ and Equation (5.151) can be replaced by

$$D_A = L_A \frac{RT}{c_A} \tag{5.157}$$

As an example, we will apply the theory to a binary alloy consisting of two pure elements A and B. Their concentrations can be expressed either as c_A and c_B (number of atoms per unit volume) or as mole fractions x_A and x_B.

The total chemical potential of the solution can be written as

$$\mu_{total} = x_A \mu_A + x_B \mu_B \tag{5.158}$$

where

$$\mu_A = \mu_A^0 + RT \ln x_A \gamma_A \tag{5.159}$$

$$\mu_B = \mu_B^0 + RT \ln x_B \gamma_B \tag{5.160}$$

In the absence of all external gradients and negligible strain energy, the fluxes of the solute and solvent are functions of the chemical potential gradients. In a one-dimensional flow in an isotropic binary material there are two concentration gradients, one for each component.

The two fluxes J_A and J_B depend on the two gradients of chemical potentials μ_A and μ_B. In the case of one-dimensional diffusion we obtain

$$J_A = -L_{AA} \frac{d\mu_A}{dy} - L_{AB} \frac{d\mu_B}{dy} \tag{5.161}$$

$$J_B = -L_{BA} \frac{d\mu_A}{dy} - L_{BB} \frac{d\mu_B}{dy} \tag{5.162}$$

where
  J_A, J_B = fluxes of components A and B
  L_{ij} = constants
  μ_A, μ_B = chemical potentials of A and B
  y = coordinate along the diffusion direction.

Owing to the interaction between the A and B atoms, 'mixed' terms appear in Equations (5.161) and (5.162). The second term in Equation (5.161) describes how the B atoms affect the diffusion rate of A atoms. The first term in Equation (5.162) is a measure of the influence of the A atoms on the diffusion rate of the B atoms.

By use of the relationship between D and L [identification of Equations (5.150) and (5.113)], the equation system can be replaced by

$$J_A = -D_{AA}\frac{dc_A}{dy} - D_{AB}\frac{dc_B}{dy} \tag{5.163}$$

$$J_B = -D_{BA}\frac{dc_A}{dy} - D_{BB}\frac{dc_B}{dy} \tag{5.164}$$

Equations (5.163) and (5.164), valid for binary alloys, are the equations which correspond to Fick's first law in systems with only one diffusing agent.

Diffusion in Binary Alloys

The diffusion in binary alloys depends strongly on the diffusion mechanisms in the alloy. On pages 255–258 we mentioned that the dominant diffusion mechanisms are the vacancy mechanism and the interstitial mechanism.

The vacancy mechanism occurs in substitutional alloys, where a fraction of the lattice atoms are replaced by other atoms of similar size. The interstitial mechanism dominates in binary alloys with lighter alloying elements than the lattice atoms. The diffusion differs widely in the two cases and completely different models of the diffusion have to be applied.

Interstitial Diffusion

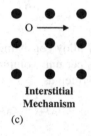

**Interstitial
Mechanism**

(c)

Figure 5.25   (c) Interstitial mechanism of diffusion.

Interstitial atoms are normally much smaller than the lattice atoms for simple space reasons. The most typical example of interstitials in binary alloys is carbon atoms in steel and cast iron. The number of interstitials is small compared with the number of lattice atoms. Hence the interstitial atoms are normally surrounded by lattice atoms and empty interstitial sites (Figure 5.25c).

The interstitials and the great number of empty interstitial sites can be regarded as a sublattice of their own and the diffusion of the interstitials occurs in this sublattice as single-component diffusion. The interaction with the crystal lattice can be entirely neglected. The crystal lattice behaves like a rigid body and the lattice atoms do not diffuse.

This model of interstitial diffusion works very well. This is the reason why the theory in the preceding sections is valid for diffusion by the interstitial mechanism. An example is impurity diffusion on solidification, which results in microsegregation in the solidified product.

Substitutional Diffusion

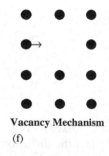

Vacancy Mechanism

(f)

Figure 5.25   (f) Vacancy mechanism of diffusion.

The dominant diffusion mechanism of lattice atoms and substitutional atoms is the vacancy mechanism (Figure 5.25f). Each lattice point is occupied by a substitutional atom A, a lattice atom B or a vacancy. The fact that two types of atoms diffuse makes substitutional diffusion much more complicated than self-diffusion and interstitial diffusion. The theory, given in the preceding sections, is not directly applicable.

Below we will discuss the theory of substitutional diffusion in binary alloys and consider the simultaneous diffusion of both components.

Theory of Substitutional Diffusion in Binary Alloys

Analysis of the Fluxes on Substitutional Diffusion

We consider two bars of metals A and B which are welded together along a cross-section plane (weld plane in Figure 5.31). The A and B atoms have in most cases different jump frequencies and different diffusion rates. Consequently the atom fluxes will also be different. As we have mentioned above and will see later, there is a strong interaction between the two types of atoms. The consequence is that the diffusion coefficients of atoms A and B vary with the composition of the alloy.

We consider the flux of atoms across an arbitrary plane in the alloy, perpendicular to the diffusion direction (Figure 5.31). We assume that

1. The solute atoms A are smaller than the solvent atoms B.
2. The diffusion rate of the A atoms is larger than that of the B atoms.

As both A and B atoms are assumed to diffuse by the vacancy mechanism, it is reasonable to study the net vacancy flux in particular. Each atom jump across the plane is connected with the movement of a vacancy across the plane in the opposite direction. Three driving forces are present, one for the A atoms, one for the B atoms and one for the vacancies.

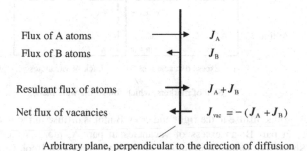

Flux of A atoms J_A

Flux of B atoms J_B

Resultant flux of atoms $J_A + J_B$

Net flux of vacancies $J_{vac} = -(J_A + J_B)$

Arbitrary plane, perpendicular to the direction of diffusion

Figure 5.31   Fluxes of A atoms, B atoms and vacancies at substitutional diffusion in the absence of edge dislocations. Driving forces act on the A atoms, the B atoms and the vacancies. Reproduced with permission from P. Shewmon, *Diffusion in Solids*. © 1963 McGraw-Hill Book Company, Inc.

Hence the resultant flux of atoms across the plane is associated with a net flux of vacancies of the same size and opposite direction. This can be expressed by the 'material balance':

$$J_A + J_B + J_{vac} = 0 \tag{5.165}$$

This relationship holds provided that there are no sources and sinks of vacancies in the alloy. In the presence of edge dislocations (page 26 in Chapter 1) the vector sum of the fluxes is no longer zero. Zero has to be replaced by the rate of vacancy production or annihilation at the edge dislocations.

The diffusion process and Equation (5.165) correspond to local conservation of vacancies, which is a necessary condition. It leads to an *uneven distribution of vacancies* in the alloy. The region to the left of the plane receives an excess of vacancies whereas the region to the right obtains a vacancy concentration below the average (Figure 5.31).

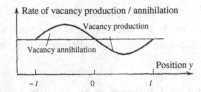

Figure 5.32 Excess of vacancies (to the right) is reduced by annihilation at edge dislocations. Lack of vacancies is reduced by vacancy creation at edge dislocations (to the left). Edge dislocations can act both as sinks and sources of vacancies. For position scale, see Figure 5.34 on page 271. Reproduced with permission from P. G. Shewnon, *Transformations in Metals*. © 1969 McGraw-Hill Book Company, Inc.

There must be a mechanism which brings the vacancy concentration back to its average value in both regions. The only realistic and energetically possible means is vacancy annihilation at edge dislocations in excess regions and vacancy production at edge dislocations in low-vacancy regions (Figure 5.32).

These processes involve deposition of new atomic planes in one region and removal of atomic planes in other regions. It has been possible to check the theory by observing markers, i.e. insoluble inclusions at the weld plane, and study their displacement of the weld plane and hence indirectly study the rate at which vacancies are created or annihilated in the alloy to maintain the equilibrium concentration x_{vac}. The positions of the markers are studied as a function of time. Figure 5.33 shows the result. A small displacement of the position of the weld plane is observed.

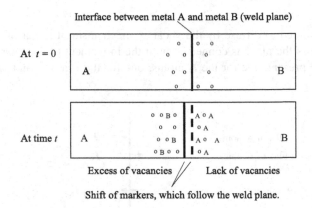

Figure 5.33 Kirkendall effect. Fast A atoms diffuse to the right and slow B atoms diffuse in the opposite direction. They cause: a vacancy concentration, lower than the average, in part B; an excess of vacancies in part A; more A atoms in part B than B atoms in part A. Reproduced with permission from P. Shewmon, *Diffusion in Solids*. © 1963 McGraw-Hill Book Company, Inc.

Experimental measurements of marker displacements have been performed for many alloys with $D_A \neq D_B$. The effect was first discovered and interpreted by Kirkendall and Smigelskas in the middle of the 20th century and is called the Kirkendall effect. It appears in all binary alloys with substitutional structure and different diffusion rates of the solute and solvent atoms.

For the special case that $D_A \approx D_B$, the diffusion process will be much simpler. There will be no flux of vacancies and no production or annihilation of vacancies at edge dislocations. The concentration of atoms will be constant:

$$c_A + c_B = c_0$$

and Equation (5.165) will be simplified to

$$J_A + J_B = 0 \tag{5.166}$$

An analysis of the motion of the diffusing atoms A and atoms B offers a possibility of a better understanding the diffusion in a binary alloy. For this reason we will discuss the theory of the Kirkendall effect below.

Chemical Diffusion Coefficient

The simplest possible description of the total diffusion process in a substitutional alloy is obtained if the concept of *chemical diffusion coefficient* is introduced. For this purpose we will discuss the simultaneous diffusion of atoms A, atoms B and vacancies from a mathematical point of view.

Chemical diffusion coefficients are marked with a special sign. The reason for this will be explained on page 272.

Theory of the Kirkendall Effect

Assume as before that the A atoms have a higher diffusion rate than the B atoms. Therefore, an excess of A atoms arises to the right of the interface (weld plane). This excess is balanced by an excess of vacancies to the left of the interface (Figure 5.34).

Owing to the mass transport, the B part expands (greatly exaggerated in the figure) at the expense of the A part. This leads to the Kirkendall shift, indicated in Figure 5.34 by short arrows.

We introduce a coordinate system with $\tilde{y}$ connected with the left end of the A part and set up an expression of the B flux through the interface at time t:

$$J_B(\tilde{y} = l, t) = -D_B \frac{\partial c_B}{\partial \tilde{y}} + v c_B(\tilde{y} = l, t) \tag{5.167}$$

where v is the velocity of the interface in the coordinate system $\tilde{y}$. The first term on the right-hand side is the usual diffusion term. The second term has to be added owing to the motion of the interface relative to the origin (the left-hand end of the A part).

An analogous expression can be set up for atoms A:

$$J_A(\tilde{y} = l, t) = -D_A \frac{\partial c_A}{\partial \tilde{y}} + v c_A(\tilde{y} = l, t) \tag{5.168}$$

$$\text{The total density of atoms is } c = c_A + c_B \tag{5.169}$$

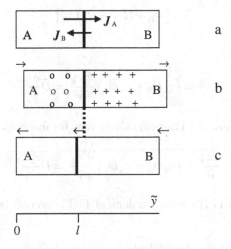

Figure 5.34 (a) The diffusion in (a) leads to the situation in (b), i.e. an excess of A atoms in part B and an excess of vacancies in part A.
(b) Part B swells and part A shrinks slightly.
(c) This does *not* result in a displacement of the whole piece of metal as indicated in (b). Instead, the volume changes result in a displacement of the weld plane as indicated in by small arrows in (c). Reproduced with permission from P. Haasen, *Physical Metallurgy*, 3rd edn. © 1996 Cambridge University Press.

In Chapter 4 [Equation (4.84) on page 202] we derived Fick's second law. We will use it here with concentrations c instead of mole fractions x and the additional vc term included. We take the partial derivatives of Equation (5.169) with respect to t and the partial derivatives of Equations (5.167) and (5.168) with respect to y and introduce them into Equation (4.84) The result is

$$\frac{\partial c}{\partial t} = \frac{\partial c_A}{\partial t} + \frac{\partial c_B}{\partial t} = \left(\tilde{D}_A \frac{\partial^2 c_A}{\partial \tilde{y}^2} - v \frac{\partial c_A}{\partial \tilde{y}} \right) + \left(\tilde{D}_B \frac{\partial^2 c_B}{\partial \tilde{y}^2} - v \frac{\partial c_B}{\partial \tilde{y}} \right) \tag{5.170}$$

or, if we assume that the total density c remains constant, we obtain

$$\frac{\partial c}{\partial t} = \frac{\partial}{\partial \tilde{y}} \left(\tilde{D}_A \frac{\partial c_A}{\partial \tilde{y}} + \tilde{D}_B \frac{\partial c_B}{\partial \tilde{y}} - vc \right) = 0 \tag{5.171}$$

Hence the expression inside the parentheses in Equation (5.171) must be constant. The value of the constant can be determined from the conditions at $\tilde{y} = 0$. It is far from the diffusion region and the gradients and v are zero. Hence the expression inside the parentheses must be zero also at $\tilde{y} = l$. This condition can be used to solve the velocity v. As c is constant, $x_A = c_A/c$, $x_B = c_B/c$ and $x_B = 1 - x_A$, we obtain from Equation (5.171)

$$v = \frac{1}{c} \left(\tilde{D}_A \frac{\partial c_A}{\partial \tilde{y}} + \tilde{D}_B \frac{\partial c_B}{\partial \tilde{y}} \right) = \tilde{D}_A \frac{\partial x_A}{\partial \tilde{y}} + \tilde{D}_B \frac{\partial x_B}{\partial \tilde{y}} = \left(\tilde{D}_B - \tilde{D}_A \right) \frac{\partial x_B}{\partial \tilde{y}} \tag{5.172}$$

Equation (5.170) is valid for arbitrary compositions. The only possibility then is that Equation (5.170) is divided into two equations of equal shape, one for each component. One of these equations is Equation (5.173), where v has been replaced by the expression (5.172):

$$\frac{\partial c_B}{\partial t} = \tilde{D}_B \frac{\partial^2 c_B}{\partial \tilde{y}^2} - v \frac{\partial c_B}{\partial \tilde{y}} = \tilde{D}_B \frac{\partial^2 c_B}{\partial \tilde{y}^2} + \left(\tilde{D}_A - \tilde{D}_B \right) \frac{\partial x_B}{\partial \tilde{y}} \frac{\partial c_B}{\partial \tilde{y}} \tag{5.173}$$

or

$$\frac{\partial c_B}{\partial t} = \frac{\partial}{\partial \tilde{y}} \left[\tilde{D}_B \frac{\partial c_B}{\partial \tilde{y}} + \left(\tilde{D}_A - \tilde{D}_B \right) x_B \frac{\partial c_B}{\partial \tilde{y}} \right]$$

which can be transformed into

$$\frac{\partial c_B}{\partial t} = \left[\tilde{D}_B + \left(\tilde{D}_A - \tilde{D}_B \right) x_B \right] \frac{\partial^2 c_B}{\partial \tilde{y}^2}$$

or

$$\frac{\partial c_B}{\partial t} = \left(x_A \tilde{D}_B + x_B \tilde{D}_A \right) \frac{\partial^2 c_B}{\partial \tilde{y}^2} = \tilde{D} \frac{\partial^2 c_B}{\partial \tilde{y}^2} \tag{5.174}$$

where $\tilde{D}$ is the *chemical diffusion coefficient*. By identical calculations for the A atoms we obtain (A and B are exchanged)

$$\frac{\partial c_A}{\partial t} = \left(x_B \tilde{D}_A + x_A \tilde{D}_B \right) \frac{\partial^2 c_A}{\partial \tilde{y}^2} = \tilde{D} \frac{\partial^2 c_A}{\partial \tilde{y}^2} \tag{5.175}$$

The differential Equations (5.174) and (5.175) are the versions of Fick's second law which are used to solve c_A and c_B as functions of time and position.

Chemical Diffusion Coefficient or Interdiffusion Coefficient

Equations (5.174) and (5.175) are identical. This means that it is possible to describe the diffusion of both components in the binary alloy by a *single* diffusion coefficient, which depends on the diffusion coefficients of the A and B atoms and the composition of the alloy. The diffusion coefficient is called the *chemical diffusion coefficient*:

$$\tilde{D} = x_A \tilde{D}_B + x_B \tilde{D}_A \tag{5.176}$$

where
$\tilde{D} = $ chemical diffusion coefficient of the binary alloy
x_A, $x_B = $ mole fraction of element A and B of the alloy
$\tilde{D}_A = $ diffusion coefficient of A atoms at the given composition of the alloy
$\tilde{D}_B = $ diffusion coefficient of B atoms at the given composition of the alloy.

The diffusion coefficients $\tilde{D}_A$ and $\tilde{D}_B$ are marked with a special sign (tilde) to indicate very clearly that they vary with the composition of the alloy (Figure 5.35). They must be distinguished from D_A and D_B, the self-diffusion coefficients of A and B (page 265).

If the solution is dilute, $x_A \approx 0$ and $x_B \approx 1$, and Equation (5.176) gives

$$\tilde{D} = D_A \tag{5.177}$$

If the solution is dilute, $x_B \approx 0$ and $x_A \approx 1$, and Equation (5.176) gives similarly

$$\tilde{D} = D_B \tag{5.178}$$

This can be summarized as follows:

- In a dilute solution, the chemical diffusion coefficient is equal to the intrinsic diffusion coefficient of the solute and not the solvent.

The diffusion coefficients of the two components of a solid binary alloy vary strongly with the composition of the alloy. This is shown in Figure 5.35 for solid CuNi alloys. For each alloy, for example the one defined by the dotted vertical line in the figure, the diffusion coefficients for Cu and Ni have been measured by tracer methods (see below). As the corresponding composition is known, the chemical diffusion coefficient can be calculated from Equation (5.176). In this way, $\tilde{D}$ as a function of composition can be derived.

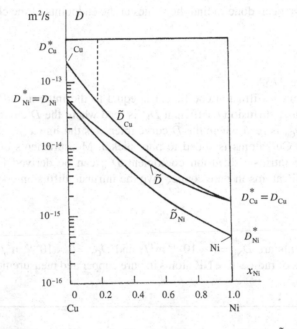

Figure 5.35 The diffusion coefficients of Cu and Ni and the chemical diffusion coefficient $\tilde{D}$ of CuNi alloys as functions of the alloy composition. Reproduced with permission from A. G. Guy, *Elements of Physical Metallurgy*, 2nd edn. © Addison-Wesley Publishing Company, Inc. (now under Pearson Education).

This is *not* true for alloy *melts*. The diffusion coefficients in pure metal melts and in binary alloy melts are approximately equal and of magnitude $10^{-9}\,\mathrm{m}^2/\mathrm{s}$. As the forces between the atoms are weaker in a melt than in a crystal, it is easy to understand that the diffusion coefficients of metal melts are much larger than those of the corresponding solid metals.

Experimental Determination of Diffusion Coefficients by Tracer Methods

Diffusion coefficients in solids can in principle be determined in the following way.

The self diffusion coefficient of a metal can be measured with the aid of radioactive atoms in question. At $t = 0$ a small amount of tracer atoms are added to one end of a bar of the metal ($y = 0$). The times $t_1, t_2, t_3, \ldots$ when the tracer atoms reach the distances y_1, y_2, y_3 from the radioactive end of the bar are measured.

The atoms start to diffuse towards the other end of the bar at $t = 0$ under the influence of a concentration gradient of the tracer atoms. Einstein's law of random walk gives

$$y^2 = 2Dt \qquad (5.179)$$

y^2 is plotted as a function of t and the self diffusion coefficient can be derived from the slope of the straight line.

The tracer atoms have exactly the same properties as the ordinary atoms because their electron orbits are identical. Hence the diffusion coefficient $\tilde{D}_{A*}$ will be equal to $\tilde{D}_A$.

In this way, the diffusion coefficient $\tilde{D}_A$ can be measured for AB alloys of arbitrary compositions. By analogous measurements on radioactive B* atoms, the diffusion coefficient $\tilde{D}_B$ can be determined for the same alloys.

For the special case that the alloy consists of pure element A, tracer measurements on A* atoms give the self-diffusion coefficient D_A or the intrinsic diffusion coefficient of A.

The self-diffusion coefficient D_B can be determined by analogous tracer measurements on B* atoms in pure element B gives the self-diffusion coefficient D_B or the intrinsic diffusion coefficient of B.

Example 5.5

Consider Figure 5.35.

(a) Read the self-diffusion coefficients for Ni and Cu from Figure 5.35.
(b) What type of measurements have been done to find the values at the endpoints of the chemical diffusion coefficient curve?

Solution:

(a) We apply the 'theorem' on page 273:

In a dilute solution, the chemical diffusion coefficient is equal to the intrinsic diffusion coefficient of the solute and not the solvent. Hence the intrinsic diffusion coefficient D_{Ni} is read where the $\tilde{D}$ curve intersects the line $x_{Ni} = 0$, and the intrinsic diffusion coefficient D_{Cu} is read where the $\tilde{D}$ curve intersects the line $x_{Ni} = 1$.

(b) A small amount of radioactive Cu* atoms is added to pure nickel. Measurements of the Cu* distribution as a function of time are performed and the intrinsic diffusion coefficient D_{Cu} can be derived from the measurements. Analogous measurements on radioactive Ni* atoms in pure copper give the intrinsic diffusion coefficient D_{Ni}.

Answer:

(a) The intrinsic diffusion coefficients are $D_{Ni} = 5 \times 10^{-14} \, \text{m}^2/\text{s}$ and $D_{Cu} = 2 \times 10^{-15} \, \text{m}^2/\text{s}$.
(b) Measurements on small amounts of radioactive Ni* atoms in pure copper and measurements on small amounts of radioactive Cu* atoms in pure nickel.

Diffusion in Ternary Alloys

Diffusion in alloys with two or more alloying elements is much more complex than diffusion in binary alloys. The mutual interaction between the two alloying elements and between each of them and the solvent in ternary alloys influences the diffusion of the components strongly.

It is difficult to give a mathematical description of the diffusion process. We will restrict the discussion to a concrete example and choose the Fe–C–Si system (Figure 5.36).

Two properties of the ternary alloy are of special importance:

1. The diffusion coefficient of carbon in iron is about 10^5 times larger than that of silicon in iron. The diffusion of the Si atoms is consequently much slower than the diffusion of the C atoms.
2. The carbon atoms are remarkably repelled by silicon atoms.

Figure 5.36a shows two pieces of steel. The left one contains about 0.45% C and 3.8% Si. The right one has the same size, shape and cross-sectional area, but it contains about 0.45% C and no Si at all. The two pieces are brought into close contact with each other and are annealed for 13 days.

At $t = 0$, before the steel pieces have been brought together, both steel pieces are at equilibrium and their alloying elements are evenly distributed. Figure 5.36b shows the concentration profile of Si and C before the annealing starts at $t = 0$.

When the pieces are brought into close contact with each other, the diffusion of the alloying elements starts. The concentration profile of C and Si at the end of the annealing time is shown in Figure 5.36c. By that time, considerable diffusion of the carbon atoms has occurred. C atoms have moved from the left, across the interface, into the right piece of steel. The distribution of the Si atoms in Figure 5.36c is essentially the same as the dotted profile in Figure 5.36b because the Si diffusion is so slow that it can hardly be noticed after the annealing time.

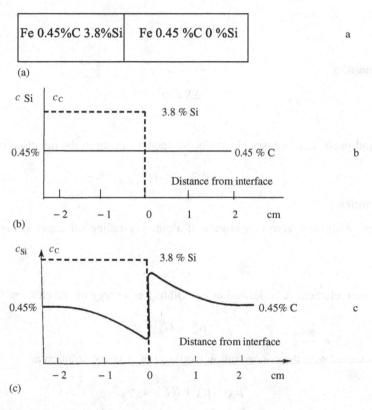

Figure 5.36   (a) The two pieces of steel are brought into close contact with each other at $t = 0$.
(b) Concentration distribution of C and Si in two pieces of steel before annealing starts.
(c) Concentration distribution of C and Si after annealing at 1050 °C for 13 days. Reproduced with permission from P. G. Shewmon, *Transformations in Metals.* © 1969 McGraw-Hill Book Company, Inc.

The carbon concentration in Figure 5.36c is perturbed about 1.5 cm on each side of the interface during the same time. The effect is entirely due to the influence of the Si atoms. In the absence of Si in the left bar, no net diffusion of C atoms at all would have occurred as the pieces would have had the same C concentration.

The explanation is that the chemical potential of carbon is not the same in the two alloys. The difference depends on the interaction between the Si and C atoms on the left side. The system is not in equilibrium. The uneven chemical potential causes a driving force of diffusion. The system tries to achieve equilibrium ($\mu_1 = \mu_2$) by diffusion in order to change the carbon concentrations in both alloys. After the annealing time there is still no equilibrium except near the interface.

This is a typical case when grad μ and not the concentration gradient represents the driving force of diffusion. At equilibrium there will be a permanent difference in carbon concentration of the two alloys.

Summary

■ *Basic Laws of Thermodynamics*

First Law of Thermodynamics

$$Q = U + A$$

Some Basic Concepts

Enthalpy:
$$H = U + pV \quad (dH)_p = (dQ)_p$$

Entropy:
$$dS = \frac{dQ}{T} \quad S = k_B \ln P$$

Gibbs free energy:
$$G = H - TS$$

Second Law of Thermodynamics

Reversible processes: $\quad \Delta S = 0$
Irreversible processes: $\quad \Delta S > 0$

At constant temperature and pressure processes always occur spontaneously in the direction of decreasing G:

$$dG_{T,p} \leq 0$$

Third Law of Thermodynamics

Nernst's theorem: the entropy at absolute zero temperature of a pure crystalline substance is zero.

Chemical Potential

The chemical potential of a pure element A is defined as the Gibbs free energy of the element:

$$\mu_A^0 = G_A^0$$

The chemical potential of A varies with the concentration (activity) of A in the solution as

$$\mu_A = \mu_A^0 + RT \ln x_A \gamma_A$$

■ *Thermodynamic Condition for Equilibrium*

A system is in equilibrium when its Gibbs free energy is a minimum. The equilibrium condition at constant temperature and pressure is $dG_{T,p} = 0$.

The driving force in chemical and metallurgical reactions is

$$-\Delta G = -\int_{\text{initial}}^{\text{final}} dG = -(G_f - G_i)$$

In liquids and solids pV and pdV are small compared with other terms and can be often neglected:

$$H = U + pV \quad \Rightarrow \quad H \approx U$$
$$dH = dU + pdV \quad \Rightarrow \quad dH \approx dU$$
$$dG = dH - TdS \quad \Rightarrow \quad dG \approx dU - TdS$$

■ *Ideal and Nonideal Solutions*

Ideal Solutions

Homogeneous forces (A–A, B–B) and heterogeneous forces (A–B) are equal. The heat of mixing is zero. The solubility is complete at all proportions.

Heat of mixing:

$$-\Delta H_{\text{mix}} = 0$$

Entropy change of mixing (solids, liquids and gases):

$$\Delta S_{\text{mix}} = -nR(x_A \ln x_A + x_B \ln x_B)$$

Nonideal Solutions

All other solutions than ideal solutions are nonideal.

$$-\Delta H_{\text{mix}}^{\text{nonideal}} = H_{\text{mix}}^{\text{Ex}} = L_{\text{mix}} X_A X_B$$

$$S_{\text{mix}}^{\text{nonideal}} = S_{\text{mix}}^{\text{ideal}} + S_{\text{mix}}^{\text{Ex}} = \Delta S_{\text{mix}} + S_{\text{mix}}^{\text{Ex}}$$

■ *Basic Concepts and Fundamental Relationships of Transformation Kinetics*

Activation Energy of a Reaction or a Transformation

The activation energy U_{act} is the energy barrier which has to be overcome before a transformation of a system can occur.

Driving Force of a Reaction or a Transformation

As a measure of the probability of the transformation, the concept of *driving force* is used:

$$\text{Driving force} = -\Delta G = -(G_f - G_i)$$

The driving force of a spontaneous process is always a *positive* quantity. The larger the driving force is, the more probable and rapid will be the transformation.

Endothermic and Exothermic Reactions

ΔG is always <0 whereas ΔU can be positive *or* negative.
  If $U_f > U_i$ the transformation is *endothermic*; heat has to be added to the system.
  If $U_f < U_i$ the transformation is *exothermic*; heat is released and transferred to the surroundings.

■ *Maxwell–Boltzmann Distribution Law*

$$N_i = \frac{N_0}{Z} g_i e^{-\frac{u_i}{k_B T}}$$

where

$$Z = g_1 e^{-\frac{u_1}{k_B T}} + g_2 e^{-\frac{u_2}{k_B T}} + \ldots + g_i e^{-\frac{u_i}{k_B T}} = \sum_i g_i e^{-\frac{u_i}{k_B T}}$$

The Maxwell–Boltzmann law can be used to calculate the fraction of thermally excited atoms and vacancies.

■ Reaction Rates of Thermally Activated Reactions and Transformations

The reaction rate is the fraction of the total number of particles which reach the final state per unit time.

The reaction rate of simple reactions and transformations is proportional to three factors:

- the frequency with which the atoms have the opportunity to transform, i.e. the collision frequency between the particles in a gas or the vibration frequency in a solid;
- the fraction of the total number of atoms in the initial state, which has sufficient thermal energy to overcome the energy barrier:

$$f^* = e^{-\frac{U_{a\,act}}{k_B T}} \frac{}{Z}$$

where Z is the partition function;
- a probability factor associated with the entropy term in the relationship

$$G_{a\,act} = U_{a\,act} - TS_{a\,act}$$

$$S_{a\,act} = k_B \ln P_{a\,act} \quad \text{or} \quad P_{a\,act} = e^{\frac{S_{a\,act}}{k_B}}$$

The reaction rate can be written as

$$k = \frac{\nu}{Z} e^{\frac{S_{a\,act}}{k_B}} e^{-\frac{U_{a\,act}}{k_B T}} = \frac{\nu}{Z} e^{-\frac{U_{a\,act} - S_{a\,act} T}{k_B T}} = \frac{\nu}{Z} e^{-\frac{G_{a\,act}}{k_B T}}$$

or if the entropy factor is included in the constant

$$k = A e^{-\frac{U_{a\,act}}{k_B T}} \quad \text{Arrhenius equation}$$

The Arrhenius equation can be written as

$$\ln k = \ln A - \frac{U_{a\,act}}{k_B T}$$

In the case of simple reactions and transformations, this equation can be used to determine reaction and transformation rates as a function of temperature by plotting $\ln k$ as a function of $1/k_B T$. The activation energy $U_{a\,act}$ and the frequency factor A can be derived graphically.

■ Kinetics of Homogeneous Reactions in Gases

A homogeneous reaction occurs simultaneous in the whole available volume.

Collision Theory of Homogeneous Chemical Reactions

$$\mathbf{A + B \rightarrow AB}$$

$$-\frac{dx_A'}{dt} = -\frac{dx_B'}{dt} = \frac{dx_{AB}'}{dt} = k_1 x_A x_B$$

$$\mathbf{A + B \leftarrow AB}$$

$$\frac{dx_A''}{dt} = \frac{dx_B''}{dt} = -\frac{dx_{AB}''}{dt} = k_2 x_{AB}$$

which can be summarized as

$$\frac{dx_A}{dt} = \frac{dx_B}{dt} = -\frac{dx_{AB}}{dt} = -k_1 x_A x_B \left(1 - \frac{k_2}{k_1} \frac{x_{AB}}{x_A x_B}\right)$$

or

$$\frac{dx_A}{dt} = \frac{dx_B}{dt} = -\frac{dx_{AB}}{dt} = -k_1 x_A x_B \left(1 - e^{-\frac{\Delta G}{RT}}\right)$$

Temperature Dependence of the Rate Constant

$$k = Ae^{-\frac{U_{a\ act}}{k_B T}}$$

A is proportional to the total number of atomic collisions in the gas per unit time and unit volume. $U_{a\ act}$ is the minimum kinetic energy of the atoms necessary for a reaction to occur.

Rate Constant as a Function of Thermodynamic Quantities

According to the kinetic theory of gases, the collision frequency equals the mean velocity divided by the mean free path:

$$\nu_{coll} = \frac{\overline{\nu_{kin}}}{l} = 16 N r^2 \sqrt{\frac{\pi R T}{M}}$$

Arrhenius model: $k = 16 N r^2 \sqrt{\dfrac{\pi R T}{M}} e^{-\frac{U_{a\ act}}{k_B T}}$

Eyring's model: $k = 16 N r^2 \sqrt{\dfrac{\pi R T}{M}} e^{-\frac{G_{a\ act}}{k_B T}}$

The rate constant is proportional to the number of collisions per unit time and unit volume which results in a chemical reaction:

$$k = \nu_{coll} e^{-\frac{U_{a\ act}}{k_B T}}$$

It is possible to evaluate the overall reaction rate in terms of concentrations and thermodynamic quantities.

■ *Driving Force and Reaction Rate of Homogeneous Chemical Reactions*

$$\mathbf{A + B \rightarrow AB}$$

$$\text{Driving force} = -\Delta G_{AB} = -\int_{initial}^{final} dG = -(G_f - G_i) = G_A + G_B - G_{AB}$$

$$k_1 = 16 N r^2 \sqrt{\frac{\pi R T}{M}} e^{-\frac{G_{a\ act}}{k_B T}}$$

$$\mathbf{AB \rightarrow A + B}$$

$$\text{Driving force} = -\Delta G_{A+B} = -\int_{initial}^{final} dG = -(G_f - G_i) = G_{AB} - (G_A + G_B)$$

$$k_2 = 16 N r^2 \sqrt{\frac{\pi R T}{M}} e^{-\frac{G_{a\ AB} + \Delta G_{a(A+B)}}{k_B T}}$$

The total reaction rate is given in the text.

■ *Activated Complex Theory*

The theory is built on the accurate calculation of the total energy of the system. It includes the interactions between the nuclei, their potential and kinetic energies and their interaction with the electrons and the interaction between the electrons. The calculations cannot be performed without some simplifying assumptions. The motions of the nuclei and of the electrons are treated independently of each other. The methods of quantum chemistry and high-capacity computers are used for the calculations.

The energy surface gives a detailed description of the reaction from beginning to end. Initially the reacting atoms A and B are far from each other. The reaction process is described by a path along the energy surface. It follows the path of lowest possible energy:

$$k_1 = \kappa \frac{k_B T}{h} \frac{Z_{(AB)*}}{Z_A Z_B} e^{-\frac{U_0}{k_B T}}$$

or

$$k_1 = \frac{k_B T}{h} e^{-\frac{G_{a\ act}*}{k_B T}} = \frac{k_B T}{h} e^{\frac{S_{a\ act}*}{k_B}} e^{-\frac{H_{a\ act}*}{k_B T}}$$

The activated complex theory gives much better agreement with experimental values of the frequency factor A than the simple hard-sphere collision theory. The reason is that the activated complex theory involves the properties of the reactants and products. Its disadvantage is the complex and extensive calculations.

■ Diffusion in Solids

Diffusion is a systematic motion of the atoms. It overlaps the random motion, which is always present. Diffusion is caused by a driving force, which normally is a concentration gradient, and implies mass transport

Basic Law of Diffusion

Fick's first law:
$$J = -D \ \ \text{grad} \ \ c$$

Diffusion along the y axis:
$$J = -D \frac{dc}{dy}$$

The concentration c is measured in numbers or kilomol per unit volume. If mole fraction is used, a factor $1/V_m$ has to be included on the right-hand side of the diffusion equation.

■ Diffusion Mechanisms

Diffusion of atoms in the interior of a crystalline solid can occur in many different ways. The most common are the *vacancy mechanism* and the *interstitial mechanism*.

■ Theory of Diffusion in Solids

Substitutional diffusion of lattice atoms is performed with the aid of the vacancy mechanism. Both the atom and the vacancy move within the lattice when a lattice atom jumps to a vacancy.

In *interstitial diffusion* only small atoms can jump from site to site within a solid for space reasons (interstitial mechanism). Vacancies and interstitials are point defects.

Formation Energy of Point Defects

The formation of point defects requires thermal energy. The fraction of available point defects depends strongly on the temperature and the formation energy.

Equilibrium fraction of vacancies: $x_{vac\ form} = e^{\frac{-\Delta G_{vac\ form}}{RT}}$

Equilibrium fraction of interstitials: $x_{i\ form} = e^{\frac{-\Delta G_{i\ form}}{RT}}$

The formation energy is considerably lower for vacancies than for lattice atom interstitials. The formation energy for small interstitials is practically zero.

Energy Barrier for Atom Jumps

The atoms in the crystal lattice vibrate around their equilibrium positions. On rare occasions an atom gains enough phonon energy to overcome the energy barrier.

A lattice atom jump corresponds to diffusion by the vacancy mechanism. An interstitial atom jump corresponds to diffusion by the interstitial mechanism.

Fraction of available vacancies with enough energy to overcome the energy barrier:

$$f_{\text{vac barrier}} = e^{\frac{-\Delta G_{\text{vac barrier}}}{RT}}$$

Fraction of available interstitials with enough energy to overcome the energy barrier:

$$f_{\text{i barrier}} = e^{\frac{-\Delta G_{\text{i barrier}}}{RT}}$$

■ Activation Energy. Jump Frequency

Activation Energy

The total energy required for a jump of an atom is the sum of the formation energy and the energy barrier. It is called the activation energy of diffusion:

$$H_{\text{a act}} = H_{\text{a form}} + H_{\text{a barrier}}$$

or

$$U_{\text{a act}} = U_{\text{a form}} + U_{\text{a barrier}}$$

Jump Frequency of Atoms

Substitutional Diffusion

$$f_{\text{jump}} = \frac{F_{\text{jump}}}{N_0} = x_{\text{vac}} Z_{\text{coord}} f_{\text{barrier}} \nu_{\text{vibr}}$$

or

$$f_{\text{jump}} = Z_{\text{coord}} \nu_{\text{vibr}} e^{\frac{-(G_{\text{a form}} + G_{\text{a barrier}})}{k_{\text{B}} T}}$$

In solids the product pV is small compared with H and U and can be neglected. Hence $H \sim U$ and $G \sim U - TS$ and

$$f_{\text{jump}} = f_0 e^{-\frac{U_{\text{a act}}}{k_{\text{B}} T}}$$

Interstitial Diffusion

Large interstitials (lattice atoms or foreign atoms) have higher formation energy than vacancies and consequently higher activation energy. Small interstitials are small dissolved foreign atoms with a given concentration, independent of temperature. The activation energy equals the energy barrier as the formation energy is zero. Interstitials may be activated by phonon collisions. Otherwise the same theory is valid for interstitial and substitutional diffusion.

Temperature Dependence of Diffusion Coefficient

$$D = D_0 e^{-\frac{U_{\text{a act}}}{k_{\text{B}} T}}$$

Relationship Between Diffusion Coefficient, Jump Distance and Jump Frequency

For cubic crystal structures:

$$D = \frac{1}{6} d_j^2 f_{j \text{ total}}$$

The general case is discussed in the text:

$$D = \frac{1}{2} \sum_{uvw} f_{uvw} d_{uvw}{}^2$$

Driving Force of Diffusion

$$-\Delta G^{\text{diff}} = \text{grad } G = \text{grad } \mu$$

Rigorous condition for diffusion:

$$J = -L \frac{d\mu}{dy}$$

A relationship between the diffusion constants L and D is derived in the text.

■ Self-diffusion

Radioactive atoms A* diffusing in pure element A:

$$J_A = -D_A \frac{dc_A}{dy}$$

The intrinsic diffusion or self-diffusion coefficient D_A can be measured experimentally by tracer methods.

■ Diffusion in Alloys

In alloys with two or more components, diffusion is in most cases more complicated than in single-component diffusion. The reasons are several diffusing components and interaction between atoms of different kinds.

■ Diffusion in Binary Alloys

Interstitial Alloys

The interstitials and the great number of empty interstitial sites can be regarded as a sublattice of their own. The diffusion of the interstitials occurs in this sublattice as single-component diffusion. The interaction with the crystal lattice can be entirely neglected. The crystal lattice behaves like a rigid body and the lattice atoms do not diffuse.

Substitutional Alloys

In substitutional alloys, both components diffuse. In most cases the diffusion coefficients of the two components differ.

Fluxes in Substitutional Diffusion

The diffusion of atoms in opposite directions and at different rates leads to a lack or an excess of atoms in different regions of the alloy. Fast A atoms and slow B atoms diffusing across a weld surface results in an excess of A atoms and a lack of vacancies in the B region and an excess of vacancies in the A region.

The mass transport leads to a slight displacement of the weld plane. This displacement can be observed experimentally by inclusions in the alloy at the weld plane. This phenomenon, called the Kirkendall effect, appears in substitutional alloys because the diffusion rates and the diffusion constants of A atoms and B atoms differ.

The deviations from the equilibrium concentration of vacancies disappear successively with time by annihilation or production of vacancies at edge dislocations in the A and B regions.

■ Chemical Diffusion Coefficient

$$\tilde{D} = x_A \tilde{D}_B + x_B \tilde{D}_A$$

$\tilde{D}$, $\tilde{D}_A$ and $\tilde{D}_B$ vary strongly with the composition of the alloy.

The diffusion coefficients $\tilde{D}_A$ and $\tilde{D}_B$ are marked with a special sign (tilde) to indicate very clearly that they vary with the composition of the alloy.

The chemical diffusion coefficient appears in Fick's second law, which can be applied on both components of the binary alloy:

$$\frac{\partial c_B}{\partial t} = \tilde{D}\frac{\partial^2 c_B}{\partial \tilde{y}^2} \quad \text{and} \quad \frac{\partial c_A}{\partial t} = \tilde{D}\frac{\partial^2 c_A}{\partial \tilde{y}^2}$$

■ Diffusion in Ternary Alloys

Diffusion in ternary alloys is strongly influenced by the interaction between the components. The diffusion depends on grad μ and not on concentration gradients.

Exercises

5.1 The temperature inside a refrigerator is kept at 4 °C. The temperature of the surroundings is 24 °C. The amount of heat per hour which has to be removed from the interior of the refrigerator is 1.0×10^6 J/h.

    The refrigerator is supposed to work like an ideal Carnot engine. What is the average minimum electrical power which is required to run it? The answer should be given both in SI units and in kWh/year, which is a common technical unit.

5.2 To keep an indoor temperature of 20 °C in a house on a cold January day in a northern country, a heat pump is used. It consists of a closed liquid–vapour system including two heat exchangers and a compressor.

    The heat pump extracts heat from a lake with a water temperature of 5 °C via a heat exchanger (evaporation of the liquid and expansion of the gas require energy). When the gas is compressed and condenses to a liquid, heat is delivered via another heat exchanger to the water-filled radiator system of the house. The temperature of the 'radiator water' is 50 °C at the heat exchanger. To maintain the indoor temperature a power of ~ 20 MJ/h is required. The overall efficiency of the heat pump is about 40% of the theoretical efficiency of an ideal Carnot engine. Calculate

(a) the average electrical power which is required to run the heat pump (in SI units and in kWh per 24 hours);
(b) the power required to heat the house by direct electrical heating (in SI units and in kWh per 24 hours);
(c) the ratio of the answers in (b) and (a).

5.3 A 1 kg amount of water of temperature 20 °C is mixed by an adiabatic and isobaric process with 1 kg of water of temperature 100 °C. Calculate the entropy change of the system.

5.4 A 1 kmol amount of ice of temperature 0 °C is heated and transferred at constant pressure (1 atm) to water vapour of temperature 100 °C. Calculate the energy required for the process and the entropy change. Material constants can be found in a standard table.

5.5 Two equal, insulated cylinders ($V_1 = V_2 = 4.0$ l) are connected with a short tube with a closed tap. One of them contains nitrogen gas of 400 mm Hg pressure and the other is evacuated. When the tap is opened the gas pressure becomes equal in both cylinders. No temperature change is observed. The temperature is constant and equal to 20 °C.
  Calculate the entropy change of the system.

5.6 Define
  (a) the enthalpy of a system
  (b) ideal and nonideal solutions of two components A and B
  (c) molar heat of mixing of pure substances A and B.

5.7 The heat of fusion of aluminium is 390 kJ/kg.

  (a) Calculate the entropy change when 1 kmol of Al is molten. Material constants can be found in a standard table.
  (b) The result in (a) can be interpreted in statistical terms. When solid Al is transformed into liquid Al, the number of possible distributions of the molecules in available energy states becomes much higher in the liquid than in the solid state. Calculate the ratio of the numbers of possible distributions in liquid Al and in solid Al.

5.8 Define
  (a) the concept of Gibbs free energy of a system and the condition for equilibrium of a system
  (b) driving force of a reaction or transformation in the case of non- equilibrium
  (c) chemical potential of a pure substance.
  (d) Why is the Gibbs free energy such an important thermodynamic quantity?

5.9 (a) Define the concepts of reaction rate and activation energy and give the relationship between them. Draw a figure and mark the activation energy and the driving force of the reaction.
  (b) The reaction rate depends on three factors. Which ones?

5.10 All the components in the chemical reaction $2NO_2 \rightleftharpoons 2NO + O_2$ are gases, moving at random in the available volume. When two NO_2 molecules collide, decomposition may occur. Therefore, it is reasonable to assume that the initial rate of decomposition of NO_2 is proportional to the square of the NO_2 concentration:

$$\frac{d[NO_2]}{dt} = -k[NO_2]^2$$

The rate constant depends on the temperature as is shown from the measured values below:

T(K)	610	620	630	640	650	660
$k(m^3/s\ kmol)$	0.82	1.30	1.86	2.72	3.60	4.95

  (a) What is the initial decomposition rate of NO_2 at $T = 650$ K if the concentration of NO_2 at $t = 0$ is $0.060\ kmol/m^3$?
  (b) Calculate the activation energy of the decomposition reaction.

5.11 In a chemical reaction an A atom reacts with two B atoms:

$$A(g) + 2\ B(g) \rightarrow products$$

The reaction kinetics were studied and resulted in the data given in the table.

Experiment No.	Initial concentration (kmol/dm³)		Initial reaction rate (kmol/dm³s)
	$[A]_{t=0}$	$[B]_{t=0}$	
1	5.0×10^{-4}	4.0×10^{-4}	18.7×10^{-3}
2	2.0×10^{-4}	4.0×10^{-4}	7.5×10^{-3}
3	4.0×10^{-4}	2.0×10^{-4}	3.7×10^{-3}
4	3.0×10^{-4}	3.2×10^{-4}	

   (a) What is the order of the reaction with respect to each reactant and the overall order of the reaction?
   (b) Determine the rate constant of the reaction.
   (c) Predict the initial reaction rate for experiment 4.

5.12 (a) Describe the two most common diffusion mechanisms in pure crystalline materials.
   (b) What is the basic equation of diffusion in solids? Define the concept of the self-diffusion constant.
   (c) Define the concepts of jump frequency and jump distance. Give the relationship between these quantities and the diffusion constant in one direction.
   (d) Give D as a function of temperature and activation energy. How can the activation energy be derived experimentally?

5.13 Even in a dense-packed FCC crystal structure it is possible to introduce small interstitial atoms between the normal sites of the lattice atoms.

   (a) Calculate the maximum size of such a foreign atom (radius r expressed as a function of the radius of the lattice atoms) in an interstitial position, for the two possible alternatives. On interstitial diffusion these holes form vacant sites for the jumping atoms.
   (b) How many neighbour atoms does the foreign atom touch in each case?

   The lattice atoms can be regarded as hard spheres with equal radius R.

5.14 Commercial hydrogen is stored at high pressure in tubes, made of a special type of hardened steel with very low hydrogen solubility. Consider such a tube with an internal volume of $50\,dm^3$ and a total internal area of $80\,dm^2$. The thickness of the tube walls is $5.0\,mm$. Owing to the pressure of 200 atm inside the wall, hydrogen gas dissolves slightly in the metal. At steady state the concentration of hydrogen atoms is 8×10^{-8} mol/kmol steel at the inner surface. The concentration of H in the steel at the outer surface is zero. The diffusion constant of H in steel at room temperature is $10^{-8}\,m^2/s$. Occasionally the tube happened to be left unopened for 10 years. Calculate

   (a) the number of H atoms which escape from the tube during 10 years
   (b) the amount of hydrogen (kg) which leaks from the tube during 10 years
   (c) the pressure decrease (atm) due to diffusion of H through the steel tube during 10 years.

5.15 The temperature dependence of the self-diffusion constant of iron has been studied experimentally, which resulted in the following data:

$T(K)$	700	900	1100	1300
$D(m^2/s)$	10^{-15}	10^{-14}	10^{-13}	10^{-12}

Derive the constant D_0 and the activation energy $U_{a\ act}$ graphically.

5.16 Diffusion of an atom consists of numerous jumps from one site to another in a medium (magnitude 10^8 jumps per second). The jumps are irregular with respect to length and direction as shown in the figure.

   Such a motion pattern is called a *random walk*. Another example is Brownian motion, which can be observed in colloidal suspensions or in air, polluted by floating small particles, for example smoke. The kinetic motion of the solvent molecules or the nitrogen and oxygen molecules results in irregular macroscopic pushes of the bigger particles. Their motions can be observed and filmed through a microscope.

The jump distances are measured, independent of direction, and the average distance L can be calculated by the following equation:

$$nL^2 = \sum_1^i d_i^2$$

where n is the number of jumps. In such an experiment, the following jump distances were measured (the unit is irrelevant here):

2, 6, 3, 8, 5, 1, 4, 9, 2, 6, 7, 2, 8, 5, 2, 6, 3, 1, 7, 12, 3, 6, 2, 5, 4, 7

Calculate the average jump distance.

5.17 The self diffusion constant of Cu is $1 \times 10^{-12}\,\mathrm{m^2/s}$ at 600 K. The jump distance has been estimated as $2 \times 10^{-9}\,\mathrm{m}$. Calculate the magnitude of the total jump frequency in Cu at this temperature.
Hint: Cu has an FCC crystal structure.

5.18 The diffusion in a material with substitutional diffusion is described by a temperature-dependent diffusion constant:

$$D = D_0 e^{-\frac{U_{a\ act}}{k_B T}}$$

(a) What factors are included in D_0?
(b) Discuss the effect of the vacancy concentration on the diffusion coefficient.
(c) Suppose that the material is heated to a high temperature and then rapidly quenched to the original temperature. Will the diffusion constant be the same after this process? If the answer is no, give the reason and give the new equation.
(d) In the case of interstitial diffusion of small atoms, the diffusion constant is described by an equation which formally appears to be identical with that valid for substitutional diffusion. What is the fundamental difference between the two equations?

5.19 Diffusion in alloys is more complicated than that in pure metals.

(a) Analyse the fluxes of atoms and vacancies on substitutional diffusion in a binary alloy. Explain the influence of vacancies.
(b) Define the concept of the chemical diffusion coefficient of a binary alloy AB with a given composition. Describe graphically how the chemical diffusion coefficient and the diffusion coefficients of A and B atoms vary with the composition of the alloy.

5.20 A diffusion couple is made by welding pure Cu to an alloy of the composition Cu, 20%Zn with inert markers in the joined surface. Zinc diffuses much faster than copper.

(a) In what direction do the markers move? Motivate your answer.
(b) Describe how the fractions of vacancies vary on both sides of the joined interface.

5.21 Consider Figure 5.35 on page 273.

(a) Verify that Equation (5.176) on page 272 is valid for the $\tilde{D}$ curve in Figure 5.35.
(b) Describe the diffusion which corresponds to the four values where $\tilde{D}_{Cu}$ and $\tilde{D}_{Ni}$ intersect the axes.

6

Mechanical, Thermal and Magnetic Properties of Solids

6.1	Introduction	287
6.2	Total Energy of Metallic Crystals	288
	6.2.1 Electron Energy	288
	6.2.2 Phonon Energy	290
6.3	Elasticity and Compressibility	290
	6.3.1 Elasticity	290
	6.3.2 Compressibility	293
	6.3.3 Deformation Energy	295
	6.3.4 Temperature Dependence of the Modulus of Elasticity	296
6.4	Expansion	296
	6.4.1 Length and Volume Expansions	296
	6.4.2 Thermal Expansion	299
6.5	Heat Capacity	303
	6.5.1 Heat Capacity at Constant Volume and at Constant Pressure	303
	6.5.2 Models of Heat Capacity	304
	6.5.3 Molar Heat Capacity at Constant Volume	309
	6.5.4 Molar Heat Capacity at Constant Pressure	312
	6.5.5 Influence of Order–Disorder Transformations on Heat Capacity	315
6.6	Magnetism	317
	6.6.1 Magnetic Moments of Atoms	317
	6.6.2 Dia- and Paramagnetism	319
	6.6.3 Ferromagnetism	323
	6.6.4 Properties of Ferromagnetic Materials	328
Summary		333
Exercises		337

6.1 Introduction

Space limitations in this book make it necessary to select only a few mechanical and thermal properties of crystalline matter of special interest. They are analysed from the atomic point of view.

We have selected elasticity and compressibility (Section 6.3) and expansion (Section 6.4), which are of special interest and will be used later.

Physics of Functional Materials   Hasse Fredriksson and Ulla Åkerlind
© 2008 John Wiley & Sons, Ltd

Temperature is one of the most important parameters in material processes. Thermal properties such as heat capacity (Section 6.5) and thermal conduction for solid materials are very important. Heat capacity is treated in this chapter. Owing to the close relationship between thermal and electrical conduction, thermal conduction is treated together with electrical conduction, which is the basis of the classification of the materials in conductors, semiconductors and insulators. The band theory of solids is applied in the theories of both thermal and electrical conductions in Chapter 7.

Diffusion is extremely important in alloys and doped semiconductors. The basic laws and the theory of diffusion have been treated in Chapter 5.

The magnetic properties of solids, particularly ferromagnetic metals, are closely related to the band theory of solids, which has been treated in Chapter 3. The magnetic properties of these materials and the theoretical background of these properties are discussed in Section 6.6.

In Chapter 4 we used the general law of gases to describe most of the properties of gases. It would be desirable to use an analogous equation of state for solids. Unfortunately, there is no such equation. Solids are much more complex than gases owing to the presence of strong interatomic forces.

To understand some of the properties of metallic solids, it is useful to consider the total energy of metallic crystals. The outlines of this topic, which is not included in Chapter 3, will be discussed below.

6.2 Total Energy of Metallic Crystals

The total energy of a crystal is closely related to the bonds between the particles which constitute the crystal. In Chapter 3, Section 3.3.1, we discussed the total energy per crystal unit in ionic crystals. In Section 3.3.2, we discussed covalent bonds starting with the energy conditions in this type of crystals. Ionic and covalent bonds in solids are very strong. Metallic bonds are also very strong and it would be desirable to find an expression for the total energy per crystal unit even in metals.

The total energy per metal crystal unit consists of several contributions. The main contributions are the formation energy of the crystal lattice, the energy of the free electrons and the phonon energy owing to vibrations in the crystal lattice:

$$E_{total} = E_{formation} + E_{electron} + E_{phonon} \qquad (6.1)$$

The last two terms are very important for the properties of solids and will be discussed below.

6.2.1 Electron Energy

The quantum mechanical model of electrons in metals and the band theory of solids were discussed extensively in Chapter 3.

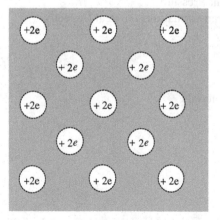

Figure 6.1 Structure of a metal with two valence electrons per atom. Reproduced with permission from W. Schatt (ed.), *Einführung in die Werkstoffwissenschaft*, 7th edn. © 1991 Deutscher Verlag für Grundstoffindustrie, Leipzig.

Metals have crystalline structures. An example of such a structure is given in Chapter 1. A metal consists of a crystal lattice of positive ions surrounded by a 'cloud' of valence electrons common to the whole crystal. Each ion consists of the positive atom nucleus plus all the shell electrons except the valence electrons. The whole crystal is electrically neutral. The dark area in Figure 6.1 illustrates the valence electrons or free electrons, which can move fairly freely inside the crystal.

There are several sorts of electrical forces between the particles in the metal, *attractive* forces between the ions and the surrounding valence electrons and *repulsive* forces between the positive ions and also between the valence electrons themselves. So far it has not been possible to find a simple general expression for the total energy of a unit of a metallic crystal.

Here we will restrict the treatment to a qualitative discussion of the relationship between the energy levels and the interatomic distances in the crystal. As an example we choose monovalent Na metal to illustrate the formation of a metallic bond.

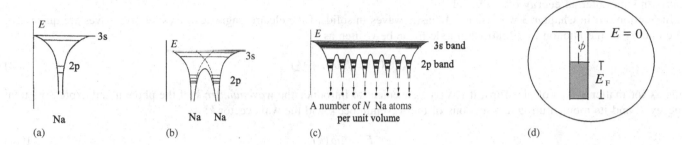

Figure 6.2   Potential energy of (a) a single Na atom, (b) two Na atoms and (c) N Na atoms. (d) Enlargement of part of (c). Reproduced with permission from W. Schatt (ed.), *Einführung in die Werkstoffwissenschaft*, 7th edn. © 1991 Deutscher Verlag für Grundstoffindustrie, Leipzig.

Figure 6.2a shows the potential energy around an Na atom and the sharp energy levels of the valence electron of an Na atom. The inner electrons are tightly bound to the Na nucleus and we will concentrate on the energy level of the 3s electron.

In Figure 6.2b, the corresponding potential energy around a pair of Na atoms is shown. The 3s energy state of the valence electron in each atom has split up into two levels owing to *exchange energy* (pages 72–73 in Chapter 2).

In Figure 6.2c, many Na atoms are present and the sharp energy states form energy bands with many narrow electron energy states. Each energy band contains N energy states per unit volume, equal to the number of Na atoms. Each energy state can accommodate two electrons, i.e. $2N$ electrons per unit volume (Chapter 3, page 129). The 3s band contains only N valence electrons and is therefore only half filled. The electrons occupy the lowest energy states in consistency with the Pauli principle.

If we choose $E_{pot} = 0$ at infinity the strong bonds between the Na nucleus and the inner electrons result in large negative values of the narrow 1s band (not shown in the figure) and the 2s and 2p subbands. The wide half-filled 3s band has the smallest negative energy and the 3s valence electrons are relatively mobile. Part of Figure 6.2c is enlarged and shown in Figure 6.2d on a smaller energy scale (eV instead of keV). This is the familiar potential well with the work function, i.e. the energy required to remove an electron from the metal.

It is easy to realize that

- The electron energy states depend on the interatomic distance a

Expansion and compression represent deviations from the equilibrium state, which corresponds to an energy minimum. Such processes require energy addition and will be studied here from a macroscopic point of view in Section 6.3.

The energy levels of the electrons of *a free atom* are independent of temperature. This is not true for the electrons in a crystal. On the contrary:

- The energies of the electron bands in a solid vary with temperature and other properties which change the interatomic distance.

This is true for all sorts of crystals, not only metals. As in ionic crystals and solids with covalent bonds there is an interatomic distance which corresponds to equilibrium, i.e. minimum energy of the crystal.

So far it has not been possible to find a quantitative expression for the complex electron energy in terms of interatomic distances, temperature and other parameters which influence the electron energy of a crystal, e.g. the structure of the crystal.

6.2.2 Phonon Energy

The internal energy of the positive metal ions in the crystal lattice of a metal is the sum of their potential and kinetic energies. This fact offers no feasible way to find a satisfactory model of the phonon energy in metals. The reason is the difficulty of finding a simple and convenient function for description of the potential energy of the metal ions, surrounded by the 'Fermi sea' of valence electrons. The only possible way to find a successful model for the description of the internal energy is to concentrate on the vibrational motion of the ions in the lattice. The total vibrational energy of the lattice or the *phonon energy* represents the internal energy of the metal.

In Section 3.6 in Chapter 3 we discussed elastic waves in solids. Like electromagnetic waves, elastic waves are quantized. The energy of the phonon in vibrational mode n can be written as

$$G(n) = \hbar\omega(n + {}^1\!/_2) \tag{6.2}$$

Just as for matter waves of electrons, it is very convenient to introduce the *wavenumber* k of the phonon and express its total energy E and its momentum $\boldsymbol{p}$ as functions of the wavenumber k and the wavevector $\boldsymbol{k}$:

$$E = \hbar\omega(k) \tag{6.3}$$

$$\boldsymbol{p} = \hbar\boldsymbol{k} \tag{6.4}$$

where the angular frequency $\omega(k) = 2\pi\nu$.

> *Energy distribution function of photons and phonons*
> Average energy per oscillator:
>
> $$\overline{u} = \frac{h\nu}{\mathrm{e}^{\frac{h\nu}{k_\mathrm{B}T}} - 1}$$

Both phonons and photons obey Bose–Einstein statistics (Chapter 3, pages 148–149). For this reason, the distribution functions for phonons and photons are the same.

The internal energy of the metal is well described as phonon energy provided that the volume of the crystal is kept constant. The model is no longer satisfactory when the volume changes.

Materials expand or contract with temperature changes or under the influence of external forces. When the average interatomic distances within the metal change, it is necessary to consider the change in potential energy of the lattice ions as a function of the interatomic distances. The positions of the electron bands also change in a way that is difficult to predict theoretically.

These circumstances make the theories of elasticity, expansion and heat capacity at constant pressure and other properties, which depend on the interatomic distances, most complicated. There is no simple, convenient and generally accepted model of the internal energy of a solid so far.

6.3 Elasticity and Compressibility

6.3.1 Elasticity

Origin of Elastic Forces

If a solid body is exposed to a mechanical load it will be deformed. Energy is required to deform the body.

The stable state of a crystal is achieved at the distance between the atoms or ions where the attraction forces balance the repulsion forces. This occurs at the equilibrium distance which corresponds to a minimum of the total potential energy of the crystal (Figure 6.3a). Changing the distance between the atoms or ions requires energy and restoring forces appear which try to carry the system back to equilibrium. The steeper the energy curve is, the stronger will be the restoring forces.

Crystals are much more complex than molecules. Figure 6.3b gives a better description than Figure 6.3a of the energy levels in metallic crystals. When the interatomic distances change, a large number of energy levels will change.

The restoring forces are the tensile or compressive forces which appear in solid matter as soon as the distance between the atoms is changed.

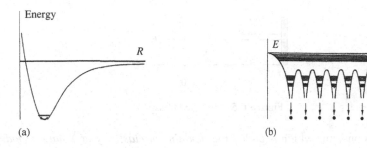

Figure 6.3 (a) Potential energy of a diatomic molecule as a function of the distance between the two atoms. (b) One-dimensional row of metal atoms. The energy bands of the free electrons are indicated. Reproduced with permission from W. Schatt (ed.), *Einführung in die Werkstoffwissenschaft*, 7th edn. © 1991 Deutscher Verlag für Grundstoffindustrie, Leipzig.

Some Basic Concepts and Definitions

Stress is defined as *force per unit area*. If the force F is perpendicular to the surface with area A the stress is called *normal stress*, which is designated by σ:

$$\sigma = \frac{F}{A} \tag{6.5}$$

The normal stress refers to the surface of a section of the material. Its sign is determined by its direction relative to the normal of the surface which is directed outwards. $\sigma < 0$ means that the stress is directed inwards and $\sigma > 0$ means that the stress is directed outwards.

The deformation of a solid under the influence of stress is expressed as strain. *Strain* is defined as the *relative length change*, i.e. the ratio of the length change Δl and the length l:

$$\varepsilon = \frac{\Delta l}{l} \tag{6.6}$$

where the strain ε is a dimensionless quantity.

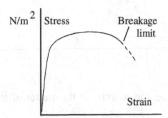

Figure 6.4 Stress as a function of strain.

If the deformation of the solid is small, it will disappear completely when the stress is removed. The deformation is *elastic*.

If the stress is increased, the deformation will no longer be elastic. It remains more or less when the stress is removed. Such a deformation is said to be *plastic*.

A further increase in the stress leads to the breakage limit when the material is ruptured (Figure 6.4).

Hooke's Law

If the relative length change or strain ε, which arises in a rod due to the tensile stress σ, is small then the tensile stress is proportional to the strain (Figure 6.5):

$$\sigma = Y\varepsilon \quad \text{Hooke's law} \tag{6.7}$$

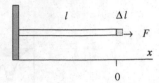

Figure 6.5 Tensile stress.

where the constant Y is a material constant, which is called the *modulus of elasticity* or *Young's modulus*. The magnitude of Y is 10^{10}–10^{11} N/m^2 for metals.

σ is *positive* for *tensile* forces and *negative* for *compressive* forces. Both are normal stresses. ε is *positive* in case of *elongation* and *negative* at *contraction*.

Lateral Contraction

When the bar in Figure 6.6 a is stretched owing to the tensile stress in its length direction, its cross-section will shrink. At small tensile stresses the relative transverse contraction ε_{trans} is proportional to the strain ε in the length direction. The proportionality constant is called *Poisson's ratio*:

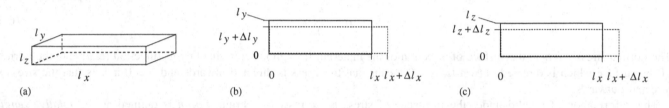

Figure 6.6 (a) Initial box before straining. The strain is applied in the x direction, $\varepsilon_x = \Delta l_x / l_x$ where Δl_x is positive. It is not shown in (a) but in (b) and (c).
(b) Strain in the y direction. $\varepsilon_y = \Delta l_y / l_y$ where Δl_y is negative. (c) Strain in the z direction. $\varepsilon_z = \Delta l_z / l_z$ where Δl_z is negative.

$$\varepsilon_{trans} = -\nu\varepsilon \tag{6.8}$$

where
  ε = strain in the length direction
  ε_{trans} = strain in a perpendicular direction
  ν = Poisson's ratio.

In Figure 6.6, the length direction is chosen as the x axis. If the material is *isotropic* the strains in the y and z directions, which both are negative, are equal:

$$\varepsilon_y = \varepsilon_z = -\nu\varepsilon_x \tag{6.9}$$

An extension of the bar in the x direction corresponds to contractions, i.e. negative strain, in the y and z directions, and vice versa. Hence Δy and Δz always have opposite signs compared with Δx. This is the explanation of the minus sign in Equations (6.8) and (6.9).

For metals, Poisson's ratio usually has a value of approximately 0.3.

Hooke's Generalized Law

Consider a parallelepiped with the volume $V = xyz$, which is exposed to the stresses σ_x, σ_y, σ_z and shown in Figure 6.7. We want to calculate the resulting strains ε_x, ε_y and ε_z of this mechanical load at constant temperature.

As the body is exposed to stress from all sides, ε_x can no longer be calculated from Equation (6.7) only. The stresses in the y and z directions, due to lateral contraction, must also be taken into consideration.

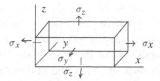

Figure 6.7   The stress σ_y on the rear surface exists but is not shown.

This can be done by application of Equation (6.9) twice. The total strain in the x direction will be

$$\varepsilon_x = \frac{\sigma_x}{Y} - \nu \frac{\sigma_y}{Y} - \nu \frac{\sigma_z}{Y} \qquad (6.10)$$

We obtain analogous equations for ε_y and ε_z by permutation of the indices x, y and z. The three equations

$$\varepsilon_x = \frac{1}{Y}\left[\sigma_x - \nu\left(\sigma_y + \sigma_z\right)\right] \qquad (6.10)$$

$$\varepsilon_y = \frac{1}{Y}\left[\sigma_y - \nu\left(\sigma_z + \sigma_x\right)\right] \qquad (6.11)$$

$$\varepsilon_z = \frac{1}{Y}\left[\sigma_z - \nu\left(\sigma_x + \sigma_y\right)\right] \qquad (6.12)$$

represent *Hooke's generalized law,* valid for isotropic solids exposed to pure tensile stresses. It is valid in case of a general stress in a body. Equations (6.10)–(6.12) can be used for calculation of the volume change associated with the strains. This will be discussed in connection with compression of solids below.

6.3.2 Compressibility

When a solid body is exposed to an external pressure Δp, it becomes slightly compressed. $\Delta p > 0$ when it is directed towards the surface. At constant temperature the volume change is negative and proportional to the applied pressure change. This can be written as

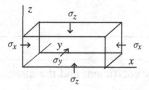

Figure 6.8   The stress σ_y on the rear surface exists but is not shown.

$$V + \Delta V = V\left(1 - \kappa \Delta p\right) \qquad (6.13)$$

where
  V = volume of the body
  Δp = applied all-round pressure on the body
  ΔV = volume change (<0) due to the pressure $\Delta p\,(>0)$
  κ = compressibility coefficient of the solid at constant temperature.

κ is a material constant. The proper definition of κ is

$$\kappa_T = -\frac{1}{V}\left(\frac{\partial V}{\partial p}\right)_T \qquad (6.14)$$

but we will exclude the subscript for simplicity as no confusion will arise.

A positive external pressure, directed towards the surface of the body, results in a decrease in volume. For this reason, there is a minus sign in Equation (6.14) to give a positive value of κ. The compressibility coefficient is generally small for liquids and even smaller for solids, owing to strong repulsive forces between the atoms in the crystal lattice when the distance between the atoms is forced to decrease. For metals it is of the magnitude $10^{-11}\,\mathrm{Pa}^{-1}$ ($1\,\mathrm{Pa} = 1\,\mathrm{N/m^2}$).

$1/\kappa$ is called the *modulus of compressibility* or *bulk modulus* and is denoted by B:

$$B = \frac{1}{\kappa} \tag{6.15}$$

The SI unit of the bulk modulus is pascal.

Relative Volume Change as a Function of Pressure Change

The volume change of the body in Figure 6.8 under the influence of pressure Δp can be written as

$$\Delta V = (l_x + \Delta l_x)\left(l_y + \Delta l_y\right)(l_z + \Delta l_z) - l_x l_y l_z \tag{6.16}$$

where ΔV, Δl_x, Δl_y and Δl_z are negative quantities.

If Equation (6.16) is divided by $V = l_x l_y l_z$, the right-hand side is reduced and second- and third-order terms of small quantities are neglected, we obtain the relationship

$$\frac{\Delta V}{V} = \frac{\Delta l_x}{l_x} + \frac{\Delta l_y}{l_y} + \frac{\Delta l_z}{l_z} \tag{6.17}$$

or, according to Equation (6.6)

$$\frac{\Delta V}{V} = \varepsilon_x + \varepsilon_y + \varepsilon_z \tag{6.18}$$

This relationship is generally valid, independent of the shape of the body.

Addition of Equations (6.10)–(6.12) for the special case $\sigma_x = \sigma_y = \sigma_z = -\Delta p$ gives the relative volume decrease:

$$\frac{\Delta V}{V} = -\frac{3}{Y}\left(1 - 2\nu\right)\Delta p \tag{6.19}$$

where $\Delta V < 0$ and $\Delta p > 0$.

Example 6.1
Derive a relationship between the compressibility coefficient and the modulus of elasticity.

Solution and Answer:

If we divide Equation (6.19) by $-\Delta p$, we obtain

$$\frac{-\Delta V}{V \Delta p} = \frac{3}{Y}\left(1 - 2\nu\right) \tag{1'}$$

The left-hand side of this equation is identical with Equation (6.14) for the definition of κ and we obtain

$$\kappa = \frac{3}{Y}\left(1 - 2\nu\right) \tag{2'}$$

Equation (2') is the desired relationship between κ and Y.

Alternatively Equation (2') is often written as a relationship between the modulus of compressibility and elasticity:

$$B = \frac{Y}{3\left(1 - 2\nu\right)} \tag{3'}$$

6.3.3 Deformation Energy

Deformation energy or *strain energy* is the work required to change the shape of a solid body.

One-dimensional Deformation

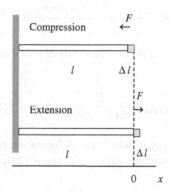

Figure 6.9 Deformation of a bar.

For a *small* deformation, compression or extension, in one dimension, of a bar made of an elastic material (Figure 6.9), the work to deform the body can be written with the aid of Equations (6.6) and (6.7):

$$W = \int_0^{\Delta l} F \, dx = \int_0^{\Delta l} \sigma A \, dx = \int_0^{\Delta l} Y \varepsilon A l \, d\varepsilon = A l \frac{Y \varepsilon^2}{2}$$

where

W = deformation work
l = length of the bar
A = cross-section of the bar.

The deformation energy per unit volume $E_{\text{deform}} = W/Al$ can in this case be written as

$$E_{\text{deform}} = \frac{Y \varepsilon^2}{2} = \frac{\sigma \varepsilon}{2} = \frac{\sigma^2}{2Y} \tag{6.20}$$

where Y is the modulus of elasticity. Equation (6.20) is valid independent of the shape of the body.

Three-dimensional Deformation

In the *case* of a *small general deformation* and *in the absence of shear stresses*, the energy per unit volume of an isotropic elastic material is obtained if Equation (6.20) is replaced by the expression

$$E_{\text{deform}} = \frac{\left(\sigma_x + \sigma_y + \sigma_z\right)^2}{18B} \tag{6.21}$$

where B is the bulk modulus.

Equation (6.21) can be transformed with the aid of Equation (3′) in Example 6.1:

$$E_{\text{deform}} = \frac{\left(\sigma_x + \sigma_y + \sigma_z\right)^2}{18 \dfrac{Y}{3\left(1 - 2\nu\right)}}$$

which can be reduced to

$$E_{\text{deform}} = \frac{1-2\nu}{6Y}\left(\sigma_x + \sigma_y + \sigma_z\right)^2 \tag{6.22}$$

If we add Equations (6.10)–(6.12), we obtain

$$\varepsilon_x + \varepsilon_y + \varepsilon_z = \frac{\sigma_x + \sigma_y + \sigma_z}{Y}\left(1-2\nu\right) \tag{6.23}$$

We express $\sigma_x + \sigma_y + \sigma_z$ in terms of $\varepsilon_x + \varepsilon_y + \varepsilon_z$ and introduce this expression into Equation (6.22). The deformation energy per unit volume will then be

$$E_{\text{deform}} = \frac{Y}{6\left(1-2\nu\right)}\left(\varepsilon_x + \varepsilon_y + \varepsilon_z\right)^2 \tag{6.24}$$

The deformation energy is independent of the way in which the deformation has been achieved.

Equations (6.22) and (6.24) show that the maximum value of Poisson's ratio is 0.5:

$$\nu < 0.5 \tag{6.25}$$

The maximum value, which never is reached by any material, corresponds to an incompressible material.

In anisotropic materials, Equations (6.22) and (6.24) are not valid and have to be replaced by other equations which contain the components of both tensile and shear stresses.

6.3.4 Temperature Dependence of the Modulus of Elasticity

Empirical Results

Experimental studies of oxides in the 1960s showed that the temperature dependence of Young's modulus Y can be described by the empirical relationship

$$Y(T) = Y(0)\left(1 - bTe^{-\frac{T_0}{T}}\right) \tag{6.26}$$

where
   Y = modulus of elasticity
   b, T_0 = constants
   $Y(0)$ = value of Y at $T = 0\,\text{K}$.

If $T \gg T_0$, series development of the exponential factor gives

$$Y(T) \approx Y(0)\left[1 - b\left(T - T_0\right)\right] \tag{6.27}$$

Since then temperature dependence of elastic constants has been observed and confirmed for many other solids and for elastic constants other than Young's modulus.

Efforts have been made to understand the empirical relationships on a theoretical basis. These theories will not be discussed here.

6.4 Expansion

6.4.1 Length and Volume Expansions

In the absence of change in structure, most solid materials expand when temperature increases and shrink with decreasing temperature. This can be described by the empirical relationship

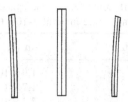

Figure 6.10   A bimetallic thermometer consists of two thin metal bars with different expansion coefficients, firmly attached to each other. At a given temperature the bar is straight and at all other temperatures it is bent.

$$l + \Delta l = l(1 + \alpha \Delta T) \tag{6.28}$$

where
l = length of specimen at temperature T
$l + \Delta l$ = length of specimen at temperature $T + \Delta T$
α = linear thermal expansion coefficient at constant pressure.

The linear thermal expansion coefficient α at constant pressure varies with the temperature of the solid. Hence it must be defined at a specified temperature. The proper definition is

$$\alpha = \frac{1}{l}\left(\frac{\partial l}{\partial T}\right)_p \tag{6.29}$$

where p is the pressure. The temperature dependence can be used for temperature measurements (Figure 6.10).

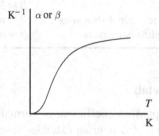

Figure 6.11   Linear expansion or volume expansion coefficient as a function of temperature.

Linear expansion is closely related to the thermal volume expansion of solids. In analogy with the one-dimensional case we have

$$V + \Delta V = V(1 + \beta \Delta T) \tag{6.30}$$

where
V = volume of the specimen at temperature T
$V + \Delta V$ = volume of the specimen at temperature $T + \Delta T$
β = thermal volume expansion coefficient at constant pressure.

The thermal volume expansion coefficient β at constant pressure varies with the temperature of the solid (Figure 6.11). The definition of β is

$$\beta = \frac{1}{V}\left(\frac{\partial V}{\partial T}\right)_p \tag{6.31}$$

where p is the pressure.

Table 6.1 Length expansion coefficients of some metals. Temperature interval 0–100 °C.

Metal	α (K^{-1})
Ag	1.9×10^{-5}
Cu	1.7×10^{-5}
Al	2.3×10^{-5}
Mg	2.6×10^{-5}
Fe	1.2×10^{-5}

Some α values are given in Table 6.1.

In the absence of phase transformations, the expansion coefficient α increases with temperature. In the temperature interval from about $-200\,°C$ to the melting point of the metal α is proportional to the heat capacity of the metal. At temperatures close to $T = 0\,K$ α drops rapidly and approaches zero.

Relationship between α and β

α and β are normally of the magnitude $10^{-5}\,K^{-1}$. In this case there is an approximate relationship between α and β. Table 6.2 shows that the relationship depends of the structure of the crystalline solid.

Table 6.2 Relationship between length and volume expansion coefficients in crystals.

Type of crystal structure	Relationship between α and β
Cubic structures	$\beta \approx 3\alpha$
Hexagonal, trigonal and tetragonal structures	$\beta \approx \alpha_x + 2\alpha_y$
Rhombic, mono- and triclinic structures	$\beta \approx \alpha_x + \alpha_y + \alpha_z$

Average Volume Expansion Coefficient in Metals

The temperature dependence of the volume expansion coefficient is normally not especially strong. For this reason, it is a reasonable approximation to use an average value of β over an extended temperature interval. The average value of β can be written as

$$\overline{\beta} = \frac{1}{V_0} \frac{\Delta V}{\Delta T} \tag{6.32}$$

where
$V_0 = $ volume of the solid at temperature T_0
$\Delta T = $ change of temperature $= T - T_0$
$\Delta V = $ change of volume $= V - V_0$.

As an application we will consider a number of metals with cubic structures (BCC and FCC). If we choose temperature T equal to the melting point T_M of the metal and $T_0 = 0\,K$, the temperature interval ΔT will be equal to T_M. It has been found empirically that the average volume expansion coefficient within the temperature interval from $0\,K$ to T_M times the melting temperature (K):

$$\overline{\beta}T_M = \frac{1}{V_0} \frac{\Delta V}{\Delta T} T_M = \frac{\Delta V}{V_0}$$

is practically *constant* (values between 0.06 and 0.07) for metals with cubic structure, *independent of the kind of metal*.

As α is approximately one-third of β, the conclusion is that

$$\overline{\alpha}T_M = \frac{\Delta l}{l_0} = constant \approx 0.02 \tag{6.33}$$

The average relative length expansion between $0\,K$ and the melting point of any metal is approximately 2%.

Equation (6.33) is called *Grüneisen's rule*. It can be verified directly from experimental data for some metals in Figure 6.12. From Equation (6.33), we can conclude that

- Metals with cubic structures and high melting points have low linear expansion coefficients and vice versa.

This statement is verified by the hyperbolic curve in Figure 6.13, which is based on experimental data.

A group of FeNi alloys, known as Invar alloys, show practically no expansion at all close to room temperature. These alloys are ferromagnetic and the magnetic effect, the so-called *magnetostriction* or change of length in a magnetic field, can be either positive or negative.

In the case of Invar alloys, the magnetostriction happens to be practically equal to the thermal expansion and of opposite sign. Hence the two effects cancel and the net effect is nearly zero.

Such alloys do not change their length when the temperature changes. The term Invar comes from the word invariance.

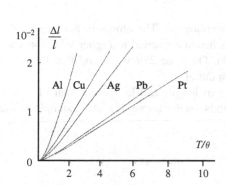

Figure 6.12 Average value of the relative length expansion coefficient within the interval $0\,K$–T_M as a function of the absolute temperature divided by the Debye temperature for some metals.

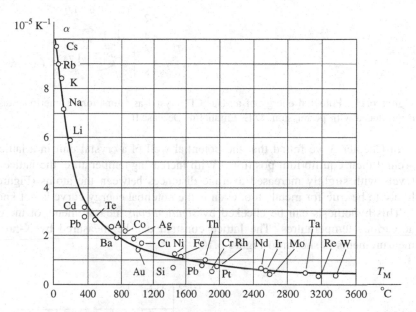

Figure 6.13 The linear expansion coefficient of some metals with cubic crystal structures as a function of their melting points. Reproduced with permission from W. Schatt (ed.), *Einführung in die Werkstoffwissenschaft*, 7th edn. © 1991 Deutscher Verlag für Grundstoffindustrie, Leipzig.

6.4.2 Thermal Expansion

Figure 6.14 shows the average volume per atom of iron crystals as a function of temperature, and indicates that two structure transformations occur in iron within the temperature interval 0–1500 °C: from α-Fe to γ-Fe at 910 °C and from γ-Fe to δ-Fe at 1390 °C. A change from one structure of a crystal to another must necessarily result in a considerable and discontinuous volume change as different crystal structures have different packings of the atoms.

Below we will discuss the reasons for the continuous volume change of crystals, due to temperature changes, from an atomic point of view.

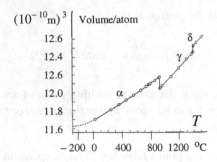

Figure 6.14 Average volume per atom in iron crystals as a function of temperature. Reproduced with permission from W. Schatt (ed.), *Einführung in die Werkstoffwissenschaft*, 7th edn. © 1991 Deutscher Verlag für Grundstoffindustrie, Leipzig.

Influence of Lattice Constant

When a crystal is heated its length L increases. An explanation is that the expansion is due to *a slight increase in the lattice constant a* with increase in temperature.

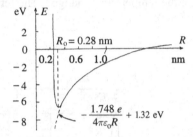

Figure 6.15 Potential energy of an $Na^+ Cl^-$ crystal as a function of interionic distance. Compare Figure 3.6 on page 108 in Chapter 3. Reproduced with permission. © E. Lindholm (Deceased).

In Chapter 3 we found that the potential well of a crystal unit in a lattice is asymmetric. The atoms in the lattice vibrate around their equilibrium positions. With increasing temperature, the lattice atoms become excited to higher vibration energy levels with slightly increased average distances between the atoms (Figure 6.15). On page 289 we found that the same is likely to be true for metals too, even if the potential energy curve is not known in detail.

This hypothesis can be checked by *simultaneous* measurements of the change of length and lattice constant of a crystal at various temperatures. The lattice constant can be measured by X-ray methods (see Chapter 1). An example of such measurements is shown in Figure 6.16.

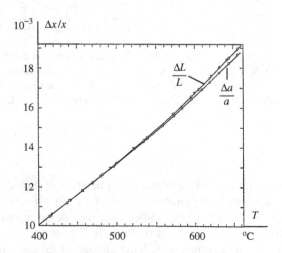

Figure 6.16 Relative length change and relative lattice parameter change for aluminium as functions of temperature. Reproduced with permission from P. Shewmon, *Diffusion in Solids*. © 1963 McGraw-Hill Book Company.

Simmons and Balluffi carried out careful experimental studies of the relative changes in length L and the lattice constant a for different metals over a wide range of temperatures. Their results for aluminium are plotted in Figure 6.16, which shows clearly that stretching of the distance between the atoms in the crystal lattice is the dominant but not only explanation of the length expansion.

Influence of Vacancies in the Crystal Lattice

The increasing discrepancy between relative length change and relative change of the lattice constant at higher temperatures is due to the *formation of vacancies in the crystal lattice.*

Consider a crystal with a cubic structure and lattice constant a at temperature T. If the crystal contains N_{total} atoms, we can write its volume as

$$V = L^3 = N_{total} a^3 \qquad (6.34)$$

After logarithmic taking the derivative of Equation (6.34), we obtain

$$\frac{3\Delta L}{L} = \frac{\Delta N_{total}}{N_{total}} + \frac{3\Delta a}{a} \qquad (6.35)$$

which can be written as

$$\frac{\Delta N_{total}}{N_{total}} = 3\left(\frac{\Delta L}{L} - \frac{\Delta a}{a}\right) \qquad (6.36)$$

Equation (6.36) gives the relative increase in the number of atom sites independent of the type of crystal defect. Vacancies increase the number of sites, i.e. the length L and hence also the volume of the crystal.

Interstitials are not expected to influence the lattice constant but there is some experimental evidence (e.g. C atoms in FCC iron at constant temperature) which indicates that they may increase the crystal volume and hence the lattice constant slightly [N_{total} is constant in Equation (6.34) at constant temperature].

For metals, $\Delta N_{total}/N_{total}$ is positive and we can neglect the formation of interstitials. In this case, $\Delta N_{total}/N_{total}$ depends mainly on vacancy formation. The number of vacancies increases rapidly (exponentially) with temperature. At each temperature there is an equilibrium concentration of vacancies in the metal.

If the metal is heated, the equilibrium concentration of vacancies increases. New vacancies are formed first at dislocations and boundaries, which act as sources. The vacancies become distributed homogeneously by diffusion. When the metal is cooled, the vacancies diffuse to the dislocations and boundaries, which act as sinks.

Hence length expansion at constant pressure and variable temperature consists of two parts:

- a contribution due to change of the lattice constant a
- a contribution due to vacancies.

The dominant effect originates from the change of the interatomic distances in the crystal lattice. The fraction of vacancies is comparatively small but increases with increase in temperature (Figure 6.16). The maximum influence of vacancies occurs close to the melting point of the metal.

Theory of Thermal Expansion

It is well known that thermal expansion of solids is caused by a change in the interatomic distances in the crystal lattice and to a minor extent vacancy formation at higher temperatures below the melting point of the solid.

A very reasonable explanation of length expansion is the anharmonicity of the vibrations of the atoms around their equilibrium positions in the lattice, which corresponds to increasing distances between the atoms with increase in temperature. This effect is *not* sufficient to explain the magnitude of the length expansion. The increase in the lattice constant a is *larger* than the anharmonicity effect on the interatomic distance.

In classical mechanics, the equilibrium is found by minimizing the total energy, which is the sum of the kinetic and potential energy, and the temperature is not involved. This approach is not possible here. The potential energy is a complex function of all the atoms with many variables as many distances are involved. A more feasible way is to use thermodynamics. In Chapter 5 we defined the internal energy U, which is the sum of the kinetic and potential energy of the system.

At constant temperature and pressure, the equilibrium condition of a nonmechanical system is a *minimum of Gibbs free energy* (Chapter 5 on page 237). This condition is widely used in chemistry and is also applied to, for example, diffusion and solidification processes. The Gibbs free energy is defined as [Equation (5.55) on page 235 in Chapter 5]

$$G = H - TS \tag{6.37}$$

which can be written as

$$G = U + pV - TS \tag{6.38}$$

where
  G = Gibbs free energy of the system (J/kmol)
  H = enthalpy of the system
  U = internal energy of the system
  p = pressure of the system
  V = volume of the system
  T = temperature of the system
  S = entropy of the system (J/kmol K).

Thermodynamic Explanation of Thermal Expansion
When the temperature increases the lattice constant increases and the whole crystal expands. Each temperature corresponds to a certain value of the lattice constant a. When the temperature increases or decreases, a new equilibrium with a different value of a is developed. The equilibrium condition is that the function $G(V, T)$ has a minimum.

The volume of a solid is small. Normal pressures, for example 1 atm, can be regarded as small and the product pV can be neglected in comparison with U. In this case, we can instead minimize the Helmholtz free energy F, which is defined as

$$F = U - TS \tag{6.39}$$

F is a function of V and T. The equilibrium volume at a given temperature (when T no longer is a variable) is given by the condition

$$\left(\frac{\partial F}{\partial V} \right)_T = 0 \tag{6.40}$$

The equilibrium volume V depends on the temperature only if F contains a term which is a function of both V and T. If, for example, the vibrations around the equilibrium positions of the atoms were completely harmonic, the equilibrium positions would not change with temperature changes and V would be independent of T. Consequently, there would be no expansion.

Hence anharmonicity of the atom vibrations in the lattice is a *necessary but not sufficient* condition for expansion. Without going into details, the explanation of expansion, in terms of thermodynamics, can be expressed as follows:

- The whole crystal expands or contracts until it finds the volume for which the total free energy is a minimum.

or, more precisely, the expansion depends on the term TS in Equation (6.39). The entropy S is also a function of V and T. At a certain change of T a new equilibrium value of V is developed. The volume change stops and the equilibrium volume is achieved when the term TS acquires such a value that the condition (6.40) is fulfilled.

The equilibrium value of the interatomic distance, obtained from the condition (6.40), does *not* coincide with the average distance caused by the anharmonicity of the potential well.

Grüneisen's Rule
In the absence of safe information about the anharmonic terms of the lattice vibrations, we will assume that the frequency of the vibrations is a function of the volume of the crystal. We assume that the relative change in the frequency is proportional to the relative change in the volume. When the volume increases the frequency decreases, which gives

$$\frac{\Delta \nu}{\nu} = -\gamma \frac{\Delta V}{V} \tag{6.41}$$

where γ is a constant.

The total free energy F can is obtained by addition of two terms:

1. the deformation energy due to the volume change of the crystal
2. the sum of the free energies of the phonons, calculated by Bose–Einstein statistics of the Einstein oscillators , i.e. the phonon distribution in the lattice.

When the derivative of the total free energy F with respect to V is set to zero, the minimum condition for F can be written as

$$\frac{1}{\kappa}\left(\frac{\Delta V}{V}\right) = \gamma \overline{E_{\mathrm{vibr}}}(T) \tag{6.42}$$

where
κ = compressibility coefficient of the crystal
γ = a constant
$\overline{E_{\mathrm{vibr}}}(T)$ = average energy of the lattice vibration mode per unit volume.

Equation (6.42) can be written as

$$\frac{\Delta V}{V} = \kappa \gamma \overline{E_{\mathrm{vibr}}}(T) \tag{6.43}$$

- The volume expansion at temperature T is proportional to the average thermal energy density.

Taking the derivative of Equation (6.43) with respect to T gives

$$\frac{\mathrm{d}\left(\dfrac{\Delta V}{V}\right)}{\mathrm{d}T} = \kappa \gamma \frac{\mathrm{d}\left(\overline{E_{\mathrm{vibr}}}(T)\right)}{\mathrm{d}T} \tag{6.44}$$

or

$$\beta = constant \times C_{\mathrm{V}} \tag{6.45}$$

- The thermal volume expansion coefficient is proportional to the molar heat capacity at constant volume.

Equation (6.45) is known as *Grüneisen's rule*. We will come back to these topics in connection with heat capacity and find a third Grüneisen's rule.

6.5 Heat Capacity

6.5.1 Heat Capacity at Constant Volume and at Constant Pressure

If an amount of heat $\mathrm{d}Q$ is added to a system, the heat is used to increase the internal energy U of the system and to perform work, according to the first law of thermodynamics:

$$\mathrm{d}Q = \mathrm{d}U + p\mathrm{d}V \tag{6.46}$$

The internal energy is a function of the temperature T and the volume V of the system:

$$\mathrm{d}U = \left(\frac{\partial U}{\partial T}\right)_{\mathrm{V}} \mathrm{d}T + \left(\frac{\partial U}{\partial V}\right)_{\mathrm{T}} \mathrm{d}V \tag{6.47}$$

This expression for dU is introduced into Equation (6.46):

$$dQ = \left(\frac{\partial U}{\partial T}\right)_V dT + \left[\left(\frac{\partial U}{\partial V}\right)_T + p\right] dV \tag{6.48}$$

We consider 1 kmol of the solid and use Equation (6.48) as the basis for definition of the molar heat capacity at constant volume C_V and the molar heat capacity at constant pressure C_p:

$$C_V = \left(\frac{\partial Q}{\partial T}\right)_V = \left(\frac{\partial U}{\partial T}\right)_V \tag{6.49}$$

$$C_p = \left(\frac{\partial Q}{\partial T}\right)_p \tag{6.50}$$

For gases and at higher temperatures even for liquids and solids it is necessary to distinguish between C_V and C_p.

The heat capacity of 1 mass unit (1 kg instead of 1 kmol) is called the *specific heat capacity* or *heat capacitivity* and is denoted by c_V and c_p. Generally we have

$$dQ = cm\,dT \tag{6.51}$$

6.5.2 Models of Heat Capacity

Classical Model

From the kinetic theory of gases (Chapter 4), it is well known that $C_V = 3R/2$ and $C_p = 5R/2$ for a monoatomic gas. According to the equipartition principle, each degree of freedom corresponds to an energy equal to $RT/2$ per kilomol. The three degrees of freedom are, in the case of a monoatomic gas, the translation motion of the molecules in three directions, x, y and z.

In the case of a crystalline solid, the atoms in the lattice have no translation movement but vibrate incessantly around their equilibrium positions. Each atom can be regarded as an oscillator. Its total vibrational energy is the sum of its potential energy and its kinetic energy, which corresponds to two degrees of freedom for each vibration direction. The total internal energy u per atom equals the sum of the vibrational energy in all three directions:

$$u = 3\left(\frac{k_B T}{2} + \frac{k_B T}{2}\right) = 3k_B T$$

The total internal energy U of 1 kmol is obtained if u is multiplied by Avogadro's number N_A. Remembering that $k_B N_A = R$, we obtain

$$U = 3RT \tag{6.52}$$

We know that $(\partial U/\partial T)_V = C_V$, and differentiate Equation (6.52) with respect to T, which gives the Dulong–Petit law, named after the scientists who discovered it:

$$C_V = 3R \tag{6.53}$$

The classical model of molar heat capacity of solids agrees well with experiments at *high* temperatures. At *low* temperatures the model is a disaster, as is shown in Figure 6.17. C_V decreases with temperature at low temperatures and becomes zero at $T = 0\,\mathrm{K}$.

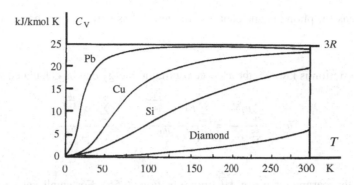

Figure 6.17 The molar heat capacities at constant volume of some solids as a function of temperature. Reproduced with permission from D. R. Gaskell, *Introduction to Metallurgical Thermodynamics*, 2nd edn. © 1981 Hemisphere Publishing Corporation (with Taylor & Francis) Taylor & Francis Group are the rightsholder.

Einstein's Model

Einstein developed a new model of the heat capacity of solids, which agreed much better with experiments than the classical model.

Einstein's Model of Phonon Energy

Einstein considered a crystal consisting of N_{total} atoms. He assumed that the N_{total} atoms can be regarded as $3N_{total}$ linear harmonic oscillators, i.e. each atom corresponds to three harmonic oscillators, which vibrate in the three perpendicular directions x, y and z, with *a single basic frequency ν* around their average positions *independently of all their neighbours*.

Einstein realised that even if the basic frequency of the vibrations is the same, the energies of the oscillators can differ. This is best understood in terms of vibrations of a string with a given length. In such a system overtones of various intensities occur. The frequencies of these overtones are multiples of the basic frequency and correspond to higher energies than the basic frequency.

Einstein applied Planck's famous proposal, used in the derivation of his radiation law, and assumed that the energies of the oscillators are not continuous but *quantized*. He assumed that the allowed phonon energies u_i are integer multiples of $h\nu$:

$$u_i = ih\nu \quad i = 1, 2, 3, \ldots, \infty \tag{6.54}$$

where

u = energy of oscillator i
h = Planck's constant
ν = frequency of the basic oscillation.

The integer quantum number i is analogous with the vibrational quantum number of the harmonic oscillator according to older quantum mechanics. In Chapter 3, the quantum number was denoted n instead of i, which is used in this special case.

Energy Distribution Function

The energy distribution of the oscillators depends on the temperature. The fraction of the atoms N_i/N_{total} which have the energy u_i can be calculated statistically with the same approach as in Chapter 4, page 175. The fraction decreases exponentially with increasing energy $ih\nu$:

$$\frac{N_i}{N_{total}} = \frac{e^{-\frac{ih\nu}{k_B T}}}{\sum_{i=1}^{\infty} e^{-\frac{ih\nu}{k_B T}}} \tag{6.55}$$

where

N_i = number of atoms which have the energy u_i
N_{total} = total number of atoms in the crystal
k_B = Boltzmann's constant
T = absolute temperature
$\sum_{i=1}^{\infty} e^{-\frac{ih\nu}{k_B T}}$ = the partition function (Chapter 4, page 175).

The energy distribution functions for phonons and photons are identical as both types of particles obey Bose–Einstein statistics (Chapter 3, pages 148–149).

Average Oscillator Energy
When the energy distribution function is known, the average oscillator energy can be calculated with the aid of Equation (6.55):

$$\overline{u} = \frac{\sum\limits_{i=1}^{\infty} u_i N_i}{\sum\limits_{i=1}^{\infty} N_i} = \frac{\sum\limits_{i=1}^{\infty} ih\nu\, e^{-\frac{ih\nu}{k_B T}}}{\sum\limits_{i=1}^{\infty} e^{-\frac{ih\nu}{k_B T}}} = h\nu \frac{\sum\limits_{i=1}^{\infty} i e^{-\frac{ih\nu}{k_B T}}}{\sum\limits_{i=1}^{\infty} e^{-\frac{ih\nu}{k_B T}}} \tag{6.56}$$

For simplicity, we introduce the parameter $p = e^{-\frac{h\nu}{k_B T}}$ into equation (6.56). For small values of p, the exponential function can be developed in series:

$$\overline{u} = h\nu\, \frac{p(1 + 2p + 3p^2 + 4p^3 + \dots)}{p + p^2 + p^3 + \dots} = h\nu\, \frac{p\dfrac{d}{dp}(p + p^2 + p^3 + \dots)}{\dfrac{p}{1-p}}$$

or

$$\overline{u} = h\nu\, \frac{p\dfrac{d}{dp}\left(\dfrac{1}{1-p}\right)}{\dfrac{p}{1-p}} = h\nu\, \frac{\dfrac{p^2}{(1-p)^2}}{\dfrac{p}{1-p}} = h\nu\, \frac{p}{1-p} = \frac{h\nu}{p^{-1} - 1}$$

or, if we replace p with the exponential expression

$$\overline{u} = \frac{h\nu}{e^{\frac{h\nu}{k_B T}} - 1} \tag{6.57}$$

Equation (6.57) is valid for both phonons and photons and closely related to Planck's radiation law, which describes the density of electromagnetic radiation at temperature T (Chapter 2, page 49).

In Chapter 3 on page 148 we mentioned that phonons obey Bose–Einstein statistics. This statement is in agreement with Equation (6.57), which can be written as

$$\overline{u} = h\nu f_{BE} \tag{6.58}$$

Total Phonon Energy
The total phonon energy is calculated as the product of the number of oscillators and the average oscillator energy per atom. It represents the internal energy of the crystal. If we let U represent the internal energy of 1 kmol, the number of oscillators will be equal to three times Avogadro's number N_A:

$$U = 3N_A \overline{u} = \frac{3 N_A h\nu}{e^{\frac{h\nu}{k_B T}} - 1} \tag{6.59}$$

Modification of Einstein's Model with Respect to Quantum Mechanics
Einstein's model for the internal energy and the heat capacity of solids was published in 1906. If his assumption of the quantized phonon energies, Equation (6.54), is modified with respect to quantum mechanics (1920s), we obtain

$$u_i = (i + \tfrac{1}{2})h\nu \quad i = 1, 2, 3, \dots, \infty \tag{6.60}$$

If we add $\tfrac{1}{2} h\nu$ to each phonon energy, then the average energy per oscillator must also increase with the same amount and we obtain

$$\overline{u} = \frac{h\nu}{2} + \frac{h\nu}{e^{\frac{h\nu}{k_B T}} - 1} \tag{6.61}$$

and

$$U = 3N_A \overline{u} = 3N_A \left(\frac{h\nu}{2} + \frac{h\nu}{e^{\frac{h\nu}{k_B T}} - 1} \right) \tag{6.62}$$

Molar Heat Capacity at Constant Volume

The molar heat capacity is by definition the derivative of U with respect to T. It can be calculated by differentiating Equation (6.59) or Equation (6.62) with respect to T:

$$C_V = 3N_A \frac{-h\nu}{(e^{h\nu/k_B T}-1)^2} e^{\frac{h\nu}{k_B T}} \frac{h\nu}{k_B} \frac{-1}{T^2} = 3N_A k_B \left(\frac{h\nu}{k_B T}\right)^2 \frac{e^{\frac{h\nu}{k_B T}}}{(e^{h\nu/k_B T}-1)^2}$$

or

$$C_V = 3R \frac{x^2 e^x}{(e^x-1)^2} \tag{6.63}$$

where

$$x = \frac{h\nu}{k_B T} = \frac{\theta_E}{T}$$

The temperature $\theta_E = h\nu/k_B$ is called the Einstein temperature.

At *high* temperatures $(x \to 0)$, Einstein's model gives the classical value $C_v = 3R$, in good agreement with experiments. At *low* temperatures, i.e. for large values of x, $e^x \gg 1$ and Equation (6.63) can be written as

$$C_V = constant \times x^2 e^{-x} = constant \times \left(\frac{1}{T}\right)^2 e^{\frac{-h\nu}{k_B T}} \tag{6.64}$$

in fairly good agreement with experimental observations, much better than the classical model. This was a great success for Einstein.

The frequency of the basic oscillation ν is a material constant. It can be estimated experimentally from measurements of C_V as a function of temperature.

Debye's Model

Einstein's model of the heat capacity of solids was a great improvement compared with the classical model. However, at extremely low temperatures there was a marked discrepancy between Einstein's theory and the experimental evidence. The latter showed that C_V is proportional to T^3 in this temperature region, whereas Einstein's theory gave a too rapid decrease of C_V due to the exponential factor in Equation (6.64), which dominates over the factor T^{-2}.

An improved model was suggested by Debye. He modified Einstein's theory in two respects:

1. Einstein assumed that all the atoms in the crystal were vibrating with the same frequency and that the vibrations of each atom or harmonic oscillator were independent of all the others.

   Debye included all the phonons in the whole crystal in his calculations, i.e. all the elastic waves with a manifold of frequencies. He realized that each atom is influenced by all the others, as the crystal can be regarded as a three-dimensional system of mass points connected by 'elastic springs'. It is impossible to set a single atom into motion without transferring kinetic energy to all the others via the 'elastic springs' between the atoms.

2. Einstein assumed, in analogy with Planck's photon theory of blackbody radiation, that there is no upper limit of the oscillators or the phonon energies [Equation (6.54) with $i = \infty$ on page 305]. This is correct for photons but not for phonons.

   Debye realized that there must be a maximum phonon energy $h\nu_{max}$. The shortest possible wavelength that makes sense is of the same magnitude as the interatomic distances or 3×10^{-10} m. This condition implied that the range of vibrational frequences must be cut off at a maximum frequency, called the *Debye frequency* ν_D. The upper limit of the phonon frequencies is equal to the Debye frequency ν_D. The magnitude of ν_D is 10^{13} Hz (velocity of sound in the crystal divided by the minimum wavelength).

In Chapter 3, Section 3.6, we discussed briefly elastic waves in solids and some of the properties of phonons. We found (Chapter 3, pages 147 and 150–151) that for each wave of frequency ν there are three possible vibrational modes, one longitudinal mode (acoustic mode) and two transversal modes (optical modes).

Debye made the approximation that the wavelengths of all the phonons are large compared with the interatomic distances and treated the crystal as continuous rather than as a crystal lattice. He found an expression for the number of vibrational modes within the crystal. The Debye frequency is derived from the condition that this number equals $3N_{total}$, where N_{total} is the total number of atoms in the crystal. This condition can be written as

$$4\pi V \left[\frac{1}{c_l^3} + \frac{2}{c_t^3}\right] \int_0^{\nu_D} \nu^2 d\nu = 3N_{total} \tag{6.65}$$

where
V = volume of the crystal
c_l = velocity of the longitudinal elastic wave
c_t = velocity of the transversal elastic waves
ν = frequency
ν_D = Debye frequency
N_{total} = number of atoms in the crystal lattice.

The Debye frequency can be derived from Equation (6.65). The result is

$$\nu_D = \sqrt[3]{\frac{9N_{total}}{4\pi V}\left(\frac{1}{c_l^3} + \frac{2}{c_t^3}\right)^{-1}} \tag{6.66}$$

The Debye frequency depends on the velocities of the longitudinal and transversal elastic waves and the number of atoms per unit volume (N_{total}/V) of the crystal.

Debye used the same energy distribution of phonons as Einstein. Integration over all occurring frequencies (up to ν_D) and vibration modes gave Debye's modified expression for the total internal vibrational energy of 1 kmol ($N_{total} = N_A$), which corresponds to Equation (6.59) in Einstein's model:

$$U = 9N_A k_B T \left(\frac{T}{\theta_D}\right)^3 \int_0^{\frac{\theta_D}{T}} \frac{x^3}{e^x - 1} dx \tag{6.67}$$

where

$$\theta_D = \frac{h\nu_D}{k_B} \tag{6.68}$$

is the *Debye temperature* of the crystal.

We obtain the molar heat capacity at constant volume by taking the derivative of Equation (6.67) with respect to T. After reduction we obtain

$$C_V = 9R \left(\frac{T}{\theta_D}\right)^3 \int_0^{\frac{\theta_D}{T}} \frac{x^4 e^x}{(e^x - 1)^2} dx \tag{6.69}$$

Equation (6.69) is the Debye equation for the molar heat capacity at constant volume. It gives excellent agreement between theory and experiments for both high and low temperatures.

An example of this is given in Figure 6.18. As a comparison, Einstein's theoretical curve has also been included.

At low temperatures, the molar heat capacity at constant volume can approximately be written as

$$C_V = \frac{12\pi^4}{5} R \left(\frac{T}{\theta_D}\right)^3 \qquad T << \theta_D \tag{6.70}$$

The values of θ_D of some metals are given in Table 6.3. Varying values of the same metal are given in the literature. One reason for this may be that ν_D, and consequently θ_D, varies slightly with temperature and is slightly anisotropic, i.e. has different values along different crystal axes.

Figure 6.11 on page 297 shows a striking resemblance with Figure 6.18. The length expansion coefficient is proportional to the molar heat capacity at low temperatures. *The mechanism behind the phenomenon is the lattice vibrations in both cases.* At higher temperatures the mechanisms differ.

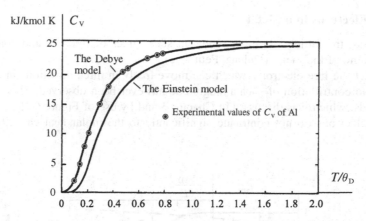

Figure 6.18 The molar heat capacity at constant volume of Al as a function of T/θ_D. Reproduced with permission from D. R. Gaskell, *Introduction to Metallurgical Thermodynamics*, 2nd edn. © 1981 Hemisphere Publishing Corporation (with Taylor & Francis) Taylor & Francis Group are the rightsholder.

Table 6.3 Debye temperatures of some metals.

Metal	θ_D (K)
Ag	225
Al	428
Au	165
Be	1460
Cu	345
Fe (α)	462
K	91
Mg	400
Mn	410
Ni	453
Zn	327

6.5.3 Molar Heat Capacity at Constant Volume

In Section 6.5.2, we discussed three models of the heat capacity of solids: the classical theory and Einstein's and Debye's models. Debye's model shows the best agreement with experimental values at both low and high temperatures:

$$C_V = 9R \left(\frac{T}{\theta_D} \right)^3 \int_0^{\frac{\theta_D}{T}} \frac{x^4 e^x}{(e^x - 1)^2} dx \tag{6.69}$$

where

$$x = \frac{h\nu}{k_B T}$$

and the Debye temperature

$$\theta_D = \frac{h\nu_D}{k_B}$$

All three models deal only with the energy of the lattice vibrations. If the solid is a metal it also contains free electrons. Below we will examine the contribution of the electrons to the heat capacity.

Heat Capacity of the Free Electrons in a Metal

According to classical theories, the lattice vibrations represent six degrees of freedom and contribute $6 \times R/2$ to the molar heat capacity at constant volume of the metal (Dulong–Petit rule).

Similarly, one would expect the free electrons, which can move freely in three directions in the metal, to contribute $3R/2$ to the molar heat capacity. No contribution of such a magnitude has ever been observed. This can be explained with the aid of the electron theory of solids, which was discussed in Chapter 3 and by use of Figure 6.19.

The reason why the free electrons do not contribute significantly to the molar heat capacity is that they obey the Pauli exclusion principle.

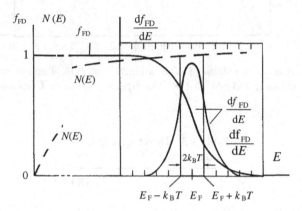

Figure 6.19 Three diagrams in one: (1) the electron energy density $N(E)$ as a function of E (dashed curve); (2) the Fermi factor f_{FD} as a function of E; (3) the derivative of f_{FD} with respect to the energy E as a function of E. Reproduced from F. Blatt, *Modern Physics*. © McGraw-Hill Inc (1992).

The electron gas is in thermal equilibrium with the crystal lattice. At temperature $T = 0$ all energy states up to the Fermi level $E_F = 0$ are filled. At temperatures $T > 0$ some electrons with energies close to E_F become excited to higher energy levels above E_F. Since the thermal excitation energy is of the magnitude $k_B T$, only the electrons within the energy range from $E_F - k_B T$ to E_F can become excited. Thermal excitation of electrons with lower energies $<< E_F$ is impossible because no unoccupied energy levels are available.

Hence only electrons within the narrow energy region $E_F \pm k_B T$ are able to contribute to the molar heat capacity. The number of electrons that can be thermally excited is the integral of the density of energy states over the energy interval when the energy distribution function f_{FD} changes from 1 to 0. This number is small and can be approximated by the product of the density of energy states at E_F and the energy interval $2k_B T$.

The number of electrons per unit volume ΔN which can be thermally excited can be written as

$$\Delta N = \int N(E) f_{FD} dE \approx N(E_F) \times 2k_B T \qquad (6.71)$$

If all the N electrons per unit volume had contributed to the molar heat capacity as in the classical theory, the contribution would have been $3 \times R/2$. As only the fraction $\Delta N/N$ contributes, the molar capacity of heat of the electrons will be

$$C_e = \frac{2k_B T \, N(E_F)}{N} \frac{3R}{2} = \frac{9k_B T}{2E_F} R$$

More exact calculations result in the expression

$$C_e = \frac{\pi^2 k_B T}{2E_F} R \qquad (6.72)$$

The magnitude of E_F is $5\,eV$ and at room temperature $k_B T \approx 0.02\,eV$. Hence the electronic contribution to the molar heat capacity is negligible at higher temperatures compared with $3R/2$. At very low temperatures, when the vibrations of the lattice are not developed, the electronic contribution dominates.

Total Molar Heat Capacity at Low Temperatures

The total molar heat capacity at low temperatures, which includes the phonon contribution [Equation (6.70) on page 308] and the electron contribution [Equation (6.72)], can be written as

$$C \approx C_p \approx C_V = \frac{12\pi^4 R}{5\theta_D^3}T^3 + \frac{\pi^2 k_B R}{2E_F}T \tag{6.73}$$

Equation (6.73) is the total molar heat capacity at low temperatures. At temperatures below 1 K the linear term dominates.

Measurements of the molar heat capacity at very low temperatures as a function of temperature are of special interest as they allow the determination of the Fermi energy E_F, the Debye temperature θ_D and indirectly the effective mass m^* of the free electrons in the metal (Chapter 3, page 146).

If C/T is plotted as a function of T^2, a straight line is obtained. The coefficients of T and T^3 can be determined graphically as the intercept and the slope of the line. By use of these values, the quantities mentioned above can be derived. An example of such calculations is given below.

Example 6.2

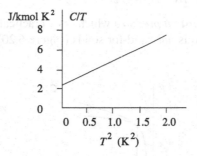

The plot in the figure is based on experimental data for the molar heat capacity of potassium in the temperature region 0.25–1.35 K (W. H. Lien and N. E. Phillips, *Phys. Rev.* 1964, **133**, A1370). The equation of the line was found to be

$$\frac{C}{T} = 2.1 \times 10^{-6} + 2.6 \times 10^{-6} \times T^2$$

Calculate (a) the Debye temperature of potassium, (b) the Fermi energy and (c) the effective mass of the free electrons of potassium.

Solution:

Equation (6.73) can be written as

$$\frac{C}{T} = \frac{\pi^2 k_B R}{2E_F} + \frac{12\pi^4 R}{5\theta_D^3}T^2 \tag{1'}$$

Identification gives

(a) $\dfrac{12\pi^4 R}{5\theta_D^3} = 2.6 \times 10^{-6} \quad \Rightarrow \quad \theta_D = \left(\dfrac{12\pi^4 R}{5 \times 2.6 \times 10^{-6}}\right)^{\frac{1}{3}} = 91\,\text{K}$

(b) $\dfrac{\pi^2 k_B R}{2E_F} = 2.1 \times 10^{-6} \quad \Rightarrow \quad E_F = \dfrac{\pi^2 k_B R}{2 \times 2.1 \times 10^{-6} \times e} = 1.7\,\text{eV}$

(c) Equation (3.67) on page 124 can be used for calculation of m^* when E_F is known:

$$m^* = \frac{h^2}{8E_F}\left(\frac{3n_{total}}{\pi}\right)^{\frac{2}{3}} \tag{2'}$$

where n_{total} is the number of valence electrons in potassium per unit volume. It can be calculated from known data for 1 kmol:

$$n_{total} = \frac{N_{total}}{V} = \frac{N_A \rho}{M} = \frac{6.02 \times 10^{26} \times 0.87 \times 10^3}{39.1} = 1.34 \times 10^{28}\,\text{m}^{-3}$$

Hence we obtain

$$m^* = \frac{h^2}{8E_F}\left(\frac{3n_{total}}{\pi}\right)^{\frac{2}{3}} = \frac{(6.62 \times 10^{-34})^2}{8 \times 1.7 \times 1.60 \times 10^{-19}}\left(\frac{3 \times 1.34 \times 10^{28}}{\pi}\right)^{\frac{2}{3}} = 1.1 \times 10^{-30}\,\text{kg} = 1.2\,m$$

Answer:

(a) $\theta_D = 91\,\text{K}$, (b) $E_F = 1.7\,\text{eV}$ and (c) $m^* = 1.2\,m$.

6.5.4 Molar Heat Capacity at Constant Pressure

Generally it is the molar heat capacity at *constant pressure* which is of practical interest. The difference $C_p - C_V$ is constant and equal to R for gases but this relationship is *not* valid for solids (Figure 6.20).

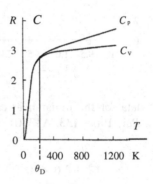

Figure 6.20 Molar heat capacities C_p and C_V for copper as functions of temperature. Reproduced with permission from A. J. Dekker, *Solid State Physics*. © 1962 Macmillan & Co Ltd.

If the temperature is below the Debye temperature, C_p and C_V are approximately equal. At temperatures higher than θ_D, the difference is roughly proportional to the temperature interval $T - \theta_D$ and the volume of the crystal.

$$C_p - C_V = \frac{\beta^2 V_A^{\theta_D}}{\kappa}(T - \theta_D) \quad \text{Grüneisen's rule} \tag{6.74}$$

where
  C_p = heat capacity of 1 kmol of the crystal at constant pressure
  C_V = heat capacity of 1 kmol of the crystal at constant volume
  β = thermal volume expansion coefficient at constant pressure (Section 6.4.1, page 296)
  κ = coefficient of compressibility (Section 6.3.2, page 293)
  $V_A^{\theta_D}$ = molar volume of the crystal at the Debye temperature
  T = absolute temperature.

Equation (6.74) is a relationship between heat capacity, expansion and compressibility, which all depend on the interatomic distance in the solid. It is derived in Example 6.3 below. Grüneisen derived Equation (6.74), which also is called Grüneisen's rule (compare pages 299 and 303).

Example 6.3

Verify Grüneisen's rule:

$$C_p - C_v = \frac{\beta^2 V_A^{\theta_D}}{\kappa}(T - \theta_D)$$

for isotropic cubic crystals.

Solution and Answer:

Consider a 1 kmol crystal with a cubic crystal structure with the molar volume $V_A{}^{\theta_D}$ at the Debye temperature θ_D. The temperature of the crystal is increased by an amount ΔT at constant pressure.

The internal energy U is a function of the volume V of the crystal and the temperature T:

$$dU = \left(\frac{\partial U}{\partial V}\right)_T dV + \left(\frac{\partial U}{\partial T}\right)_V dT \tag{1'}$$

The first law of thermodynamics (Equation (5.3) on page 220 in Chapter 5) is applied to the crystal. In combination with Equation (1') we obtain

$$dQ = \left(\frac{\partial U}{\partial T}\right)_V dT + \left[\left(\frac{\partial U}{\partial V}\right)_T + p\right] dV \tag{2'}$$

By definition [Equations (6.49) and (6.50)] on page 304, we obtain

$$C_p dT = C_V dT + \left[\left(\frac{\partial U}{\partial V}\right)_T + p\right] dV \tag{3'}$$

At normal pressures p can be neglected in comparison with the term $(\partial U/\partial V)_T$ and Equation (3') can be written as

$$C_p dT = C_V dT + \left(\frac{\partial U}{\partial V}\right)_T dV \tag{4'}$$

The second term on the right-hand side of Equation (4') is equal to the deformation energy change dE_{deform} associated with an infinitesimal volume change dV. According to Equation (6.24) on page 296, the deformation energy of 1 kmol of an isotropic cubic crystal can be written as

$$E_{deform} = \frac{Y}{6(1-2\nu)}(3\varepsilon)^2 V_A^{\theta_D} \tag{5'}$$

where $\varepsilon = \Delta a/a$ and a is the interatomic distance.

The infinitesimal change dE_{deform} can be obtained by taking the derivative of Equation (5'):

$$dE_{deform} = \frac{Y}{6(1-2\nu)} \times 2 \times 3\varepsilon \times d(3\varepsilon) \times V_A^{\theta_D}$$

or

$$dE_{deform} = \frac{Y}{(1-2\nu)} V_A^{\theta_D} \times 3\varepsilon \, d\varepsilon \tag{6'}$$

The expression in Equation (6') is introduced into Equation (4'):

$$(C_p - C_V) dT = \frac{Y}{(1-2\nu)} V_A^{\theta_D} \times 3\varepsilon \, d\varepsilon \tag{7'}$$

We use Equation (2′) between E_Y and κ in Example 6.1 on page 294:

$$\kappa = \frac{3}{Y}(1-2\nu) \quad \Rightarrow \quad \frac{Y}{1-2\nu} = \frac{3}{\kappa} \tag{8′}$$

Equation (8′) is introduced into Equation (7′), which gives

$$(C_p - C_V)\,dT = \frac{3}{\kappa}V_A^{\theta_D} \times 3\varepsilon\,d\varepsilon$$

or

$$(C_p - C_V)\,dT = V_A^{\theta_D}\frac{9\varepsilon d\varepsilon}{\kappa} \tag{9′}$$

The strain ε is equal to the relative lattice constant of the crystal:

$$\varepsilon = \frac{\Delta a}{a} \tag{10′}$$

One way to verify Grüneisen's rule is to express ε as a function of T and use the method of identification to express $C_p - C_V$ in terms of κ and β.

The total length change Δa would corresponds to the temperature change ΔT if the expansion were caused by thermal expansion. Similarly, an infinitesimal length change da would require a temperature change dT. We have

$$a + \Delta a = a(1 + \alpha\Delta T) \quad \text{and} \quad a + da = a(1 + \alpha dT)$$

which give

$$\varepsilon = \frac{\Delta a}{a} = \alpha\Delta T \quad \text{and} \quad d\varepsilon = \frac{da}{a} = \alpha dT \tag{11′}$$

If we replace ε and $d\varepsilon$ by the expressions (11′) in Equation (9′) and use the relationship $3\alpha = \beta$ between the length and volume dilatation coefficients (page 298) twice, we obtain

$$(C_p - C_V)\,dT = \frac{\beta\Delta T\beta dT}{\kappa}V_A^{\theta_D} \tag{12′}$$

ΔT corresponds to $T - \theta_D$. If this value is introduced into Equation (12′), we obtain the desired relationship after division with dT:

$$C_p - C_V = \frac{\beta^2}{\kappa}V_A^{\theta_D}(T - \theta_D)$$

The difference $C_p - C_V$ increases linearly with $T - \theta_D$, which is shown in Figure 6.20 on page 312. The values of the coefficients β and κ are difficult to derive theoretically but values derived from experiments may be available.

Heat Capacity at Constant Pressure in Various Temperature Intervals

Low-temperature Region, $T < \theta_D/10$

$$C_p = C_V = \underbrace{\frac{12\pi^4 R}{5}\left(\frac{T}{\theta_D}\right)^3}_{\substack{\text{lattice (phonon)}\\\text{contribution}}} + \underbrace{\frac{\pi^2 k_B R}{2E_F}T}_{\substack{\text{free electron}\\\text{contribution}}} \tag{6.75}$$

At *very low* temperatures, practically all phonons have the lowest possible energy and very few become excited to higher energies when the temperature is slightly increased. Hence the phonon contribution [Equation (6.70), page 308] to the heat capacity is negligible at extremely low temperatures and the free electron contribution [Equation (6.72), page 310] dominates.

When the temperature increases, the phonon contribution to the heat capacity gradually grows and becomes part of the molar heat capacity.

Intermediate-temperature Region, $\theta_D/10 < T < \theta_D$

$$C_p = C_V = 9R\left(\frac{T}{\theta_D}\right)^3 \int_0^{\frac{\theta_D}{T}} \frac{x^4 e^x}{(e^x - 1)^2} dx \qquad (6.69)$$

In this region the free electron contribution can be neglected in comparison with the phonon contribution [Debye Equation (6.69) on page 309].

Equation (6.23) represents the molar heat capacity approximately up to the temperature θ_D. It approaches asymptotically the value $3R$ at high temperatures, which corresponds to the Dulong–Petits law.

High-temperature Region, $T > \theta_D$
In this region C_p and C_V differ. Approximately we have

$$C_V = 3R \qquad (6.53)$$

$$C_p = 3R + \frac{\beta^2 V_A^{\theta_D}}{\kappa}(T - \theta_D) \qquad (6.76)$$

Equation (6.76) is valid only if no structural transformations occur in the solid. We will discuss the shape of the C_p–T curve in the presence of phase transformations below.

6.5.5 Influence of Order–Disorder Transformations on Heat Capacity

The heat capacity at constant pressure C_p depends, like other material properties, on the structure of the material. Typical structure changes are melting and phase changes, i.e. changes of the *structure* of the crystal lattice.

Even if the structure of a crystal lattice of a solid remains unchanged, discontinuous changes in its C_p–T curve may appear. Such changes are due to transformations from an ordered to a disordered state or vice versa.

Order and disorder refer to the *distribution* of atoms in a crystal lattice. It is important to realize that the *lattice structure* remains unchanged. It is only the *occupancy of the sites* which is concerned. A and B atoms can be ordered or distributed at random in the sites.

The matter of order and disorder will be discussed further in connection with the theory of ferromagnetism in Section 6.6.3. Here we will discuss order–disorder transformations and give two examples, which are manifested by an abrupt change in the heat capacity at a specific critical temperature.

Order–Disorder Transformations in Binary Alloys

Ordered and Disordered Binary Alloys
Consider a crystal lattice which consists of two types of atoms, A and B. At low temperature the alloy has an *ordered structure*, i.e. the atoms are arranged in a regular pattern and the atoms can only vibrate around their equilibrium positions.

If the temperature is increased to a certain critical temperature, the kinetic motion will be violent enough to break the ordered structure in the sense that the atoms A and B can exchange lattice sites at random. A completely *disordered structure* is characterized by the equal probability of finding, for example, an A atom at any lattice site. Similarly, there is an equal probability of finding an atom B at any site. If the alloy has the composition AB, both probabilities are $\frac{1}{2}$.

Measure of the Degree of Order–Disorder of a Binary Alloy
Figures 6.21a and b and 6.22 represent the two extremes: complete order and complete disorder. Intermediate cases are possible.

The degree of order is described by the long-range order parameter S:

$$S = \frac{f_A - x_A}{1 - x_A} \qquad (6.77)$$

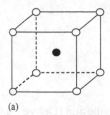

(a)

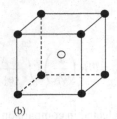

(b)

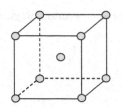

Figure 6.21 (a) Ordered structure. Black sphere = atom A. White spheres = atoms B.

Figure 6.21 (b) The same ordered structure as in (a) Black spheres = atoms A. White sphere = atom B.

Figure 6.22 The same structure as in Figure 6.21a but disordered. Grey spheres = average 'AB' atoms.

where

S = long-range order parameter

f_A = fraction of A sites occupied by A atoms

x_A = mole fraction of A atoms.

In the completely ordered state, all A atoms are in their proper positions, which means that $f_A = 1$. Hence $S = 1$ for the completely ordered state. In the disordered state, the probability of finding an A atom in an A site equals x_A. Hence $S = 0$ for the completely disordered state. For intermediate cases, $0 \leq S \leq 1$. The degree of order is manifested by the intensities of the X-ray superlattice lines. The X-ray lines have maximum intensity for $S = 1$, fade away with decreasing S values and disappear completely at $S = 0$.

Order–Disorder Transformations in CuZn

As the properties of solids depend strongly on their structure, a sudden transformation from one structure to another will result in abrupt changes of the properties of the solid, for example its heat capacity and its thermal and electrical conductivity. As a concrete example we will discuss the alloy β brass, which consist of 46–50% Zn dissolved in Cu. Its chemical composition is close to CuZn.

At room temperature, β brass is an ordered alloy. The unit cell corners are occupied by Cu atoms and the centre by a Zn atom (Figure 6.21b) or reversed (Figure 6.21a). The structure can be described as two simple cubic structures of either type of atoms. The structures are displaced relative each other by $a/2$, where a is the lattice constant.

The order is maintained when the temperature changes and the properties, for example the heat capacity, change gradually until the temperature reaches a critical value T_{cr}. At the critical temperature the kinetic motion is violent enough to let the atoms change sites fairly freely and the structure becomes disordered. It can be described as a BCC structure with a completely random distribution of the Cu and Zn atoms (Figure 6.22).

The critical temperature T_{cr} when the order–disorder transformation occurs is approximately 740 K. The change in heat capacity is shown in Figure 6.23. Energy is required to change the system from an ordered to a disordered state. The increased

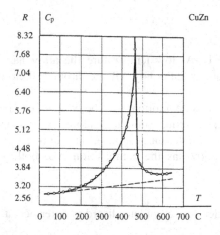

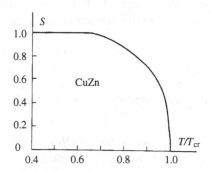

Figure 6.23 Molar heat capacity at constant pressure as a function of temperature for β brass (CuZn). The critical transformation temperature from an ordered to a disordered structure occurs at $T_{cr} = 740$ K. Reproduced with permission from C. Kittel, *Introduction to Solid State Physics*, 6th edn. © 1986 John Wiley & Sons, Inc.

Figure 6.24 Long-range order parameter S as a function of the ratio T/T_{cr} for β brass (CuZn). Reproduced with permission from B. D. Cullity, *Elements of X-Ray Diffraction*. © Addison-Wesley Publishing Company, Inc. (now under Pearson Education).

need for energy appears as an abrupt increase in C_p in the vicinity of the critical temperature. When the transformation is carried through, the C_p value decreases rapidly to the 'normal' value of the disordered state.

A plot of the variation of the long-range parameter S as a function of T/T_{cr} is given in Figure 6.24. S was obtained from X-ray measurements of the type described above, performed within the temperature range 300–740 K $(0.4T_{cr} - T_{cr})$. The resemblance between Figure 6.24 and Figure 6.38 on page 328, which concerns the Curie point in ferromagnetic materials, is striking.

Theoretical Background of Order–Disorder Transformations

Heat capacity is closely related to elastic vibrations in solids. In Chapter 3 (page 149) we found that elastic waves in a one-dimensional row of equal atoms have an upper angular frequency limit equal to

$$\omega_{max} = 2\sqrt{\frac{\beta}{M}} \tag{6.78}$$

where β is a material constant. The phonons obey Bose–Einstein statistics. This is the background to Debye's model of heat capacity which results in Equations (6.67) and (6.69):

$$U = 9N_A k_B T \left(\frac{T}{\theta_D}\right)^3 \int_0^{\frac{\theta_D}{T}} \frac{x^3}{e^x - 1} dx \tag{6.67}$$

$$C_p = C_V = 9R \left(\frac{T}{\theta_D}\right)^3 \int_0^{\frac{\theta_D}{T}} \frac{x^4 e^x}{(e^x - 1)^2} dx \tag{6.69}$$

where $x = h\nu/k_B T$ and the Debye temperature $\theta_D = h\nu_D/k_B$.

As β is a material constant, the Debye temperature θ_D, which is a function of the maximum frequency of the elastic waves, must be specific for each type of atom.

In the ordered structure, the Cu atoms form their own simple cubic crystal lattice and the Zn atoms form another simple cubic crystal lattice, which is displaced half an atomic distance relative to the Cu lattice. The Cu and Zn atoms lie in different planes. As a rough approximation, we will neglect the interaction between the two lattices.

In the ordered structure, we can approximately describe the average phonon energy as the average of the phonon energies of the two lattices:

$$U = \frac{U_{Cu} + U_{Zn}}{2} \tag{6.79}$$

Referring to Equations (6.67), (6.69) and (6.79), we can conclude that the heat capacity of the ordered structure is a function of both θ_D^{Cu} and θ_D^{Zn}.

When the temperature increases, the number of phonons and their average energy increase and the phonons collide with electrons and atoms. Some of the atoms gain sufficient energy for an exchange of position with neighbouring atoms. When the temperature increases further, the collisions become more and more frequent. At the critical temperature a great number of atoms change place and the ordered distribution in the two lattices collapses. When the completely disordered structure is a fact, the frequent collisions give no macroscopic changes.

In the disordered state, the Cu and Zn atoms are distributed at random in the crystal lattices and θ_D^{Cu} and θ_D^{Zn} are no longer defined as the Cu and Zn planes do not exist. They are replaced by another value of θ_D related to the elastic waves in a single and denser crystal structure.

6.6 Magnetism

6.6.1 Magnetic Moments of Atoms

Magnetism has been discussed in Chapter 2, where the origin and basis of magnetism were briefly discussed in connection with the theory of atoms, particularly the Zeeman effect.

In Chapter 2 (page 60), we found that the angular momentum of an electron in an atom is always firmly coupled to a magnetic moment. The same is true for all orbital electrons in an atom and also for their resultant, the total orbital angular momentum

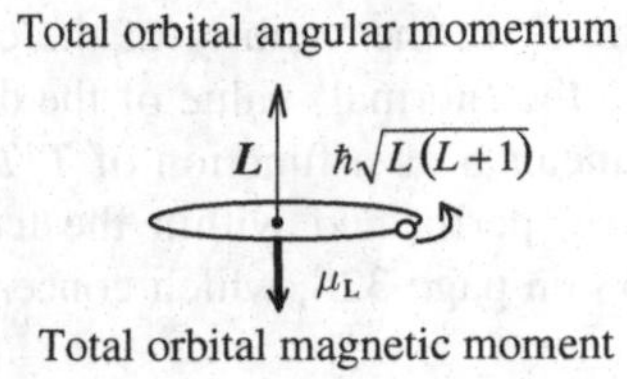

Total orbital angular momentum

L $\hbar\sqrt{L(L+1)}$

μ_L

Total orbital magnetic moment

Figure 6.25 Total orbital angular momentum L of an atom and its corresponding magnetic moment μ_L.

L. The vector L is rigidly coupled to the total magnetic moment μ_L. L and μ_L have opposite directions (Figure 6.25). In analogy with Equation (7′) in the box on page 70 in Chapter 2, we have

$$|\mu_L| = \mu_B\sqrt{L(L+1)} \tag{6.80}$$

The constant μ_B [Equation (2.47) on page 71] is called the *Bohr magneton* and has the value

$$\mu_B = \frac{e\hbar}{2m} \tag{6.81}$$

It is a very useful 'unit' for magnetic moments.

According to quantum mechanics, L and its corresponding magnetic moment μ_L are space quantized. In the general case, L is coupled to the resulting electron spin vector S, which gives a resulting vector J. Its resulting magnetic moment can be written as

$$|\mu_J| = \mu_B\sqrt{J(J+1)} \tag{6.82}$$

J and μ_J are also space quantized and the corresponding energy state splits into $2J+1$ energy levels with different energies in a magnetic field. The reason is the space quantization of the magnetic moment (page 60 in Chapter 2).

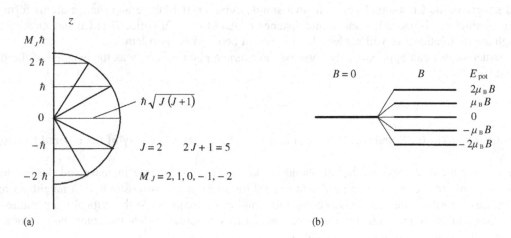

Figure 6.26 Splitting of a ^{1}D state in a magnetic field. (a) Space quantization of J in the magnetic field. (b) Splitting of energy levels without and with a magnetic field B in the z direction.

The components of μ_J on the z axis of a coordinate system can be written as

$$\mu_{J,z} = -\mu_B M_J \tag{6.83}$$

where M_J is the magnetic quantum number.

In connection with the treatment of the normal Zeeman effect on pages 69–71 in Chapter 2, it was shown that the energy levels in the presence of a magnetic field B in the z direction can be written as [Equation (2.48) on page 71]

$$E_{pot} = \mu_B B M_J \tag{6.84}$$

where M_J is the component of the J vector on the direction of the B field (z axis). A concrete example of the energy splitting of a ^1D state ($S = 0$ and $L = J = 2$) is given in Figure 6.26.

It was clearly pointed out in Chapter 2 that the splitting described by Equation (6.84) is valid *only for singlet states*, i.e. when the quantum number $S = 0$ and $J = L$. The reason for this will be explained below.

Electron Spin and Intrinsic Magnetic Moment of Electrons

In addition to the orbital angular momentum and the orbital magnetic moment, the electron also has an intrinsic angular momentum s and an intrinsic magnetic moment μ_s (Chapter 2, page 61).

The z component of the magnetic moment is *twice* the size which might be expected from an analogy with Equation (6.83):

$$\mu_{sz} = -2\mu_B m_s \tag{6.85}$$

As $m_s = \pm^1/_2$, we obtain

$$\mu_{sz} = \pm\mu_B \tag{6.86}$$

According to Figure 2.19 on page 61, we have

$$|\boldsymbol{\mu}_s| = 2\mu_B|s| = 2\mu_B\sqrt{s(s+1)} = \sqrt{3}\mu_B \tag{6.87}$$

The resulting angular momentum L of all the electrons is obtained by vector addition of all the l_i vectors. Similarly, all the s_i vectors give the resultant S. The corresponding magnetic moments also form a resultant. In the general case it can be written as

$$|\boldsymbol{\mu}_S| = 2\mu_B\sqrt{S(S+1)} \tag{6.88}$$

The vectors L and S are coupled to each other. The corresponding magnetic moments also form a resultant which in the general case can be written as

$$\boldsymbol{\mu}_{total} = \boldsymbol{\mu}_L + \boldsymbol{\mu}_S \tag{6.89}$$

which is the total magnetic moment of the atom.

The factor 2 on the right-hand side in Equation (6.88) is responsible for the anomalous Zeeman effect (page 71 in Chapter 2).

J and $\boldsymbol{\mu}_{total}$ are *not* antiparallel unless $S = 0$. In all other cases the upper and lower energy levels no longer have the same splitting as in Figure 2.30a on page 71, which results in many more lines than three when the selection rules are applied in the general case.

6.6.2 Dia- and Paramagnetism

The net magnetic moment of an atom depends on the orientation of its electron orbits. If L and S are zero, for example for symmetry reasons in filled shells or subshells (Chapter 2, Section 2.7.1), the atomic magnetic moment will be zero. Even if the atom possesses a net magnetic moment, the magnetization of the solid is zero for most materials owing to random orientation of the atoms in the absence of an external magnetic field.

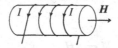

Figure 6.27 Magnetizing field.

In the presence of an external magnetic field, all materials become influenced. Consider a long current-carrying solenoid which gives a fairly homogeneous magnetic field (Figure 6.27). If it contains a solid cylinder the material becomes magnetized. The *magnetizing field H* is defined by the relationship

$$H = nI \tag{6.90}$$

where
H = magnetizing field (A/m)
n = the number of coils per unit length of the solenoid
I = current through the coils.

The magnetization of the solid is described by the *magnetization vector M*, which is defined as the *magnetic moment per unit volume of the solid.*
The magnetization vector of a solid is proportional to the magnetizing field:

$$M = \chi H \tag{6.91}$$

where
M = magnetic moment per unit volume (Am^2/m^3)
χ = magnetic susceptibility of the solid
H = magnetizing field (A/m).

The *magnetic susceptibility χ* of the solid is a dimensionless material constant.
Owing to their behavior in a magnetic field, solids are classified as

- diamagnetic
- paramagnetic
- ferromagnetic
- antiferromagnetic
- ferrimagnetic.

The influence of a magnetic field on the atoms of all materials depends on the interaction between the magnetic field and the atoms. There are two effects. One is the influence of the field on the orbits of the electrons. This effect, which is described below under the heading Diamagnetism, appears in *all* atoms. The other effect is an orientation effect, which appears in such atoms that have a permanent magnetic moment.

The magnetic phenomena have been the object of much research in theoretical physics. It is beyond the scope of this book to penetrate the theories of dia-, para- and ferromagnetism but some brief outlines will be given without demands upon either derivation or completeness.

Ferromagnetism, antiferromagnetism and ferrimagnetism belong to the same type of magnetic materials. They will be discussed in Section 6.6.3.

Diamagnetism

When a diamagnetic substance is exposed to a magnetizing field, it obtains a weak magnetization in a direction *opposite* to the magnetizing field. The magnetic susceptibility of diamagnetic solids is *negative* and $<< 1$. In most cases it is of the magnitude 10^{-5}.

$$M = \chi H \qquad \chi < 0 \tag{6.92}$$

Diamagnetic effects appear in all solids but are generally completely screened by other magnetic effects unless the resulting magnetic moment of the atoms is zero.

Theory of Diamagnetism

The electrons in the shells and subshells of the atoms appear in pairs. Consider a pair of electrons with equal angular momentum l. The two electrons 'rotate' in opposite directions, i.e. their magnetic quantum numbers are $\pm m_l$. Their electron spins have opposite directions (spin up and spin down), which can be expressed as $m_s = \pm \frac{1}{2}$. Such electron pairs have no resulting magnetic moment. All filled shells and subshells consist of such electron pairs and have no resulting magnetic moment.

The atoms of all diamagnetic solids have a resultant magnetic moment equal to zero. Hence diamagnetic solids have filled electron shells or subshells. Diamagnetic metals are, for example, Zn, Cd and Hg, all with an electron structure with no unpaired electron spins.

Under the influence of a magnetizing field, the electron orbits become distorted in such a way that a resulting magnetic moment arises, opposite to the H field. It is maintained as long as the H field is present.

The negative direction of the magnetization can be understood as follows. Consider a single closed coil which suddenly is exposed to a perpendicular magnetic field. An electric field is induced which causes an induction current in the coil in such a direction that it gives a magnetic field opposite to the external magnetic field (Lentz's law). If the electric resistance in the coil is zero, the current continues to flow. The coil corresponds to a magnet with its magnetic moment in the opposite direction to the magnetizing field.

Alternatively, we can say that the time-dependent magnetic field induces an electric field, which accelerates or retards the electrons in their orbits slightly. The result is that their magnetic moments no longer cancel but have a resultant in the direction opposite to the magnetizing field.

The susceptibility of diamagnetic solids, which consist of atoms or ions with filled electron shells, can be written as

$$\chi = -\frac{Ze^2 N}{6m} \overline{r^2} \tag{6.93}$$

where

χ = magnetic susceptibility of the solid
Z = number of electrons of the atom
e = charge of the electron
m = mass of the electron
N = number of atoms per unit volume
$\overline{r^2}$ = root mean square radius of the electron cloud around the atom.

Owing to its origin, it can be named *orbital magnetic susceptibility*. It should be noted that χ is independent of temperature for diamagnetic substances.

Paramagnetism

When a paramagnetic substance is exposed to a magnetizing field, it obtains a weak magnetization in the same direction as the magnetizing field. The magnetic susceptibility (page 320) of paramagnetic solids is *positive* and $\ll 1$. It is proportional to $1/T$. In most cases it is of magnitude 10^{-4} for solids at room temperature.

$$M = \chi H \qquad \chi > 0 \tag{6.94}$$

The atoms of paramagnetic solids have a resulting magnetic moment. It originates partly from the orbital motion of the electrons and partly from their spins.

The interaction between paramagnetic atoms is negligible.

Theory of Paramagnetism

The paramagnetism in solids consists of two effects. One of them originates from the orbital electrons of all lattice atoms. The other emanates from the free electrons which are common for the crystal lattice in metals.

Paramagnetism of the Permanent Resultant Atomic Orbital Magnetic Dipole Moment

Atoms with a resulting magnetic moment behave like small magnets. Under the influence of the external magnetic field, they orient themselves in directions closer to the magnetic field than their original positions.

The reorientation is counteracted by the scattering of the atoms and their magnetic moments, owing to interaction with the electron gas and electron–phonon collisions, until equilibrium is achieved. The stronger the external magnetic field is, the stronger will be the orientation of the small magnets in the direction of the external field. Hence the weak total magnetic moment of a piece of paramagnetic material is proportional to the external magnetizing field.

The diamagnetic effect appears in all materials. If the induced magnetic moment and the permanent magnetic moment of the atoms are equal, a magnetically indifferent substance is obtained.

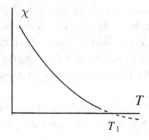

Figure 6.28 Susceptibility of paramagnetic solids versus temperature. A negative values of χ means that the diamagnetic effect, which is always present, dominates over the paramagnetic effect.

The alignment of the permanent magnetic moments of the atoms is counteracted by the kinetic motion of the atoms. Hence the paramagnetic susceptibility depends on temperature. The higher the temperature is, the smaller will be χ (Figure 6.28).

Only those materials in which the paramagnetic effects dominate over the diamagnetic effects are called paramagnetic and show paramagnetic properties. Examples of such paramagnetic solids are the alkali metals with single electrons in their outer shells and rare earth metals with unfilled shells.

In Chapter 2, we discussed the orbital electrons with a magnetic spin quantum number m_s (page 61). In a magnetic field, space quantization of the electron spin occurs.

In a magnetic field, the degeneracy between the orbital electrons, which share the same orbital state but have opposite spins, is broken and the electrons have slightly different energy levels (Figure 6.29).

The orbital electrons with positive spin relative to the direction of the magnetic field B *lower* their energy in the magnetizing field by the amount $-\boldsymbol{\mu} \cdot \boldsymbol{B}$, where $\boldsymbol{\mu}$ is the total magnetic moment of the electron, which includes both $\boldsymbol{\mu}_L$ and $\boldsymbol{\mu}_S$ [Equation (6.89) on page 319]. Orbital electrons with negative spin relative to B *increase* their energy by the amount $\boldsymbol{\mu} \cdot \boldsymbol{B}$ in the magnetic field.

Paramagnetism of the Free Conduction Electrons in Metals

The Fermi distribution of valence electrons has been discussed extensively in Chapter 3 (pages 121–124). If we exchange the axes of Figure 3.26c on page 123 and draw two distribution functions, one for electrons with spin up and the other for electrons spin down, we obtain the dotted curves in Figure 6.29.

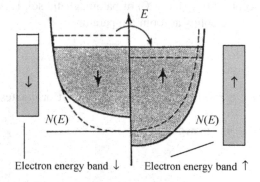

Figure 6.29 Density of electron energy states of spin-up and spin-down electrons of paramagnetic atoms in a magnetic field. The tops of the energy bands have the same energy. The energy difference between the bottoms of the electron bands is $2\mu_B B$. Adapted with permission from J. M. Ziman, *Principles of the Theory of Solids*, 2nd edn. © Cambridge University Press.

The shaded areas in Figure 6.29 represent the number of spin-down ($n_\downarrow$) and spin-up electrons ($n_\uparrow$) per unit volume. Obviously there are more spin-up than spin-down electrons. The excess spin gives rise to a resulting magnetic moment. This is the explanation of paramagnetism of the conduction electrons.

A necessary condition for equilibrium is that the maximum energy level is the same for both spin-down and spin-up electrons. This is achieved by transfer of some spin-down electrons to the spin-up side.

In analogy with Equation (6.86) on page 319, the magnetic moment of a free electron is $2\mu_B \times 1/2 = \mu_B$. Hence the net magnetic moment of the free electrons per unit volume will be

$$M = \mu_B \left(n_\uparrow - n_\downarrow \right) \tag{6.95}$$

where

M = net magnetic moment per unit volume due to the free electrons

μ_B = magnetic moment of a free electron, i.e. the Bohr magneton

$n_\uparrow$ = number of electrons per unit volume with spin up

$n_\downarrow$ = number of electrons per unit volume with spin down.

The positions of the spin-up and spin-down electron energy bands depend, of course, on the type of metal and are also strongly influenced by alloying or impurity atoms, which change the total number of free electrons in the solid.

6.6.3 Ferromagnetism

Ferromagnetism, Antiferromagnetism and Ferrimagnetism

A characteristic and common feature of ferromagnetic, antiferromagnetic and ferrimagnetic materials is that there exists *strong coupling between the magnetic moments of neighbouring atoms*, which is strong enough to persist in spite of thermal excitation. The nature of this coupling will be discussed below.

In ferromagnetic materials, the atomic magnetic moments of equal size are oriented roughly in the same direction (Figure 6.30a). In antiferromagnetic materials the atomic magnetic moments are coupled roughly in opposite directions (Figure 6.30b) and cancel, which makes the material unmagnetic. Ferrimagnetic materials are a type of antiferromagnetic material where the two types of magnetic moments have different sizes and do not cancel (Figure 6.30c). Consequently, such materials are magnetic.

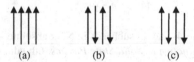

(a)   (b)   (c)

Figure 6.30   Coupling between the magnetic moments in case of (a) ferromagnetism, (b) antiferromagnetism and (c) ferrimagnetism.

Below we will restrict the discussion to ferromagnetism and the dominating ferromagntic materials.

Ferromagnetism. Spin Interaction

The difference between paramagnetic and ferromagnetic solids is that that the ferromagnetic materials have much higher values of the magnetic susceptibility (page 320), which is of magnitude 1000.

$$M = \chi H \qquad \chi >> 0 \tag{6.96}$$

Examples of ferromagnetic solids are Fe, Ni, Co, some of their alloys and some alloys which contain Mn and Cr.

The magnetic moments of ferromagnetic atoms are of the same magnitude as those of paramagnetic atoms. Another similarity is that the magnetic susceptibility depends on temperature in both paramagnetic and ferromagnetic solids. Above a certain temperature, characteristic for each material, the ferromagnetism disappears and the material becomes paramagnetic. This temperature is called the *Curie point* or *Curie temperature* (page 328).

Electron Configurations and Magnetic Moments of Ferromagnetic Atoms

It is striking that ferromagnetic solids consist of atoms with unfilled shells or subshells (the 3d subshell can accommodate a maximum of 10 electrons) and high resulting angular momentum and spin vectors. This is shown in Table 6.4.

Table 6.4 Electron configurations and ground states of Fe, Co and Ni atoms.

Element	Electron configuration of atoms	Ground state	L	S
Fe	$1s^2 2s^2 2p^6 3s^2 3p^6 3d^6 4s^2$	$^5D_4$	2	2
Co	$1s^2 2s^2 2p^6 3s^2 3p^6 3d^7 4s^2$	$^4F_{9/2}$	3	3/2
Ni	$1s^2 2s^2 2p^6 3s^2 3p^6 3d^8 4s^2$	$^3F_4$	3	1

However, reality is more complex than Table 6.4 indicates. The magnetic moments of the ferromagnetic atoms have been measured carefully and turn out to be non-integer instead of integer numbers of the Bohr magneton [Equation (6.81) on page 318]. The unavoidable conclusion is that the numbers of electrons in the subshells are also non-integers. The explanation is *hybridization* of the 4s and 3d electrons in ferromagnetic atoms. This phenomenon, which is a consequence of quantum mechanics, was discussed in Chapter 3 in connection with covalent bonds in diamond.

Therefore, it is in fact *not* correct to keep the names 3d and 4s for the bands of the outer electrons in the transitions metals. Instead we have hybrid bands. However, for the sake of simplicity and tradition, the notations are retained.

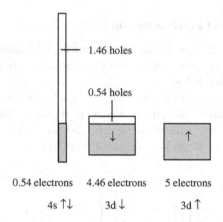

Figure 6.31 Schematic electron distribution in the 4s and 3d subbands in Ni at absolute zero temperature. The figure is not drawn to scale. Reproduced with permission from C. Kittel, *Introduction to Solid State Physics*, 6th edn. © 1986 John Wiley & Sons, Inc.

According to Table 6.4, a nickel atom has two electrons in the 4s and eight electrons in the 3d subshell. It can accommodate a maximum of 10 electrons, five with spin up and five with spin down. In Figure 6.31 we have schematically separated the 3d sub-band with respect to the spin direction (compare Figure 6.29 on page 322). Owing to quantum mechanical interaction between the wave functions of the electrons, the two sub-bands have different energies, as indicated in Figure 6.31 (*exchange energy*, compare the H_2^+ molecule in Chapter 2, page 73).

The 3d sub-band with spin up is assumed to be filled, i.e. contains five electrons, whereas the sub-band with spin down on average contains 4.46 electrons and the 4s subband 0.54 electrons. This electron configuration corresponds to the lowest possible energy of the Ni atom.

Alternatively, we can plot the density of electron energy states as a function of energy to show the electron distribution in the crystal. This type of diagram was frequently used in Chapter 3 (page 123). It is used in Figure 6.32.

Figure 6.32 shows that the density of electron states in the 4s band is low and that the band is very wide. The band can accommodate $2N$ electrons per unit volume but is only filled with electrons up to the Fermi level. The number of electrons with spin up is equal to the number of spin-down electrons and the net magnetic moment of the 4s electrons is zero.

The total number of electrons in the 4s and 3d shells equals 10. Hence the resulting magnetic moment of all the 3d electrons is $(5.0 - 4.46)\mu_B = 0.54\mu_B$ (Bohr magnetons). The measured value is $0.60\mu_B$. The deviation $0.06\mu_B$ can be explained by other effects.

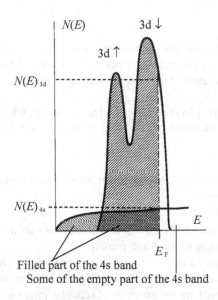

Figure 6.32   Density of the electron energy states in the 4s and 3d bands as a function of the energy E in Ni at absolute zero temperature. The shaded areas of the 4s and 3d bands between $E = 0$ and $E = E_F$ represent the number of electrons per unit volume in the 4s and 3d bands (0.54 and 9.46 electrons, respectively). The figure is not drawn to scale. Reproduced with permission from A. G. Guy, *Elements of Physical Metallurgy*, 2nd edn. © Addison-Wesley Publishing Company, Inc. (now under Pearson Education).

The 3d band is narrow compared with the 4s band and the density of electron states per energy unit is much higher in the 3d band than in the 4s band. The two 3d bands overlap strongly. The peak at lower energy corresponds to spin up. Both the spin-up and spin-down electrons contribute to the second peak.

On the whole, the spin-up electrons have lower energy than the spin-down electrons. As the lowest energy states always become filled first there is an excess of spin-up electrons. The 3d band is not completely filled, which gives a resulting magnetic moment. The distribution represents the lowest possible total energy.

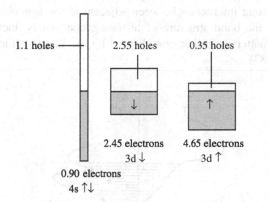

Figure 6.33   Electron distribution in the 4s and 3d subbands in Fe at absolute zero temperature. The total number of electrons in the 4s and 3d subbands is eight. The figure is not drawn to scale. Reproduced with permission from C. Kittel, *Introduction to Solid State Physics*, 6th edn. © 1986 John Wiley & Sons, Inc.

The conditions are analogous in iron (Figure 6.33). Iron has two electrons in the 4s and six electrons in the 3d subband according to Table 6.4. Owing to hybridization, the electron distribution can roughly be described by 4.65 electrons with spin up, 2.45 electrons with spin down in the 3d subband and 0.90 electrons in the 4s subband, which do not contribute to the magnetic moment. This distribution corresponds to a resulting magnetic moment of all the 3d electrons equal to $(4.65 - 2.45)\mu_B = 2.20\mu_B$, in good agreement with the measured value $2.22\mu_B$.

Theory of Ferromagnetism

In the preceding section, the magnetic moments per atom of the ferromagnetic elements Ni and Fe were derived. As there is no difference in the magnitude of the magnetic moments per atom between paramagnetic and ferromagnetic elements, the reason for the strong magnetism of ferromagnetic materials must be an *alignment* of the magnetic moments in ferromagnetic solids, which is absent in paramagnetic solids.

Much effort has been devoted by many physicists to finding a theory which can explain the puzzling phenomenon of ferromagnetism. The magnetic dipole moment of an atom is firmly coupled to its spin (Chapter 2). It is generally accepted that

- The strong coupling of the magnetic moments of neighbouring atoms in ferromagnetic materials is caused by the coupling of the spin vectors of neighbouring atoms.

Spin coupling has been discussed in Chapter 2. Electrons within the *same* atom with equal values of the quantum numbers n, l and m_l must have antiparallel spins owing to the Pauli principle.

An example of coupling of electron spins in *different* atoms is the H_2 molecule. Quantum mechanical calculations show that the two electrons have antiparallel spins in the ground state (Figure 2.37 on page 75). This leads to a *binding* state.

The coupling of electron spins is described by the so-called exchange integral, which is a function of the symmetric and antisymmetric combinations of the wave functions of the two electrons. The exchange integral represents the difference between the two energy states which correspond to parallel and antiparallel spins. Conclusions regarding the equilibrium state can be drawn from the sign and magnitude of the exchange integral.

The unpaired electrons in an atom form a resultant spin vector S. The study of the exchange integral shows that

- The energy state with lowest possible energy is that with highest possible value of S.

This statement (Hund's rule) explains why atoms with incompletely filled d shells, for example the ferromagnetic transition metals, have high resulting spins and consequently large magnetic moments.

Two entirely different models have been used to give a theoretical explanation of ferromagnetism. One is the quantum mechanical approach and the other is a statistical model.

Quantum Mechanical Model of Ferromagnetism

In contrast to the H_2 molecule, the strong interaction between adjacent ferromagnetic atoms is caused by electron spins with *parallel* orientation. To explain this, the band structure of ferromagnetic metals, including the Fermi–Dirac distribution of electrons, must be considered. The matter is further complicated by the fact that the electrons in the 4s and 3d bands are hybridized (Figures 6.29, 6.31 and 6.33).

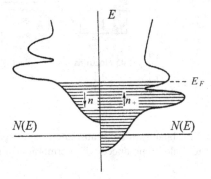

Figure 6.34 Density of electron energy states in a ferromagnetic metal as a function of energy. Adapted with permission from J. M. Ziman, *Principles of the Theory of Solids*, 2nd edn. © Cambridge University Press.

Stoner set up a collective electron model of the 'magnetic' electrons. The energies of the electrons were expressed with the aid of Bloch functions. The density of electron energy states per unit volume of the ferromagnetic metal as a function of energy is shown in Figure 6.34, which is analogous to Figure 6.32 on page 325. Expressions for the energy of an electron with spin up and spin down are set up. The total energy is calculated and an expression for the susceptibility is derived. This expression contains the temperature T.

The Ising Model

A totally different approach was introduced by Ising in the middle of the 20th century. He suggested a general statistical 'digital' model with two alternatives at each site. This is shown in Figure 6.35. The same model was used in connection with discussion of the influence of order–disorder transformations on heat capacity in Section 6.5.5 (page 315) and will be discussed in more general terms below.

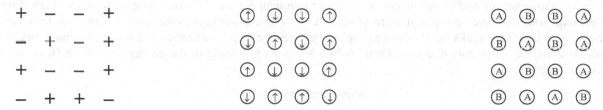

Figure 6.35   The Ising model.

Figure 6.36   Atoms with spin-up and spin-down electrons in a ferromagnetic material.

Figure 6.37   Atoms A and B in a binary alloy.

The Ising model has several applications. It can be used to study and explain ferromagnetism (Figure 6.36) and can also be applied to alloys (Figure 6.37). There is one fundamental difference in the two cases, however. Spin-up electrons can be converted into spin-down electrons and vice versa. A corresponding transformation of an A atom into a B atom or vice versa in an alloy is impossible. In spite of this, some analogous properties of the two types of systems have been found.

The advantage of the Ising model is that the difficult problem of finding the exact eigenfunctions of a given assembly of spins is avoided. The Ising model concentrates on finding the statistical distribution of energy states of such a system. It is no longer a quantum mechanical problem but a statistical problem and concerns order–disorder transformations.

The interaction between neighbouring atoms in a binary alloy has been studied experimentally. At low temperature the alloy tends to be ordered, i.e. with a regular pattern of A and B atoms. At a certain transition temperature the order was found to be broken and the distribution of atoms to be random.

Application of the Ising Model to Ferromagnetism

N atoms per unit volume of a ferromagnetic material are considered and all the number of different configurations of spin-up and spin-down atoms of the N sites per unit volume of the crystal lattice are counted. An energy function that corresponds to the exchange integral is set up and the condition for minimum free energy is derived. From this condition, it can be concluded that the minimum energy corresponds to a *strong coupling of parallel spins in neighbouring atoms*.

We can also make a comparison between alloys and ferromagnetic materials. We assume, according to the Ising model and Figures 6.36 and 6.37, that A atoms correspond to spin-up atoms and B atoms to spin-down atoms. Just as in alloys:

- There is a certain specific transition temperature when all order disappears. This transition temperature in ferromagnetic materials corresponds to the temperature, when they suddenly lose their ferromagnetic properties, the so-called Curie point (page 328).
- There is a sudden change in the heat capacity of alloys at the transition temperature from an ordered to a disordered state (for example the CuZn alloy, Section 6.5.5 on page 316) and also a sudden change in heat capacity of ferromagnetic materials at the Curie point (page 328).

These properties of ferromagnetic materials will be discussed more extensively below.

The assumptions connected with the quantum mechanical theory of ferromagnetism are very rough but the results agree on the whole with those derived from the Ising model. Hence it seems reasonable to accept the idea of alignment of the spins of neighbouring atoms. A large number of atoms with parallel spins are supposed to be coupled strongly to each other in ferromagnetic materials.

Weiss Domains

The alignment of the spins of a large number of neighbouring atoms, predicted by theory, can easily be verified experimentally. When a ferromagnetic material is studied under a microscope, small grains, so-called *Weiss domains*, can be observed. The *Weiss domains* consist of a 10^{12}–10^{20} atoms with parallel magnetic moments. The extension of the domains are of magnitude 10^{-3}–10^{-4} cm. Each crystal may contain a number of domains of magnitude 10^5.

6.6.4 Properties of Ferromagnetic Materials

Curie Point

The properties of ferromagnetic and paramagnetic solids depend on temperature. At higher temperatures, the thermal motion in a ferromagnetic solid becomes more and more violent. At a certain temperature characteristic for each ferromagnetic material, the domain structure breaks up, the strong coupling between the magnetic moments of the atoms disappears and the ferromagnetic material becomes paramagnetic (Figure 6.38). A concrete example of the energy changes in Ni at the Curie point is given in Figure 6.39.

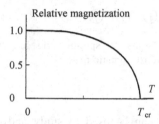

Figure 6.38 Saturation magnetization as a function of temperature for a ferromagnetic material.

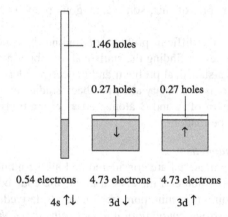

Figure 6.39 The 3d energy band in Ni at temperatures above the Curie point. The figure should be compared with Figure 6.31 on page 324. Above the Curie point the energy difference between the 3d↓ and 3d↑ electron bands in the Ni crystal, which is identical with the *exchange energy*, has disappeared. The 3d↓ and 3d↑ bands have the same energy independent of spin direction. The figure is not drawn to scale. Reproduced with permission from C. Kittel, *Introduction to Solid State Physics*, 6th edn. © 1986 John Wiley & Sons, Inc.

The transition temperature T_{cr} is called the *Curie point*. Some values for ferromagnetic metals are given in Table 6.5.

Table 6.5 Curie points of some common ferromagnetic metals.

Metal	Curie point (K)
Fe	1043
Co	1390
Ni	631

It should also be remembered that the positions of the energy bands are functions of the lattice constant, i.e. of the interatomic distances.

Order–Disorder Transformations in Ferromagnetic Materials

Transformations in ferromagnetic solids at the Curie point represent a special type of order–disorder transformation. Two examples of such transformations will be discussed below.

Figure 6.40 shows C_p for nickel as a function of T. The curve has a typical anomaly at the Curie point of Ni, 627 K. The discontinuity of C_p appears at the transition from an ordered structure with Weiss domains to a random, disordered structure without coupling between the spins of the atoms.

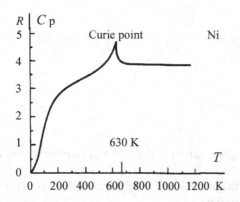

Figure 6.40 The molar heat capacity at constant pressure of Ni as a function of temperature. Reproduced with permission from A. J. Dekker, *Solid State Physics*. © 1962 Macmillan & Co Ltd.

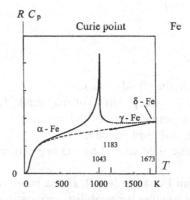

Figure 6.41 The molar heat capacity at constant pressure of Fe as a function of temperature within the interval 0–1800 K.

Figure 6.41 shows experimental values of C_p for iron as a function of T. The upper curve corresponds to α-Fe, which has a BCC structure (Chapter 1, page 21). The lower curve corresponds to γ-Fe, which has an FCC structure. At temperatures up to 1183 K the stable form of iron is the α-phase. In the temperature interval 1183–1673 K the stable form with lowest energy is γ-Fe.

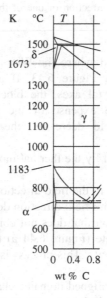

Figure 6.42 Low-carbon part of the phase diagram of the Fe–C system. Reproduced with permission from M. Hansen and K. Anderko, *Constitution of Binary Alloys*, 2nd edn. © 1958, McGraw-Hill Book Company, Inc.

Figure 6.41 shows that discontinuities appear in the transitions $\alpha \rightarrow \gamma$ (1183 K) and $\gamma \rightarrow \delta$ (1673 K). These transformations are true structure changes of the crystal lattice and cause an abrupt change in the heat capacity of Fe but no peak in the curve.

The anomalous peak in the α-curve at 1043 K corresponds to the transformation from an ordered magnetic structure of α-Fe to a disordered nonmagnetic structure, the Curie point of α-Fe. The disorder depends on random orientation in space of the magnetic moments of the atoms when the Weiss domains disappear.

Ferromagnetic Materials in Magnetic Fields

Magnetization of Ferromagnetic Materials

If ferromagnetic materials are exposed to a magnetizing field H they become magnetized, i.e. the Weiss domains align themselves more or less in the direction of the H field and contribute to the magnetic B field. The magnetic B field is defined by the relationship

$$B = (1 + \chi)\,\mu_0 H = \mu\mu_0 H \tag{6.97}$$

where

H = magnetizing field (A/m)
χ = magnetic susceptibility (of magnitude 10^3 for ferromagnetic materials)
μ = relative permeability of the material = $1 + \chi$
μ_0 = permeability in vacuum
B = magnetic field caused by the magnetizing field and the varying polarization in the material (ferromagnetic material).

As is seen from Equation (6.97), B is a measure of the strength of the magnetization of the material. It is not proportional to H because the relative permeability varies strongly with the magnetizing field.

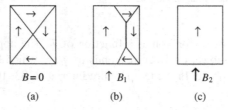

Figure 6.43 Magnetizing a ferromagnetic crystal in the direction of one of the crystal axes. The arrows represent the directions of the magnetic moments of the Weiss domains.

Consider a single crystal of a ferromagnetic solid where the Weiss domains have random orientations (Figure 6.43a). For simplicity only a few of them are shown in Figure 6.43. If a weak magnetizing field ($B = \mu\mu_0 H$) is applied in a direction that coincides with one of the crystal axes, the block walls of the Weiss domains become displaced. The domains in the field direction grow at the expense of the others (Figure 6.43b). The process is reversible for small displacements. When the magnetic field decreases, the domain walls become displaced in the reverse directions.

The magnetic order of the domains is counteracted by the thermal motion. The equilibrium depends on the strength of the magnetic B field.

When the magnetic field is increased, the domains in the field direction continue to grow by domain wall displacements. If the magnetic field is strong enough, the domains in the field direction dominate completely (Figure 6.43c).

If a weak magnetic B field is applied in a direction that does not coincide with one of the crystal axes, the directions of magnetic domains first change from the random state (Figure 6.44 a) to a state as close as possible to the magnetic field direction (Figure 6.44 b) by domain wall displacements. This process is the same as that described above and reversible for small displacements.

When the magnetic field is further increased, the aligned domains change direction discontinuously by sudden rotation in steps. This process is irreversible. If the magnetic field is strong enough, the directions of the magnetic moments and the B field coincide (Figure 6.44 c). The magnetization is said to be *saturated*.

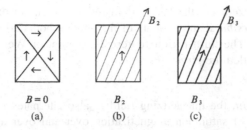

Figure 6.44 Magnetizing a ferromagnetic crystal in an arbitrary direction other than the crystal axes. The arrows represent the directions of the magnetic moments of the Weiss domains.

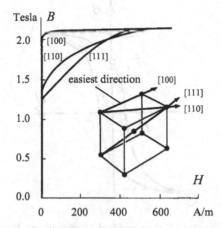

Figure 6.45 Magnetization of a single crystal of Fe in three different crystallographic directions. Reproduced with permission from A. G. Guy, *Elements of Physical Metallurgy*, 2nd edn. © Addison-Wesley Publishing Company, Inc. (now under Pearson Education).

It is easier to magnetize ferromagnetic crystals in the directions of some of its crystallographic axes than in other directions. This is shown in Figure 6.45, which shows that full magnetization is easily obtained in the <100> direction in iron.

Hysteresis

The easiest way to magnetize a ferromagnetic specimen is to place it in a homogeneous magnetizing field H caused by a coil (see Figure 6.27 on page 319). The H field is easily changed by varying the current I in the coil ($H = nI$, where n = number of turns per unit length).

The specimen is magnetized by increasing the H field from zero until saturation magnetization has been achieved, i.e. becomes constant. The process is shown in Figure 6.46. B and H are no longer proportional.

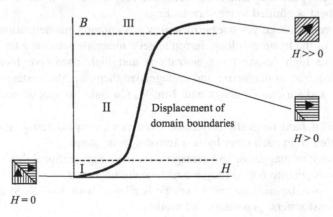

Figure 6.46 Virgin curve. The parallel lines inside the Weiss domains indicate the directions of their magnetic moments. Reproduced with permission from A. G. Guy, *Elements of Physical Metallurgy*, 2nd edn. © Addison-Wesley Publishing Company, Inc. (now under Pearson Education).

At $H = 0$ the specimen is unmagnetized, i.e. the Weiss domains have random orientations. Region I in Figure 6.46 of the curve represents the process of *reversible* displacements of the domain boundaries. Further displacements of the domain boundaries are *irreversible* (region II). The third region represents rotation of Weiss domains from easy directions to the direction of the H field. The curve is called the *virgin curve*.

Hysteresis Loop

If the coil is fed with *alternating current*, the magnetizing field H also alternates with the same frequency as the current. The specimen has to alter its direction of saturation magnetization over and over again. The remagnetization is shown in Figure 6.47.

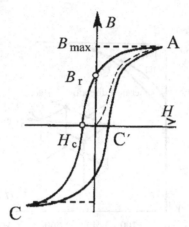

Figure 6.47 Hysteresis loop of a ferro-magnetic material. B_r = residual magnetization; H_c = coercive force. Reproduced with permission from A. G. Guy, *Elements of Physical Metallurgy*, 2nd edn. © Addison-Wesley Publishing Company, Inc. (now under Pearson Education).

The dashed curve represents the virgin curve discussed above. At point A the specimen is fully magnetized in the direction of the H field. When the H field decreases, becomes zero and then negative, the B field does not follow the virgin curve but the AB_rH_cC curve. When the H field decreases again and changes direction once more, the B field follows the curve $CC'A$ back to the starting point. The closed curve is known as the *hysteresis loop*.

B_{max} is the *saturation magnetization*. The curve intersects the B axis at $B = B_r$ when $H = 0$ and B decreases from B_{max}. B_r is the *residual magnetization*. For negative H values B continues to decrease. The value of H which corresponds to $B = 0$ is H_c, which is called the *coercive force*. If the magnetic material is exposed to H_c the residual magnetization B_r disappears.

The area inside the hysteresis loop represents the heat losses or *hysteresis losses*. So-called *eddy currents* are induced in the metal by the incessant changes of the magnetic field during the magnetization–remagnetization cycles. The eddy currents generate heat in the metal. The heat is emitted to the surroundings.

Depending on their magnetic properties, i.e. relative permeability μ, saturation magnetization B_{max}, residual magnetization B_r, coercive force H_c and the area of the hysteresis loop, ferromagnetic materials are characterized as hard or soft.

Hard magnetic materials have high resistant magnetizations and high coercive forces and large hysteresis losses (Figure 6.48). Much energy is required to magnetize and remagnetize them, i.e. the hysteresis losses are large. Examples of such materials are carbon steels and various Fe alloys with Mn, W, Co and also special non-iron alloys based on rare earth metals.

Permanent magnets are made of hard magnetic materials. Clusters of coupled ferromagnetic atoms form tiny domains, so-called single domains, separated from each other by non-ferromagnetic atoms.

Soft magnetic materials are easy to magnetize and remagnetize, have high permeabilities and low coercive forces and, particularly, small hysteresis losses (Figure 6.49). They are technically important.

Pure iron is often used. To reduce the eddy current losses Fe is alloyed with 3–4.5% Si. Such alloys are very convenient for use in iron cores in power transformers, generators and motors.

Undesired residual magnetism in specimens, for example iron ships, can be removed by surrounding the specimen with coils and running an alternating current through the coils, with an amplitude which gradually decreases to zero. The process causes smaller and smaller hysteresis loops, which finally end at the origin.

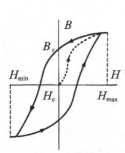

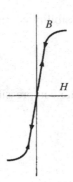

Figure 6.48 Hysteresis loop for a magnetically hard material. Reproduced with permission from A. G. Guy, *Elements of Physical Metallurgy*, 2nd edn. © Addison-Wesley Publishing Company, Inc. (now under Pearson Education).

Figure 6.49 Hysteresis loop for a magnetically soft material. In the ideal case the enclosed area of the loop is zero. Reproduced with permission from A. G. Guy, *Elements of Physical Metallurgy*, 2nd edn. © Addison-Wesley Publishing Company, Inc. (now under Pearson Education).

Summary

■ *Elasticity and Compressibility*

Origin of elastic forces:

   Restoring forces in the atoms and their electron shells when the equilibrium in the crystal lattice is disturbed.

Basic Concepts

$$\text{Stress}: \quad \sigma = \frac{F}{A}$$

$$\text{Strain}: \quad \varepsilon = \frac{\Delta l}{l}$$

Lateral contraction:

$$\varepsilon_{\text{trans}} = -\nu\varepsilon$$

where $\nu =$ Poisson's ratio

Elasticity

$$\text{Hookes's law}: \quad \sigma = Y\varepsilon$$

Hooke's general law is given in the text.

Compressibility

$$V + \Delta V = V\left(1 + \kappa\Delta p\right)$$

$$\kappa = -\frac{1}{V}\left(\frac{\partial V}{\partial p}\right)_T$$

where κ is the compressibility coefficient.

$$B = \frac{1}{\kappa} = \text{bulk modulus}$$

Relationship between the compressibility coefficient and the modulus of elasticity:

$$\kappa = \frac{3}{Y}(1 - 2\nu)$$

Deformation Energy

Linear deformation energy per unit volume: $E_{com} = \dfrac{Y\varepsilon^2}{2} = \dfrac{\sigma\varepsilon}{2} = \dfrac{\sigma^2}{2Y}$

Three-dimensional deformation energy per unit volume: $E_{deform} = \dfrac{Y}{6(1-2\nu)}(\varepsilon_x + \varepsilon_y + \varepsilon_z)^2$

■ Expansion

$$\begin{aligned}
\text{Length expansion}: \quad & l + \Delta l = l(1 + \alpha\Delta T) \\
& \alpha = \frac{1}{l}\left(\frac{\partial l}{\partial T}\right)_p \\
\text{Volume expansion}: \quad & V + \Delta V = V(1 + \beta\Delta T) \\
& \beta = \frac{1}{V}\left(\frac{\partial V}{\partial T}\right)_p
\end{aligned}$$

Relationship between α and β for cubic structures: $\quad \beta \approx 3\alpha$

Origin of Length and Volume Expansion

One contribution is due to the lattice vibrations. One contribution is due to vacancies. The dominant effect comes from the vibrational lattice part. The fraction of vacancies is comparatively small at room temperature but increases with increase in temperature.

Full agreement with experiments is achieved when the following thermodynamic principle is applied:

The whole crystal expands or contracts until it finds the volume for which the total free energy is a minimum.

■ Heat Capacity

$$\mathrm{d}Q = cm\mathrm{d}T$$

where c is the specific heat capacity.

Molar heat capacity at constant volume:

$$C_V = \left(\frac{\partial Q}{\partial T}\right)_V = \left(\frac{\partial U}{\partial T}\right)_V$$

Molar heat capacity at constant pressure:

$$C_p = \left(\frac{\partial Q}{\partial T}\right)_p$$

Debye's Model of Heat Capacity

In analogy with Planck's radiation law, Einstein calculated the energy of the elastic waves in a crystal lattice and derived C_V by taking the derivative of U with respect to T. Einstein's model was improved by Debye:

$$U = 9N_A k_B T \left(\frac{T}{\theta_D}\right)^3 \int_0^{\frac{\theta_D}{T}} \frac{x^3}{e^x - 1} \, dx$$

where θ_D = Debye temperature and ν_D = Debye frequency.

$$C_V = 9R \left(\frac{T}{\theta_D}\right)^3 \int_0^{\frac{\theta_D}{T}} \frac{x^4 e^x}{(e^x - 1)^2} \, dx$$

$$\theta_D = \frac{h\nu_D}{k_B}$$

At high temperature both models give $C_V = 3R$ (Dulong–Petit law).
At low temperatures Debye's model gives

$$C_V = \frac{12\pi^4}{5} R \left(\frac{T}{\theta_D}\right)^3 \qquad T \ll \theta_D$$

■ *Heat Capacity of Electrons*

At high temperatures the heat capacity of the valence electrons can be neglected in comparison with the phonon contribution.
Molar heat capacity of electrons at low temperatures:

$$C_e = \frac{\pi^2 k_B T}{2E_F} R$$

Total molar heat capacity at low temperatures:

$$C \approx C_p \approx C_V = \frac{12\pi^4 R}{5\theta_D^3} T^3 + \frac{\pi^2 k_B R}{2E_F} T$$

The electron contribution dominates at very low temperatures.

Molar Heat Capacity at Constant Pressure at High Temperatures

Grüneisen's rule:

$$C_p - C_V = \frac{\beta^2 V_A^{\theta_D}}{\kappa} (T - \theta_D)$$

Influence of Order–Disorder Transformations on Heat Capacity

Order and disorder refer to the *distribution* of atoms in a crystal lattice. Its *structure* remains unchanged. It is only the *occupancy of the sites* which is concerned. A and B atoms can be ordered or distributed at random in the sites
An order–disorder transformation influences many properties of a solid, among them the heat capacity.
Order–disorder transformations occur at the transformation temperature and require *energy*. A steep increase in C_p is followed by an even more rapid decrease after the transformation.

■ *Magnetism*

Magnetic Moment of Atoms

Orbital magnetic moment of an atom:

$$|\boldsymbol{\mu}_{\mathrm{L}}| = \mu_{\mathrm{B}}\sqrt{L(L+1)}$$

Bohr magneton: $\mu_{\mathrm{B}} = \dfrac{e\hbar}{2m}$

Component of the total magnetic moment of an atom:

$$\mu_{J,z} = -\mu_{\mathrm{B}}M_{\mathrm{J}}$$

Component of the total spin magnetic moment of an atom:

$$|\boldsymbol{\mu}_{\mathrm{S}}| = 2\mu_{\mathrm{B}}\sqrt{S(S+1)}$$

Total magnetic moment of an atom:

$$\boldsymbol{\mu}_{\mathrm{total}} = \boldsymbol{\mu}_{\mathrm{L}} + \boldsymbol{\mu}_{\mathrm{s}}$$

■ *Dia- and Paramagnetism*

All solids become magnetized in a magnetizing field:

$$H = nI$$

Magnetization vector:

$\boldsymbol{M}$ = magnetic moment per unit volume
$\boldsymbol{M} = \chi\,\boldsymbol{H}$

where χ is the magnetic susceptibility

  $\chi < 0$ diamagnetic solid; magnitude 10^{-5}
  $\chi > 0$ paramagnetic solid; magnitude 10^{-4}
  $\chi \gg 0$ ferromagnetic solid; magnitude 1000.

Diamagnetic atoms have no resulting spin and no magnetic moment. Paramagnetic atoms have a resulting spin and a magnetic moment, firmly coupled to the spin.

■ *Ferromagnetism*

Ferromagnetic atoms have unfilled 3d shells and hybridization of their 3d and 4s electrons. The atoms have a resulting spin owing to the unfilled 3d shell. The resulting spin is firmly coupled to a magnetic moment of the atom, just as for paramagnetic atoms.

The two groups of 3d bands with spin-up and spin-down electrons in ferromagnetic solids have different energies owing to exchange energy.

Theory of Ferromagnetism

The reason for the strong magnetism of ferromagnetic materials is an *alignment* of the magnetic moments in ferromagnetic solids, which is absent in paramagnetic solids.

The strong coupling between the magnetic moments of neighbouring atoms in ferromagnetic materials is caused by the coupling of the spin vectors of neighbouring atoms.

A large number of atoms with parallel spins are coupled strongly to each other in ferromagnetic materials and form so-called Weiss domains.

Curie Point

The structures of both ferromagnetic and paramagnetic solids depend on temperature. At a certain temperature characteristic for each ferromagnetic material, the domain structure breaks up, the strong coupling between the atoms disappears and the ferromagnetic material becomes paramagnetic. The transition temperature is called the *Curie point*.

Order–Disorder Transformations in Magnetic Materials

Ferromagnetic materials have an ordered structure owing to interaction of the spins of the atoms. Above the Curie point the ferromagnetism disappears and the materials become disordered.

The magnetic order–disorder transformation at the Curie point in ferromagnetic materials is accompanied by a similar anomalous maximum of the molar heat capacity C_p at the transition temperature. The order–disorder transformation is the reason in both cases.

Magnetization of Ferromagnetic Materials

$$B = (1 + \chi)\,\mu_0 H = \mu \mu_0 H$$

where μ is the relative permeability.

If a weak magnetizing field ($B = \mu \mu_0 H$) is applied in a direction that coincides with one of the crystal axes, the domain walls become displaced. The domains in the field directions grow at the expense of the others. The process is reversible for small displacements, i.e. the domain walls become displaced in the reverse directions when the magnetic field decreases.

The magnetic order is counteracted by the thermal motion. The equilibrium depends on the strength of the magnetic B field.

When the magnetic field is increased the domains in the field direction continue to grow by domain wall displacements. If the magnetic field is strong enough, the domains in the field direction dominate completely.

If a weak magnetic B field is applied in a direction that does not coincide with one of the crystal axes, the sizes the magnetic domains first change from the random state to a state as close as possible to the magnetic field direction by domain wall displacements. The process is reversible for small displacements.

When the magnetic field is further increased, the aligned domains change direction discontinuously by sudden rotation in steps. This process is irreversible. If the field is strong enough, the magnetization becomes saturated.

Hysteresis

Ferromagnetic materials can be magnetized and remagnetized. Then B and H are no longer proportional but follow the so-called hysteresis loop.

Soft magnetic materials have narrow hysteresis loops and are easy to magnetize. The heat losses are small. Hard magnetic materials have broad hysteresis loops. The heat losses are large.

Exercises

6.1 A steel bar is heated to $200\,°C$ and fixed safely at both ends at this temperature. Calculate the stress in the bar when it has been cooled to room temperature, $20\,°C$. Material constants for steel are $\alpha = 1.15 \times 10^{-5}\,K^{-1}$ and $Y = 2.0 \times 10^{11}\,N/m^2$.

6.2 A homogeneous metal tube has a length of $2.0\,m$ and an external diameter of $8.0\,mm$. The thickness of the metal is $2.0\,mm$. When the tube was exposed to a pulling force of $1.47 \times 10^3\,N$ in its length direction, the internal volume was increased by $4.0\,mm^3$ and the length of the tube by $0.80\,mm$.

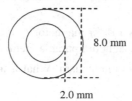

8.0 mm

2.0 mm

Calculate Poisson's ratio of the metal.

6.3 A steel spring is compressed 0.10 m and is kept in this compressed mode by the force of 1.0×10^4 N.

(a) What is the increase in internal energy of the spring in its compressed mode?
(b) What happens with the potential energy of the spring when the force suddenly is removed and the spring rapidly returns to its initial equilibrium position?
(c) Does the temperature of the spring change? If the answer is yes, calculate the initial temperature change. The mass of the spring is 10 kg. The specific heat capacity of steel is 0.46×10^3 J/kg K.

6.4 An elastic body floats on a water surface. To what depth under the water surface must the body at least be moved to sink by itself?

The modulus of elasticity of the body is 7.0×10^5 Pa (N/m^2) and its Poisson's ratio is 0.30. The density of the water is assumed to be 1.0 kg/dm^3, independent of the depth. The thermal expansion of the body can be neglected. The density of the elastic body in air is 700 kg/m^3.
Hint: Calculate the modulus of compressibility with the aid of the modulus of elasticity (page 294).

6.5 A sphere of solid brass has a radius of 1.0 cm when exposed to atmospheric pressure.

(a) When the sphere is placed in vacuum it expands. Calculate the increase in its radius.
(b) Calculate the decrease in the radius when the sphere is placed in a liquid in which the pressure is 10 atm.
(c) Calculate the deformation energy required for the compression in (b).

Poisson's ratio for brass is 0.22. The modulus of elasticity for brass is 10.5×10^{10} Pa(N/m^2).
Hint: Calculate the modulus of compressibility with the aid of the modulus of elasticity.

6.6 Consider a metal with length expansion coefficient α within a medium temperature interval ($\sim$20–200 °C). What is the relative change of the density of the metal (in %) when the temperature is changed by 1 K?

6.7 (a) Mention two effects that contribute to thermal expansion in crystals.
(b) Discuss briefly the thermodynamic explanation of thermal expansion in crystals.

6.8 (a) Describe briefly the Debye model of heat capacity for solids and define the concept of Debye temperature.
(b) Give the relationship between the thermal volume expansion coefficient, the compressibility coefficient and the molar heat capacities at constant pressure and constant volume, respectively.

6.9 Test the validity of the Dulong–Petit law for some common metals, for example Be, Ag, Zn, Fe, Cu and Al. One of these metals shows a striking deviation from the classical rule. Explain why this particular metal deviates so strongly from the other ones.

Heat capacities of the metals can be found in standard tables. The Debye temperatures are listed in Table 6.3 on page 309.

6.10 The heat capacities at constant volume C_V of diamond at different temperatures are

T (K)	100	150	200
C_V (kJ/kmol K)	0.29	1.06	2.34

(a) Verify that Debye's T^3 law is valid for diamond within this temperature range.
(b) Calculate the Debye temperature θ_D and compare it with those of other elements.

6.11 Calculate the electron contribution to the heat capacity of copper at 750 K. What fraction of the total heat capacity is this contribution? The Debye temperature of Cu is 343 K and the Fermi level of Cu is 7.04 eV.

6.12 (a) In what way does an order–disorder transition in a solid influence its heat capacity?
(b) Define the long-range order parameter S. Sketch S as a function of T/T_{cr} for a transformation from an ordered to a completely disordered state of a system.

6.13 (a) Discuss the origin of magnetism. What is a Bohr magneton?
(b) What circumstances make a material diamagnetic, paramagnetic or ferromagnetic?

6.14 Give a short review of modern theories of ferromagnetism. What is a Weiss domain?

6.15 (a) Explain the concept of Curie point.
(b) Explain the anomalous C_p values of ferromagnetic materials at their Curie points.

6.16 Consider Figures 6.31 on page 324 and Figure 6.39 on page 328 [shown below as Figures (a) and (b), respectively].

   (a) How can the number of electrons in each energy state be non-integer?

   (b) Describe the two mechanisms behind the electron distributions in the figures.

   (c) Explain the numerical electron and hole distributions in the 4s and 3d bands in Figure (a).

   (d) Explain the numerical electron and hole distributions in the 4s and 3d bands in Figure (b) and the reason why the two figures differ.

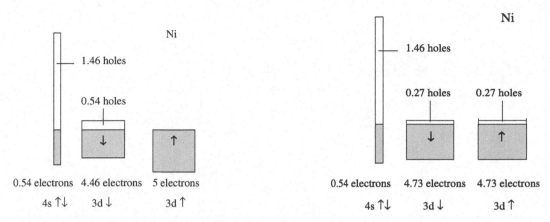

Sub-bands 4s, 3d ↑ and 3d ↓ of Ni at (a) $T = 0\,\mathrm{K}$ and (b) above the Curie point temperature.

6.17 Ferromagnetic materials can be magnetized in various crystallographic directions by an external magnetic field. Verify the statement that the [100] direction is the easiest direction in iron.

6.18 When a ferromagnetic specimen is exposed to an alternating magnetizing field from an alternating current in a coil it becomes magnetized and remagnetized periodically. This is illustrated by the hysteresis loop, which is a relationship between the magnetizing field H and the resulting magnetic field B. Describe the 'micro-changes' in the material during the hysteresis loop.

7

Transport Properties of Solids. Optical Properties of Solids

7.1	Introduction	342
7.2	Thermal Conduction	342
	7.2.1 Thermal Conductivities of Nonmetals	343
	7.2.2 Thermal Conductivities of Metals and Alloys	345
7.3	Electrical Conduction	347
	7.3.1 Resistivity and Conductivity	347
	7.3.2 Conductors, Insulators and Semiconductors	348
7.4	Metallic Conductors	350
	7.4.1 Resistivities of Pure Metals	350
	7.4.2 Temperature Dependence of Resistivity	351
	7.4.3 Resistivities of Alloys	352
	7.4.4 Theory of Electrical Conduction in Metals. Mobility	353
	7.4.5 Ratio of Thermal and Electrical Conductivities in Pure Metals. Wiedemann–Franz Law	357
7.5	Insulators	357
	7.5.1 Theory of Electrical Conduction in Insulators	358
	7.5.2 Conductivities of Insulators	361
7.6	Semiconductors	362
	7.6.1 Pure Semiconductors	362
	7.6.2 Theory of Electrical Conduction in Pure Semiconductors	364
	7.6.3 Doped Semiconductors	369
	7.6.4 Theory of Electrical Conductivity in Doped Semiconductors	369
	7.6.5 Types of Doped Semiconductors. Acceptor and Donor Energy Levels	370
7.7	Optical Properties of Solids	375
	7.7.1 Optical Properties of Metals	375
	7.7.2 Optical Properties of Semiconductors	376
	7.7.3 Optical Properties of Insulators	378
	7.7.4 Polarized Light and Insulators	379
Summary		388
Exercises		394

Physics of Functional Materials   Hasse Fredriksson and Ulla Åkerlind
© 2008 John Wiley & Sons, Ltd

7.1 Introduction

The phenomena, that are called *transport phenomena* as a common name involve the transport of one or several physical quantities simultaneously. Table 7.1 gives a survey of these processes.

Table 7.1 Transport phenomena in solids.

Property	Involves transport of
Diffusion	Mass
Thermal conduction	Energy
Electrical conduction	Charge

Diffusion has been discussed in Chapter 5. The mechanisms of thermal and electrical conductivities of solids will be treated in this chapter. The basis of the theories is the band theory of solids and the phonon theory, of both which have been discussed in Chapter 3.

As mentioned in Chapter 6, heat conduction is one of the properties which is very important for control of the rate of crystallization processes and hence for the quality and properties of the solidified products.

The electrical properties of solids are determined by their electrical conductivities, which decide the applications of the solidified products, for example insulating materials, metals, alloys and semiconductors.

7.2 Thermal Conduction

Thermal conduction plays a very important role in crystallization processes. The heat which is developed during the solidification must be transported away with the same rate as it is produced, otherwise the process stops.

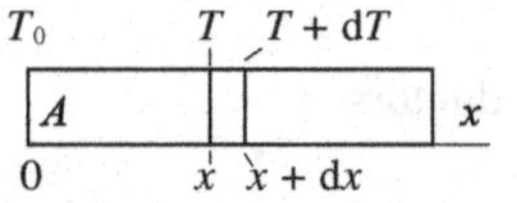

Figure 7.1 Heat conduction in a bar.

Thermal conduction implies transport of heat, i.e. kinetic energy. Consider a bar in which the temperature changes linearly (Figure 7.1). The amount of heat dQ which passes a cross-sectional area A in time dt is proportional to the cross-sectional area A and the temperature gradient dT/dx:

$$\frac{dQ}{dt} = -\lambda A \frac{dT}{dx} \tag{7.1}$$

Heat flows from higher to lower temperature ($dT < 0$). As the temperature decreases with increasing distance x, a minus sign has to be included in Equation (7.1). We divide the equation by A and introduce the heat flux dq/dt instead of dQ/Adt into Equation (7.1), which results in Equation (7.2). The thermal conductivity coefficient λ is defined by the equation

$$\frac{dq}{dt} = -\lambda \frac{dT}{dx} \tag{7.2}$$

where
  dq/dt = thermal flux (amount of heat per unit area and unit time)
  λ = thermal conductivity
  T = temperature.

There are two possibilities to carry heat in a crystalline solid, by free electrons or by phonons (lattice vibrations) of the crystal lattice:

$$\lambda = \lambda_e + \lambda_{lattice} \tag{7.3}$$

The number of available free electrons and the phonon transport through the crystal lattice control the thermal conduction in all solid materials.

It has been known for a long time that conduction of electricity and heat in *pure metals* are related. Pure metals such as copper and silver are excellent conductors of both electricity and heat. Lead conducts both electricity and heat rather poorly. These facts indicate that the mechanisms of conduction in metals are the same in the two cases.

Pure metals conduct electricity excellently or very well compared with non-metallic solids. Pure metals have many free electrons. As the phonon contribution to the electrical conduction in pure metals is negligible compared with the contribution from the free electrons, we can conclude that this it is true also for thermal conduction.

- The phonon contribution to the thermal conduction in a pure metal is small in comparison with that of the free electrons.

Other solids except pure metals, such as semiconductors, insulators and alloys, show no correlation between heat conduction and electrical conduction. Some materials are good thermal conductors but transport electricity poorly. This is true for diamond, for example (page 345).

Ionic crystals and solids with covalent bonds have practically no free electrons. As the free electron contribution is close to zero, we can conclude that

- The phonon contribution is responsible for the total thermal conduction in ionic crystals and solids with covalent bonds.

In other cases, for example in alloys, both the free electrons and the phonons contribute to the total thermal conduction.

The theory of thermal conduction in nonmetallic solids will be discussed in Section 7.2.1 and the theory of thermal conduction in metals in Section 7.2.2.

7.2.1 Thermal Conductivities of Nonmetals

In Chapter 3, Section 3.6, we discussed lattice vibrations in crystals and introduced phonons, particles associated with the lattice vibrations. In a crystal lattice of homogeneous constant temperature, the phonons move at random within the solid. The phonons collide with each other and the crystal faces. They change directions and energies incessantly in the lattice. If there is a temperature gradient in the crystal, the irregular motion of the phonons is overlapped by a systematic resulting motion in the direction from higher towards lower temperature.

The phonons in the crystal can be regarded as particles in a phonon gas container. One difference is that phonons, but not gas molecules, can be created or disappear in collisions. The phonon gas model will be used to derive an expression of the thermal conductivity $\lambda_{lattice}$ in terms of atomic quantities.

Theory of Thermal Conductivity in Nonmetals

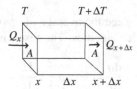

Figure 7.2   Heat balance.

Consider the heat balance of the volume element $A\Delta x$ in Figure 7.2. If the heat capacity (J/kg K) of the crystal is c, the heat capacity (J/K) of the volume element will be $c_V\rho A\Delta x$. A change in temperature from T to $T+\Delta T$ corresponds to the energy change $c_V\rho A\Delta x\Delta T$. This energy is equal to the net amount of heat into and out from the volume element:

$$\Delta Q = -Q_{x+dx} + Q_x = c_V\rho A\Delta x\Delta T$$

or by use of the relationships $\Delta x = v_x dt$ and $\Delta q = \Delta Q/A$ we obtain (minus sign because Δq is positive and ΔT is negative)

$$\Delta q = -c_V \rho v_x \Delta t \Delta T \tag{7.4}$$

where

Δq = heat amount per unit cross-sectional area
Δx = length of volume element
c_V = heat capacity per unit mass of the crystal (J/kg K)
ρ = density of the crystal (kg/m^3)
$c_V \rho$ = heat capacity per unit volume of the crystal (J/m^3 K)
v_x = phonon velocity in the x direction
Δt = time interval
ΔT = temperature difference.

After division by dt and a limit transformation, we obtain

$$\frac{dq}{dt} = -c_V \rho v_x \Delta T \tag{7.5}$$

We choose the length of the volume element equal to the mean free path l of the phonons. The temperature difference ΔT can be expressed in terms of the temperature gradient and the mean free path:

$$\Delta T = T(x + \Delta x) - T(x) = T(x) + \frac{dT}{dx}\Delta x - T(x) = \frac{dT}{dx}\Delta x = \frac{dT}{dx}l$$

Inserting this expression of ΔT into Equation (7.4) gives

$$\frac{dq}{dt} = -c_V \rho v_x \frac{dT}{dx}l$$

or with $l = v_x \tau$

$$\frac{dq}{dt} = -c_V \rho v_x \frac{dT}{dx}v_x \tau = -c_V \rho v_x^2 \frac{dT}{dx}\tau \tag{7.6}$$

where

l = mean free path of the phonons
τ = average time interval between two consecutive phonon collisions.

Equation (4.1) on page 170 is valid for the phonons both in the solid and for molecules in an ideal gas.

$$v^2 = v_x^2 + v_y^2 + v_z^2$$

As $v_x^2 = v_y^2 = v_z^2$ for symmetry reasons we can replace the average of v_x^2 of all the phonons by $\overline{v^2}/3$ in Equation (7.6). $\overline{v^2}$ is the average value of the sum of the squared particle velocities. The result is

$$\frac{dq}{dt} = -c_V \rho \frac{\overline{v^2}}{3} \tau \frac{dT}{dx}$$

If we assume that all the phonons have approximately equal velocities, the root mean square velocity equals this velocity. We can replace $v\tau$ by the mean free path of the particles and obtain

$$\frac{dq}{dt} = -\frac{c_V \rho v l}{3}\frac{dT}{dx} \tag{7.7}$$

Equation (7.7) can be compared with the definition Equation (7.2) of thermal conductivity on page 342. Identification gives

$$\lambda = \frac{c_V \rho v l}{3} \tag{7.8}$$

where

λ = thermal conductivity of the solid

c_V = heat capacity per unit mass of the solid (J/kg K)

ρ = density of the solid (kg/m³)

$c_V\rho$ = heat capacity per unit volume of the solid (J/m³ K)

v = phonon velocity, independent of direction

l = mean free path of the phonons.

Equation (7.8) is the total thermal conductivity in nonmetals such as ionic crystals and solids with covalent bonds because there are no free electrons in such solids.

Equation (7.8) formally has the same shape as Equation (4.68) on page 198 in Chapter 4, which represents the thermal conductivity for gases. From a more atomistic point of view, a different but equivalent expression for the thermal conductivity in nonmetal solids can be derived. The heat capacity per unit volume can be written in two ways:

$c_V \times \rho$	$= \quad n \times c_V^a$
Heat capacity per unit mass × density	number of atoms per unit volume × heat capacity per atom in the solid
$J/kg\,K \times kg/m^3 = J/m^3 K$	$m^{-3} \times J/K = J/m^3\,K$

Hence the thermal conductivity of the nonmetal solid can alternatively be written as

$$\lambda = \lambda_{\text{lattice}} = \frac{n c_V^a v l}{3} \tag{7.9}$$

The expression in Equation (7.9) will be used in Section 7.2.2 on page 346.

Table 7.2 gives the thermal conductivities for some different materials.

Table 7.2 Thermal conductivities of some nonmetallic materials.

Material	Conductivity (W/m K)
Mica	0.5
Fused SiO_2 (~ sand)	0.2
Dry concrete	0.4–1.7
Brick	0.6–0.8
Diamond	1000
Graphite	150

7.2.2 Thermal Conductivities of Metals and Alloys

Thermal Conductivities of Metals

The earliest theory of metallic conduction is ascribed to Drude. He considered the metal as a box filled with n_e electrons per unit volume in a homogeneous distribution of positive charge of equal total amount. As the Pauli principle was unknown at that time, he treated the conduction electrons of the metal as a classical gas, which gave puzzling discrepancies between theory and experiments.

In Chapter 6, Section 6.5.2 [Equation (6.63) on page 307], we found that the molar heat capacity at temperatures above the Debye temperature equals $3R$ owing to the lattice vibrations. The electron gas has three degrees of freedom and would therefore give a contribution of $3R/2$. No such contribution has ever been observed.

The explanation is that quantum mechanics has to be taken into account:

1. The conduction electrons of the metal have to obey the Pauli exclusion principle.
2. The electrons do not obey the classical Maxwell–Boltzmann distribution law. An electron has a half-integer spin and all particles with half-integer spin obey the Fermi–Dirac distribution law, which is based on the assumption that the particles are indistinguishable (Chapter 3, pages 117 and 148–149).

Figure 7.3 Electron–hole pair created in the valence band by thermal excitation. The valence band is only half filled for metal atoms with the valence 1.

At $T > 0$ and with the electron gas in thermal equilibrium with the crystal lattice, the valence electrons easily gain enough kinetic energy to be excited to the empty part of the valence band (Figure 7.3). It leaves a vacancy in the occupied part of the band.

Such vacancies are called *holes*. They can be treated as positive particles with an effective mass of the same magnitude as the electrons. Both electrons and holes can easily move and both contribute to the transport of electrical charge and heat.

Theory of Thermal Conductivity in Pure Metals

In metals, the phonon contribution can be neglected in comparison with the electron contribution. If we assume that the relationship between the thermal conductivity and the heat capacity is the same in nonmetals as in metals, we can easily calculate the thermal conductivity of a metal.

In Chapter 6 on page 310 we found that the molar heat capacity of the free electrons is

$$C_e = \frac{\pi^2 k_B T}{2 E_F} R \tag{7.10}$$

where
k_B = Boltzmann's constant
E_F = the Fermi energy.

We obtain the heat capacity per atom by dividing Equation (7.10) by Avogadro's number, N_A, and replacing R/N_A by Boltzmann's constant, k_B:

$$c_e = \frac{\pi^2 k_B^2 T}{2 E_F} \tag{7.11}$$

We let Equation (7.11) replace c in Equation (7.9) and obtain the thermal conductivity of the metal:

$$\lambda_e = \frac{n_e c_e v l}{3} = \frac{n_e v l}{3} \frac{\pi^2 k_B^2 T}{2 E_F} \tag{7.12}$$

In addition, the velocity v of the electrons in Equation (7.12) has to be replaced by the velocity v_F of those electrons of the Fermi–Dirac distribution which contribute to the heat conduction. We also replace the Fermi energy E_F by $m^* v_F^2/2$ and the mean free path l of the electrons by $v_F \tau$. After reduction we obtain

$$\lambda_e = \frac{n_e \pi^2 k_B^2 T \tau}{3 m_e^*} \tag{7.13}$$

where
λ_e = contribution to the thermal conductivity from the free electrons of the metal
k_B = Boltzmann's constant
n_e = number of free electrons per unit volume of the metal
T = temperature (K)
τ = average time interval between two consecutive particle collisions (mean free path divided by the average velocity of the particles)
m^* = effective mass of the electron in the conduction band.

The thermal conductivities of some metals are given in Table 7.3.

Table 7.3 Thermal conductivities of some metals at 293 K.

Metal	Conductivity (W/m K)
Ag	4.2×10^2
Cu	4.0×10^2
Fe	0.82×10^2
Pb	0.35×10^2
Sn	0.65×10^2
Zn	1.2×10^2

Thermal Conductivities of Alloys

In pure metals, the electronic contribution to the total thermal conductivity dominates at all temperatures. In impure metals and disordered alloys, collisions with vacancies or impurity atoms occur. This reduces the electron contribution considerably (Figure 7.4) and the phonon and electron contributions may be of the same magnitude. In this case the total thermal conductivity has to be written as

$$\lambda_{\text{alloy}} = \lambda_{\text{lattice}} + \lambda_e = \frac{nc_A^a vl}{3} + \frac{n_e \pi^2 k_B^2 T\tau}{3m_e^*} \tag{7.14}$$

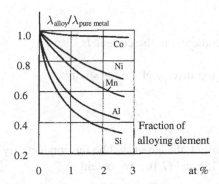

Figure 7.4 Influence of alloying elements on the thermal conductivity of iron at room temperature. Reproduced with permission from W. Schatt (ed.), *Einfuhrung in die Werkstoffwissensc* (7th edn). © 1991 Deutscher Verlag für Grundstoffindustrie, Leipzig.

7.3 Electrical Conduction

In Section 7.2 we showed that thermal energy in solids is transported by the free valence electrons of the atoms and by phonons in the crystal lattice. The mechanism of the transport and the contributions of the two sources depend strongly on the structure and composition of the solid.

In Sections 7.3–7.6 we will analyse the electrical conductivities of different types of solids, especially metals and semiconductors.

7.3.1 Resistivity and Conductivity

The electrical properties of matter can be described in two ways, either by *resistivity* or by *conductivity*.

If a voltage U is applied to a solid in a closed circuit (Figure 7.5), an electric current I starts in the circuit. It is customary to introduce the resistance R of the solid and write the relationship between the three quantities as

$$U = RI \tag{7.15}$$

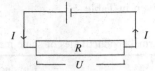

Figure 7.5 Simple electrical circuit.

The resistance R of the solid body depends of its shape. In order to extract the material property electrical conductivity and eliminate the influence of the shape of the solid body, we introduce the *resistivity* ρ of the solid (Figure 7.6):

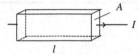

Figure 7.6 Rectangular bar of length l and cross-sectional area A.

$$R = \rho \frac{l}{A} \tag{7.16}$$

where
 ρ = resistivity of the solid
 R = resistance of the solid body
 l = length of the bar
 A = cross-sectional area of the bar perpendicular to the current I.

The conductivity σ of a solid is defined as the inverse of the resistivity:

$$\sigma = \frac{1}{\rho} \tag{7.17}$$

The better the solid conducts the electric current, the larger will be the conductivity σ and the smaller will be the resistivity ρ.
 If we eliminate R between Equations (7.15) and (7.16), we obtain

$$U = \rho \frac{l}{A} I \tag{7.18}$$

Equation (7.18) can be written as

$$j = \sigma E \tag{7.19}$$

where
 j = current density I/A through the solid
 E = electric field U/l in the solid.

7.3.2 Conductors, Insulators and Semiconductors

The classification of solids is based on their *electrical conductivities*. No other property shows such an enormous variation in magnitude as electrical conductivity. The ratio of the resistivities of the best insulators to the best conductors is more than 10^{40}.
 A material is characterized as an *insulator* if its resistivity $\rho > 10^9\,\Omega\,\text{m}$. The electric current through a solid is transported by the valence electrons. Hence the conductivity of a solid is closely related to the widths and positions of the energy bands of the solid. Metallic *conductors* have resistivities within the interval $10^{-12} < \rho < 10^{-5}\,\Omega\,\text{m}$ (Table 7.4).

Table 7.4   Resistivities of some materials.

Type of material	Resistivity $(\Omega\,\mathrm{m})$
Conductors	$10^{-12} - 10^{-5}$
Semiconductors	$10^{-5} - 10^{9}$
Insulators	$> 10^{9}$

As the name indicates, pure *semiconductors* have resistivities within the intermediate interval $10^{-5} < \rho < 10^{9}\,\Omega\,\mathrm{m}$. The boundaries between the three groups are not sharp and the values given above are to be regarded as merely representative for the three types of materials.

The resistivity of a solid varies with its temperature. Experiments show that

- The resistivity of metals *increases* with *increase* in temperature.
- The resistivities of insulators and semiconductors *decrease* with *increase* in temperature.

The resistivity variations of metals and semiconductors are shown in Figures 7.7 and 7.8.

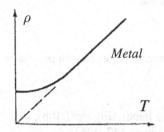

Figure 7.7   Resistivity of a metal as a function of temperature. At low temperatures $\rho = $ constant (Ohm's law). Reproduced with permission from H. Benson, *University Physics.* © John Wiley & Sons, Inc.

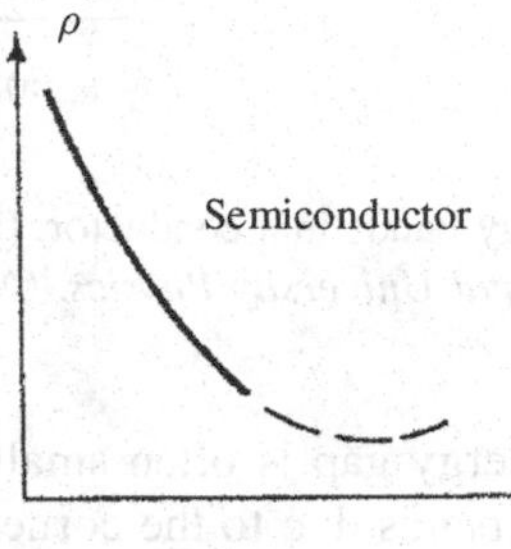

Figure 7.8   Restistivity of a semiconductor as a function of temperature. Reproduced with permission from H. Benson, *University Physics.* © John Wiley & Sons, Inc.

These properties can easily be explained in terms of the band theory of solids, which was discussed in Chapter 3. The lowest energy band, which is the ground state, is called the *valence band.*The first excited non-filled band is called the *conduction band* (Figure 7.9).

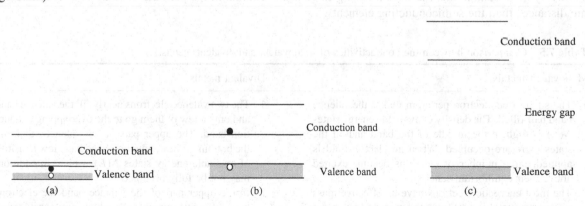

Figure 7.9   (a) Conductor. In monovalent atoms the valence band is only half filled. The conduction band is close to the valence band or overlaps the valence band. Hence the electrons can move rather freely within the metal.
(b) Semiconductor. In a semiconductor the valence band is filled. The energy gap is of the magnitude $\leq 1\,\mathrm{eV}$. Thermal excitation of electrons across the energy gap is possible.
(c) Insulator. In an insulator the valence band is filled. The energy gap between the conduction band and the valence band is very high (magnitude $5\,\mathrm{eV}$). Very few electrons have enough thermal energies to reach the conduction band.

7.4 Metallic Conductors

The main reason why *metals* and *alloys* are good conductors is the presence of free electrons. Their conductivities varies considerably (by a factor of 10^7), depending on the size of the energy gap between the valence band and the conduction band, the structure of the crystal lattice and the number of valence electrons per atom.

The shape and energies of the Brillouin zones and the way in which they become filled with electrons, according to the general principle of lowest possible total energy of the system, influence the conductivity strongly. These topics have been discussed in Chapter 3 on pages 140–145 and in Section 3.7.2 on page 153.

In most cases there is no energy gap between occupied and empty energy states, as is shown in Figures 7.9a and 7.10a. This is particularly evident in monovalent metals, where *the valence band is only half filled.* Hence the empty sites are easily available and the excited electrons can easily move through the metal.

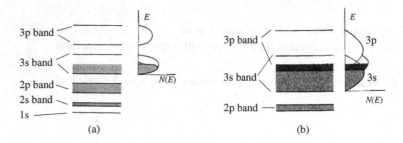

Figure 7.10 (a) Energy bands in a conductor. (b) Overlapping energy bands in a conductor. Reproduced with permission from M. Alonso and E. Finn, *Fundamental University Physics.* © Addison-Wesley.

In addition, the energy gap is often small between the valence and conduction bands (Figure 7.10b), which may result in overlapping of these bands due to the corners of the Brillouin zones (Chapter 3, pages 142–145).

7.4.1 Resistivities of Pure Metals

Monovalent metals usually are better conductors than divalent metals. This is explained in Table 7.5 by use of discussion in Chapter 3 (pages 121–124) on the energy distribution in the electron gas in a metal and the effective mass of the free electrons (Chapter 3, pages 145–146).

The monovalent metals Cu and Ag are excellent conductors, in agreement with the conclusions in Table 7.5. Tin and lead are comparatively poor conductors. Sn and Pb belong to the same column in the periodic table as Si and Ge and have larger interatomic distances than the semiconducting elements.

Table 7.5 Comparison between the conductivities of monovalent and divalent metals.

Monovalent metals	Divalent metals
1. The single free electron per atom makes the valence band half filled. The density of available energy states $N(E)$ is high in the middle of the band and all the states there are occupied. When an electric field is applied, large number of electrons become excited above their equilibrium states.	1. The two valence electrons nearly fill the valence band and only a few of them go to the overlapping conduction band. The upper part of the valence band and the bottom of the conduction band have low densities of available energy states $N(E)$, which, in addition, may not be fully occupied.
2. The most energetic electrons have an effective mass m^* approximately equal to m, the mass of an electron in free space. Compare pages 145–146 in Chapter 3.	2. In the upper part of the valence band the electrons have effective masses m^* which are considerably larger than m. Compare pages 145–146 in Chapter 3.

Normally the conductivity of a metal decreases when it melts owing to the disappearance of the regular crystal lattice. One exception is Bi, due to its special Brillouin zone structure.

The resistivity of a single crystal of a noncubic metal varies with the direction of the current. For example, in a crystal with HCP structure (Chapter 1) the resistivities parallel to the c axis and perpendicular to this axis are different.

Table 7.6 gives the resistivities of some common metals.

Table 7.6   Resistivities of some common metals at 300 K.

Metal	Resisistivity ($10^{-8}\,\Omega\,m$)
Ag	1.59
Cu	1.67
Fe (α)	9.7
Pb	21
Sn	11

7.4.2 Temperature Dependence of Resistivity

The resistivities of metals and alloys vary with temperature. Figure 7.11 shows the resistivity of Al and some of its alloys as an example.

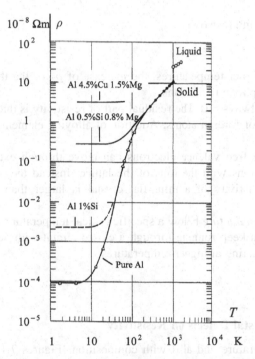

Figure 7.11   The resistivity of Al and some Al alloys as a function of temperature (double logarithmic scales). At low temperatures $\rho(T) = \rho_r + AT^2 + BT^5$, where the T^2 term describes the electron–electron collisions and the T^5 term describes the electron–phonon collisions. Reproduced with permission from J. E. Hatch (ed.), *Aluminum: Properties and Physical Metallurgy.*© 1984 American Society for Metals.

At room temperature and over a wide temperature range, the resistivity can be described empirically by a linear function of the type

$$\rho = \rho_{293}\left[1 + \alpha\left(T - 293\right)\right] \tag{7.20}$$

where

ρ_{293} = resistivity at 20° C (293 K)
α = temperature coefficient of resistivity
T = absolute temperature.

The temperature coefficients are of the magnitude $10^{-3}\,\mathrm{K^{-1}}$.

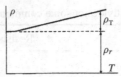

Figure 7.12 Resistivity of an alloy as a function of temperature. Reproduced with permission from A. G. Guy, *Elements of Physical Metallurgy*, 2nd edn. © Addission-Wesley Publishing Company, Inc. (now under Pearson Education).

Alternatively, it is convenient to write the resistivity of an alloy as a sum of two terms (Figure 7.12):

$$\rho = \rho_T + \rho_r \qquad (7.21)$$

where
   ρ_T = temperature-dependent part of the resistivity
   ρ_r = residual part of the resistivity.

At $T = 0$, ρ_T is practically zero. At higher temperatures, the increase of ρ_T or the thermal disturbance of the crystal lattice can be described in terms of electron–phonon collisions.

In the presence of impurities, ρ_r is always > 0. The residual part of resistivity is due to disturbances in the crystal lattice. It is extremely sensitive to the presence of foreign atoms. Addition of alloying elements to a pure metal increases the value of ρ_r strongly.

At extremely low temperatures, the free valence electrons can move through extremely pure crystal lattices containing few defects (vacancies) without collisions with the ions of the lattice. Instead they experience collisions against the walls of the solid. For this reason, the resistivity of a thin, flat crystal is larger than that of such a crystal with a thicker shape.

A number of pure metals are *superconducting* below a specific critical temperature near absolute zero. The resistance drops to zero and strong electric currents can keep running through a closed circuit with negligible effect losses. Efforts are made to find materials which are superconducting at higher temperatures.

7.4.3 Resistivities of Alloys

Influence of Foreign Atoms and Crystal Defects on Resistivity

The conductivity σ varies with temperature and also with composition (Figures 7.13 and 7.14), i.e. it depends on *foreign atoms*, alloying elements or impurities, in the crystal lattice and *crystal defects*, for example vacancies or interstitials.

Resistivities of Alloys

Foreign atoms and crystal defects increase the scattering of the electrons and reduce the time τ between consecutive collisions, which increases the resistivity. The higher the disorder in the crystal lattice is, the higher will be the resistivity. This is demonstrated in Figure 7.13.

The resistivity of an alloy increases with concentration of the alloying element. An example is given in Figure 7.13, which shows the resistivities of Cu and Cu–Ni alloys as functions of temperature. The time τ between two collisions decreases with increase in temperature. This is the reason for the inclination of the curves. The dominant reason for the change in ρ is the concentration of the alloying element.

Another example is Cu–Au alloys, which show four deep resistivity minima for the 'ordered' structures of pure Cu, Cu_3Au, CuAu and pure Au and resistivity peaks for maximum disordered structure for intermediate concentration values

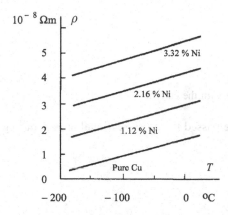

Figure 7.13 Resistivities of Cu–Ni alloys as functions of temperature. Reproduced with permission from A. J. Dekker, *Solid State Physics*. © 1962 Macmillan & Co. Ltd.

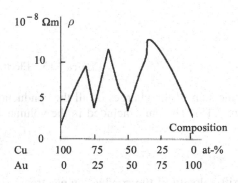

Figure 7.14 Resistivities of Cu–Au alloys as functions of composition. Reproduced with permission from A. J. Dekker, *Solid State Physics*. © 1962 Macmillan & Co. Ltd.

(Figure 7.14). Compare also Chapter 3, page 154. Disordered structures have been discussed in Chapter 6, Section 6.5.5 (page 315).

The resistivities of some common alloys are given in Table 7.7. Some of them have low temperature coefficients. Such alloys are used in resistors.

Table 7.7 Resistivities of some alloys at 293 K.

Alloy	Resistivity ($10^{-8}\,\Omega\,m$)	Temperature coefficient ($10^{-3}\,K^{-1}$)
Kanthal	145	0.03
Constantan	50	0.03
Manganin	43	0.02
Ni–Cr 80:20	105	0.18
Brass	6.5	1.5
Invar	10	2
Steel	16	3.3

Some alloys, for example manganin (84% Cu, 12% Mn, 4% Ni) constantan (55% Cu, 45% Ni) and others have very low temperature coefficients and the variations of their resistances with temperature are very small.

7.4.4 Theory of Electrical Conduction in Metals. Mobility

Classical Theory

As a first coarse and classical approximation, we regard the free conduction electrons in a metal as an electron gas in the volume occupied by the metal. The electrons are supposed to have a random motion just like the molecules in a gas. They seldom collide with each other because of the mutual electrostatic repulsion. Owing to electrostatic attraction, the electrons collide with the positive metal ions in the crystal lattice.

Drift Velocity and Current Density

The average velocity of the free electrons is high and their kinetic energy is considerable but the resulting random motion is zero. An external electric field changes the velocity distribution slightly as each free electron is accelerated in the field. In the frequent collisions, the electrons change both the direction and magnitude of their high velocities incessantly.

The collisions are equivalent to a friction force and the resulting motion is an average constant velocity in the opposite direction to the electric field. This velocity is the so-called *drift velocity* v_e of the electrons. The current in a metal wire is related to the drift velocity.

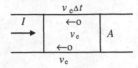

Figure 7.15 Electron motion in a metal wire with the current I.

ΔQ is the sum of the charges of all the conduction electrons which have passed the cross-sectional area A during the time Δt (Figure 7.15). They are included in the volume $Av_e\Delta t$:

$$\Delta Q = I\Delta t = n_e e A v_e \Delta t \tag{7.22}$$

where
v_e = drift velocity of the conduction electrons
n_e = number of conduction electrons per unit volume
e = charge of the electron
A = cross-section area of the wire.

If Equation (7.22) is divided by Δt, we obtain

$$I = n_e e A v_e \tag{7.23}$$

The *current density* j is defined as the current per unit area:

$$j = \frac{I}{A} \tag{7.24}$$

Relationship Between Drift Velocity and Current Density
Equation (7.23) is divided by A and Equation (7.24) is applied. The result is

$$j = n_e e v_e \tag{7.25}$$

Example 7.1

The current density in a pure copper wire is $10\,A/mm^2$. The density of Cu is $8.93 \times 10^3\,kg/m^3$. Its molar weight is 63.6.

(a) Calculate the density of conduction electrons in the metal if we assume that there is one conduction electron per Cu atom.
(b) Calculate the drift velocity at the given current density.

Solution:

(a) Consider 1 kmol of Cu and calculate the volume of 63.6 kg of Cu.

$$V = \frac{M}{\rho} \tag{1'}$$

1 kmol contains $N_A = 6.02 \times 10^{26}$ Cu atoms. Hence the electron density will be

$$n_e = \frac{N_A}{V} = \frac{N_A}{\dfrac{M}{\rho}} = \frac{N_A \rho}{M} \tag{2'}$$

or in this case

$$n_e = \frac{6.02 \times 10^{26} \times 8.93 \times 10^3}{63.6} = 8.5 \times 10^{28}\,m^{-3}$$

(b) Equation (7.25) can be written as

$$v_e = \frac{j}{n_e e} = \frac{10 \times 10^6}{8.5 \times 10^{28} \times 1.60 \times 10^{-19}} = 0.7 \times 10^{-3}\,\text{m/s}$$

Answer:

(a) The electron density is 8.5×10^{28} electrons/m^3.
(b) The drift velocity is 0.7 mm/s.

Example 7.1 shows that the drift velocity is surprisingly low. The reason is that the number of charge carriers is very high.

Conductivity
The electric field accelerates the conduction electrons between the collisions. The electrons are retarded by the collisions, which can be regarded as a friction force. A force balance for the electron motion can be set up:

$$\underset{\substack{\text{Resulting} \\ \text{force}}}{m\frac{dv}{dt}} = \underset{\substack{\text{Electrical} \\ \text{accelerating} \\ \text{force}}}{eE} - \underset{\substack{\text{Retarding} \\ \text{friction} \\ \text{force}}}{\frac{mv - 0}{\tau}} \tag{7.26}$$

The friction force balances the electric force exactly, because the average velocity is constant, i.e. the net acceleration must be zero. In this case the velocity is equal to the drift velocity:

$$eE = \frac{mv_e}{\tau} \tag{7.27}$$

or

$$v_e = \frac{eE\tau}{m} \tag{7.28}$$

If we combine this equation with Equation (7.25), we obtain

$$j = \frac{n_e e^2 \tau}{m} E \tag{7.29}$$

It is obvious from Equation (7.26) that τ is the average time between two collisions. Ohm's law (page 348) can be written as

$$j = \sigma E \tag{7.30}$$

where σ is the electric conductivity. Identification of Equations (7.29) and (7.30) gives

$$\sigma = \frac{n_e e^2 \tau}{m} \tag{7.31}$$

The conductivity can easily be found experimentally and Equation (7.31) can be used for the calculation of τ. In the case discussed in Example 7.1, the time between two collisions is of the magnitude 10^{-14} s.

Relaxation Time
The time τ also has another significance. If the electric field is suddenly switched off [insert $E = 0$ in Equation (7.26)], the friction force gradually retards the electrons from their average velocity to zero:

$$\frac{dv}{dt} = -\frac{v}{\tau} \tag{7.32}$$

or after integration:

$$\int_{v_e}^{v} \frac{dv}{v} = \int_{0}^{t} -\frac{dt}{\tau}$$

which can be written as

$$v = v_e e^{-t/\tau} \qquad (7.33)$$

where

  v = velocity at time t
  v_e = drift velocity, i.e. velocity at time $t = 0$
  τ = average time between two consecutive collisions or *relaxation time*, the time constant in Equation (7.33).

Mobility

If we divide Equation (7.25) with the electric field E, we obtain

$$\frac{j}{E} = n_e e \frac{v_e}{E}$$

which can be written (page 348) as

$$\sigma = n_e e \mu_e \qquad (7.34)$$

where μ_e is the *mobility* of the conduction electrons. It is defined as the *drift velocity per unit electric field*:

$$\mu_e = \frac{v_e}{E} \qquad (7.35)$$

Equation (7.34) is a very useful and central equation, which will be frequently used in the rest of this chapter. It can be used to determine the mobility for different metals. Mobility is a material constant, i.e. it has a characteristic value for each metal.

Mean Free Path

If the time τ between two collisions and the average kinetic velocity of the conduction electrons are known, the mean free path can be calculated from the relationship

$$l = \tau v \qquad (7.36)$$

where

  l = mean free path of the conduction electrons
  τ = average time between two collisions
  v = average kinetic velocity of the conduction electrons.

The magnitude of the mean free path is of the magnitude 100 distances between the atoms in the lattice. For Cu the value is about 40 nm.

Band Theory

However, there are serious discrepancies between the theoretical and experimental values of σ and the classical theory has to be modified.

The error lies in the fact that the conduction electrons cannot be treated like a classical gas. Instead, we have to take the band theory of metals into account.

The band theory requires the introduction of the 'effective mass' of the electron (Chapter 3, pages 145–146) and the resulting equation will be, in analogy with Equation (7.31):

$$\sigma = \frac{n_e e^2 \tau}{m_e^*} \tag{7.37}$$

The resistivity is equal to the inverted conductivity:

$$\rho = \frac{m_e^*}{n_e e^2 \tau} \tag{7.38}$$

These equations agree well with experimental values.

7.4.5 Ratio of Thermal and Electrical Conductivities in Pure Metals. Wiedemann–Franz Law

The theory of thermal conductivity of a pure metal shows that the thermal conductivity can be written as Equation (7.13) on page 346. Equation (7.37) above is the analogous expression for the electrical conductivity of a pure metal. We form the ratio λ_e/σ by dividing the Equations (7.13) and (7.37):

$$\frac{\lambda_e}{\sigma} = \frac{n_e \pi^2 k_B^2 T \tau}{3 m_e^*} \bigg/ \frac{n_e e^2 \tau}{m_e^*} \Rightarrow \frac{\lambda_e}{\sigma} = \frac{\pi^2 k_B^2}{3 e^2} T = constant \times T \tag{7.39}$$

The value of the constant is $2.45 \times 10^{-8}\,\mathrm{W\Omega/K^2}$. Equation (7.39) is Wiedemann–Franz law, which states that

- The ratio of the thermal and electrical conductivities of a pure metal is proportional to the absolute temperature.

The law is illustrated in Table 7.8, where the values of $\lambda_e/\sigma T$ are given for some metals at $T = 273$ K.

Table 7.8 The ratio $\lambda_e/\sigma T$ of some metals at 273 K.

Metal	$\lambda_e/\sigma T \ (\mathrm{W\Omega/K^2})$
Ag	2.31×10^{-8}
Au	2.35×10^{-8}
Cd	2.42×10^{-8}
Cu	2.23×10^{-8}
Mo	2.61×10^{-8}
Pb	2.47×10^{-8}
Sn	2.52×10^{-8}
Zn	2.31×10^{-8}

The fact that the ratio of the thermal and electrical conductivities has almost the same value for all pure metals and is independent of their material constants strongly supports the theory that the carriers of energy and charge are the free electrons of the metals in both cases. The phonon energy transport represents only a minor contribution to the thermal conductivity in metals.

It should be strongly emphasized that the Wiedemann–Franz law is valid only for *pure* metals. A striking example of the contrary is diamond, which is a good thermal conductor (page 345) and an excellent insulator (page 358), e.g. a very poor electrical conductor. The reason is that the phonon contribution to the thermal conductivity of diamond is large.

7.5 Insulators

The characteristic features of *insulators* are that they have

- filled valence bands
- empty conduction bands
- large energy gaps.

The electrons in a filled band cannot move as it contains no empty states (Figure 7.9 c on page 349). If the energy gap is large, few electrons have energies enough to jump to the empty conduction band with plenty of available energy states. The resulting current will be very low.

As an example of a good insulator we will choose carbon in the configuration of diamond. The four covalent bonds of carbon are symmetrical in space and end in the corners of a tetrahedron due to sp³ hybridization (Chapter 3, page 109).

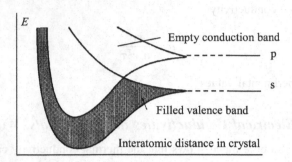

Figure 7.16 Energy bands of diamond (C). The large energy gap between the valence band and the conduction band makes diamond to a very good electrical insulator. Reproduced with permission from M. Alonso and E. Finn, *Fundamental University Physics*. © Addison-Wesley Publishing Company, Inc. (now Pearson Education).

Figure 7.16 shows the simplified band scheme of diamond. One s electron and three p electrons per C atom result in two binding states and two anti-binding states per atom in diamond in analogy with Si and Ge (Figure 3.8 on page 110). The binding states, i.e. the valence band, are filled with electrons whereas the upper band, the conduction band, is empty. At the equilibrium distance between the atoms the energy gap is about 5 eV, which makes diamond to a very good insulator.

Most covalent solids which are composed of atoms that have an even number of valence electrons are insulators.

7.5.1 Theory of Electrical Conduction in Insulators

In Chapter 3, Section 3.4.3, Sommerfeld's quantum mechanical model of the electron gas was treated extensively. It is the basis for the band theory of solids. The theory, which includes concepts such as Fermi level and Fermi distribution of electrons, was successfully applied to metals. Of special interest here is the calculation of the *density of available electron energy states per energy unit* in a solid [Equation (3.62) on page 122]:

$$N(E) = \frac{(2m^*)^{3/2}}{4\pi^2\hbar^3} E^{1/2} \tag{7.40}$$

and the density of occupied electron energy states per energy unit [Equation (3.64) on page 123]:

$$N(E)f_{FD} = \frac{(2m^*)^{3/2}}{4\pi^2\hbar^3} E^{1/2} f_{FD} \tag{7.41}$$

These concepts and equations will be most useful when we discuss a simple model for the electrical conductivity of insulators below and later an improved model for semiconductors.

Simple Model of Electrical Conductivity in Insulators

We want to calculate the electron density, i.e. the number of electrons per unit volume in the conduction band and the valence band, respectively, in order to find an expression for the conductivity of an insulator. For this purpose we must know the position of the Fermi level.

The distribution laws (7.40) and (7.41) have been applied to metals and cannot be used uncritically for the electron energy distribution in all types of solids. The Fermi level (Chapter 3, page 116) has a very specific physical significance for metals. In insulators and semiconductors there is no corresponding meaning. The Fermi level E_F is located in the energy gap between the valence and the conduction bands in these types of solids and cannot accommodate any electrons.

According to the theory given in Chapter 3, the electron concentration $n(E)\mathrm{d}E$ (the number of electrons per unit volume within the energy range E and $E+\mathrm{d}E$) is equal to *twice* the number of occupied energy levels $N(E)f_{FD}\mathrm{d}E$ within the energy interval:

$$n(E)\,\mathrm{d}E = 2N(E)\,f_{FD}\mathrm{d}E \qquad (7.42)$$

where

$n(E)\mathrm{d}E$ =concentration of electrons (number of electrons per unit volume) with energies between E and $E+\mathrm{d}E$
$N(E)\mathrm{d}E$ =number of *available* energy states per unit volume with energies between E and $E+\mathrm{d}E$
f_{FD} =Fermi factor (Fermi–Dirac distribution function).

According to Equation (3.43) on page 117 in Chapter 3, we have

$$f_{FD} = \frac{1}{e^{\frac{E-E_F}{k_B T}} + 1} \qquad (7.43)$$

The Fermi level E_F can be calculated by integration of Equation (7.42). According to Equation (3.65) on page 123, we have

$$n = \int n(E)\,\mathrm{d}E = \int 2N(E)\,f_{FD}\mathrm{d}E \qquad (7.44)$$

provided that E_F and the function $N(E)$ are known. The electron energy distribution function $N(E)\mathrm{d}E$ of an insulator is a more complicated function of E than that for a metal. n is the concentration of valence electrons (number of electrons per unit volume), which often is known or can be calculated from known data.

Calculation of the Electron Concentration in the Valence and Conduction Bands

To simplify the calculations, we will make the approximation that the widths of all inner bands, the valence band and the conduction band are small compared with the energy gap between the latter bands. If this condition is fulfilled, the valence band can be assumed to have a single energy E_v and the conduction band the single energy E_c (Figure 7.17).

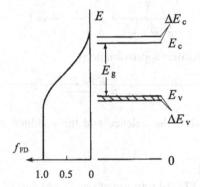

Figure 7.17 Energy levels in an insulator together with the Fermi distribution function. $\Delta E_c \ll E_g$ and $\Delta E_v \ll E_g$. Reproduced with permission from A. J. Dekker, *Solid State Physics*. © 1962 Macmillan & Co. Ltd.

In Chapter 3 on page 129, we found that the total number of energy states in each band (Brillouin zone) in a solid equals N_{total}, where N_{total} is the number of atoms in the crystal lattice. The total number of available energy levels per unit volume in a band is obtained by dividing N_{total} by the volume of the crystal:

$$N = \frac{N_{total}}{V} \qquad (7.45)$$

Each energy state can accommodate two electrons (one with spin up and one with spin down). Hence the total number of electrons per unit volume which can be accommodated in each band (Brillouin zone) is $2N$.

At temperature $T = 0$, the valence band and all lower bands are completely filled ($f_{FD} = 1$) and the conduction band is completely empty ($f_{FD} = 0$).

At temperature T, some electrons in the valence band are excited up to the conduction band. The electron concentration (number of electrons per unit volume) in the conduction band can be written as [Equations (7.42) and (7.43)]

$$n_c = 2Nf_{FD} = \frac{2N}{e^{\frac{E_c - E_F}{k_B T}} + 1} \tag{7.46}$$

The number of electrons per unit volume in the valence band is

$$n_v = 2Nf_{FD} = \frac{2N}{e^{\frac{E_v - E_F}{k_B T}} + 1} \tag{7.47}$$

where

n_c = concentration of electrons (number of electrons per unit volume) with energy E_c in the conduction band
n_v = concentration of electrons (number of electrons per unit volume) with energy E_v in the valence band
N = number of energy levels per unit volume in each of the valence and conduction bands
k_B = Boltzmann's constant
T = absolute temperature.

When the temperature approaches $T = 0$ K no electrons can be thermally excited, i.e. the conduction band is empty and all the electrons are in the filled valence band, i.e. $n_v = 2N$. This is in agreement with Equations (7.46) and (7.47).

Calculation of the Position of the Fermi Level

It is reasonable to assume that all free electrons per unit volume in the conduction band come from the valence band and that no valence electrons disappear elsewhere. This condition can be written as

$$n_v + n_c = 2N \tag{7.48}$$

By combining equations (7.46)–(7.48), we obtain after division by $2N$

$$\frac{1}{e^{\frac{E_v - E_F}{k_B T}} + 1} + \frac{1}{e^{\frac{E_c - E_F}{k_B T}} + 1} = 1 \tag{7.49}$$

Reduction of Equation (7.49) gives without further approximations

$$E_F = \frac{E_v + E_c}{2} \tag{7.50}$$

- The Fermi level is located half way between the valence and the conduction bands in an insulator, independent of the temperature

according to the simple model we have used.

If we introduce the value of E_F [Equation (7.50)] into Equations (7.46) and (7.47), we obtain

- the number of electrons in the conduction band at temperature T:

$$n_c = \frac{2N}{e^{\frac{E_c - E_v}{2k_B T}} + 1} = \frac{2N}{e^{\frac{E_g}{2k_B T}} + 1} \tag{7.51}$$

- the number of electrons in the valence band at temperature T:

$$n_v = \frac{2N}{e^{-\frac{E_c - E_v}{2k_B T}} + 1} = \frac{2N}{e^{-\frac{E_g}{2k_B T}} + 1} \tag{7.52}$$

where E_g is the width of the energy gap between the conduction and valence bands.

7.5.2 *Conductivities of Insulators*

Classical Theory of Conductivity in Metals

The classical theory of electrical conductivity in metals has been treated on page 356. The concept of mobility was introduced and defined as the drift velocity per unit electric field in Equation (7.35). The relationship between the conductivity σ and the mobility was found to be according to Equation (7.34).

Band Theory of Conductivity in Insulators

The classical theory derived for metals is *not* valid for pure insulators and pure semiconductors. It is only valid for metals with plenty of free electrons.

Insulators have filled valence bands, which cannot transport charge, have very few free electrons and are poor conductors. The charge transport is performed by excited electrons and holes.

Another objection to the classical theory is that no consideration is taken of the effective masses of the electrons and holes when they move within the lattice environment.

Conductivity of Insulators

The transport of charge is performed not only by electrons in the conduction band but also to the same extent by holes, which move in the direction of the electric field in the valence band. Therefore, Equation (7.34) on page 356 has to be replaced by the general equation

$$\sigma = n_e e \mu_e + n_h e \mu_h \tag{7.53}$$

where

σ = conductivity of the insulator
n_e = concentration of electrons (number of electrons per unit volume) in the conduction band
μ_e = mobility of electrons, i.e. their drift velocity per unit electric field
n_h = concentration of holes (number of missing electrons per unit volume) in the valence band
μ_h = mobility of holes, i.e. their drift velocity per unit electric field.

The theory of electrical conductivity for pure insulators is the same as that of pure semiconductors. The only difference is the magnitude of the energy gap E_g. Hence the Equations derived for pure semiconductors in Section 7.6.2 on pages 364–368 are also valid for pure insulators. The revised band theory is more accurate than the simple model given in Section 7.5.1 on pages 358–360.

An adequate expression for n_e (number of electrons in the conduction band per unit volume) is given by Equation (7.59) on page 366:

$$n_e = \frac{(2\pi m_e^* k_B T)^{3/2}}{4\pi^3 \hbar^3} e^{-\frac{E_c - E_F}{k_B T}} \tag{7.59}$$

The corresponding expression for n_h (the number of holes in the valence band per unit volume) in given by Equation (7.67) on page 367:

$$n_h = \frac{(2\pi m_h^* k_B T)^{3/2}}{4\pi^3 \hbar^3} e^{-\frac{E_F - E_v}{k_B T}} \tag{7.67}$$

For an insulator, $n_e = n_c$ and $n_h = 2N - n_v$. From now on n_e and n_h will be used instead of n_c and $2N - n_v$ in most cases for both insulators and semiconductors.

On page 368, it is shown that the conductivity for an intrinsic insulator can be written as

$$\sigma = n_i e (\mu_e + \mu_h) \tag{7.73}$$

where

$$n_i = \sqrt{n_h n_e} = \frac{(2\pi m k_B T)^{3/2}}{4\pi^3 \hbar^3} \left(\frac{m_e^*}{m} \frac{m_h^*}{m} \right)^{3/4} e^{-\frac{E_g}{2k_B T}} \tag{7.71}$$

The revised theory implies that the Fermi level no longer is the average of energies of the conduction and valence bands [Equation (7.70) on page 368].

Determination of the Energy Gap of an Insulator

Equations (7.73) and (7.71) show that the conductivity is proportional to $T^{3/2} \times \exp(-E_g/2k_BT)$. The other factors are not very strongly temperature dependent.

If we plot $\ln \sigma$ as a function of $1/T$, we can therefore derive $-E_g/2k_B$ from the slope of the straight line as $\ln T^{3/2}$ changes fairly slowly.

7.6 Semiconductors

Semiconductors are a very special and important group of solids. Owing to a rather small energy gap between the valence and conduction bands (magnitude 1 eV), their electrical conductivity is higher than that of the insulators but lower than the metals.

Semiconducting solids acquired enormous technical and industrial importance during the last part of the 20th century. The reason is that it is possible to control the conductivity of semiconducting materials and produce large amounts of cheap complex components. Two technical processes have made this development possible:

- production of extremely pure semiconducting materials
- doping of extremely pure semiconducting materials.

Without these vital premises, the development would never have been possible. Refining methods, originally invented by Pfann at Bell Telephone Laboratories at the end of the 1950s, have been developed nearly to perfection. Semiconducting layers are designed for innumerable special purposes and produced simultaneously as multicopies at low cost.

The properties and the basic theory of pure and doped semiconductors are given in this section.

7.6.1 Pure Semiconductors

The dominant semiconductors are silicon and germanium. For this reason, Si and Ge have been extensively studied and are among the best understood solid elements.

Figure 7.18 shows a sketch of the energy levels of an Si or Ge atom in the crystal lattice and its correlation with the energy levels of the orbitals of a free Si or Ge atom.

A free Si or Ge atom has two s electrons and two p electrons. Owing to hybridization in the solid state, one of the s electrons is excited up to a p state (Chapter 3, page 109). Each of the four sp^3 orbitals can accommodate two electrons, i.e. a total of eight electrons, due to opposite spins. Si and Ge have only four outer electrons. They all occupy the lowest energy states in the crystal lattice, which is the valence band with four electrons per atom.

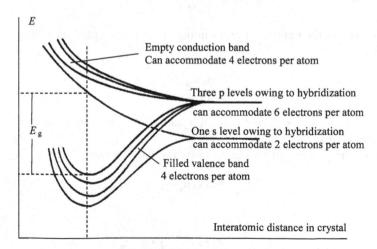

Figure 7.18 Correlation between energy levels in a free Si or Ge atom and a corresponding atom in the crystal lattice of Si or Ge.

The conduction band can also accommodate four electrons. The valence band is filled and the conduction band is empty at $T = 0$.

When the influence of all the valence electrons and the lattice atoms in the Si or Ge crystals is taken into consideration, it is easy to understand that the energy levels become wide bands in analogy with diamond, which has the same crystal structure as Si and Ge.

The interatomic distances in Si and Ge are larger and the energy gaps are considerably smaller than that in diamond (Table 7.9). Characteristics of all semiconductors are

- small energy gaps
- nearly filled valence bands at room temperature
- nearly empty conduction bands at room temperature.

Table 7.9 Energy gaps of some solids.

Material	Energy gap (eV)
Diamond	5.3
Si	1.1
Ge	0.7

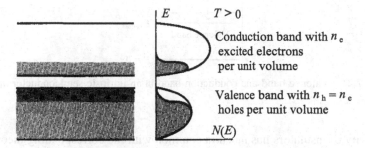

Figure 7.19 Energy bands and electron distribution in a pure semiconductor. Reproduced with permission from M. Alonso and E. Finn, *Fundamental University Physics*. © Addison-Wesley Publishing Company, Inc. (now Pearson Education).

At $T = 0$, all semiconductors are insulators. At temperatures $T > 0$, some electrons are excited up to the conduction band due to thermal excitation (Figure 7.19). In the conduction band many empty energy states are available for the electrons, which can easily move through the metal. They leave vacancies in the valence band that are called *holes*. The holes are also mobile and contribute to the current. They can be regarded as particles with a positive charge $+e$ and an effective mass similar but not equal to that of the electron.

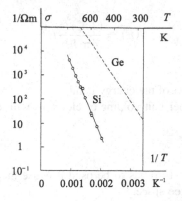

Figure 7.20 σ as a function of T and $1/T$ for silicon and germanium. Logarithmic scale. The resistivity of a semiconductor, which is the inverse of the conductivity, decreases with increasing temperature (Figure 7.8, page 349). C. Wehrt, *Physics of solids*, © McGraw-Hill, 1964.

The conductivity of an extremely pure semiconductor increases strongly with temperature (Figure 7.20). The reason is that the number of electrons in the conduction band and the corresponding vacancies or holes in the valence band increase exponentially as many electrons become thermally excited up to the conduction band at higher temperatures.

Pure semiconductors with the same number of electrons in the conduction band and holes in the valence band are called *intrinsic semiconductors*.

When semiconducting materials are *doped*, their electrical conductivities change. The possibility of controlling the conductivity is one of the two main conditions for the technical importance of semiconductors. Doped semiconductors will be discussed on page 369.

7.6.2 Theory of Electrical Conduction in Pure Semiconductors

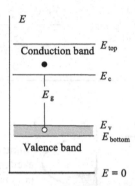

Figure 7.21   Valence band and conduction band in an intrinsic semiconductor. $n_e = n_h$.

The theory of the conductivity of insulators has much in common with the corresponding theory of semiconductors, but the simple model for insulators cannot be uncritically applied to semiconductors.

The reason is that the assumption we made on page 359 that the widths of the valence and conduction bands are much smaller than the width of the energy gap is *not* valid for a semiconductor. The theory has to be modified with respect to this fact in order to be applicable on an intrinsic semiconductor (Figure 7.21).

Calculation of Density Distribution of Electron Energy States and Electron Concentration in the Conduction Band

According to Equation (7.40) on page 358, the number of available energy states with energies between E and $E + dE$ at the bottom of a band can be written as

$$N(E) = \frac{(2m_e^*)^{3/2}}{4\pi^2\hbar^3} E^{1/2} \tag{7.40}$$

where
  E = energy of the electron above the bottom of the energy band
  $N(E)$ = number of available energy states per unit volume of electrons with energies between E and $E + dE$
  m_e^* = effective electron mass
  h = Planck's constant.

The effective mass m_e^* (Chapter 3, pages 145–146) has to be used because the mass of the free electrons in the solid is not the same as the mass of a single electron in free space.

$N(E)$ represents the number of available energy states per unit volume within the energy interval dE. The number of occupied energy states per unit volume within the interval dE is obtained by multiplying $N(E)$ by the Fermi factor f_{FD}.

The total number of electrons per unit volume in the conduction band is obtained by integration over an energy range equal to the width of the conduction band:

$$n_e = \int\limits_{E_c}^{E_{top}} 2N(E) f_{FD} dE \approx \int\limits_{E_c}^{\infty} 2N(E) \frac{1}{e^{\frac{E-E_F}{k_B T}} + 1} dE \tag{7.54}$$

Near the bottom of the conduction band (marked in the right-hand part of Figure 7.22), $N(E)$ is proportional to $(E - E_c)^{1/2}$. In Equation (7.54), we have replaced E_{top} with infinity as the upper integration limit. This is a reasonable approximation as the integrand is very small because the Fermi factor rapidly approaches zero in the upper part of the band, as shown in Figure 7.23. Combining Equations (7.40) and (7.54), we obtain

$$n_e = \int\limits_{E_c}^{\infty} 2 \frac{(2m_e^*)^{3/2}}{4\pi^2 \hbar^3} (E - E_c)^{1/2} \frac{1}{e^{\frac{E-E_F}{k_B T}} + 1} dE \tag{7.55}$$

The Fermi level E_F is roughly half way between E_c and E_v. Hence it is reasonable to assume that $E - E_F > 4k_B T$. In this case, the second term in the denominator of the Fermi factor can be neglected in comparison with the exponential term and we obtain

$$n_e = \int\limits_{E_c}^{\infty} 2 \frac{(2m_e^*)^{3/2}}{4\pi^2 \hbar^3} (E - E_c)^{1/2} e^{-\frac{E-E_F}{k_B T}} dE \tag{7.56}$$

If we replace $-(E - E_F)$ by $-(E - E_c) - (E_c - E_F)$ and place constant factors outside the integral sign, Equation (7.56) can be written as

$$n_e = 2 \frac{(2m_e^*)^{3/2}}{4\pi^2 \hbar^3} e^{-\frac{E_c - E_F}{k_B T}} \int\limits_{E_c}^{\infty} (E - E_c)^{1/2} e^{-\frac{E-E_c}{k_B T}} dE \tag{7.57}$$

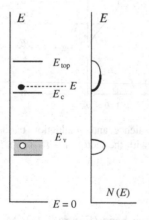

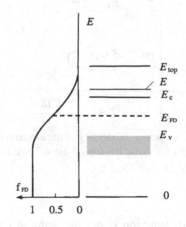

Figure 7.22 Sketch of the density of energy states in a semiconductor. Near the bottom of the conduction band $N(E)$ is proportional to $(E - E_c)^{1/2}$.

Figure 7.23 Valence and conduction bands of a semiconductor together with the Fermi factor f_{FD} as a function of E.

A change of dE to $d(E - E_c)$ gives

$$n_e = 2 \frac{(2m_e^*)^{3/2}}{4\pi^2 \hbar^3} e^{-\frac{E_c - E_F}{k_B T}} \int\limits_0^{\infty} (E - E_c)^{1/2} e^{-\frac{E-E_c}{k_B T}} d(E - E_c) \tag{7.58}$$

The integral in Equation (7.58) is of the type

$$\int_0^\infty x^{1/2}e^{-x}dx = \frac{\sqrt{\pi}}{2}$$

We introduce $x = (E - E_c)/k_BT$ and $dx = d(E - E_c)/k_BT$ into the integral in Equation (7.58). Its value is calculated as $(k_BT)^{3/2}\pi^{1/2}/2$ and after reduction we obtain the *electron concentration*(number of electrons per unit volume) *in the conduction band*:

$$n_e = \frac{(2\pi m_e^* k_BT)^{3/2}}{4\pi^3\hbar^3}e^{-\frac{E_c-E_F}{k_BT}} \tag{7.59}$$

The larger the difference $(E_e - E_F)$ is, the smaller will be n_e.

Calculation of Density Distribution of Hole Energy States and Hole Concentration in the Valence Band

To find the density distribution of hole energy states and the hole concentration in the valence band, we will perform analogous calculations to those for the electron concentration in the conduction band.

The holes can be regarded as missing electrons in the filled valence band and can be calculated as the difference between the *available* electron energy levels and the *occupied* electron energy levels within the valence band. If the electron distribution in the top of the valence band is denoted $N'(E)$, we obtain

$$dn_h = 2N'(E)dE - 2N'(E)f_{FD}dE = 2N'(E)(1 - f_{FD})dE \tag{7.60}$$

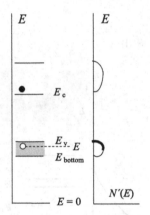
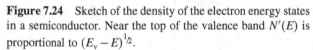

Figure 7.24 Sketch of the density of the electron energy states in a semiconductor. Near the top of the valence band $N'(E)$ is proportional to $(E_v - E)^{1/2}$.

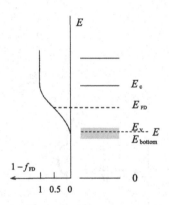

Figure 7.25 Valence and conduction bands of a semiconductor together with the factor $(1 - f_{FD})$ as a function of E.

The electron distribution is not the same at the top and bottom of the valence band (Figure 7.24). Most of the holes are located to the top of the band, where the distribution of electron energy states per unit volume $N'(E)$ is proportional to $\sqrt{E_v - E}$, in analogy with the corresponding value for the conduction band $(E - E_c)^{1/2}$ for symmetry reasons. Hence the total number of holes per unit volume in the valence band is obtained from the integral

$$n_h = 2\int_{E_{bottom}}^{E_v} N'(E)(1 - f_{FD})dE \approx 2\int_{E_{bottom}}^{E_v} N'(E)\left(1 - \frac{1}{e^{\frac{E-E_F}{k_BT}} + 1}\right)dE \tag{7.61}$$

The number of holes decreases rapidly towards the bottom of the band. Hence a change of the lower integral limit from E_{bottom} to $-\infty$ is a reasonable approximation as the integrand is very small because the factor $1 - f_{FD}$ rapidly approaches zero in the lower part of the band, as can be seen in Figure 7.25. Hence Equation (7.61) can be written approximately as

$$n_h = 2 \int_{-\infty}^{E_v} \frac{(2m_h^*)^{3/2}}{4\pi^2 \hbar^3} (E_v - E)^{1/2} \left(1 - \frac{1}{e^{\frac{E-E_F}{k_B T}} + 1}\right) dE \tag{7.62}$$

A change of dE to $-d(E_v - E)$ gives

$$n_h = 2 \int_{\infty}^{0} \frac{(2m_h^*)^{3/2}}{4\pi^2 \hbar^3} (E_v - E)^{1/2} \frac{e^{\frac{E-E_F}{k_B T}}}{1 + e^{\frac{E-E_F}{k_B T}}} [-d(E_v - E)] \tag{7.63}$$

As $(E - E_F)T < -4k_B T$, the second term in the denominator of the 'Fermi factor' is very small in comparison with the first term and can be neglected. The minus sign is cancelled by change of the integration limits:

$$n_h = 2 \int_{0}^{\infty} \frac{(2m_h^*)^{3/2}}{4\pi^2 \hbar^3} (E_v - E)^{1/2} e^{\frac{E-E_F}{k_B T}} d(E_v - E) \tag{7.64}$$

If we replace $E - E_F$ by $-(E_F - E_v) - (E_v - E)$ and place constant factors outside the integral sign, Equation (7.64) can be written as

$$n_h = 2 \frac{(2m_h^*)^{3/2}}{4\pi^2 \hbar^3} e^{-\frac{E_F - E_v}{k_B T}} \int_{0}^{\infty} (E_v - E)^{1/2} e^{-\frac{E_v - E}{k_B T}} d(E_v - E) \tag{7.65}$$

We introduce $x = (E_v - E)/k_B T$ and $dx = d(E_v - E)/k_B T$, which gives

$$n_h = 2 \frac{(2m_h^*)^{3/2}}{4\pi^2 \hbar^3} e^{-\frac{E_F - E_v}{k_B T}} (k_B T)^{3/2} \int_{0}^{\infty} (x)^{1/2} e^{-x} dx \tag{7.66}$$

The integral in Equation (7.66) is the same as that in Equation (7.58). It has the value $\sqrt{\pi}/2$ and after reduction the final result, i.e. the *concentration of holes* (number of holes per unit volume) *in the valence band* can be written as

$$n_h = \frac{(2\pi m_h^* k_B T)^{3/2}}{4\pi^3 \hbar^3} e^{-\frac{E_F - E_v}{k_B T}} \tag{7.67}$$

The larger the difference $E_F - E_v$ is the smaller will be n_h.

Calculation of the Fermi Level in an Intrinsic Semiconductor

In an intrinsic semiconductor, the number of holes in the valence band and excited electrons in the conduction band are equal:

$$n_e = n_h \tag{7.68}$$

When the expressions in Equations (7.59) and (7.67) are introduced into Equation (7.68), we obtain

$$\frac{(2\pi m_e^* k_B T)^{3/2}}{4\pi^3 \hbar^3} e^{-\frac{E_c - E_F}{k_B T}} = \frac{(2\pi m_h^* k_B T)^{3/2}}{4\pi^3 \hbar^3} e^{-\frac{E_F - E_v}{k_B T}} \tag{7.69}$$

After division with common factors and taking the logarithm of both sides of Equation (7.69), we obtain

$$\frac{3}{2} \ln m_e^* - \frac{E_c - E_F}{k_B T} = \frac{3}{2} \ln m_h^* - \frac{E_F - E_v}{k_B T}$$

E_F is solved from the last equation:

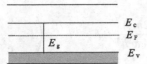

Figure 7.26 Energy levels in an intrinsic semiconductor. $m_h^* > m_e^*$ and $E_F > (E_v + E_g)/2$.

$$E_F = \frac{E_c + E_v}{2} + \frac{3}{4} k_B T \ln \frac{m_h^*}{m_e^*} \tag{7.70}$$

If the effective masses of the electrons and holes are equal, the last term in Equation (7.70) will be zero and the Fermi level lies exactly half way between the top of the valence band and the bottom of the conduction band, i.e. in the middle of the energy gap $E_g = (E_c - E_v)/2$.

In general, the effective mass of the holes in the valence bands is larger than that of the electrons in the conduction band and E_F is somewhat higher than half way between E_c and E_v (Figure 7.26).

Electrical Conductivities of Intrinsic Semiconductors

A comparison between Equations (7.59) and (7.67) shows, that the expressions for n_e and n_h are not exactly equal. If we form the expression $n_i = \sqrt{n_h n_e}$ and insert the expressions (7.59) and (7.67), we obtain after reduction

$$n_i = \sqrt{n_h n_e} = \frac{(2\pi k_B T m)^{3/2}}{4\pi^3 \hbar^3} \left(\frac{m_e^*}{m} \frac{m_h^*}{m} \right)^{3/4} e^{-\frac{E_g}{2k_B T}} \tag{7.71}$$

where m is the mass of a free electron outside the metal.

Provided that $m_e^* \approx m_h^*$, then

$$n_e \approx n_h \approx n_i \tag{7.72}$$

and it is reasonable to replace n_e and n_h by n_i in the general expression for σ [Equation (7.53) on page 361]:

$$\sigma = n_i e (\mu_e + \mu_h) \tag{7.73}$$

This theory is valid for both pure semiconductors and pure insulators and has to be used when the widths of the valence and conduction bands cannot be neglected in comparison with the energy gap. Equations (7.73) and (7.71) are very useful for the calculation of electron and hole concentrations from measurements of conductivities.

Calculation of the Energy Gap of an Intrinsic Semiconductor

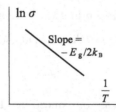

Figure 7.27 Derivation of E_g.

If the expression (7.71) for n_i is inserted into Equation (7.73), we obtain

$$\sigma = constant \times (\mu_e + \mu_h) T^{3/2} e^{-\frac{E_g}{2k_B T}} \tag{7.74}$$

If $\ln \sigma$ is plotted as a function of $1/T$, a straight line will probably be obtained as $\ln T^{3/2}$ changes slowly in comparison with the function $1/T$ and the mobilities are not very temperature dependent. E_g can be derived from the slope of the straight line.

7.6.3 Doped Semiconductors

Extremely pure semiconductors are called intrinsic semiconductors because their electrical properties are a consequence of the inherent nature of the elements themselves.

If impurities are added to intrinsic semiconductors, their electrical properties change radically. Semiconductors which contain impurities are called *extrinsic semiconductors* because their electrical properties are controlled by the nature and amounts of impurities added.

For this reason, is is necessary that the basic semiconductor is extremely free from impurities. If this condition is fulfilled then controlled amounts of impurities can be added to the semiconductor in order to design its electrical properties in a desired way. The semiconductor is said to be *doped* and the impurity is called *dopant*.

If the semiconductor is doped with elements in group 5 in the periodic table there will be an *excess* of valence electrons. The semiconductor is said to be *n-doped* (n = negative). If the dopant comes from group 3 it is said to be *p-doped*. In this case there will be a *lack* of valence electrons, which we call holes. The holes behave like positive charges (p = positive).

7.6.4 Theory of Electrical Conductivity in Doped Semiconductors

Even if the amounts of dopants added to a pure semiconductor are very small the concentrations of the excess electrons n_e or excess holes n_h will be much larger than the concentration n_i in a pure semiconductor. Hence the intrinsic contribution to the electrical conductivity can normally be neglected. In this case the general Equation (7.53) on page 361 is valid.

The concentrations of electrons and holes are *not* equal in a doped semiconductor but obey the relationship

$$n_e n_h = n_i^2 \tag{7.75}$$

Normally the concentration of the dopant, for example n_e, in a doped semiconductor is known. Then the concentration of holes can be calculated with the aid of Equation (7.75). The higher the dopant concentration is, the lower will be the concentration of the holes.

Provided that $n_e \gg n_h$, the latter can be neglected and Equation (7.53) will be simplified to

$$\sigma = n_e e \mu_e + n_h e \mu_h \implies \sigma = n_e e \mu_e \tag{7.76}$$

If the indices e and h are exchanged we realize that the same arguments are valid for a p-doped semiconductor, which gives

$$\sigma = n_e e \mu_e + n_h e \mu_h \implies \sigma = n_h e \mu_h \tag{7.77}$$

If the condition $n_e \gg n_h$ or the reverse is not fulfilled, both contributions must be included and Equation (7.53) has to be used.

Influence of Temperature and Dopant Concentration on the Number of Charge Carriers

Temperature and dopant concentration have a strong influence on the number of current carriers, i.e. the sum of free electrons and holes, in a doped semiconductor. To illustrate this, we choose B-doped Si semiconductors with various degrees of doping as an example.

Figure 7.28 shows the total number of current carriers as a function of $1/T$ for silicon doped with boron. The B concentrations are given in at-%. From Figure 7.28, we can conclude that

1. The number of charge carriers per unit volume is remarkably even at very low temperatures. Even extremely pure semiconductors contain impurities.
2. The steep line of the lowest curve above 300 °C corresponds to an intrinsic excitation of electrons. The energy gap can be derived from the slope of the line. It amounts to about 1.1 eV.
3. At low dopant concentrations, all the electrons and holes are not excited at low temperatures. When the temperature increases, more and more electrons become excited into the conduction band and leave holes in the valence band. Hence the number of charge carriers that contribute to the current increases strongly.

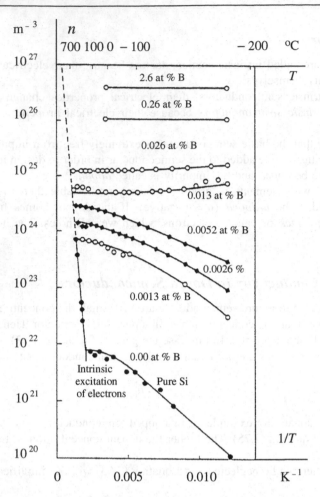

Figure 7.28 Number of charge carriers per m³ as a function of temperature in a B-doped Si semiconductor. The percentages on the curves represent the degree of doping in at-%. The lowest curve shows that impurities are present even in pure Si. C. Wehrt, *Physics of Solids*, © McGraw-Hill, 1964.

4. At high temperatures, all the charge carriers are activated and the curves becomes horisontal, i.e. independent of the temperature.

5. At high degrees of doping, the number of dopants is much greater than the number of intrinsic, thermally activated pairs of electrons and holes. The number of current carriers depends only on the dopant concentration and not on the temperature. The number of charge carriers is roughly proportional to the dopant concentration.

7.6.5 Types of Doped Semiconductors. Acceptor and Donor Energy Levels

n-Doped or n-Type Semiconductors

If the semiconductor is doped with very small amounts of an element in group 5 of the periodic table, the conductivity increases drastically because the dopant supplies additional electrons in the conduction band of the semiconductor in addition to the available thermally excited electrons. Simultaneously the lattice becomes ionized to some extent when the dopant atoms are dissolved substitutionally, i.e. replace semiconductor atoms. An example is given in Figure 7.29.

Elements in group 5 in the periodic table have five outer electrons. If an Si atom, for example, is replaced by a P atom, four of its outer electrons replace the normal valence electrons of the missing Si atom and the fifth electron is excited up to the conduction band and is able to move freely within the lattice if the available thermal energy is large enough. If the energy is too low, the fifth electron is instead trapped in an orbit around the site with the P atom (Figure 7.29a). At very low temperatures the fifth electrons of the impurity atoms do not increase the conductivity of the semiconductor.

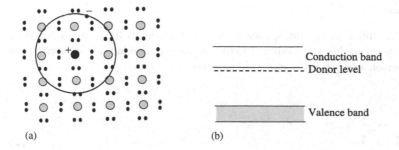

(a)   (b)

Figure 7.29   (a) The covalent bonds in an Si lattice are indicated schematically by electron pairs. The P site in the crystal lattice has a charge +e compared with the Si sites. Four of the five outer electrons of the P atom are involved in covalent bonds and the fifth electron is bound to the P site (as in the figure) or excited into the conduction band as a free electron.
(b) Donor level in an n-doped semiconductor. The donor level has excited states but these are not indicated in the figure for the sake of simplicity.

The excitation energy required for exciting the fifth electron has been found to be *much lower* than that which corresponds to the energy gap in the semiconductor. The extra electrons obviously stay in a special energy level close to the lower part of the conduction band (Figure 7.29b). This energy level is called *donor level*.

p-Doped or p-Type Semiconductors

If the semiconductor is doped with very small amounts of an element in group 3 in the periodic table, the element dissolves substitutionally in the crystal lattice of the semiconductor.

In an intrinsic semiconductor, the number of thermally excited electrons in the conduction band is equal to the number of holes in the valence band. In an n-doped semiconductor there are many more mobile electrons than holes; in a p-doped semiconductor the opposite is true. Elements in group 3 in the periodic table have three outer electrons. If an Si atom, for example, is replaced by a B atom the three outer electrons replace the normal valence electrons of the missing Si atom and the absent fourth electron corresponds is a lack of an electron and behaves as a hole with a charge +e (Figure 7.30a).

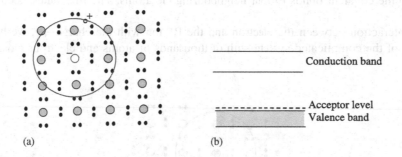

(a)   (b)

Figure 7.30   (a) The covalent bonds in an Si lattice are indicated schematically by electron pairs. The B site in the crystal lattice has a charge −e compared with the Si sites. The three outer electrons of the B atom are involved in covalent bonds and the hole is bound to the B site (as in the figure) or excited down into the valence band as a free hole.
(b) Acceptor level in a p-doped semiconductor. The acceptor level has excited states but these are not indicated in the figure for the sake of simplicity.

The holes can be excited down to the valence band and be able to move freely within the lattice if the available thermal energy is large enough. If the energy is too low, the hole is instead trapped in an orbit around the site with the B atom which has a lower charge (−e) than the surrounding Si atoms. At very low temperatures, the holes of the impurity atoms do not increase the conductivity of the semiconductor.

The excitation energy required for exciting a hole in a p-doped semiconductor into the valence band has been found to be of the same magnitude as the energy required to excite an electron from the donor level into the conduction band in an n-doped semiconductor. Hence the holes obviously stay in a special energy level close to the valence band (Figure 7.30b). This energy level is called *acceptor level*. Electrons in the upper part of the valence band become exited into the acceptor level and leave mobile holes in the valence band.

Compound Semiconductors

Some semiconductors are chemical compounds, usually of one component from group 3 and one element from group 5 in the periodic table. Common examples are InSb and GaAs. Such semiconductors are ionized as is shown in Figure 7.31a. Figure 7.32 gives a comparison with an ordinary semiconductor.

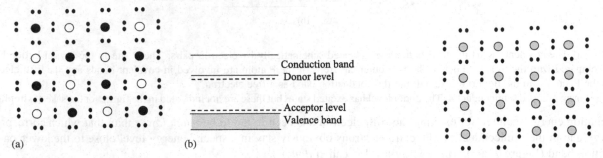

(a)   (b)

Figure 7.31   (a) Compound semiconductor. 50% of the sites are positive and 50% are negative. (b) Energy levels in a compound semiconductor. Both acceptor and donor levels are present.

Figure 7.32   Intrinsic Si or Ge semiconductor.

Compound semiconductors have both donor levels and acceptor levels (Figure 7.31b). If GaAs is doped with Si, a Si atom can occupy a Ga site and act as an n-dopant. On the other hand, if a Si atom ocupies an As site, it is a p-dopant.

Calculation of Donor and Acceptor Energy Levels in Doped Semiconductors

Consider an impurity atom in a semiconductor, for example a P atom in a Ge lattice. Phosphorus has five outer electrons and four of them contribute to the covalent bonds to four neighbouring Ge atoms. The fifth outer electron is bound to the P^+ ion at low temperatures.

In order to study the interaction between the electron and the P^+ ion with the charge $+e$, we have to make considerable and rough simplifications of the complicated system with of thousands of atoms and electrons closely involved.

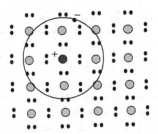

Figure 7.33   A bound phosphorus–electron system.

1. The bound system is analogous with the hydrogen atom. The simple Bohr model will be used as an approximation for calculation of its energy levels. The only difference is that the space within the bound system is filled with a dielectric medium.
2. The Ge sites in the lattice and the electron pair bonds are considered as an electrically neutral continuum with a relative dielectric capacitivity ε_r. The value of ε_r is specific for the semiconductor material.

The Bohr Model of the Hydrogen Atom

Derivation of the Radius of the Electron Orbit

The attraction force between the electron and the proton is

$$\frac{1}{4\pi\varepsilon_0}\frac{e^2}{r_n^2} = \frac{mv^2}{r_n} \tag{1'}$$

The quantization condition of the orbital angular momentum can be written as

$$L = mvr_n = n\frac{h}{2\pi} \tag{2'}$$

Combining Equations (1') and (2'), we obtain

$$\frac{1}{4\pi\varepsilon_0}e^2 r_n = \frac{m^2 v^2 r_n^2}{m} \quad \Rightarrow \quad r_n = \frac{4\pi\varepsilon_0}{me^2}\left(\frac{h}{2\pi}\right)^2 n^2 \quad \text{or} \quad r_n = r_0 n^2 \tag{3'}$$

where

$$r_0 = \frac{4\pi\varepsilon_0}{me^2}\left(\frac{h}{2\pi}\right)^2 \tag{4'}$$

Kinetic, Potential and Total Energy of the Hydrogen Atom

The proton is supposed to be at rest. The kinetic energy of the electron can be written [from Equation (1)] as

$$E_{kin} = \frac{mv^2}{2} = \frac{1}{4\pi\varepsilon_0}\frac{e^2}{2r_n} \tag{5'}$$

The potential energy of the system is

$$E_{pot} = (-e)\frac{1}{4\pi\varepsilon_0}\frac{e}{r_n} \tag{6'}$$

The total energy of the electron–proton system is equal to the sum of Equations (5') and (6'). In combination with Equations (3') and (4'), we obtain

$$E_n = -\frac{1}{8\pi\varepsilon_0}\frac{e^2}{r_n} = -\frac{1}{8\pi\varepsilon_0}\frac{e^2}{n^2}\frac{me^2}{4\pi\varepsilon_0}\left(\frac{2\pi}{h}\right)^2$$

or

$$E_n = -\frac{E_0}{n^2} \tag{7'}$$

where

$$E_0 = \frac{me^4}{8h^2\varepsilon_0^2} \tag{8'}$$

Analogous calculations to those for the hydrogen atom (given in the box) give the following equations of the energy levels and the smallest radius of the bound system:

$$E_n^* = -\frac{m_e^* e^4}{8h^2 \varepsilon_r^2 \varepsilon_0^2} \frac{1}{n^2} = -E_n \frac{m_e^*/m}{\varepsilon_r^2} \tag{7.78}$$

$$r_n^* = \frac{4\pi\varepsilon_r\varepsilon_0}{m_e^* e^2} \left(\frac{h}{2\pi}\right)^2 n^2 = r_n \frac{\varepsilon_r}{m_e^*/m} \tag{7.79}$$

where E_n and r_n refer to the energy level n and the Bohr radius n of the hydrogen atom, which have been derived in the box.

The relative dielectric capacity ε_r for the semiconductor is given in Table 7.10. It is more complicated to find an adequate value of the effective mass of the electron m_e^* because it is not a constant but depends on the direction in space. All directions in solids are not equivalent. An example is graphite, where π-binding electrons are free to move along the planar layer of carbon rings but not perpendicular to them. In this case the orbit of the electron is an ellipsis rather than a circle.

Numerical calculations of the ground state, which involve the anisotropic mass of the electron, have been performed for both Si and Ge. The obtained values of the binding energies correspond to weighted scalar average values of the longitudinal and transverse effective masses of the electron in Si and Ge. These values are also used for the calculation of the radius of the ground state orbit. The values obtained are given in Table 7.10.

Table 7.10 Data for a bound H-like system of P atoms in a semiconductor.

Semiconductor	Relative dielectric constant $\varepsilon_r = \varepsilon/\varepsilon_0$	Ratio m_e^*/m	Energy of ground state (eV)	Radius of ground state orbit (nm)
Si	12	0.31	−0.0092	2.3
Ge	16	0.17	−0.029	5.4

The radius of the electron in its ground state is surprisingly large. Within the orbit thousands of semiconductor sites are included. The model of the semiconductor as a continuous dielectric medium instead of a point lattice is obviously very reasonable.

The binding energies, which represent the energy of the donor level in Si and Ge, close to the bottom of the conduction band in the semiconductor, are very small. The calculated values agree comparatively well with experimental values.

The values of the experimental binding energies are similar for all group 5 impurity atoms, for example P, As and, Sb. For Si, the values are about 0.011 eV and for Ge about 0.045 eV (Tables 7.11 and 7.12).

Table 7.11 Parameters of doped silicon.

Semiconductor	Dopant	Experimental binding energy (eV)
Si	Donors (group 5):	
	P	0.0120
	As	0.0127
	Sb	0.0096
$\varepsilon_r = 12$	Acceptors (group 3):	
	B	0.0104
	Al	0.0102
	Ga	0.0108
	In	0.0112

Similar calculations have been performed for group 3 elements as dopants in semiconductors with similar results (Tables 7.11 and 7.12). In this case the H-like bound system consists of a negative nucleus, for example B, Ga or In, and a hole (Figure 7.34). The binding energies correspond to the energies of the acceptor levels close to the top of the valence band in

Table 7.12 Parameters of doped germanium.

Semiconductor	Dopant	Experimental binding energy (eV)
Ge	Donors (group 5):	
	P	0.044
	As	0.049
	Sb	0.039
$\varepsilon_r = 16$	Acceptors (group 3):	
	B	0.045
	Al	0.057
	Ga	0.065
	In	0.160

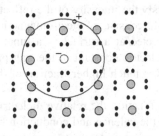

Figure 7.34 A bound boron–hole system.

the semiconductor. The experimental values for group 3 dopants in silicon are about 0.050 eV for B, Ga and higher for In (0.16 eV). The experimental acceptor levels for B, Al, Ga and In in germanium are all close to 0.011 eV.

The experimental values are derived from measurements of the temperature dependence of the conductivity of doped semiconductors at low temperatures or by optical absorption methods.

The binding energies of the bound systems are very small and hence the donor and acceptor levels are very close to the allowed conduction and valence bands.

The conductivity of a semiconductor varies strongly with the concentration of the dopants, as was shown in Figure 7.28 on page 370.

7.7 Optical Properties of Solids

The optical properties of solids vary considerably depending on the type of solid. The reason is that the properties of a solid depend entirely of its structure. This statement has already been shown in this and earlier chapters, for example when magnetic properties were discussed in Chapter 6 and when the colour centres in ionic crystals were treated in Chapter 3.

Below we will briefly discuss the properties of different types of solids when they are exposed to light or other electromagnetic radiation. The solids will be classified according to their electrical conduction properties, i.e. as metals, insulators and semiconductors. The section ends with a discussion about the interaction between polarized light and insulators.

7.7.1 Optical Properties of Metals

Metals are opaque and many metals reflect light very well. These topics can be explained with the aid of the properties of metals which have been discussed in earlier chapters.

Transparency and Absorption

In Section 7.4 on page 350, the positions of the valence and conduction bands of solid metals were discussed. Figures 9.9a and 7.10b show sketches of the positions of some of the outer energy bands in a metal. Each metal is characterized by its own widths and positions of the energy bands, but for a general discussion the figures are sufficient.

Figure 7.9a shows that in monovalent metals the valence band is only half filled and that the nearest conduction band is very close to the valence band or the two bands overlap. Figure 7.10b shows that the conduction bands become wider and wider with increasing energy and often overlap.

In addition, the figures show that photons of many energies can be absorbed, extending from the top of the valence band to the bottom of the conduction band, from the bottom of the valence band to the top of the conduction band and all energies between the two.

The photons of visible light have energies of 1.8–3.2 eV, which is of the magnitude of the gap between the energy bands in all metals. Hence it is easy to realize that all photons within this energy interval can be absorbed. This is the reason why metals are opaque and not transparent.

Reflection

Most metals reflect a shiny or grayish non-coloured light. Exceptions are the monovalent metals copper and gold, which show a selective reflection in red and yellow. Obviously the intensity of the reflected light depends on the wavelength.

The most useful model in this case is that light consists of electromagnetic waves. As is discussed in next section on page 379, the fraction of the radiation which is reflected at a surface is a function of the angle of incidence and the refractive index of the material, which depends on the dielectric constant of the transparent material. These equations are not valid for conductors. As no light is transmitted in metals, a much higher fraction of the incident light is reflected than for transparent materials.

The reflection of light at a metal surface is a function the relative dielectric constant ε_r, which varies with the wavelength of the light. This variation and the positions of the valence and conduction bands are the reasons why the reflected light has different composition for different metals. Copper has an intensity maximum in the red spectral region. The dominant yellow region gives gold its colour. Most metals reflect all colours and the reflected light has no specific colour.

7.7.2 Optical Properties of Semiconductors

Like metals, semiconductors have a wide valence band and a still wider conduction band. The difference is that the conduction band is separated from the valence band by an energy gap of the magnitude 1 eV (Figure 7.9b on page 349) without the overlap that is common in metals.

Different types of semiconductors have been described in Sections 7.6.3 and 7.6.5. Semiconductors such as Si and Ge doped with elements in groups 3 and 5 and many semiconductors of combined group 3 and 5 elements in the periodical table are used in industry.

Transparency and Absorption

Many semiconductors are transparent in the visible region of the electromagnetic spectrum but show absorption in the infrared region.

Determination of the Energy Gap of a Semiconductor

The energy gap of a semiconductor can be determined experimentally by studying its electrical conductivity as a function of temperature (page 368). One of the best methods for accurate determination of the energy gaps of semiconductors is measurement of the absorption coefficient as a function of wavelength or photon energy.

A semiconductor specimen is exposed to monochromatic infrared radiation and intensity measurements of the light before and after the passage through the specimen are performed:

$$I = I_0 e^{-\alpha x} \tag{7.80}$$

where
  I = intensity after passage of the specimen
  I_0 = intensity before passage of the specimen
  α = absorption coefficient
  x = thickness of the specimen.

The wavelength is varied and for each value of photon energy the absorption coefficient α is calculated from the experimental measurements and plotted as a function of λ. Figure 7.35b gives an example of such a diagram. Obviously α varies strongly with the wavelength of the radiation. The temperature is kept low to avoid thermal excitation.

The appearance of the resulting diagrams depends on the positions and shape of the bands. Two examples of band shapes are shown in Figures 7.35a and 7.36a. Figures 7.35b, 7.36b and 7.37 show some typical examples of absorption curves. The band shapes and the absorption curves depend strongly on the temperature.

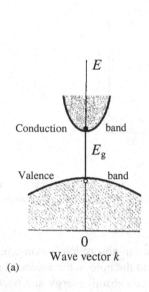

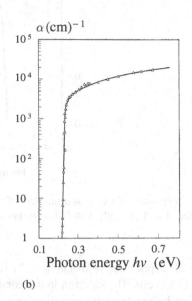

Figure 7.35   (a) Sketch of the valence and conduction bands of a semiconductor with a direct energy gap. (b) Absorption coefficient as a function of the photon energy for InSb at 5 K (after G. W. Gobeli and H. Y Fan). Adapted with permission from C. Kittel, *Introduction to Solid State Physics*, 6th edn. © 1986 John Wiley & Sons, Inc.

In the simplest case (Figure 7.35a), the energy gap E_g is the minimum energy and the transition is vertical, i.e. involves no change of the **k** vector at the excitation of the electron. Figure 7.35b shows the absorption curve for InSb, which has a *direct* energy gap, i.e. is of the type shown in Figure 7.35a.

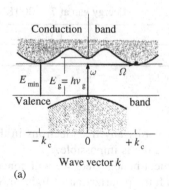

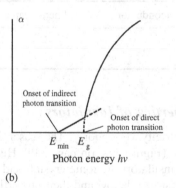

Figure 7.36   (a) Sketch of the valence and conduction band in a semiconductor with an indirect energy gap. (b) The absorption constant as a function of the photon energy in case of an indirect energy gap. Adapted with permission from C. Kittel, *Introduction to Solid State Physics*, 6th edn. © 1986 John Wiley & Sons, Inc.

A type of absorption curve of a semiconductor with an *indirect* energy gap is sketched in Figure 7.36a. Transitions with lower energy than E_g are possible only if the electron changes its **k** value, i.e. an electron, a photon ω and a phonon Ω are involved and the energy and momentum (right part of Figure 7.36a) are conserved at the excitation of the electron. Such absorption processes result in energy absorption at lower energies than E_g. The energy and momentum of the phonon are *much* smaller than those of the photon.

The value of the energy gap can be extrapolated from the photon absorption part of the curve. Germanium belongs to this type of semiconductor with an indirect energy gap.

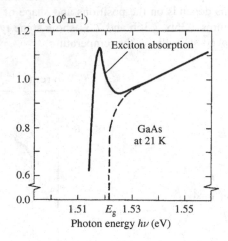

Figure 7.37   The absorption constant of GaAs as a function of the photon energy near the band gap region, measured at 21 K. Reproduced with permission from H. Ibach and H. Luth, *Solid-State Physics – An Introduction to Theory and Experiment.* © Springer-Verlag, Berlin Heidelberg 1991.

Figure 7.37 shows an absorption curve of GaAs at 21 K. The curve has an absorption maximum, at lower energy than E_g, before it decreases steeply to zero. The electron in the conduction band and the hole in the valence band form a bound system, an exciton, bound together by electrostatic attraction forces. The exciton can absorb energy and become excited. Excitons are identical with the bound systems which have been described on pages 371–375.

If the effect of the exciton absorption is subtracted, the normal photon absorption remains (the dashed curve in Figure 7.37) from which E_g can be determined. GaAs is a semiconductor with an indirect energy gap.

All absorption curves are influenced by the temperature. The derived E_g values become smaller when the temperature increases. An example is given in Table 7.13.

Table 7.13   Values of the energy gap for some semiconductors at $T = 0$ and 300 K.

Semiconductor	Energy gap at $T = 0\,K$ (eV)	Energy gap at $T = 300\,K$ (eV)
Si	1.17	1.11
Ge	0.74	0.66

7.7.3 *Optical Properties of Insulators*

Insulators conduct electricity very poorly. The reason is that they have hardly any free electrons in their lattices and that the energy gap is very high (Figure 7.9c on page 349). Hence thermal excitation is impossible.

The main groups of insulators are ionic crystals and solids with covalent bonds. Here we will concentrate on ionic crystals and discuss their transparency below and their properties connected with the polarization of light in Section 7.7.4.

Transparency and Absorption

Ionic crystals are transparent in the visible and infrared regions of the electromagnetic spectrum. This is fully explained by their high energy gaps between the valence band and empty conduction band. As no absorption is possible in the visible region, all crystals ought to be colourless, but this is not always the case.

The observed intense colours of some types of crystals can easily be explained. They are caused by lattice defects. Special wavelengths in the visible region, characteristic of each type of crystals, are absorbed by so-called colour centres, which results in strongly coloured crystals (page 152–153 in Chapter 3).

7.7.4 *Polarized Light and Insulators*

Reflection of Light at a Planar Surface

When electromagnetic radiation strikes the surface of a solid material, part of it will be reflected and the rest enters the interior of the solid. The fraction of the radiation which is reflected depends on the material properties and the angle of incidence.

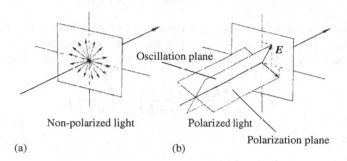

Figure 7.38 (a) Nonpolarized light: The *E* vector oscillates in all directions perpendicularly to the propagation direction. (b) The *E* vector oscillates in the *oscillation plane*. The *polarization plane* is perpendicular to the oscillation plane. Reproduced with permission. © O. Beckman.

The intensity of the incident non-polarized beam (Figures 7.38 and 7.39) is I_0. The intensity of the reflected light is given by the so-called Fresnel's equations

The intensity of the components of the *reflected* light is

$$I_{r\perp} = \frac{I_0}{2} \frac{\sin^2(i-b)}{\sin^2(i+b)} \tag{7.81}$$

$$I_{r\|} = \frac{I_0}{2} \frac{\tan^2(i-b)}{\tan^2(i+b)} \tag{7.82}$$

where

I_0 = intensity of the incident beam

$I_{r\perp}$ = intensity of the part of the light which (the *E* vector) oscillates perpendicularly to the plane of incidence (Figure 7.39)

$I_{r\|}$ = intensity of the part of the light which (the *E* vector) oscillates in the plane of incidence (Figure 7.39)

i = angle of incidence (Figure 7.39)

b = angle of refraction (Figure 7.39).

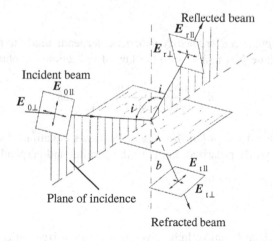

Figure 7.39 Reflection and refraction of an incident beam. $E_0 = E$ vector of incident beam; $E_r = E$ vector of reflected beam; $E_t = E$ vector of transmitted (refracted) beam. Symbols $\perp$ and $\|$ refer to the plane of incidence. Reproduced with permission. © O. Beckman.

Special Case 1: Angle of Incidence = 0

If the incident beam is perpendicular to the surface $i = b = 0$. In this case $\sin i \sim i$ and $\sin b \sim b$ (the angles are measured in radians) and we obtain

$$I_\perp = I_{\parallel} = \frac{I_0}{2} \frac{(i-b)^2}{(i+b)^2} = \frac{I_0}{2} \left(\frac{i-b}{i+b} \right)^2 \tag{7.83}$$

Provided that light in vacuum or air is reflected towards the surface of a medium with a refractive index n, Snell's law of refraction can be written: $\sin i = n \sin b$. If the angle i is small, the angle b is also small, and the relationship can be written $i = nb$. The total intensity of the fraction R of the incident light which is reflected can be written as

$$R = \frac{I_{r\perp} + I_{r\,\parallel}}{I_0} = \left(\frac{i-b}{i+b} \right)^2 = \left(\frac{n-1}{n+1} \right)^2 \tag{7.84}$$

where n is the refractive index.

Special Case 2: Angle of Incidence = Polarizing Angle

If the reflected and the refracted beams are perpendicular, i.e. $i + b = \pi/2$, Equations (7.81) and (7.82) can be written as

$$I_{r\perp} = \frac{I_0}{2} \sin^2(i - b) \tag{7.85}$$

$$I_{r\,\parallel} = 0 \tag{7.86}$$

as $\sin(i + b) = \sin \pi/2 = 1$ and $\tan(i + b) = \tan \pi/2 = \infty$.

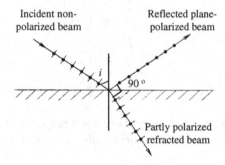

Figure 7.40 Oscillation directions of the reflected and refracted (transmitted) beams when i is the polarizing angle. Reproduced with permission. © O. Beckman.

Hence the *reflected* beam is *plane polarized* or *linearly polarized* perpendicularly to the plane of incidence, which is shown in Figure 7.40. If we introduce the value $b = \pi/2 - i$ into the law of refraction, we obtain

$$n = \frac{\sin i}{\sin b} = \frac{\sin i}{\sin \left(\frac{\pi}{2} - i \right)} = \tan i \tag{7.87}$$

This special angle of incidence is called the *polarizing angle*. Fresnel's equations for the refracted beam (not given here) show that the refracted beam is only partly polarized, i.e. both the parallel and perpendicular components are present. This is indicated in Figure 7.40.

Double Refraction

We have seen above that a nonpolarized beam of light gives a plane-polarized reflected beam if the angle of incidence is equal to the polarizing angle. Plane-polarized light can be obtained in simpler and better ways, for example with the aid of double-refracting crystals or with so-called polaroids.

Double refraction means that an incident beam of light is separated into two transmitted beams, one ordinary and one extraordinary, in all crystalline media except those with a cubic structure. The two beams are plane polarized and their E vectors oscillate in perpendicular planes:

- the 'extraordinary' E vector in a plane through the optical axis and the extraordinary beam
- the 'ordinary' E vector in a plane perpendicular to the plane through the optical axis and the 'ordinary' beam.

The direction where the ordinary and extraordinary beams have the same velocity (vertical in Figure 7.41) is called the *optical axis*. It is in fact a *direction* and not an axis, but the name is traditional and it is difficult to change it.

The origin of double refraction is an anisotropic velocity of light. The ordinary beam has the same velocity in all directions whereas for the extraordinary beam the velocity varies with the direction. The wave fronts propagate as is shown in Figure 7.41a and b. The ordinary beam has a spherical wave front and the extraordinary wave front is a rotation ellipsoid.

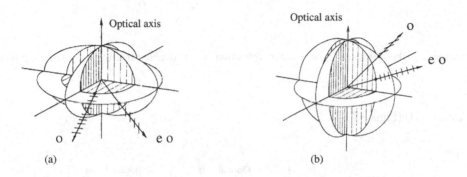

(a)   (b)

Figure 7.41   (a) Propagation of light in a *negative* crystal, e.g. Iceland spar ($CaCO_3$).
(b) Propagation of light in a *positive* crystal, e.g. quartz (SiO_2). Reproduced with permission. © O. Beckman.

If the velocity of light in air (or rather vacuum) is c_0, the velocity of the ordinary beam in the crystal is c_0/n_o, where n_o is the refractive index in the crystal (the subscript 'o' stands for 'ordinary'). The radius of the sphere in the crystal will then be $c_0 dt/n_o$, where dt is the propagation time from the origin (at $t = 0$) to the spherical wave front at time $t = dt$.

Analogously, the ellipse has the axes $c_0 dt/n_o$ and $c_0 dt/n_{eo}$, where the subscripts 'o' and 'eo' stand for 'ordinary' and 'extraordinary', respectively:

- If $n_{eo} < n_o$ the sphere is included in the ellipsoid and the crystal is said to be *negative*.
- If $n_{eo} > n_o$ the ellipsoid is included in the sphere and the crystal is said to be *positive*.

All crystal structures except the cubic structures are double refracting. A review of the crystal systems is given on pages 13–14 in Chapter 1. Hexagonal, tetragonal and trigonal crystals have only one optical axis. Iceland spar and quartz are both trigonal crystals. Orthorhombic, monoclinic and triclinic crystals are more complicated and have two optical axes. They will not be discussed here.

Refraction in Double-refracting Crystals

When parallel beams of light reach a planar surface, they are refracted and the angle of refraction can be constructed if the angle of incidence is known or calculated from the law of refraction.

The law of refraction, $\sin i = n \sin b$, is valid only for the ordinary beam and is *not valid for the extraordinary beam*. In the latter case, it is necessary to construct the angle b with the aid of wave fronts and perform calculations to find the tangent of the ellipse in the plane of incidence. The calculations will be omitted here but the construction of the wave front of the extraordinary beam is shown below for two special cases. The direction of the beam is then the direction of the line between the origin of the ellipse and the tangential point.

1. *The optical axis is in the plane of incidence. The angle of incidence > 0 (Figure 7.42).*

   The *E* vector of the ordinary beam is always perpendicular to the plane of incidence. The *E* vector of the extraordinary beam lies in the plane of incidence and its oscillation plane varies with the direction of the beam.

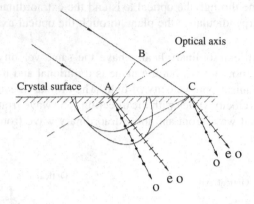

Figure 7.42 Construction of the refracted beams after refraction in a double-refracting crystal. Reproduced with permission. © O. Beckman.

2. *The angle of incidence = 0 (Figure 7.43).*

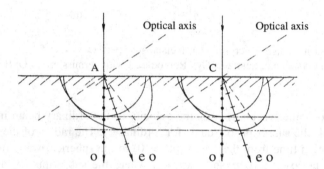

Figure 7.43 Construction of the refracted beams after refraction in a double-refracting crystal. Reproduced with permission. © O. Beckman.

This case shows very clearly that the law of refraction is not valid for the extraordinary beam. It is refracted in spite of normal incidence.

Methods to Produce Plane-polarized Light

Three methods are used in practice to produce plane-polarized light: *double refraction*, *dichroism* and *alignment of crystals*.

Double Refraction
If a prism of Iceland spar is made in such a way that the optical axis is parallel with the refracting edge of the prism (Figure 7.44a), it can be used to separate the ordinary beam and the extraordinary beam (Figure 7.44b).

Two such prisms are glued together with a substance called Canada balsam, which has a refractive index between n_o and n_{eo}. The ordinary beam will be totally reflected and only the plane-polarized extraordinary beam will be transmitted. Such a kit is called a *Nicol prism* or just a *Nicol*.

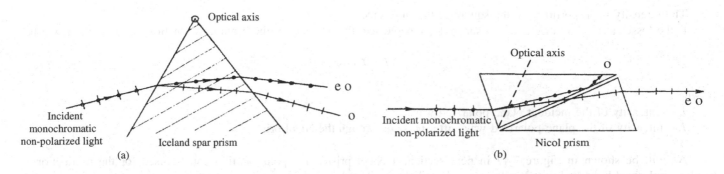

Figure 7.44 (a) Iceland spar prism. (b) A Nicol prism consists of two prisms of Iceland spar. Reproduced with permission. © O. Beckman.

Dichroism

Dichroism occurs in some double-refracting materials and means that one of the plane-polarized beams is strongly absorbed whereas the other is transmitted. The most common example is tourmaline (Figure 7.45).

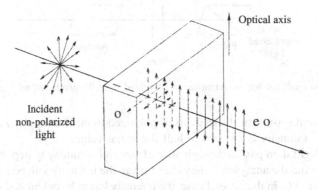

Figure 7.45 Dichroism in tourmaline. Reproduced with permission. © O. Beckman.

Alignment of Crystals

Another way to produce plane-polarized light is to use a polariod film, which consists of microscopic parallel 'needle' crystals, which only transmits light oscillating in a plane parallel with the crystals. The most common example is herepatite.

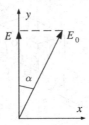

Figure 7.46 Relationship between E and E_0. Reproduced with permission. © O. Beckman.

Consider an incident plane-polarized wave with the amplitude E_0 which passes through a Nicol prism. After passage through the Nicol prism, the leaving wave is plane polarized and has the amplitude E. The angle between the oscillation planes of the incident and leaving waves is α. Figure 7.46 shows that the relationship between E and E_0 will be

$$E = E_0 \cos \alpha \tag{7.88}$$

The intensity is proportional to the square of the amplitude.

If the losses at the entrance and exit surfaces are neglected, the intensity of the beam as a function of the angle α will be

$$I = I_0 \cos^2 \alpha \qquad (7.89)$$

where

I_0 = intensity of the incident plane-polarized wave

I = intensity of the plane-polarized wave after passage through the Nicol prism.

As will be shown in Figure 7.47 in next section, a Nicol prism or a polariod film can be used for the production of plane-polarized light (polarizer) or for analysis of light (analyser).

Analysis of Polarized Light

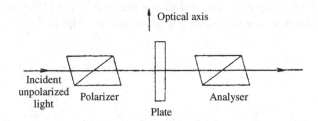

Figure 7.47 Polarizer and analyser for examination of polarized light. Reproduced with permission. © O. Beckman.

Figure 7.47 shows an experimental set-up for the analysis of polarized light. If nonpolarized light passes through a polarizer, it becomes plane polarized and its intensity is reduced to half the initial value.

In the absence of the plate, the light then passes through an analyser and its intensity depends on the relative directions of the polarization planes of the polarizer and the analyser. If they are *crossed*, the intensity will be *zero* as $\alpha = \pi/2$. If they are parallel, the intensity will be unchanged as $\alpha = 0$. In the general case, the intensity has to be calculated with the aid of Equation (7.89).

If a double-refracting plate with the *optical axis in the 'paper' plane* is inserted between the crossed polarizer and analyser in Figure 7.47, the propagation direction of monochromatic light is perpendicular to the optical axis. In this case, the propagation velocities of the ordinary and extraordinary beams are c_0/n_o and c_0/n_{eo}, respectively, as is mentioned on page 381.

If the optical axis lies in the plane of the crystal surface (Figure 7.48), the ordinary and extraordinary beams propagate with different velocities. A phase difference arises successively in the plate and remains after the exit of the plate.

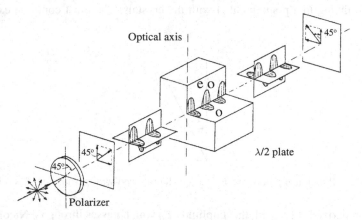

Figure 7.48 Change of the direction of plane-polarized light with the aid of a $\lambda/2$ plate. Reproduced with permission. © O. Beckman.

If monochromatic light is used and the plate is a so-called $\lambda/2$ plate, i.e. has a suitable thickness, there will be a displacement of half a wavelength between the ordinary and extraordinary beam, as shown in Figure 7.48. The light is still plane polarized but the oscillation plane is changed. Hence light will be transmitted in spite of the crossed Nicols. The analyser has to be turned 90° to obtain zero intensity after the analyser.

If the plate has another thickness, the light will normally be elliptically polarized and the intensity of the light will not be zero after the analyser. This property is used for the control of normally isotropic materials which are exposed to mechanical stress and is briefly discussed on page 387.

Rotation of the Plane of Polarization. Optical Activity

Rotation of the Plane of Polarization

If a quartz plate with *the optical axis in the direction of propagation of the light* is inserted between the crossed polarizer and analyser (Figure 7.49), the velocities of the ordinary and extraordinary monochromatic beams will be the same. Hence no displacement effect as in the case of a λ/2 plate (Figure 7.48) is to be expected.

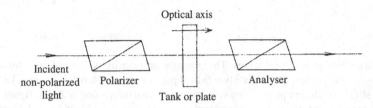

Figure 7.49   Equipment for analysis of optical activity in solid and liquid materials.

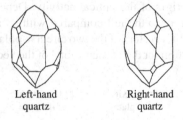

Figure 7.50   Mirror crystals of quartz. Reproduced by permission. © Ulf Ringström.

However, some materials show so-called *optical activity*, i.e. a plate of those materials turns the plane of polarization and light is transmitted through the crossed Nicols. The analyser has to be turned through the same angle to achieve darkness.

Many optically active materials have been found. Quartz is one of them. Some turn the polarization plane to the left, others to the right. There are two types of quartz: left-hand quartz and right-hand quartz (Figure 7.50). The same mirror optical activity has been found for other materials.

The angle of rotation of the polarization plane is proportional to the thickness of the optically active plate:

$$\alpha = \alpha_0 d \tag{7.90}$$

where
α = angle of rotation of the polarization plane (degrees)
α_0 = angle of rotation of the polarization plane caused by a plate of thickness 1 mm
d = thickness of the optically active plate (mm).

The proportionality constant α_0 varies with the wavelength of the light. For quartz it is about twice as large for violet light as for yellow Na light. If white light is used instead of monochromatic light, the exit light will be coloured because of the different rotation angles of the polarization plane for different wavelengths.

Optical Activity

The phenomenon of optical activity can be described as follows. After passage through the polarizer, the monochromatic light is plane polarized. The plane-polarized light can be regarded as composed of two simultaneous circular polarized waves, one with an *E* vector rotating clockwise and the other with an *E* vector rotating counterclockwise (Figure 7.51). The two rotating vectors have the same *angular* velocity but not the same *propagation* velocities. As the frequency is the same, different propagation velocities correspond to different wavelengths of the two components.

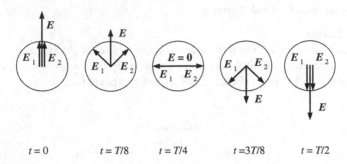

$t = 0 \qquad t = T/8 \qquad t = T/4 \qquad t = 3T/8 \qquad t = T/2$

Figure 7.51   Cooperation of two circular-polarized waves. The propagation velocity is directed perpendicularly to the plane of the paper towards the reader. The tops of the two *E* vectors move along two helical paths (right- and left-hand paths), i.e. in a circular motion at the same time as they move forwards with slightly different propagation velocities. The time for a round is *T*. The figure describes the resulting *E* vector, projected on the paper plane, as a function of time during half a period (*T*/2) of the motion or half a round. Obviously the resulting *E* vector is a plane-polarized wave.

The two wave fronts move with different velocities within the plate, which results in a *phase difference* between the rotating vectors at the exit of the plate. They give a resulting plane-polarized wave with a changed direction compared with the entrance wave. This phase difference is the origin of the optical activity. Depending on the sign of the phase difference, the plane of polarization rotates either to the right or to the left compared with its initial direction (Figures 7.52).

Figure 7.52 shows the positions of the E_1 and E_2 vectors of the two circular-polarized waves and their resultant *E* at the entrance and exit of the optically active plate. The condition for these 'snapshots' of the vector positions is given in the caption.

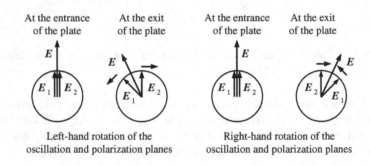

Figure 7.52   Positions of the E_1 and E_2 vectors of the two circular-polarized waves and their resultant *E* at the beginning and at the end of the optically active plate, provided that the length of the plate is a multiple of the wavelength λ_2 of one of the circular-polarized waves.

The oscillation plane and the perpendicular polarization plane are firmly connected to each other. Therefore, the oscillation plane (the *E* vector plane) and the polarization plane rotate the same angle in an optically active medium. It is customary to draw figures with the *E* vector and always talk of the rotation of the polarization plane, which is inconsequent but not wrong.

Symmetry

Optical activity seems to be coupled to *symmetry*, or rather asymmetry. Quartz with the mirror configurations is an example when crystals are involved. In organic chemistry, asymmetric carbon atoms are closely related to optical activity.

Optical activity occurs in many organic compounds. A common feature is that these compounds have an *asymmetric carbon atom*, i.e. a C atom with bonds to four different atoms or groups of atoms.

The four atom groups can be arranged in two ways, which gives two different configurations. They are mirror images of each other and cause rotation of the polarization plane to the left and right, respectively, *independent of the orientation in space* of the molecules in the solution. Molecules of one configuration can never be brought to coincidence with molecules of the other configuration by translation and/or rotation.

In Nature, both right- and left-hand quartz crystals are found. The SiO_2 molecule is not asymmetric. This is the reason why fused quartz with its disordered molecules is optically inactive. In Nature, both right- and left-hand organic molecules are present with dominance of the latter.

Applications of Polarized Light

Polarized light has a manifold of applications in many different fields. Some examples will be mentioned here.

Polaroid sunglasses are available for car drivers who want to eliminate or reduce the reflected light from wet roads. Figure 7.40 shows that the reflected beam is plane polarized in special case 2. The glasses have a 'crossed' polarization direction, which reduces the risk of dazzle and increases safety and may save lives.

The equipment described on page 384 offers an important method for *testing the stress distribution in transparent materials.* The specimen is inserted between *two* crossed Nicols. If the specimen is free from stresses, no light will be transmitted.

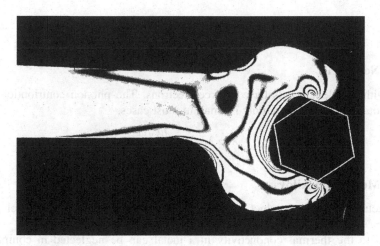

Figure 7.53 Interference pattern of light around an adjustable plastic spanner which has been exposed to external stress and is placed between crossed Nicols. Reproduced by permission. © ulf Ringström.

If the specimen contains stressed regions, the light becomes more or less elliptically polarized and part of the light passes through the analyser. Regions with a phase difference of 2π will remain dark. A striped pattern such as that in Figure 7.53 gives an idea of the stress distribution in the material.

Optically active organic materials can be studied with the aid of the equipment shown in Figure 7.49. Examples are qualitative and quantitative measurements of concentrations of optically active compounds, identification of the type of rotation of the polarization plane (left or right) of compounds and determination of reaction rates.

A classical example (1850) of the last application is the hydrolysis of sugar (saccharose) into glucose and fructose in a diluted acid solution when each molecule of saccharose splits into one molecule of glucose and one molecule of fructose with the same chemical formula but different configurations:

$$C_{12}H_{22}O_{11} \rightarrow C_6H_{12}O_6 + C_6H_{12}O_6$$

saccharose   glucose   fructose
strong right-   weak right-   strong left-
hand rotation   hand rotation   hand rotation

Initially the solution turns the polarization plane to the right as the water solution contains only saccharose ($t = 0$). During the reaction, a mixture of glucose and fructose, which rotates the polarization plane to the left as the fructose component dominates, is formed. Hence the total angle of rotation decreases with time during the reaction.

The angle of rotation is measured as a function of time. The reaction continues until an equilibrium state has been established. The reaction rate can be calculated with the aid of the relationship

$$k = \frac{1}{t} \ln \left(\frac{\alpha_0 - \beta_0}{\alpha - \beta_0} \right) \tag{7.91}$$

where
k = reaction rate
α_0 = angle of rotation at time $t = 0$
α = angle of rotation at time t
β_0 = angle at of rotation at equilibrium (time $t = \infty$).

Summary

■ Thermal Conductivity

$$\frac{dq}{dt} = -\lambda \frac{dT}{dx} \qquad \lambda = \text{thermal conductivity}$$

Total thermal conductivity:

$$\lambda = \lambda_e + \lambda_{\text{lattice}}$$

Thermal Conductivity of Nonmetals

Ionic crystals and solids with covalent bonds have no free electrons. The phonon contribution is responsible for the total thermal conductivity. The total thermal conductivity is small in most cases.

$$\lambda = \lambda_{\text{lattice}} = \frac{n c_V^a v l}{3}$$

Thermal Conductivity of Metals and Alloys

Metals have many free electrons. Metals conduct heat and electricity excellently or very well compared with nonmetallic solids.

The phonon contribution to the thermal conductivity in a metal can be neglected in comparison with that of the free electrons. The free electron contribution is responsible for the total thermal conductivity.

Thermal conductivity of pure metals:

$$\lambda_e = \frac{n_e \pi^2 k_B^2 T \tau}{3 m_e^*}$$

Thermal conductivity of alloys:

$$\lambda_{\text{alloy}} = \lambda_{\text{lattice}} + \lambda_e = \frac{n c_V^a v l}{3} + \frac{n_e \pi^2 k_B^2 T \tau}{3 m_e^*}$$

■ Electrical Conductivity

Resistivity:

$$R = \rho \frac{l}{A}$$

Electrical conductivity:

$$\sigma = \frac{1}{\rho}$$

ρ *increases* with temperature for *metals* and ρ *decreases* with temperature for *pure semiconductors*.

Relationship between current density and electric field:

$$j = \sigma E$$

Conductors, Insulators and Semiconductors

In a *metal*, the energy gap between the empty conduction band and the valence band is small or zero. The free electrons can move easily within the metal.

In a *semiconductor*, the energy gap is of the magnitude $1\,\text{eV}$. Electrons can be excited fairly easily up to the empty conduction band and leave holes in the valence band.

In an *insulator*, there are practically no free electrons. The valence band is filled and the energy gap is large. Few electrons are excited up to the conduction band.

■ *Metallic Conductors*

The shape and energies of the electron bands determine the conductivities of metals.

The valence and conduction bands may overlap. The valence of a monovalent metal is half-filled.

Monovalent metals are normally better electrical conductors than divalent metals.

Temperature Dependence of Resistivity

The resistivity of metals and alloys varies with temperature. At room temperature and over a wide temperature range, the resistivity can be described empirically by a linear function:

$$\rho = \rho_{293}\left[1 + \alpha\left(T - 293\right)\right]$$

Influence of Foreign Atoms and Crystal Defects on Resistivity. Resistivity of Alloys

σ varies with temperature and also depends on foreign atoms (alloying elements or impurities in the crystal lattice) and crystal defects, for example vacancies or interstitials.

Foreign atoms and crystal defects increase the scattering of the electrons and reduces the time τ between collisions, which increases the resistivity. The higher the disorder in the crystal lattice is, the higher will be the resistivity.

Some alloys have very low temperature coefficients and the variation of their resistances with temperature is very small.

Theory of Electrical Conductivity in Metals

The charge carriers are electrons. Drift velocity v_e is the systematic average velocity of the free electrons in a metal exposed to an electric field.

Current density:

$$j = n_e e v_e$$

Mobility:

$$\mu_e = \frac{v_e}{E}$$

Conductivity:

$$\sigma = n_e e \mu_e$$

Classical expression:

$$\sigma = \frac{n e^2 \tau}{m}$$

Band theory:

$$\sigma = \frac{ne^2\tau}{m^*}$$

Wiedemann–Franz law:

The ratio of the thermal and electrical conductivity of a pure metal is proportional to the absolute temperature.

This law strongly supports the idea that heat and charge both are transported through a pure metal by the same mechanism, i.e. by valence electrons.

■ *Pure Insulators*

Insulators are characterized by filled valence bands, empty conduction bands and large energy gaps.
Density of *available* electron energy states per energy unit:

$$N(E) = \frac{(2m^*)^{3/2}}{4\pi^2\hbar^3} E^{1/2}$$

Density of *occupied* electron energy states (number of states per energy unit and volume unit:

$$N(E)f_{\text{FD}} = \frac{(2m^*)^{3/2}}{4\pi^2\hbar^3} E^{1/2} f_{\text{FD}}$$

Fermi factor:

$$f_{\text{FD}} = \frac{1}{e^{\frac{E-E_F}{k_B T}} + 1}$$

■ *Simple Model of Electrical Conduction in Pure Insulators*

The widths of the valence and conduction bands are much less than the forbidden gap. E_v and E_c are constant.
Electron concentration:

$$n = \int 2N(E) f_{\text{FD}} dE = \int 2N(E) \frac{1}{e^{\frac{E-E_F}{k_B T}} + 1} dE$$

Electron concentration in the valence band:

$$n_v = \frac{2N}{e^{-\frac{E_c-E_v}{2k_B T}} + 1} = \frac{2N}{e^{-\frac{E_g}{2k_B T}} + 1}$$

Electron concentration in the conduction band:

$$n_c = \frac{2N}{e^{\frac{E_c-E_v}{2k_B T}} + 1} = \frac{2N}{e^{\frac{E_g}{2k_B T}} + 1}$$

Relationship between the hole and electron concentrations per unit volume and the number of atoms per unit volume:

$$n_v + n_c = 2N$$

Fermi level:

$$E_F = \frac{E_v + E_c}{2}$$

■ *More Accurate Theory of Electrical Conduction in Pure Insulators and Pure Semiconductors*

In the case of semiconductors, the valence and conduction bands are not narrow compared with the forbidden gap. It is not possible to make similar simplifications of the calculations as in the case of insulators.
Electron concentration in the conduction band:

$$n_e = \frac{(2\pi m_e^* k_B T)^{3/2}}{4\pi^3 \hbar^3} e^{-\frac{E_c - E_F}{k_B T}}$$

Hole concentration in the valence band:

$$n_h = \frac{(2\pi m_h^* k_B T)^{3/2}}{4\pi^3 \hbar^3} e^{-\frac{E_F - E_v}{k_B T}}$$

Fermi level of an intrinsic semiconductor:

$$E_F = \frac{E_c + E_v}{2} + \frac{3}{4} k_B T \ln \frac{m_h^*}{m_e^*}$$

■ *Conductivity of Pure Insulators and Pure Semiconductors*

The following relationships have to be used when the widths of the bands cannot be neglected in comparison with the forbidden gap.

Concentrations of Electrons in the Conduction Band and Holes in the Valence Band

$$n_e = n_h = n_i = \sqrt{n_e n_h}$$

$$n_e = n_h = n_i = \frac{(2\pi m k_B T)^{3/2}}{4\pi^3 \hbar^3} \left(\frac{m_e^*}{m} \frac{m_h^*}{m} \right)^{3/4} e^{-\frac{E_g}{2k_B T}}$$

General relationship:

$$\sigma = n_e e \mu_e + n_h e \mu_h$$

For intrinsic semiconductors and insulators:

$$\sigma = n_i e \, (\mu_e + \mu_h)$$

Determination of the Energy Gap of an Insulator

ln σ is plotted as a function of $1/T$, giving a straight line. The slope of this line is $-E_g/2k_B$.

■ *Conductivity of Doped Semiconductors*

$$n_e n_h = n_i^2$$

General relationship:

$$\sigma = n_e e \mu_e + n_h e \mu_h$$

n-Doped semiconductor when $n_e \gg n_h$:

$$\sigma = n_e e \mu_e$$

p-Doped semiconductor when $n_h \gg n_e$:

$$\sigma = n_h e \mu_h$$

■ *Doped Semiconductors*

Semiconductors are a very special and important group of solids. Due to a rather small forbidden gap between the valence and conduction bands (magnitude 1 eV), their electrical conductivity is higher than that of the insulators but lower than that of the metals. Semiconducting solids acquired enormous technical and industrial importance during the last part of the 20th century.

Two technical processes that have made this development possible are the production of extremely pure semiconducting materials and doping of extremely pure semiconducting materials.

The dominant semiconductors are Si and Ge. They are characterized by small energy gaps, nearly filled valence bands and nearly empty conduction bands.

n-Doped Semiconductors

If the semiconductor is doped with elements in group 5 in the periodic table, there will be an *excess* of valence electrons it is said to be *n-doped* (n = negative). In an n-doped semiconductor, there are many more mobile electrons than holes. The extra electrons obviously stays in a special energy level close to the lower part of the conduction band. This energy level is called the *donor level*.

p-Doped Semiconductors

If the dopant comes from group 3 in the periodic table, the semiconductor is said to be *p-doped*. In this case there will be a *lack* of valence electrons, which we call holes. The holes behave like positive charges (p = positive). In a p-doped semiconductor, there are many more mobile holes than electrons. The excess holes stay in a special energy level close to the valence band. This energy level is called the *acceptor level*.

Compound Semiconductors

Some semiconductors are chemical compounds, usually of one component from group 3 and one element from group 5. Common examples are InSb and GaAs. Such semiconductors are ionized.

Compound semiconductors have both donor levels and acceptor levels. If GaAs is doped with Si, an Si atom can occupy a Ga site and act as an n-dopant. On the other hand, if an Si atom ocupies an As site, it is a p-dopant.

Donor and Acceptor Energy Levels in Doped Semiconductors

The bound system of a positive group 5 ion and an excess electron is roughly analogous with the hydrogen atom. The simple Bohr model is used as an approximation for calculation of its energy levels. The only difference is that the space within the bound system is filled with a dielectric medium with the relative dielectric capacity ε_r.
Donor energy level in doped semiconductors:

$$E_n^* = -\frac{m_e^* e^4}{8\,h^2 \varepsilon_r^2 \varepsilon_0^2}\frac{1}{n^2} = -E_n\frac{m_e^*/m}{\varepsilon_r^2}$$

Radius of excess electron orbit:

$$r_n^* = \frac{4\pi\varepsilon_r\varepsilon_0}{m_e^* e^2}\left(\frac{h}{2\pi}\right)^2 n^2 = r_n\frac{\varepsilon_r}{m_e^*/m}$$

Analogous calculations can be performed for acceptor levels. In this case the H-like bound system consists of a negative nucleus and a hole. The binding energies correspond to the energies of the acceptor levels close to the top of the valence band in the semiconductor.

The binding energies of the bound systems are very small and hence the donor and acceptor levels are very close to the allowed conduction and valence bands.

■ *Optical Properties of Solids*

Optical Properties of Metals

Metals are opaque owing to absorption of valence electrons up to the empty conduction band. The absorption covers the visible part of the electromagnetic spectrum.

Metals reflect light. The intensity of the reflected light depends on the shapes and positions of the energy bands and on the dielectric constant, which varies with the wavelength of the light. The light is in most cases greyish. Exceptions are copper and gold.

Optical Properties of Semiconductors

Many semiconductors are transparent to visible light but show absorption in the infrared region.

The energy gap is of the magnitude 1 eV. It can be calculated from absorption measurements:

$$I = I_0 e^{-\alpha x}$$

Semiconductors have a direct or indirect energy gap. In the latter case, a phonon is involved in the absorption process. In some cases an exciton may absorb energy.

Optical Properties of Insulators

Ionic crystals have hardly any free electrons and high energy gaps. They are transparent and do not absorb visible and infrared electromagnetic radiation.

Pure ionic crystals are normally uncoloured. The intense colour of some types of crystals is caused by lattice defects, so-called colour centres.

■ *Polarized Light and Insulators*

Reflection

Fresnel's equations for reflected light:

$$I_{r\perp} = \frac{I_0}{2} \frac{\sin^2(i-b)}{\sin^2(i+b)}$$

$$I_{r\,\text{II}} = \frac{I_0}{2} \frac{\tan^2(i-b)}{\tan^2(i+b)}$$

Special case 1: $i = b = 0$
Fraction of reflected light:

$$R = \frac{I_{r\perp} + I_{r\,\text{II}}}{I_0} = \left(\frac{i-b}{i+b}\right)^2 = \left(\frac{n-1}{n+1}\right)^2$$

Special case 2: if $i + b = \pi/2$ the reflected beam is plane polarized perpendicularly to the plane of incidence.

Relationship between the angle of incidence i (polarizing angle) and the refractive index:

$$n = \tan i$$

Double Refraction

Double refraction means that an incident beam of light is separated into two beams, one ordinary and one extraordinary, in all crystalline media except those with a cubic structure. The two beams are plane polarized and their E vectors oscillate in perpendicular planes.

Optical axis: direction in a crystal where the ordinary and extraordinary beams move with the same velocities.

The *wave front of the ordinary beam* is a sphere. The law of refraction is valid for the ordinary beam. The *wave front of the extraordinary beam* is a rotation ellipsoid. The law of refraction is not valid.

The directions of the refracted ordinary and extraordinary beams are constructed with the aid of the wave fronts in the crystal.

If $n_{eo} < n_o$, the sphere is included in the ellipsoid and the crystal is said to be *negative*. If $n_{eo} > n_o$, the ellipsoid is included in the sphere and the crystal is said to be *positive*.

Methods to Produce Plane-polarized Light

Double Refraction
A Nicol consists of two Iceland spars glued together with Canada balsam.

Dichroism
The ordinary or extraordinary beam is strongly absorbed in the material and the other is transmitted.

Alignment of Crystals
Use of polaroid film, i.e. parallel needle crystals. Only light parallel with the needle crystals can pass the polaroid.

Analysis of Polarized Light

If the optical axis lies in the plane of the crystal surface, the ordinary and extraordinary beams propagate with different velocities perpendicular to the surface. A phase difference arises successively in the plate and remains after the exit of the plate.

Stress distribution in transparent materials can be examined with a plate with the material in the shape of a plate located between two crossed Nicols used as polarizer and analyser.

Optical Activity

Optical activity is the property of a material to rotate the plane of polarization when it is passed by plane-polarized light.

Optical activity occurs in crystals with mirror configuration and in organic compounds with asymmetric carbon atoms, i.e. C atoms bound to four different atoms or groups of atoms. Optical activity is independent of the orientation in space of the crystals or molecules.

Optical activity can be examined with the specimen located between two crossed Nicols used as polarizer and analyser.

Exercises

7.1 Solids are divided into two main groups with respect to their thermal and electrical conductivity properties – insulators and conductors.

   (a) What is the main difference in structure between insulators and conductors?
   (b) Discuss the theory of thermal conductivity in insulators.
   (c) Discuss the theory of thermal conductivity in pure metals and alloys.

7.2 In order to determine the thermal conductivity of glass, a thick spherical shell with an inner radius of 90 mm and a thickness of 12 mm was made of the material. A small electric heater with a power of 150 W was placed in the centre of the sphere. The heater was switched on and the temperatures at the inner and outer surface of the sphere were measured with the aid of thermo-elements at regular time intervals. When a steady state had been established, the temperature difference between the inner and outer glass surfaces was measured to 17 °C.
Calculate the thermal conductivity of the glass material.

7.3 Electrical conductivity has the widest variation in magnitude of all material properties. The ratio of the resistivities of the best insulators to the best conductors is more than 10^{40}.

   (a) Solids are divided into three groups – conductors, semiconductors and insulators. Characterize them by means of their valence and conduction bands and the magnitude of the energy gap between them.
   (b) Relate the valence and the conduction band of a pure metal or an alloy to the theory of solids in Chapter 3.
   (c) The resistivities of solids vary with temperature. Describe in what way for the three groups of solids.

7.4 (a) Discuss the theory of electrical conductivity in metals. Define the concepts drift velocity, relaxation time and mobility and give the relationships between them.

   (b) Explain the significance of the Wiedemann–Franz law.

7.5 (a) Calculate of the thermal conductivity of Al at 150 °C. Required material constants can be found in a standard table.

   (b) How does the thermal conductivity change if Al is alloyed with Si?

   Hint: Compare the thermal conductivities of Al and silumin listed in a standard table.

7.6 (a) The Fermi energy of copper is 7.04 eV. Calculate the maximum velocity of the conduction electrons in Cu.

   (b) A current of 10 A is sent through a copper wire of cross-sectional area 1.0 mm^2. Calculate the drift velocity of the free electrons in the wire.

   (c) The electrical conductivity of Cu is $5.8 \times 10^7 \, (\Omega \, \text{m})^{-1}$. Calculate the mobility of the electrons in the metal.

7.7 The resistivity of silver is $0.016 \, \Omega \, \text{mm}^2/\text{m}$. The Fermi energy of Ag is 5.51 eV. Assume that $m^* = m$.

   (a) Calculate the relaxation time of the free electrons.

   (b) Calculate the free mean path of the conduction electrons in the metal.

7.8 Discuss the simple theory of electrical conductivity in insulators and calculate the approximate position of the Fermi level. Give also the expressions for n_e, n_h and σ resulting from a more accurate theory.

7.9 The experimental value of the conductivity σ of diamond at room temperature ($T = 300$ K) is $1 \times 10^{-12} \, (\Omega \, \text{m})^{-1}$.

   (a) Calculate a theoretical value of σ for pure diamond by use of the following data: $E_g = 5.3$ eV, $\mu_e = 0.18 \, \text{m}^2/\text{V s}$ and $\mu_h = 0.12 \, \text{m}^2/\text{V s}$. Assume that $m_e^* = m_h^* = m$.

   (b) Does this value agree with the experimental value for intrinsic diamond? If not, what concentration of an impurity with the valence 5, which is completely ionized, is required to give the experimental value. Is the calculated impurity concentration reasonable?

7.10 (a) Compare the properties of an intrinsic semiconductor and a doped semiconductor.

   (b) Prove the relationship between the effective mass of the valence electrons in a metal and their kinetic energy E_{kin} [Equation (3.108) on page 145].

$$m^* = \frac{\hbar^2}{\dfrac{d^2 E_{\text{kin}}}{dk^2}}$$

   where k is the wavevector of a valence electron.

   Hint: Look for a relationship between k and E_{kin} in Chapter 3.

   (c) Sketch the energy levels of an n-doped and a p-doped semiconductor.

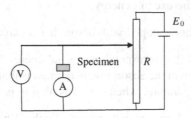

7.11 The energy gap of a pure semiconductor can be determined by measurement of the voltage U over a specimen at constant current as a function of the temperature. At such an experiment the following values were obtained:

T(K)	237.0	250.0	263.0	275.0	287.0	300.0
U(mV)	14.30	10.28	7.77	6.30	5.00	4.10

Calculate the energy gap of the semiconductor in eV.

7.12 (a) Calculate the concentration of holes and electrons in intrinsic silicon at 20 °C by use of the following data: $E_g = 1.1\,\text{eV}$, $m_e^* = 0.70\,m$ and $m_h^* = m$.

(b) Calculate the fraction of occupied energy levels in the conduction band. The density of crystalline silicon at 20 °C is $2.42 \times 10^3\,\text{kg/m}^3$.

7.13 The electrical resistivity of an intrinsic semiconductor is 0.5 Ωm at 300 K. Determine the band gap by use of the following data: $\mu_e = 0.40\,\text{m}^2/\text{Vs}$, $\mu_h = 0.20\,\text{m}^2/\text{Vs}$, $m_e^* = 0.10\,m$ and $m_h^* = 0.50\,m$.

7.14 Consider an intrinsic semiconductor with the following properties: $E_g = 1.30\,\text{eV}$, $m_e^* = 0.070\,m$ and $m_h^* = 0.69\,m$. Calculate

(a) the position of the Fermi level at $T = 0$ and 300 K
(b) the electron concentration in the conduction band and the hole concentration in the valence band at $T = 0$ and 300 K.

7.15 A single crystal of germanium, with a mass of 0.100 kg, has been doped with 2.0 μg of Sb. All impurity atoms are ionized and the mobility of the electrons is $0.36\,\text{m}^2/\text{Vs}$. A bar has been cut out of the single crystal with a length of 20 mm and a cross-sectional area of 4.0 mm². Calculate the resistance of the bar. The density of Ge is $5.35 \times 10^3\,\text{kg/m}^3$.

7.16 Germanium of n-type contains 1.0×10^{23} ionized donors per m³. Calculate the ratio of the conductivity of this doped crystal and the conductivity of extremely pure Ge at 300 K based on the following data for Ge: $E_g = 0.67\,\text{eV}$, $m_e^* = 0.60\,m$, $m_h^* = 0.30\,m$, $\mu_e = 0.39\,\text{m}^2/\text{Vs}$ and $\mu_h = 0.19\,\text{m}^2/\text{Vs}$.

7.17 Germanium is doped with 0.01 at-% aluminium. Calculate the resistivity at 300 K of the doped crystal by use of the following data for Ge: the lattice constant $a = 0.5658\,\text{nm}$, $E_g = 0.67\,\text{eV}$, the mobilities $\mu_e = 0.39\,\text{m}^2/\text{Vs}$ and $\mu_h = 0.19\,\text{m}^2/\text{Vs}$ and the intrinsic carrier concentration $n_i = 1.0 \times 10^{19}\,\text{m}^{-3}$. The impurity atoms are completely ionized at 300 K.

7.18 Calculate the concentrations of electrons and holes in a p-doped semiconductor if its conductivity σ is 10 $(\Omega\,\text{m})^{-1}$, the intrinsic charge carrier concentration n_i is $2.2 \times 10^{19}\,\text{m}^{-3}$ and the mobilities are $\mu_e = 0.40\,\text{m}^2/\text{Vs}$ and $\mu_h = 0.20\,\text{m}^2/\text{Vs}$.

7.19 In order to give an idea of how weakly bound the 'fifth' valence electron is in a phosphorus atom, which is included in a silicon crystal, you can perform the following calculations. Calculate

(a) the binding energy E_B^* of the orbit of the fifth electron in a P atom and compare it with the energy gap 1.1 eV of Si
(b) the radius r_0^* of the orbit of the fifth electron in a P atom and compare it with the lattice constant 0.54 nm of Si.

Use the following data: the dielectric constant of Si is 11.8 and the effective mass m^* of the free electrons in the Si lattice is $0.19\,m$. The binding energy and radius of the ground-state orbit in the hydrogen atom are 13.6 eV and $5.29 \times 10^{-11}\,\text{m}$, respectively.

7.20 Figure 7.35b on page 377 shows the absorption curve of pure InSb as a function of the photon energy at 5 K.

(a) Calculate the energy gap expressed in eV and the corresponding value of the wavelength of the photon.
(b) What happens if the photon energy is lower than 0.2 eV?
(c) Figure 7.37 on page 378 shows the absorption curve of pure GaAs as a function of the photon energy at 21 K. Calculate an approximate value of the exciton energy.

7.21 The transmission curve of a silicon specimen, doped with boron, has a wavelength minimum at 31.2 μm.

(a) Calculate and sketch the position of the corresponding special energy level in relation to the nearest band edge.
(b) A necessary condition for observation of the absorption maximum is that the specimen must be cooled to an extremely low temperature (liquid helium temperature) when the transmission is measured. Explain why.

7.22 Incident nonpolarized light with intensity I_0 passes two crossed Nicols. What will the intensity of the light be

(a) after the passage of the first Nicol?
(b) after passage of both Nicols?
(c) Another Nicol is placed between the two crossed Nicols. What will the intensity of the exit light be after passage of all the three Nicols if the oscillation plane of the middle Nicol forms an angle of 45° with the other two? Figures and motivated equations are required.

Assume that the reflection losses can be neglected.

7.23 Calculate the thickness of a so-called $\lambda/2$ plate made of Iceland spar and made for monochromatic Na light of wavelength 589.3 nm. Iceland spar has the following material constants: $n_{eo} = 1.4864$ and $n_o = 1.6584$. The oscillation plane of the incident plane-polarized light forms an angle of 45° with the vertical plane (see Figure 7.48 on page 384). Motivate your equation.

Draw a figure that shows the ordinary and extraordinary wave fronts and the orientation of the optical axis.

7.24 Quartz shows optical activity and rotates the polarization plane of plane-polarized Na light when it passes a quartz plate.

(a) Draw a figure, that shows the equipment used for analysis of optical activity. What is the initial relative position of the polarizer and the analyser before the quartz plate has been introduced?

  What is the criterion for the final position of the analyser, i.e. how do you adjust it?

(b) Calculate the specific angle of rotation, i.e. the angle of rotation per mm, for quartz when you know that the angle of rotation is 53° when the thickness of the plate is 2.4 mm.

(c) A quartz plate rotates the plane of plane-polarized light through 72°. What is the thickness of the plate?

(d) What difference does it make if the quartz plate is made of either of the two mirror types of quartz crystals?

8

Properties of Liquids and Melts

8.1	Introduction	400
8.2	X-ray Spectra of Liquids and Melts	400
	8.2.1 X-ray Analysis of Pure Liquids and Melts	400
	8.2.2 Temperature Dependence of the Atomic Distribution in Pure Metal Melts	401
	8.2.3 X-ray Spectra of Binary Alloys	402
8.3	Models of Pure Liquids and Melts	402
	8.3.1 Early Models	402
	8.3.2 Theory of Liquid Structure. Pair Distribution Function. Pair Potential Models	402
8.4	Melting Points of Solid Metals	406
	8.4.1 Lindemann's Melting Rule	406
8.5	Density and Volume	407
	8.5.1 Volume Change on Fusion in Metals and Semiconductors	407
	8.5.2 Volume Change and Heat of Mixing on Mixing Metal Melts	408
8.6	Thermal Expansion	409
	8.6.1 Thermal Volume Expansion	409
	8.6.2 Temperature Dependence of Density	410
8.7	Heat Capacity	412
	8.7.1 Theory of Heat Capacity of Liquids	412
	8.7.2 Heat Capacity at Constant Volume	413
	8.7.3 Heat Capacity at Constant Pressure	413
8.8	Transport Properties of Liquids	415
8.9	Diffusion	416
	8.9.1 Simple Model of Diffusion in Liquids. Diffusion in Metal Melts at Temperatures Close to the Melting Point	416
	8.9.2 Theories of Diffusion in Liquids	417
	8.9.3 Diffusion of Interstitially Dissolved Atoms in Metal Melts	421
	8.9.4 Diffusion in Alloys	421
	8.9.5 Experimental Methods of Diffusion Measurements on Metal Melts	422
8.10	Viscosity	425
	8.10.1 Basic Theory of Viscosity	426
	8.10.2 Newtonian and Non-Newtonian Fluids	427
	8.10.3 Viscosity in Pure Metal Melts	429

Physics of Functional Materials   Hasse Fredriksson and Ulla Åkerlind
© 2008 John Wiley & Sons, Ltd

8.10.4	Temperature Dependence of Viscosity of Pure Metal Melts. Relationship between D and η	432
8.10.5	Viscosity of Alloy Melts	435
8.10.6	Experimental Methods of Viscosity Measurements on Metal Melts	437
8.11	Thermal Conduction	438
8.11.1	Thermal Conductivity of Pure Metal Melts	438
8.11.2	Thermal Conductivity of Alloy Melts	439
8.12	Electrical Conduction	439
8.12.1	Electrical Conductivity of Pure Metal Melts	440
8.12.2	Electrical Conductivities of Alloy Melts	442
8.12.3	Ratio of Thermal and Electrical Conductivities in Pure Metal Melts	443
Summary		443
Exercises		449

8.1 Introduction

The properties and structures of all substances are functions of the forces between their atoms. In a *gas*, the forces between the atoms are very weak (Chapter 4), whereas they normally are very strong in *solids* (Chapter 5). In a *melt*, the average forces between the atoms are strong but weaker than those in a solid.

The atoms or molecules in a *gas* move freely and have long mean free paths and low collision frequencies. A gas has no given volume but fills the whole available space, independent of its shape.

In a *solid*, the atoms are located to permanent sites in a crystal lattice and are only able to vibrate around their equilibrium positions with small amplitudes. A solid has a rigid shape.

The interatomic forces in liquids are more similar to the forces in a solid than in a gas, where they often can be totally neglected. Unlike a gas, it does not fill the whole available space. Like a solid, a liquid or melt has a given *volume* but adapts its *shape* to the shape of the container in which it is kept.

The properties of liquids are functions of the interatomic forces. It is extremely difficult to derive reliable expressions for the dynamic properties of liquids as the atomic motion in a liquid cannot be accurately described as a function of time. This makes it very hard to find satisfactory theories of transport phenomena in liquids. Viscosity implies transport of momentum. Thermal conduction is accompanied by energy transport, electrical conduction transports electrical charge. Diffusion involves mass transport. These phenomena will be discussed in this chapter.

Below we will discuss some models of melts which may help in understanding the properties of melts in comparison with those of gases and solids. The boundary between a liquid and a gas of high density is uncertain. At the critical point (Chapter 4, page 185) there is no difference between the gas and the corresponding liquid.

8.2 X-ray Spectra of Liquids and Melts

The fundamental experimental basis for knowledge of the structure of liquids and melts is their X-ray spectra. This topic has been discussed in Section 1.2.2 on pages 4–9. Figures 1.8–1.12 indicate a short-range order of the atoms in pure metal melts. This is also true for other liquids.

8.2.1 X-ray Analysis of Pure Liquids and Melts

As was described in Chapter 1, the interpretation of the X-ray spectra of pure metal melts permits the derivation the average distances between the atoms and the coordination numbers, which are specific to the metal atoms in the melt. The short-range order means that each atom is surrounded by a number of nearest neighbour atoms. These numbers have been derived from the X-ray spectra for many metals. In Table 8.1, the coordination numbers for some common metal melts are given and compared with the distances between the atoms in the solid metal.

The table shows that metals with close-packed structures (BCC, FCC and HCP) in the solid state change their nearest neighbour distances only slightly on melting. Other metals with more complicated structures and several nearest neighbour distances do not show the same behaviour.

Table 8.1 Nearest neighbour distances for some pure solid and liquid metals at their melting points.

Metal	Position in the periodic table	Melting point (°C)	Crystal structure	Nearest neighbour distance for the solid metal (nm)	Nearest neighbour distance for the liquid metal (nm)
Na	1a	98	BCC	0.372	0.370
K	1a	64	BCC	0.452	0.470
Cu	1b	1083	FCC	0.256	0.257
Ag	1b	961	FCC	0.289	0.286
Au	1b	1063	FCC	0.288	0.285
Mg	2a	651	HCP	0.320	0.335
Hg	2b	−39	Rhombohedral	0.301 or 0.347[a]	0.307
Al	3a	660	FCC	0.286	0.296
Ge	4a	959	Diamond (two FCC)	0.245	0.270
Sn	4a	232	Tetrahedral	0.302 or 0.318[a]	0.327
Pb	4a	327	FCC	0.350	0.340
Sb	5a	631	Rhombohedral	0.291 or 0.335[a]	0.312
Bi	5a	272	Rhombohedral	0.309 or 0.353[a]	0.332

[a] Different directions. The unit cell is asymmetric.

8.2.2 Temperature Dependence of the Atomic Distribution in Pure Metal Melts

The shapes of the atomic distribution diagrams of metal melt change with temperature. This is shown in Figures 8.1 and 8.2, which show the diagrams for potassium and tin, respectively.

The temperature figures close to the curves in both diagrams represent the temperatures above the melting point. In both cases the peaks become wider and lower when the temperature is increased. The area under the first peak also changes with temperature. For potassium, the number of nearest neighbours decreases from approximately 9.5 to 9.0 when the temperature increases from 70 to 395 °C above the melting point.

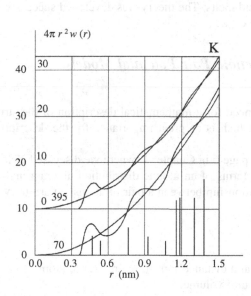

Figure 8.1 Atomic distribution diagram for liquid potassium. The vertical lines represent the distribution for solid potassium. Reproduced with permission from N. S. Gingrich, The diffraction of X-rays by liquid elements, *Rev. Mod. Phys.* **15**, 90–110. © 1943 The American Physical Society.

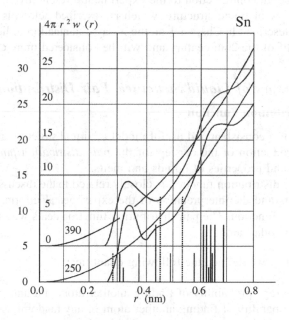

Figure 8.2 Atomic distribution diagram for liquid tin. The vertical lines represent the distribution for solid white tin (full lines) and solid grey tin (dotted lines). Reproduced with permission from N. S. Gingrich, The diffraction of X-rays by liquid elements, *Rev. Mod. Phys.* **15**, 90–110. © 1943 The American Physical Society.

Figure 8.2 shows the same effect for tin as for potassium in Figure 8.1: the number of nearest neighbours decreases with increasing temperature. Solid tin has two different structures. Grey tin has a lattice of diamond type and the structure of white tin is tetragonal. Figure 8.2 shows that the structure of liquid tin is more reminiscent of white than grey tin.

8.2.3 X-ray Spectra of Binary Alloys

X-ray spectra of binary liquid alloys have been discussed on pages 31–32 in Chapter 1. Normally the first peaks of both components are seen. In some cases the X-ray spectrum reveals the existence of an intermediate chemical compound.

8.3 Models of Pure Liquids and Melts

As is evident from Chapters 4–7, there are good and generally accepted models for gases and solids. This is not the case for melts and liquids. Attempts have been made, but unfortunately the search for a general model that can be successfully used in practice has been unfruitful so far.

8.3.1 Early Models

In the 1930s, Eyring proposed the vacancy model of liquids and melts which could explain some of the properties of liquids in a semiquantitatively way. Eyring and co-workers assumed that a liquid has the same structure as the corresponding solid with the important distinction that the vacancy concentration in the liquid is much larger than that in the solid. This difference in vacancy concentration between melts and solids was assumed to be responsible for the great differences in the properties of the melt compared with the corresponding solid.

If Eyring's model were true, the X-ray plots of melts should show the presence of long-range order structures in the liquid states like those found in solids. A serious objection to the model is that no such effects have been found.

In the early 1960s, Bernal suggested a hard sphere model of liquids. He interpreted the structure of liquids as a random distribution of close-packed spheres (atoms). The atoms were situated at the corners of five different types of rigid polyhedra. The model is described in Section 1.3.2 on pages 11–12. This model involves the fact that there exists a short-range order in liquids but it is static. The model with its fixed proportions of crystalline bodies describes a structure that does not vary with temperature, in contradiction to the experimental results from X-ray diffractions of metal melts discussed in Section 8.2.1.

The theory of liquid structure, which is described below, is based on and closely related to the fundamental experimental evidence described in Chapter 1 on the X-ray examination of liquids and melts. The theory was developed successively during the last half of the 20th century and will be considered more closely below.

8.3.2 Theory of Liquid Structure. Pair Distribution Function. Pair Potential Models

Pair Distribution Function

The lack of a consistent and useful model of liquid structure and the need for a mathematical description of the structure led to the introduction of the concept of the *pair distribution function*, which is of great importance for the description of the structures and properties of liquids and melts.

The pair distribution function is closely related to the discussion on page 6 in Chapter 1, where we discussed X-ray plots of metal melts and the interpretation of the experimental information in terms of an atomic distribution diagram and the radial distribution function. The pair distribution function deals with the relationship between a pair of atoms, an arbitrary reference atom and another atom.

In Chapter 1, we defined the following quantities:

w_r = average probability of finding another atom in a unit volume at a distance r from the reference atom
w_0 = probability of finding another atom in any randomly selected unit volume.

The pair distribution function $g(r)$ is defined as *the relative probability of finding another atom in a unit volume at a distance r from the reference atom*:

$$g(r) = \frac{w_r}{w_0} \tag{8.1}$$

The probability of finding the second atom within the volume element dV equals w_r times dV. If we choose the volume element to be a spherical shell with radius r and thickness dr, we obtain the probability of finding the second atom at a distance r from the reference atom (origin), independent of direction:

$$dW = w_r dV = w_r \times 4\pi r^2 dr \tag{8.2}$$

or

$$dW = g(r)dV = g(r)w_0 \times 4\pi r^2 dr \tag{8.3}$$

where $g(r)w_0 \times 4\pi r^2$ is the *radial distribution function*.

Each element has its own characteristic pair distribution and radial distribution functions. An example of a typical pair distribution curve is given in Figure 8.3.

The shape of the $g(r)$ value reveals interaction between the atoms. If there were no interaction at all, the probability would be the same everywhere and $g(r)$ would be equal to 1. This corresponds to the dotted line in Figure 8.4.

The deviations of the $g(r)$ value from the average value indicate interaction between the atoms. The figure shows that this interaction corresponds to a *short-range order* in the liquid as the curve approaches the value 1 after a few deviation cycles.

The $g(r)$ curve intersects the r axis at a given r value and is zero for smaller r values. The intersection point corresponds to the diameter of the atoms. The second atom cannot come closer to the reference atom than the double radius of the atoms.

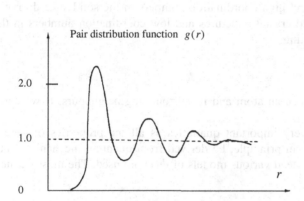

Figure 8.3 A typical pair distribution function curve for a simple melt. Reproduced with permission from T. Iida and R. I. L. Guthrie, *The Physical Properties of Liquid Metals.* © 1988 (Clarendon Press/Oxford University Press, NY-academic.permissions@oup.com)

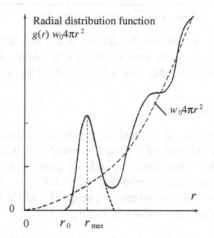

Figure 8.4 Radial distribution function $g(r)w_0 \times 4\pi r^2$ as a function of r. Reproduced with permission from T. Iida and R. I. L. Guthrie, *The Physical Properties of Liquid Metals.* © 1988 (Clarendon Press/Oxford University Press, NY-academic.permissions@oup.com)

In principle it is possible to calculate the pair distribution function $g(r)$ with the aid of statistical mechanics. In practice, it is impossible in most cases as the calculations require knowledge of the total potential energy Φ of the system. This is a function of all the *pair potential functions* ϕ_{ij} (pages 404–405). Approximate equations have been suggested, such as the Born–Green equation and others, which connect the pair distribution function $g(r)$ and the pair potential function $\phi(r)$ (pages 404–405), but mathematical complexity and poor information on the pair potential functions ϕ_{ij} have given unsatisfactory results so far.

In Chapter 1, we found that the $g(r)$ curve could be determined experimentally. A considerable amount of experimental structural information was obtained during the 1970s and 1980s.

Figure 8.4 shows a typical radial distribution function curve of a metal. It represents the product of the $g(r)$ curve times $4\pi r^2 w_0$. The latter represents the dotted curve in Figure 8.4.

The area under the first peak of the curve represents the coordination number Z_{coord}:

$$Z_{coord} \approx 2 \int_{r_0}^{r_{max}} 4\pi r^2 g(r)dr \tag{8.4}$$

Table 8.2 Coordination numbers for some pure solid and liquid metals at their melting points. Reproduced with permission from F. D. Richardson, *Physical Chemistry of Melts in Metallurgy*, Vol. 1. © 1974 Academic Press Inc. (London), now Elsevier.

Metal	Position in the periodic table	Melting point (°C)	Crystal structure	Coordination number for the solid metal	Coordination number for the liquid metal
Na	1a	98	BCC	8	9.5
K	1a	64	BCC	8	9.5
Cu	1b	1083	FCC	12	11.5
Ag	1b	961	FCC	12	10.0
Au	1b	1063	FCC	12	8.5
Mg	2a	651	HCP	12	10.0
Hg	2b	−39	Rhombohedral	6 or 6[a]	10.0
Al	3a	660	FCC	12	10.6
Ge	4a	959	Diamond (two FCC)	4	8.0
Sn	4a	232	Tetrahedral	4 or 2[a]	8.5
Pb	4a	327	FCC	12	8.0
Sb	5a	631	Rhombohedral	3 or 3[a]	6.1
Bi	5a	272	Rhombohedral	3 or 3[a]	7–8

[a] Different directions. The unit cell is asymmetric.

Table 8.2 shows that metals with close-packed structures and high coordination numbers in the solid state do not change the coordination numbers very much. Metals with complicated crystal structures and low coordination numbers in the solid state increase their coordination numbers considerably on melting.

Pair Potential Models

The pair potential ϕ_{ij} is defined as the potential energy between an atom and its surrounding neighbours. If we choose the atom as the reference atom the pair potential is denoted $\phi(r)$.

The pair distribution function and the pair potential are very important quantities as all the properties of pure liquids and melts are functions of $g(r)$ and $\phi(r)$. Pair potentials can, in principle, be derived from quantum mechanics, but this is practically impossible for all atoms except H and He so far. Instead various models of $\phi(r)$ are used. The most frequent ones are listed below and are shown in Figure 8.5.

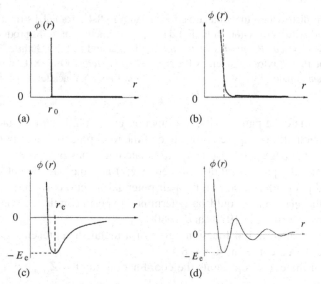

Figure 8.5 (a) Hard sphere potential; (b) Inverse power potential; (c) Lennard-Jones potential; (d) Ion–ion potential. Reproduced with permission from T. Iida and R. I. L. Guthrie, *The Physical Properties of Liquid Metals*. © 1988 (Clarendon Press/Oxford University Press, NY-academic.permissions@oup.com)

Hard Sphere Potential

$$\phi(r) = +\infty \qquad \text{for } r < r_0 \text{ (radius of the hard spheres)}$$
$$\phi(r) = 0 \qquad \text{for } r \geq r_0$$

The hard sphere model of liquids has been briefly described in Chapter 1.

Inverse Power Potential

$$\phi(r) = E\left(\frac{r_0}{r}\right)^n \tag{8.5}$$

where E, r_0 and n are parameters. This model of $\phi(r)$ is identical with the hard sphere model if $n = \infty$.

Lennard-Jones Potential

This model has been used to describe real gases in Chapter 4 (page 182):

$$\phi(r) = 4E_e\left[\left(\frac{r_0}{r}\right)^n - \left(\frac{r_0}{r}\right)^m\right] \tag{8.6}$$

where E_e is the depth of the potential well. This model is used for insulating liquids

Ion–Ion Potential

This potential can be used for metals only.

$$\phi = \frac{A}{r^3}\cos(2k_F r) \tag{8.7}$$

where
  $A = $ a parameter
  $r = $ distance between the atoms
  $k_F = $ 'radius' of the Fermi sphere (Chapter 3, page 124).

Quasi-empirical Model

Alternatively, $\phi(r)$ can be calculated from the approximate equations (page 403) which connect the pair distribution function $g(r)$ and the pair potential function $\phi(r)$. In this case an experimental $g(r)$ curve is used for the numerical calculations.

Properties such as the total internal energy and the heat capacity of liquids depend on the functions $g(r)$ and $\phi(r)$. Such relationships will be treated in connection with the property in question.

The pair potential model discussed above has the disadvantage that it is practically impossible to calculate the function $\phi(r)$ and it is necessary to use experimental values instead of a mathematical function, which makes the model semiempirical and limits its use considerably.

In addition, the model shows poor agreement between theory and experiments concerning the heat capacity for liquids. The pair potential model is hardly the desired generally accepted model of liquids and melts. It is urgent to find another and better model.

Need for a New Model for Liquids and Melts

The new model must be able to explain the known properties of liquids and melts, especially the short-range order observed in the X-ray spectra, which implies stable nearest neighbours of the liquid atoms and their ability to move practically freely relative to each other. This contradiction disappears if the liquid or melt is assumed to consist of a great number of molten grains or droplets, which we will call *clusters*.

A cluster in a pure metal melt may be assumed to be a stable unit consisting of a central atom surrounded by the number of atoms, which corresponds to the coordination number Z_{coord}, characteristic of the type of atoms in question. The interatomic forces between the atoms inside a cluster may be strong but weaker than the forces between the atoms in the corresponding solid metal. If the forces between the clusters are assumed to be weak, the ability of the melt to adapt its shape to the shape of the container can easily be explained.

However, no theory which discusses the properties of liquids and melts in terms of clusters has been developed yet and there is no experimental evidence which proves the existence of clusters so far, other then the short-range order observed by the X-ray analysis, especially in binary alloys.

The fundamental problem with the theory of liquids and melts is the difficulty of expressing the positions of the atoms as a function of time. This topic is briefly discussed in connection with the theory of heat capacity of liquids on page 412.

8.4 Melting Points of Solid Metals

At a given pressure, the melting point of an element is the only temperature at which the element can exist in stable form both as a solid and a liquid. Table 8.3 shows that the melting points of metals vary widely, and change slightly with pressure. They depend on the structure of the crystal lattice of the metal and the strength of the interatomic forces.

All matter is in constant motion. In the solid state, the atoms are not fixed to their exact sites in the crystal lattice. The sites are the centre of the atomic vibrations. At the melting point the vibration energy is larger than the binding energy and the crystal structure splits up.

Table 8.3 Melting points of some metals.

Metal	Melting point (°C)	Metal	Melting point (°C)
Ag	961	Mo	2607
Al	660	Na	98
Au	1063	Ni	1453
Be	1278	Pb	328
Co	1495	Pt	1769
Cr	1887	Si	1407
Cu	1083	Sn	232
Fe	1535	Ta	2996
Ga	30	Ti	1675
Ge	938	U	1132
Hg	−39	V	1887
In	157	W	3380
Li	179	Zn	420
Mg	651	Zr	1852

In the melt there is still a short-range order but no long-range order. As was verified in Chapter 1, there remains a certain short-range order in the melt between an atom and its nearest neighbours up to a distance of 3–4 interatomic distances.

8.4.1 Lindemann's Melting Rule

In 1910, Lindemann suggested that crystals melt when the thermal vibrational amplitude of the atoms exceeds a certain fraction of the average interatomic distance. When this fraction was chosen as 10% he obtained the relationship

$$\nu = C \left(\frac{T_M}{M V_m^{2/3}} \right)^{1/2} \tag{8.8}$$

where

C = constant, roughly $9 \times 10^8 \, \text{s}^{-1} kq^{1/2} mk^{-1/2}$

T_M = melting point of the solid

ν = average vibrational frequency of the atoms

M = molar weight

V_m = molar volume of the solid.

The constant is approximately the same for all elements. Theoretical treatments of the equation in the 1970s indicated that the constant depends on the crystal structure of the solid.

8.5 Density and Volume

Density is a very important parameter in the theory of liquids. It is easy to understand that the packing of atoms and the interatomic distances influence the properties of liquids strongly.

The volume of a given amount of a solid is closely related to density. A change in volume implies a change in density. Because mass is independent of temperature, there is a direct coupling between the relative density and volume changes of a homogeneous melt via the relationship $\rho = m/V$:

$$\frac{d\rho}{\rho} = -\frac{dV}{V} \tag{8.9}$$

The hard sphere model has proved to be a useful tool to explain properties of liquids such as viscosity and diffusion. Model experiments with computer simulations have been performed. Alder and Wainwright studied the motions of a great number of particles as a function of the packing fraction p. The *packing fraction* is defined as the ratio of the total particle volume to the total volume of the crystal. If we apply this definition to 1 kmol of atoms with equal radii r we obtain

$$p = \frac{\sum_i V_i}{V_{total}} = \frac{N_A \times \frac{4}{3}\pi r^3}{V_m} \tag{8.10}$$

where

$p =$ packing fraction
$V_i =$ volume of atom i
$r =$ radius of atom
$V_{total} =$ total volume of the crystal
$V_m =$ volume of 1 kmol of the crystal.

As a result of their studies, Alder and Wainwright could predict accurate transport coefficients (page 415) of the 'hard sphere liquid' as a function of packing fraction.

8.5.1 Volume Change on Fusion in Metals and Semiconductors

Wittenberg and DeWitt studied the melting points of metals and semiconductors in 1972 and identified some empirical rules, which are roughly valid for metals and semiconductors:

- The volume changes of metals and semiconductors on fusion can be divided into two classes: class 1 elements increase and classs 2 elements decrease their volume on melting.
- Close-packed class 1 metals with FCC and HCP structures increase their average volume by roughly 4.6%. Close-packed metals with BCC structure show a lower relative volume increase, approximately 2.7%.
  The rare earth metals with their configurations of nonfilled inner shells are not included in these observations.
- The semiconductors Ga, Si, Ge and Bi belong to class 2.

Some examples of metals in these classes are given in Table 8.4.

Wittenberg and DeWitt explained the anomalous contraction of the semiconductors on fusion in the following way. In the solid state, these elements have strong tetrahedral space symmetrical covalent bonds (Chapter 3, page 109) owing to sp^3 hybridization. On fusion, these bonds break and the orbitals become more spherical shaped and the structures become more close-packed. This process results in a volume reduction.

Elements with more complex structures are not included in the simple rules given above.

Table 8.4 Relative volume change of some pure metals at their melting points and 1 atm.

Metal	$\Delta V/V(\%)$
FCC or HCP structures:	
Cu	4.0
Ag	3.5
Au	5.5
Al	6.9
Pb	3.8
Ni	6.3
Mg	3.0
Zn	4.1
BCC structures:	
Na	2.6
K	2.5
Fe	3.6
Complex structures:	
Sn	2.4
Ga	−2.9
Si	−9.5
Ge	−5.1
Bi	−3.9

8.5.2 Volume Change and Heat of Mixing on Mixing Metal Melts

The molar volume V_m is defined as the volume of 1 kmol of a substance. For a pure metal it equals the ratio of the molar weight and the density:

$$V_m = \frac{M}{\rho} \tag{8.11}$$

Figure 8.6 shows the molar volumes for some metals at their melting points. The figure shows a striking periodicity and relationship to the periodic table of the elements.

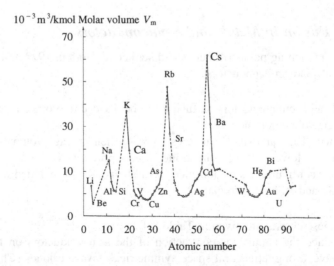

Figure 8.6 Molar volumes of some liquid metals. ○, experimental values; Δ, estimated values. Reproduced with permission from T. Iida and R. I. L. Guthrie, *The Physical Properties of Liquid Metals*. © 1988 (Clarendon Press/Oxford University Press, NY-academic.permissions@oup.com)

The densities of pure ideal metal melts can be derived if the molar volumes (Figure 8.6) are known. Experimental evidence shows that the real molar volumes of many binary alloys (no compound-forming alloys) are functions of the composition of the alloy and can be described by the following relationship:

$$V_{\text{m alloy}} = \frac{x_1 M_1}{\rho_1} + \frac{x_2 M_2}{\rho_2} \tag{8.12}$$

where
$V_{\text{m alloy}}$ = molar volume of alloy
$x_{1,2}$ = mole fractions
$M_{1,2}$ = molar weights
$\rho_{1,2}$ = alloy densities.

When two non-ideal liquids mix, their total volume changes. The mixing process is accompanied by a change in the energy of the system. The volume change can be positive or negative. The heat of mixing can also be either positive or negative. All combinations of signs occur (Table 8.5). The *excess volume* on mixing is defined by the relationship

$$V_{\text{m alloy}}^{\text{excess}} = V_{\text{end}} - V_{\text{beginning}} = \frac{x_1 M_1 + x_2 M_2}{\rho_{\text{alloy}}} - \left(\frac{x_1 M_1}{\rho_1} + \frac{x_2 M_2}{\rho_2} \right) \tag{8.13}$$

Table 8.5 Signs of excess volume and heat of mixing.

$V_{\text{m alloy}}^{\text{excess}}$	$H_{\text{m alloy}}^{\text{excess}}$
+	+
+	−
−	+
−	−

Efforts have been made to explain the volume change on mixing at constant pressure. In 1937, Scatchard presented qualitative description of V^{excess} based on thermodynamics. On this basis, Kleppa and co-workers (1960) suggested an approximate relationship between the heat of mixing and the excess volume:

$$H_{\text{m alloy}}^{\text{excess}} = T \frac{\alpha}{\kappa_T} V_{\text{m alloy}}^{\text{excess}} \tag{8.14}$$

This relationship was later questioned. Several studies of alloy systems show the absence of correlation between the signs of the excess volume and the heat of mixing. Studies of other alloy systems do show correlations between $V_{\text{m alloy}}^{\text{excess}}$ and $H_{\text{m alloy}}^{\text{excess}}$.

Negative $V_{\text{m alloy}}^{\text{excess}}$ values indicate attractive interactions between dissimilar kinds of atoms and occur particularly in compound-forming alloys. Positive excess volumes appear when there are repulsive interactions between the two kinds of atoms. Alloy systems with an immiscibility gap seem to have positive $V_{\text{m alloy}}^{\text{excess}}$ and $H_{\text{m alloy}}^{\text{excess}}$ values.

At present there is neither enough reliable and systematic experimental data nor a general complete theory which can explain the process of mixing metal melts in a satisfactory way.

8.6 Thermal Expansion

8.6.1 Thermal Volume Expansion

The definition of the volume thermal expansion coefficient at constant pressure is the same for liquids as for solids (Chapter 6, page 297). The definition of β is

$$\beta = \frac{1}{V} \left(\frac{\partial V}{\partial T} \right)_p \tag{8.15}$$

where p is the pressure.

In Chapter 6, we found that there are two reasons for thermal expansion in *solids*: an increase in interatomic distances with temperature and vacancy formation, which increases rapidly with temperature. The latter is a minor effect (Figure 6.16 on page 300), but is important close to the melting point.

In melts and liquids, the atoms move comparatively freely relative to each other. In both cases the interatomic distances are not constant but the average interatomic distances increase with temperature. Melts normally have higher volume expansion coefficients than solids.

Experimental evidence shows that β is practically constant within the temperature range from the melting point T_M to the boiling point T_b. In this case we can write

$$V_m(T) = V_m^M [1 + \beta (T - T_M)] \tag{8.16}$$

where
$V_m(T) = $ molar volume at temperature T
$V_m^M = $ molar volume at melting point
$T_M = $ melting point
$T = $ temperature
$\beta = $ volume thermal expansion coefficient at constant pressure.

8.6.2 Temperature Dependence of Density

The density ρ of a liquid or melt in the temperature range from the melting point T_M to the boiling point T_b can be derived from Equation (8.16):

$$\rho = \frac{M}{V_m} = \frac{M}{V_m^M [1 + \beta (T - T_M)]} \tag{8.17}$$

where M is the molar weight of the liquid. If β is *small*, Equation (8.17) can be transformed by series development to

$$\rho = \frac{M}{V_m} = \frac{M}{V_m^M} [1 - \beta (T - T_M)] \tag{8.18}$$

or

$$\rho = \rho_M - \frac{M\beta}{V_m^M} (T - T_M) \tag{8.19}$$

where $\rho_M = M/V_m^M$ is the density at the melting point.

Empirically, it has been found that the temperature dependence at constant pressure of the density of most metal and alloy melts is approximately linear, and can be written as

$$\rho = \rho_M + \Lambda (T - T_M) \tag{8.20}$$

where Λ is a proportionality factor.

A thermal increase in the volume corresponds to a decrease of the density of a melt or a liquid. Consequently, the proportionality factor Λ is negative. Identification of Equations (8.19) and (8.20) confirms this conclusion as β is positive:

$$\Lambda = - \frac{M\beta}{V_m^M} \tag{8.21}$$

Experimental evidence shows that the factor Λ is constant over the whole liquid temperature range from melting point to boiling point for most metals.

If β is not small enough for series development, β and Λ are no longer proportional and only one of them can be regarded as a constant.

Example 8.1

Based on careful measurements, the densities of pure liquid and solid Al are

$$\rho_L = 2380 - 0.35(T - 660)$$

$$\rho_s = 2702 - 0.228(T - 25)$$

where the temperatures are given in °C. The information comes from *Recommended Values of Thermophysical Properties for Selected Commercial Alloys* (K. C. Mills, National Physical Laboratory).

(a) Calculate the molar volume at the melting point V_m^M and the thermal expansion coefficient β for liquid Al from the given relationship.

(b) Plot V_m as a function of T for liquid Al.

Solution:

(a) The given relationship is identical with Equation (8.20), i.e.

$$\rho_L = \rho_M + \Lambda(T - T_M) \tag{1'}$$

Identification gives

$$\rho_M \equiv 2380\,\text{kg/m}^3 \quad \text{and} \quad \Lambda \equiv \frac{-M\beta}{V_m^M} = -0.35\,\text{kg/m}^3\text{K}$$

The molar volume at the melting point temperature is

$$V_m^M = \frac{M}{\rho_M} = \frac{27}{2380} = 0.0113\,\text{m}^3/\text{kmol}$$

$$\Lambda \equiv \frac{-M\beta}{V_m^M} \quad \text{gives} \quad \beta = \frac{-\Lambda V_m^M}{M} = \frac{0.35 \times 0.0113}{27} = 1.7 \times 10^{-4}\,\text{K}^{-1}$$

(b) Equations (8.18) and (8.19) give

$$V_m = \frac{M}{\rho} = \frac{M}{\rho_m + \Lambda(T - T_M)}$$

This function is plotted in the answer.

Answer:

(a) $V_m^M = 0.0113\text{m}^3$, $\quad \beta = 1.7 \times 10^{-4}\,\text{K}^{-1}$.

(b) See the figure on next page.

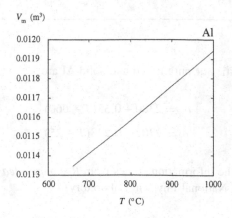

8.7 Heat Capacity

If an amount of heat dQ is added to a system, it is used to increase the internal energy U of the system and to perform work according to the first law of thermodynamics. The increase in the internal energy dU of the system is manifested by an increase dT in its temperature:

$$dU = cm\,dT \qquad (8.22)$$

where
    c = heat capacitivity (J/kg)
    m = mass of the system
    T = temperature.

Equation (8.22) is valid both for solids and liquids but the values of c differs.

For gases and for liquids and solids at high temperatures it is necessary to distinguish between the molar heat capacities at constant pressure and constant volume:

$$C_V = \left(\frac{\partial Q}{\partial T}\right)_V = \left(\frac{\partial U}{\partial T}\right)_V \text{ and } C_p = \left(\frac{\partial Q}{\partial T}\right)_p = \left(\frac{\partial H}{\partial T}\right)_p \qquad (8.23)$$

8.7.1 Theory of Heat Capacity of Liquids

Heat capacities of gases and solids have been treated in Chapters 4 and 6, respectively. In both cases, generally accepted theories have been developed, which agree very well with extensive experimental evidence.

The situation is totally different concerning the heat capacity of liquids. The motion of the atoms in the liquid state is extremely complex. It has not been possible to describe the motion of the atoms as a function of time. The reason is that the atoms move and vibrate relative to each other simultaneously. Hence it is difficult to determine the frequency of the vibrations since the distance between the atoms changes incessantly. The number of energy levels in a liquid is consequently larger than that in a solid.

One common way to proceed is to use the theory of pair distribution function and some approximate model of the pair potential and derive an expression for the internal energy expressed with the aid of these functions. Therefore we will use the functions discussed in Section 8.3.2 on page 402.

Molar Internal Energy

The total internal energy of 1 kmol of a monoatomic liquid can be written as

$$U = \frac{3}{2}RT + \overline{\Phi} \qquad (8.24)$$

where

R = general gas constant
T = absolute temperature
$\overline{\Phi}$ = average total potential energy of the liquid.

The first term on the right-hand side of Equation (8.24) represents the total kinetic energy of the atoms in the liquid (three degrees of freedom times $R/2$). The second term has to be derived.

Consider a reference atom at the origin. According to the theory in Section 2.6.1, the average number of atoms in a spherical shell with radius r and thickness dr at distance r from origin is $w_0 \times 4\pi r^2 dr g(r)$ [Equation (8.3) on page 403]. The average potential energy caused by the interaction between the reference atom and its neighbours in the spherical shell will be $w_0 \times 4\pi r^2 dr g(r) \phi(r)$. The function $\phi(r)$ is assumed to be independent of temperature.

The total potential energy per kilomol in the liquid is obtained by integrating the average potential energy over r and dividing the integral by two to avoid counting the interaction between each pair of atoms twice:

$$\overline{\Phi} = \frac{w_0 N_A}{2} \int_0^\infty g(r)\, \phi(r) \times 4\pi r^2 dr \qquad (8.25)$$

where

w_0 = probability of finding another atom in any randomly selected unit volume
N_A = Avogadro's number (number of atoms per kilomol).

Inserting Equation (8.25) for the average total potential energy into Equation (8.24) and using the relationship $w_0 = N_A/V_m$, where V_m is the molar volume, we obtain

$$U = \frac{3}{2}RT + \frac{2\pi N_A^2}{V_m} \int_0^\infty g(r)\phi(r)\, r^2 dr \qquad (8.26)$$

8.7.2 Heat Capacity at Constant Volume

Using Equation (8.26) and the definition of C_V, we obtain the molar heat capacity at constant volume of the liquid:

$$C_V = \left(\frac{\partial U}{\partial T}\right)_V = \frac{3}{2}RT + \frac{2\pi N_A^2}{V_m} \int_0^\infty \left[\frac{\partial g(r)}{\partial T}\right]_V \phi(r)\, r^2 dr \qquad (8.27)$$

where V_m is the molar volume (m^3/kmol).

Since $\phi(r)$ is assumed to be independent of temperature, the derivative of $\phi(r)$ with respect to T is zero.

Various $\phi(r)$ functions have been tested but the result has not been very successful so far. Nor has any empirical equation that describes C_V in a satisfactory way been found. Experimental results have to be reported in tables and diagrams. Unfortunately, sufficient experimental data for heat capacities of liquids that are accurate enough for testing theoretical models are not available.

8.7.3 Heat Capacity at Constant Pressure

Most metallurgical and chemical processes occur at constant pressure and hence the molar heat capacity at constant pressure C_p is of special interest.

The heat capacities of liquids resemble those of solids in some respects. The difference between C_p and C_V is described by the same relationship for both liquids and solids [Equation (6.74) on page 312], at least in the vicinity of the melting point:

$$C_p - C_V = \frac{\beta^2 V_A^{\theta_D}}{\kappa} (T - \theta_D) \qquad \text{Grüneisen's rule} \qquad (8.28)$$

where

$\beta =$ coefficient of volume expansion of the liquid

$\kappa =$ isothermal compressibility of the liquid

$\theta_D =$ Debye temperature of the liquid

$V_A^{\theta_D} =$ molar volume (kmol) of the liquid at the Debye temperature.

As $\theta_D = h\nu_D/k$ [Equation (6.68) on page 308], Equation (8.28) is of limited interest because of the difficulty in estimating vibrational frequencies in a liquid (page 408).

Temperature Dependence of Heat Capacities of Metal Melts

In general, the heat capacities of liquids vary with temperature. Experimental evidence, for example measurements by Hultgren, indicates that the heat capacities decrease slightly with increase in temperatures. The temperature dependences of some metal melts as functions of temperature are shown in Figure 8.7.

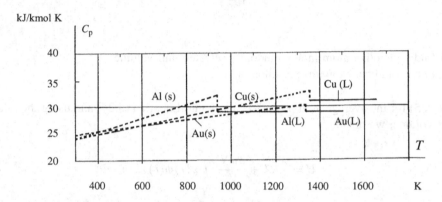

Figure 8.7 Molar heat capacities at constant pressure of solid and liquid Al, solid and liquid Cu, solid and liquid Au as functions of temperature. The dashed curves represent the solid. Reproduced with permission from D. R. Gaskell, *Introduction to Metallurgical Thermodynamics*, 2nd edn. © 1981 Hemisphere Publishing Corporation, now Taylor & Francis.

Figure 8.7 shows the heat capacities at constant pressure of some solid and liquid metals as functions of temperature. The heat capacities of metal melts are of the same magnitude as those of the corresponding solids. Figure 8.7 shows clearly that the structure of solid elements influences the heat capacity strongly. There is a discontinuous change in the heat capacities at the melting points on the Al, Cu and Au curves. The heat capacities of the solid metals vary linearly with temperature.

The heat capacities at constant pressure of the corresponding metal melts seem to be constant over a considerable temperature range. The temperature dependence is small and an average value is used within the given temperature intervals. The functions of the three metals are given in Table 8.6. Other metals, for example iron, show a similar appearance.

Table 8.6 Molar heat capacities at constant pressure as a function of temperature for some solid metals and metal melts.

Metal	C_p (kJ/kmol K)	Temperature range (K)
Al (solid)	$20.7 + 12.4 \times 10^{-3}T$	298–933
Al (liquid)	29.3	933–1273
Cu (solid)	$22.6 + 6.28 \times 10^{-3}T$	298–1356
Cu (liquid)	31.4	1356–1600
Au (solid)	$23.7 + 5.19 \times 10^{-3}T$	298–1336
Au (liquid)	29.3	1336–1600

In other cases, especially for metals with low melting points, when the C_p–T curve is inclined or bent (Figures 8.8 and 8.9), an empirical function is used to describe the temperature dependence of C_p:

$$C_p = a + bT + \frac{c}{T^2} + dT^2 \tag{8.29}$$

where a, b, c and d are parameters, which have to be determined from experimental data. Some examples of such empirical functions are given in Table 8.7. They are shown in Figures 8.8 and 8.9.

Table 8.7 Molar heat capacities at constant pressure for some metal melts.

Metal	a (kJ/kmol K)	b (kJ/kmol K^2)	c (kJ K/kmol)	d (kJ/kmol K^3)	Temperature range $T_M - T$(K)
Li	24.5	$+5.5 \times 10^{-3}$	8.7×10^5		454–1200
Na	37.5	-19.2×10^{-3}		10.6×10^{-6}	371–1200
K	37.2	-19.1×10^{-3}		12.3×10^{-6}	336–1037
Ga	26.4		1.3×10^5		303–1200
Hg	30.4	-11.5×10^{-3}		10.1×10^{-6}	298–630
In	30.3	-1.4×10^{-3}			430–800
Pb	32.4	-3.1×10^{-3}			600–1200
Bi	20.0	$+6.1 \times 10^{-3}$	21×10^5		544–820
Sn	34.7	-9.2×10^{-3}			510–810

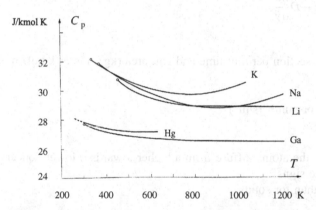

Figure 8.8 Molar heat capacities C_p of liquid Na, K, Li, Hg and Ga as functions of temperature. The curves start at the melting points of the metals, indicated by small circles. Reproduced with permission from T. Iida and R. I. L. Guthrie, *The Physical Properties of Liquid Metals*. © 1988 (Clarendon Press/Oxford University Press, NY-academic. permissions@oup.com)

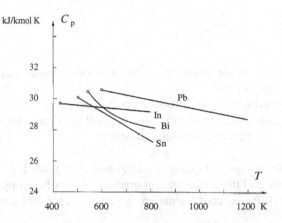

Figure 8.9 Molar heat capacities C_p of some liquid metals as functions of temperature. The curves start at the melting points of the metals, indicated by small circles. Reproduced with permission from T. Iida and R. I. L. Guthrie, *The Physical Properties of Liquid Metals*. © 1988 (Clarendon Press/Oxford University Press, NY-academic. permissions@oup.com)

8.8 Transport Properties of Liquids

Chapter 7 deals with transport phenomena in solids. Corresponding phenomena appear also in liquids with an important complement. In addition to diffusion, thermal and electrical conduction, *viscosity* or *internal friction* occurs in liquids and melts but not in solids.

Viscosity appears in flowing liquids. Momentum is transported from one liquid layer to another.

A survey of the different types of transport phenomena in liquids and the transported physical quantities is given in Table 8.8. They will be discussed in the remaining sections of this chapter.

Table 8.8 Transport phenomena in liquids.

Transport phenomenon	Involves transport of
Diffusion	Mass
Viscosity	Momentum
Thermal conduction	Energy
Electrical conduction	Charge

8.9 Diffusion

Diffusion in melts is a phenomenon that controls the microsegregation in solidifying metal melts and casting processes. Diffusion is also important in other industrial processes. For this reason, it is of great interest to make efforts to understand the mechanisms of diffusion in liquids.

The macroscopic laws of diffusion (Fick's laws) in liquids are analogous to those in solids. The diffusion coefficient is defined by Equation 5.112) on page 254:

$$J = -D \text{ grad } c \quad \text{Fick's first law} \tag{8.30}$$

In the one-dimensional case, Equation (8.30) can be written as

$$J = -D \frac{dc}{dy} \tag{8.31}$$

where

$J =$ flux or net amount of diffusing atoms passing a cross-section per unit time and unit area ($kg/m^2 s$ or $kmol/m^2 s$ or numbers/$m^2 s$)
$D =$ diffusivity or diffusion coefficient (m^2/s)
$c =$ concentration of diffusing atoms (kg/m^3 or $kmol/m^3$ or numbers/m^3)
$y =$ coordinate in diffusion direction.

The minus sign in Equations (8.30) and (8.31) indicates that the atoms diffuse from a higher towards a lower concentration. The flux and the concentration gradient always have opposite signs.

The diffusion coefficients are generally larger for liquids than for solids.

8.9.1 Simple Model of Diffusion in Liquids. Diffusion in Metal Melts at Temperatures Close to the Melting Point

Experimental evidence shows that the self-diffusion coefficient D is of the magnitude $10^{-9} m^2/s$ for most pure metal melts at temperatures close to the melting point of the metal. The values of D are of the same magnitude in spite of the fact that both the strength and type of forces involved in the solidification processes vary widely between different metals.

In this respect, the behaviour of metal melts differs strongly from that of solid metals. In solid metals the self-diffusion coefficient at temperatures close to the melting point varies considerably from one metal to another.

Diffusion Coefficient as a Function of Activation Energy and Temperature

The temperature dependence of the diffusion coefficient D in liquids is very strong and can, at constant pressure, be represented by the same equation as in solids [Equation 5.146), page 264]:

$$D = D_0 e^{-\frac{\overline{U_{a \text{ act}}^{diff}}}{k_B T}} \tag{8.32}$$

where

D = diffusion coefficient

D_0 = a temperature-independent constant

$\overline{U_{a\ act}^{diff}}$ = average activation energy of diffusion per atom

k_B = Boltzmann's constant.

$\overline{U_{a\ act}^{diff}}$ increases with the melting point for metal melts, which is shown in Figure 8.10 and Table 8.9.

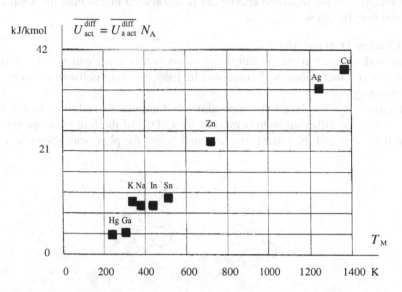

Figure 8.10 Average activation energy of diffusion for some pure metal melts as a function of their melting points. Avogadro's number $N_A = 6.02 \times 10^{26} \text{ kmol}^{-1}$. Reproduced with permission from F. D. Richardson, *Physical Chemistry of Melts in Metallurgy*, Vol. 1. © 1974 Academic Press Inc. (London), now by Elsevier.

Table 8.9 Self-diffusion coefficients and related data for some pure liquid metals. Reproduced with permission from F. D. Richardson, *Physical Chemistry of Melts in Metallurgy*, Vol. 1. © 1974 Academic Press Inc. (London), now Elsevier.

Metal	Temperature range (°C)	Melting point (°C)	$D_0 \text{(m}^2/\text{s)}$	$\overline{U_{a\ act}^{diff}} N_A$ (kJ/kmol)	Typical D values (m²/s)	Measured at temperature T(°C)
Hg	0–100	−38.9	1.0×10^{-8}	4.2	0.92×10^{-9}	30
Na	99–227	97.5	1.1×10^{-7}	10.2	4.19×10^{-9}	98
K	67–217	62.3	1.7×10^{-7}	10.7	5.44×10^{-9}	100
In	175–628	156.4	3.1×10^{-8}	10.5	1.91×10^{-9}	175
Sn	267–683	231.9	3.0×10^{-8}	10.8	3.74×10^{-9}	299
Pb	333–657	327.5	9.15×10^{-8}	18.6	2.50×10^{-9}	343
Zn	420–600	419.5	1.2×10^{-7}	23.5	3.16×10^{-9}	500
Ag	975–1350	960.5	5.8×10^{-8}	32.1	3.22×10^{-9}	1060
Cu	1140–1260	1083	1.46×10^{-7}	40.7	4.16×10^{-9}	1100
Fe 4.6%C	1240–1360	1535 (Fe)	4.3×10^{-7}	51.1	10.0×10^{-9}	1360
Fe 2.5%C	1340–1400	1535 (Fe)	1.0×10^{-6}	65.8	9.0×10^{-9}	1400

8.9.2 Theories of Diffusion in Liquids

Models of Diffusion in Liquids

In Chapter 5, Section 5.6, we discussed diffusion in solids and derived expressions for the diffusion coefficient in terms of atomic quantities.

The atomic theory of diffusion in liquids is far more complicated and less established than that in solids. It is difficult to treat diffusion in liquids theoretically because the atoms in a liquid move easily relative to each other and there is no unique activation energy for the atoms. The symptom of this is poor agreement between theoretical and experimental values of the diffusion coefficient of melts and liquids.

Instead, there is a whole spectrum of activation energies and it is customary to use an average value of them all. This corresponds to the simple model discussed above.

Because of the importance of diffusion in liquids, intensive efforts have been made to understand the diffusion process and develop better theoretical models for the diffusion coefficient in disordered media than the simple one. Four models will be mentioned and are discussed briefly below.

Fluctuation Theory of Diffusion in Metal Melts

Diffusion in liquids is assumed to result from the motion of atoms across small and variable distances in the liquid. These movements are caused by random local density fluctuations. In 1959, Swalin published a model of diffusion in pure metal melts, based on fluctuation theory.

From geometric considerations, he concluded that approximately four atoms in addition to the diffusing atom are involved in a fluctuation. In Figure 8.11a the diffusing atom is marked by x. Two of the four accompanying atoms (marked 1 and 2) are located in the plane of the paper and the other two above and below the plane (not shown in Figure 8.11).

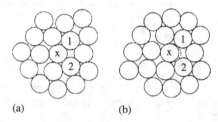

(a) (b)

Figure 8.11 (a) Schematic drawing of a liquid metal. (b) Schematic drawing of the same liquid metal showing a slight density fluctuation after a jump of the diffusing atom. Reproduced with permission from R. A. Swain, On the theory of self-diffusion in liquid metals, *Acta Metall.* **7**, 736–740. © 1959 Elsevier.

In crystals, there are vacancies in the crystal lattice (used by substitutionally dissolved atoms) and empty spaces between the lattice atoms (used by interstitially dissolved atoms). The vacancies and empty spaces enable a diffusing atom to jump from one site into a vacant site if it has enough energy to overcome the energy barrier, e.g. the activation energy.

In liquids, the atoms can move fairly freely relative to each other and there is no energy barrier to overcome for the diffusing atom. In liquid and melts, there are no vacancies such as those in a solid. Instead, holes or voids of variable size appear. Energy is required to create these voids. The voids must exceed a certain critical size, big enough for the diffusing group of atoms. The energy of void formation is assumed to be proportional to the heat of vaporization.

Swalin studied the probability forming voids and the size distribution of fluctuations. For the energy calculations of the creation of voids he used the Morse function (Chapter 2, page 83), which represents the energy between two atoms in a diatomic molecule as a function of the distance between them. It is reasonable to assume that the Morse function can be applied to liquids also.

The calculations will be omitted here but the resulting diffusion coefficient was found to be

$$D = \frac{3Z_{coord}^2 N_A k_B^2 T^2}{96 h (-\Delta H_v) \alpha^2} \qquad (8.33)$$

where

Z_{coord} = coordination number of the atoms
N_A = Avogadro's number (number of atoms/kmol)
k_B = Boltzmann's constant
T = absolute temperature of the melt
h = Planck's constant
$-\Delta H_{vap}$ = molar heat of vaporization of the melt
α = material constant.

The material constant α is related to the force constant between the atoms and can be written as

$$\alpha = \sqrt{\frac{Z_{\text{coord}}N_A k}{4(-\Delta H_{\text{vap}})}} \tag{8.34}$$

where k is a force constant characteristic of the atoms of the melt.

Comparison between Theory and Experiments

Swalin compared self-diffusion data obtained for the metals Hg, Na, In, Sn and Ga. The experimental valued were compared with values of the diffusion coefficient calculated for the five metals in question. The result is shown in Figure 8.12, where $\log D$ is plotted as a function of the inverted absolute temperature.

If the universal constants are introduced into Equation (8.33), we can write

$$D = 1.29 \times 10^{-10} \times \frac{Z_{\text{coord}}{}^2 T^2}{(-\Delta H_{\text{vap}})\alpha^2} \ \text{m}^2/\text{s} \tag{8.35}$$

If we disregard the variation in coordination number for the five metals and regard Z_{coord} as a constant, we can derive an expression for the slopes of the straight lines in Figure 8.12 as follows.

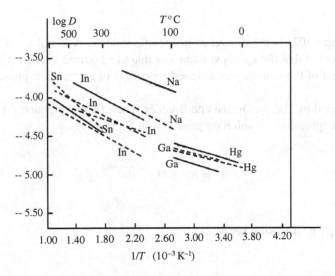

Figure 8.12 Comparison between experimental and theoretical diffusion data (dashed lines) for Hg, Na, In (several experimental values), Sn and Ga. Reproduced with permission from R. A. Swalin, On the theory of self-diffusion in liquid metals, *Acta Metall.* **7**, 736–740. © 1959 Elsevier.

If we replace $1/T$ by x and take the logarithm of Equation (8.35), we obtain

$$\log D = \log constant - 2\log x$$

or

$$\log D = \log constant - 2\log e^{\ln x}$$

or

$$\log D = \log constant - 2\ln x \log e \tag{8.36}$$

Taking the derivative of Equation (8.36) with respect to x gives

$$\frac{\mathrm{d}\log D}{\mathrm{d}x} = -\frac{2\log e}{x} = -2 \times 0.4343T \tag{8.37}$$

If we extend the right-hand side of Equation (8.37) with the gas constant R, we obtain

$$\frac{d(\log D)}{d\left(\dfrac{1}{T}\right)} = -\frac{2RT}{2.3R} = -\frac{Q}{2.3R} \tag{8.38}$$

where $Q = 2RT$ represents the formation energy of the voids.

The higher the temperature is, the higher will be the formation energy, i.e. the more difficult will be the formation of a void.

The Q values for the metal melts can be calculated from the slope of respective line. The value of the diffusion coefficient extrapolated to $1/T = 0$ in the diagram is called D_0 and corresponds to the diffusion coefficient at very high temperature.

From the comparison between theoretical and experimental values in Figure 8.12, two conclusions can be drawn:

- The extrapolated values of D_0 for the theoretical and experimental lines show surprisingly good agreement, especially when the long extrapolation distance is taken into consideration.
- According to Equations (8.37) and (8.38), the slopes of the lines should be independent of the element and depend only on the temperature. This statement agrees fairly well for the lines in Figure 8.12.

Significant Structure Theory

The base of Eyring's theory (page 402) is the difference in specific volume. A liquid has a higher specific volume than the corresponding solid. Eyring assumed that the excess volume was due to 'fluidized vacancies', with properties similar to those of gas molecules, whereas the rest of the volume was assigned properties of a crystalline phase. Hence the liquid was regarded as a mixture of gas and solid.

His model was further developed by Hicter, Durand and Bonnier. Their final expression for D can approximately be written as (compare the corresponding expression for solids on page 264 in Chapter 5)

$$D = BT + D_0 e^{-\frac{\Delta H_{a\ act}^{diff}}{k_B T}} \tag{8.39}$$

where

$-\Delta H_{a\ act}^{diff} =$ activation enthalpy of diffusion per atom
$k_B =$ Boltzmann's constant
$T =$ temperature.
D_0 and B are material constants.

Random Barrier Theory

The effects of disorder in the liquid are considered by replacing it by a fictive 'effective' medium. Hörner assumed a statistical distribution of the effective activation enthalpy $(-\Delta H_{a\ act}^{eff})$ from zero up to a maximum value $(-\Delta H_{a\ act}^{max})$. His final expression of the diffusion coefficient is

$$D = \frac{2D_0}{Z_{coord} - 2} \frac{1 - e^{-\left(1 - \frac{2}{Z_{coord}}\right)\frac{\Delta H_{a\ act}^{max}}{k_B T}}}{e^{\frac{2\Delta H_{a\ act}^{max}}{Z_{coord} k_B T}} - 1} \tag{8.40}$$

where

$Z_{coord} =$ number of nearest neighbours
$D_0 =$ a pre-exponential, temperature-independent factor
$-\Delta H_{a\ act}^{max} =$ maximum activation enthalpy of diffusion
$k_B =$ Boltzmann's constant
$T =$ absolute temperature.

The theory has been fairly successful and can among other results explain the self-diffusion in sodium, which deviates from the simple Arrhenius equation.

Theory of Dense Gases

Diffusion in disordered media has been fairly successfully explained by Enskog. He treated the liquid particles as a number of equal hard spheres with a temperature-dependent diameter σ. His final expression for the hard sphere diffusion coefficient is

$$D_{\text{hardsphere}} = \frac{3}{8} \frac{V}{\sigma^2} \sqrt{\frac{k_B T}{\pi m}} \frac{1}{g(\sigma)} \tag{8.41}$$

where
  V = volume of a hard sphere
  σ = temperature-dependent diameter of a hard sphere
  m = mass of a hard sphere
  k_B = Boltzmann's constant
  $g(\sigma)$ = the pair distribution function.

8.9.3 Diffusion of Interstitially Dissolved Atoms in Metal Melts

Most metal melts contain impurities and dissolve gases when they are exposed to gases, for example air at high temperature.

Gases such as H_2, O_2 and N_2 are interstitially solved as atoms in metal melts. The lighter the interstitial atoms are, the easier they can move in the solvent and the larger will be their diffusion coefficients in the melt. This is confirmed by Table 8.10.

Table 8.10 Diffusion coefficients of some gases dissolved at low concentrations in metal melts.

Solvent	Diffusing element	$T(°C)$	$D(m^2/s)$
Fe	H	1600	132×10^{-9}
Fe	N	1600	$(6-11) \times 10^{-9}$
Fe	O	1560–1660	2.5×10^{-9}
Ni	H	1500	$(66-320) \times 10^{-9}$
Ni	O	1511	12.6×10^{-9}
Cu	H	1200	$(126-525) \times 10^{-9}$
Cu	O	1400	12×10^{-9}

8.9.4 Diffusion in Alloys

Diffusion in solid alloys has been treated extensively in Section 5.6 in Chapter 5. Experiments on liquid alloys indicate that the same final equations can be applied to liquid and solid alloys. By use of Equation (5.176) on page 272 and Equation (5.156) on page 266, we obtain

$$\tilde{D} = x_A \tilde{D}_B + x_B \tilde{D}_A \tag{8.42}$$

and

$$\tilde{D}_A = D_A \left(1 + \frac{d\ln\gamma_A}{d\ln x_A}\right) \tag{8.43}$$

$$\tilde{D}_B = D_B \left(1 + \frac{d\ln\gamma_B}{d\ln x_B}\right) \tag{8.44}$$

where

$\tilde{D}$ = chemical diffusion coefficient of the binary alloy
x_i = mole fraction of element i of the alloy (i = A, B)
γ_i = activity coefficients of element i of the alloy
$\tilde{D}_i$ = diffusion coefficient of atoms i at the given composition of the alloy (i = A, B)
D_i = self-diffusion coefficient of atoms i (i = A, B).

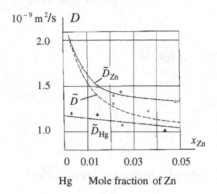

Figure 8.13 Chemical diffusion coefficient and diffusion coefficients of the system Hg–Zn at 30 °C (after Schadler and Grace). The $\tilde{D}_{Zn}$ curve intersects the D axis at D_{Zn}, the self-diffusion coefficient of Zn (not shown). Compare Figure 5.35 on page 273. Reproduced with permission from F. D. Richardson, *Physical Chemistry of Melts in Metallurgy*, Vol. 1. © 1974 Academic Press Inc. (London), now by Elsevier.

Figure 8.13 shows a simple example, the variation of $\tilde{D}$, $\tilde{D}_{Zn}$ and $\tilde{D}_{Hg}$ with composition of very dilute solutions of zinc in mercury. In other cases $\tilde{D}_i$ can have a maximum or a minimum as a function of composition.

The ratio of the self-diffusion coefficients D_{Zn} and D_{Hg} is approximately 2. The corresponding ratio for any pair of liquid metals lies approximately within the interval 0.5–2. It is very small compared with the ratio of the self-diffusion coefficients of solid metals, which may be of the magnitude 10^2–10^3.

8.9.5 Experimental Methods of Diffusion Measurements on Metal Melts

Iida and Guthrie in 1988 gave a wide review of the theories and experimental methods to determine various properties of metal melts in their book *The Physical Properties of Liquid Metals*, among them diffusion and viscosity. Theoretical models, semiempirical theories and empirical descriptions of measurements are included, and also short descriptions of the experimental methods. This book illustrates, among other things, the diverging experimental results and manifold of theories of diffusion in metal melts.

There is no generally accepted theory of diffusion in liquid metals that is valid independent of experimental method. Lack of high accuracy and varying experimental results, depending on experimental method, are a difficulty for the development of the theories. The environment of the experiments influences not only the results but also the very laws of diffusion. This is confirmed by space experiments performed by a Canadian research group during the 1990s, which will be discussed on page 424.

Influence of Experimental Method on Diffusion Results

Figure 8.14 shows the diverging results of diffusion measurements close to the melting point of some metals reported by many different authors. The line in the diagram corresponds to the empirical function

$$D_M = 0.35 \times 10^{-9} \left(\frac{T_M}{M}\right)^{1/2} V_m^{1/3}$$

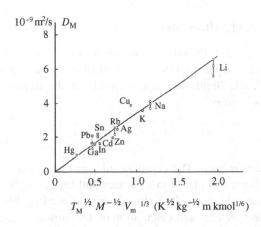

Figure 8.14 Self-diffusion coefficients in liquids at their melting point temperatures as a function of $T_M^{1/2} M^{-1/2} V_m^{1/3}$ ($K^{1/2}$ kg$^{-1/2}$ m kmol$^{1/6}$). There are several experimental values for some metals. Reproduced with permission from T. Iida and R. I. L. Guthrie, *The Physical Properties of Liquid Metals*. © 1988 (Clarendon Press/Oxford University Press, NY-academic.permissions@oup.com)

Experimental Methods for Measurement of Diffusivities in Liquid Metals

There are several methods for measurements of diffusivities in metal melts, including capillary methods and neutron scattering and nuclear magnetic resonance techniques. The latter methods give equivalent values within the frame of experimental errors.

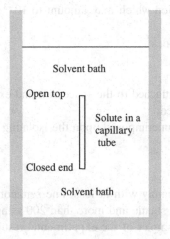

Figure 8.15 The capillary reservoir method. Reproduced with permission from T. Iida and R. I. L. Guthrie, *The Physical Properties of Liquid Metals*. © 1988 (Clarendon Press/Oxford University Press, NY-academic.permissions@oup.com)

One of the most common methods is the *long capillary reservoir method*. A long, thin capillary tube, which is closed in one end, is filled with the solute and placed in a large reservoir with solvent metal of a given constant temperature. The solvent atoms diffuse into the capillary tube with solute. After a certain time t, before solvent atoms have reached the closed end of the tube, the diffusion is interrupted when the tube is removed and quenched. The distribution of the solvent atoms in the solid tube is measured as a function of the distance y from the top (Figure 8.15) with the aid of chemical analysis. The diffusion coefficient can be derived from the equation of random walk:

$$\overline{y^2} = 2Dt \tag{8.45}$$

The advantage of a narrow capillary is that turbulence in the solute is avoided. The diameter of the tube must be > 1 mm otherwise the walls will influence the diffusion process.

The method described above is used for alloys. For measurements of self-diffusivities, the solute is replaced by solvent and mixed with a low concentration of radioactive solvent atoms (see pages 272–273 for diffusion measurements in solids).

Influence of Gravity on Measurements of Diffusivities

R. W. Smith and his research group measured the diffusivities of selected molten metals at various temperatures by using MIM (Microgravity Isolation Mount, developed by CSA, the Canadian Space Agency) as service platform in low Earth orbiting laboratories, first on NASA shuttle flights and more recently on the Russian MIR Space Station. The experiments were performed at the beginning of this century.

Experimental

A variant of the long capillary method was used. The capillary tube had a diameter of 1.5–3.0 mm and a length of 40 mm with a closed end. It was filled with solvent and placed in a furnace and kept at a given constant temperature. A 2 mm solute slug was attached to the entrance of the tube and solute atoms were allowed to diffuse into the tube during a given time. Quenching and chemical analysis of the specimen and calculation of D followed as described above.

Measurements were performed at different temperatures to give information on the diffusion coefficient as a function of temperature and to allow comparison with corresponding results on Earth.

Gravitation Environment

The gravitation in space orbits is very low compared with the gravitation on the surface of the Earth, $g_0 = 9.81$ m/s^2, but is not equal to zero.

The gravitation in the space laboratory consists of two terms, a d.c. component of magnitude $5 \times 10^{-6} g_0$, which depends on the orbit, and an a.c. component, which varies all the time and arises mainly from the momentum changes of the space station, caused by the normal functioning of the control systems and the astronaut/cosmonaut activity. This a.c. component may involve a short-period excitation which may amount to $10^{-3} g_0$. The a.c. component is commonly named 'g jitter'.

The MIM facility could be used in three modes:

1. 'Latched', in which the platform is firmly attached to the spacecraft and exposed to the d.c. and a.c. gravitation fields.
2. 'Isolating', in which the g jitter is suppressed.
3. 'Forcing', in which a forced oscillation is superimposed upon the isolating condition on purpose.

Results

The diffusion of a number of binary alloys, mainly with at least one semiconductor component, has been examined. About 50 samples have been processed on the NASA shuttle and more than 200 samples on MIR during 2 years at the beginning of this century. Figures 8.16–8.19 show some examples of the experimental results.

The results can be summarized as follows:

- The D value is much lower in a low earth orbit than on the surface of the earth. (Figures 8.16 and 8.19).
- The D value depends on the gravitational environment in space (Figure 8.17).
- A reduction in the g jitter reduces the measured D value considerably (Figures 8.16–8.19).
- All the experimental data on binary alloys indicate that the relationship between D and T is linear when the g jitter is suppressed.

In addition, the latched experiments (Figures 8.17 and 8.19) indicate that D is proportional to T^2 in the presence of g jitter. Hence the laws of the temperature dependence change with the gravitational environment:

- On Earth: $D = D_0 e^{-\frac{A}{T}}$ (constant $A > 0$)
- Microgravity in a low Earth orbit: $D = constant \times T^2$
- Microgravity in a low Earth orbit with suppressed g jitter: $D = constant \times T$.

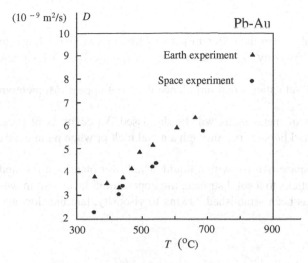

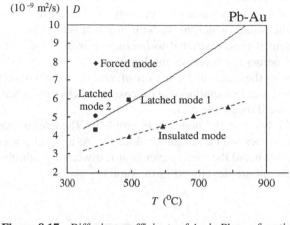

Figure 8.16 Diffusion coefficients of Au in Pb as a function of temperature. Space- and ground-based experiments. Comparison between Earth and space measurements.
Reproduced with permission from R. W. Smith *et al.*, The influence of G. Jitter on the measurement of solute diffusion in dilute liquid metals and metalloids in a low Earth orbiting laboratory. *Proc. Int. Conf. Spacebound 2000.* © Canadian Space Agency.

Figure 8.17 Diffusion coefficients of Au in Pb as a function of temperature. Measurements with the MIM facility in a forced, two latched and an isolating mode.
Reproduced with permission from R. W. Smith *et al.*, The influence of G. Jitter on the measurement of solute diffusion in dilute liquid metals and metalloids in a low Earth orbiting laboratory. *Proc. Int. Conf. Spacebound 2000.* © Canadian Space Agency.

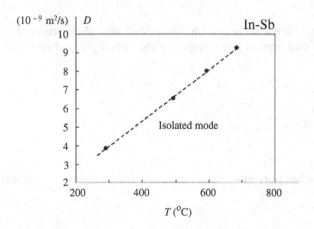

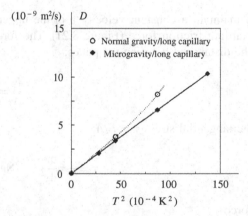

Figure 8.18 Diffusion coefficients of Sb in In as a function of temperature.
Reproduced with permission from R. W. Smith *et al.*, The influence of G. Jitter on the measurement of solute diffusion in dilute liquid metals and metalloids in a low Earth orbiting laboratory. *Proc. Int. Conf. Spacebound 2000.* © Canadian Space Agency.

Figure 8.19 Diffusion coefficients of In in Sn as a function of temperature. Comparison between Earth and space measurements.
Reproduced with permission from R. W. Smith *et al.*, The influence of G. Jitter on the measurement of solute diffusion in dilute liquid metals and metalloids in a low Earth orbiting laboratory. *Proc. Int. Conf. Spacebound 2000.* © Canadian Space Agency.

8.10 Viscosity

Viscosity is of great importance in the metal industry. The viscosity in metal melts influences the flux of the molten metal, for example during casting.

Viscosity also influences the solidification rates of metal melts indirectly and strongly. They are determined by the rate of heat transport away from the solidifying metal. The heat flow depends on conduction, radiation and convection. The convection in the melt depends strongly on its viscosity and influences many processes, among them all casting processes.

8.10.1 Basic Theory of Viscosity

Viscosity is a *dynamic* property of liquids, which manifests itself only when the different parts of the liquid move relative to each other, i.e. in the presence of a velocity gradient in the liquid. Viscosity can be described as a friction force which acts on layers of the liquid, which move with different velocities.

The treatment will be restricted to *laminar* flow of liquids, i.e. at velocities when turbulence does not appear. Momentum is transported across the layers in the liquid.

In addition to the general discussion of viscosity, the viscosities of metal melts will be discussed. Viscosity is of great practical importance in metallurgical processes, for example when small bubbles rise through a metal melt or when nonmetallic inclusions move through the melt.

Consider the two parallel plates in Figure 8.20. The intermediate space is filled with a liquid. The lower plate is at rest and the upper one moves with a constant velocity v. As a liquid always sticks to a solid surface, the upper liquid layer also moves with the velocity v and the lower layer is at rest when equilibrium has been established. Owing to viscosity, laminar flow and a velocity gradient are developed in the liquid.

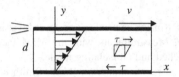

Figure 8.20 Laminar liquid flow between two parallel plates. The friction forces exert shear stress on each layer in the liquid. Adapted with permission from D. R. Gaskell, *An Introduction to Transport Phenomena in Materials Engineering.* © 1992 Macmillan Publishing Company.

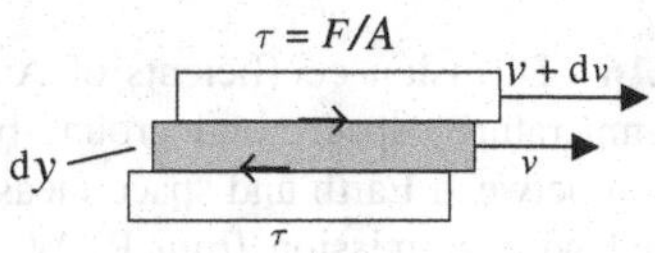

Figure 8.21 The friction forces exert shear stress on each layer in the liquid in Figure 8.20.

If we want to maintain a constant velocity v of the upper plate, a force F, acting on the contact area A, is required to overcome the retarding friction force (Figure 8.21). The force per unit area is proportional to the velocity v and inversely proportional to the distance d between the plates:

$$\frac{F}{A} = -\eta \frac{v}{d}$$

or, in terms of the tangential stress $\tau = F/A$:

$$\tau = -\eta \frac{dv}{dy} \qquad \text{Newton's viscosity law} \qquad (8.46)$$

where

F = friction force
η = dynamic[1] viscosity coefficient of the liquid
v = velocity
dv/dy = velocity gradient.

The dynamic viscosity coefficient is simply *the coefficient of viscosity* of the liquid.

Equation (8.46) can be written as

$$\tau = -\eta \ \text{grad} \ v \qquad (8.47)$$

The tangential stress, which is a shear stress, *is proportional to the velocity gradient.*

Equation (8.46) was proposed by Newton in 1687. Most liquids obey this law and are called Newtonian liquids.

---

[1] The kinetic viscosity coefficient is defined as η/ρ.

Nature of Friction Forces in Liquids

We discussed viscosity in gases in Chapter 4 on page 196. The basic law for gases is formally identical with Equation (8.46) above. However,

- The nature of the friction forces between the layers is entirely different for gases and liquids.

In the case of *gases*, the intermolecular forces are zero and the friction forces are caused by molecules which jump between the gas layers and become retarded or accelerated. In the case of *liquids*, the intermolecular forces in the liquid are responsible for the frictional adhesion forces.

It is well known that the viscosity varies strongly with temperature. The higher the temperature is, the smaller will be the coefficient of viscosity. This topic will be treated in Section 8.10.4 on page 432.

8.10.2 Newtonian and Non-Newtonian Fluids

Liquids that obey Newton's viscosity law [Equation (8.46)], are called *Newtonian* liquids. They have been studied in the preceding sections. Most liquids are Newtonian fluids but there are many liquids that do not belong to this group.

Liquids that do *not* obey Newton's viscosity law are called *non-Newtonian* fluids with a common name. Examples of such fluids are molten plastics, high polymers and slurries of clay and lime.

Below we will analyse and compare the fluid flows through a tube of given length under the influence of a pressure difference for Newtonian fluids and one type of non-Newtonian fluid.

Newtonian Liquids

Stationary Fluid Flow in a Cylindrical Tube Under the Influence of a Pressure on the Newtonian Liquid
Consider a cylindrical tube of length L and a circular cross-section with radius R (Figure 8.22). A fluid flows through the tube under the influence of a pressure difference:

$$\Delta p = p_2 - p_1 \tag{8.48}$$

We choose a volume element in the shape of a hollow cylinder with length L, radius r and thickness dr.

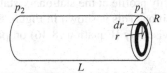

Figure 8.22   Volume element.

The element is exposed to

1. Forces in the flow direction caused by the pressure acting on the flat circular cross-section areas. The net force is

$$F_p = (p_2 - p_1)\pi r^2 \tag{8.49}$$

2. A friction force caused by the viscosity of the liquid acting along the cylindrical surface of the volume element:

$$F_f = -\eta \frac{dv}{dr} \times 2\pi rL \tag{8.50}$$

The retarding friction force is equal to the pressure force and the net force is zero. This condition gives, after separation of the variables and integration, the velocity of the cylindrical layers as a function of r:

$$v = \frac{\Delta p}{4\eta L}(R^2 - r^2) \tag{8.51}$$

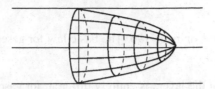

Figure 8.23 Velocity profile in the tube. Velocity v as a function of r. Adapted with permission from D. R. Gaskell, *An Introduction to Transport Phenomena in Materials Engineering*. © 1992 Macmillan Publishing Company.

The velocity is a parabolic function of the radius, which is shown in Figure 8.23.

It can also be shown that the total flux (m^3/s) through the tube is

$$\frac{dV}{dt} = \frac{\pi R^4 \Delta p}{8 \eta L} \qquad \text{Hagen–Poiseuille's law} \qquad (8.52)$$

Non-Newtonian Liquids

There are several subgroups of non-Newtonian fluids. The so-called *Ostwald's law* deals with the relationship between the shear stress τ and the velocity gradient dv/dy:

$$\tau = constant \times \left(\frac{dv}{dy}\right)^n \qquad (8.53)$$

where n is a constant.

Newton's law [Equation (8.46) on page 426] is a special case of Ostwald's law. It is obtained when the constant is replaced by η and $n = 1$.

Fluids that obey Ostwald's law are called *generalized Newtonian fluids*. Below we will discuss the case when $n > 1$. Such liquids are called *dilatant fluids*.

Stationary Fluid Flow in a Cylindrical Tube under the Influence of a Pressure on a Non-Newtonian Dilatant Liquid

In the preceding section, we discussed the velocity profile at the stationary state in a Newtonian liquid. The parabolic velocity profile is the same independent of time and position. This is shown in Figure 8.24a, which is representative for all Newtonian liquids, i.e. liquids that obey Newton's law of viscosity [Equation (8.46) on page 426].

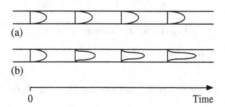

Figure 8.24 Velocity profiles of (a) a Newtonian liquid and (b) a non-Newtonian dilatant liquid flowing in a tube.

In the case of a so-called dilatant fluid, the velocity profile is as shown in Figure 8.24b. The velocity increases successively with time and position. The increase in the velocity continues until the end of the tube is reached or until the velocity equals the limit for laminar flow, e.g. the laminar flow collapses and becomes turbulent.

When the velocity increases, the velocity profile becomes extended in the direction of the flow. This is true also for the volume element which is marked in Figure 8.24b. Its central parts become elongated as more liquid flows out from the front surface than from the back surface per unit time. A lack of atoms arises in the interior of the volume element.

This lack of atoms does not result in pore formation in the fluid. Instead, the lack can be compensated by lateral diffusion of atoms into the volume element from outside if that is possible. In this case, no supply of atoms is possible because of the tube. The only possibility to keep the liquid volume constant without void formation is a deformation of the velocity profile.

The deformation occurs in such a way that the velocity decreases in the outer parts of the tube. This is indicated schematically in Figure 8.24b.

8.10.3 Viscosity in Pure Metal Melts

Theory of Viscosity in Pure Metal Melts

In Chapter 4, we found that the kinetic theory of gases provides a model of viscosity. The coefficient of viscosity for gases was, for example, shown to be a function of the mean free path of the gas and the temperature dependence of η could be derived.

It is desirable to find an analogous model for the viscosity of liquids for a similar analysis. This is extremely difficult, for both theoretical and experimental reasons. The knowledge of the interatomic forces between the atoms in a liquid is poor as the atomic motion in liquids cannot be exactly described as a function of time. The lack of a consistent single model for liquid structure has been discussed on page 402. Approximate models have to be introduced. Many empirical and approximate, semitheoretical models η have been suggested.

The predictions of the various theoretical models have to be compared with reliable experimental results. Experimental methods are discussed briefly in Section 8.10.6 on page 437. Unfortunately, accurate viscosity experiments on metal melts are very difficult to perform because of the high temperatures.

An illustration to this is shown in Figure 8.25, where Guthrie and Iida in 1988 reported experimental measurements of 21 different investigations of the coefficient of viscosity for iron as a function of temperature. It is obviously difficult to test theoretical models on such diverging experimental evidence. For the same reason, the experimental values of coefficient of viscosity, given in tables and figures in this chapter, cannot be taken for granted.

The reason for the manifold of models is the poor knowledge of the interatomic forces in liquids. A few of them will be discussed below.

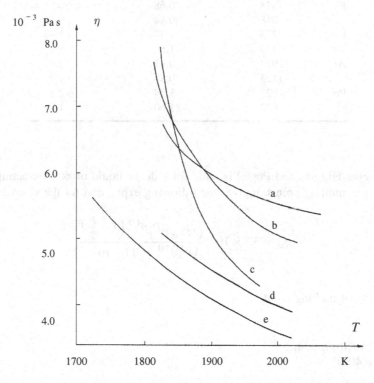

Figure 8.25 The coefficient of viscosity for iron as a function of temperature. Guthrie and Iida collected the curves from 21 different investigations and a few are shown here. The discrepancies may depend on different experimental methods and variation of the carbon content of the specimen. (a) Barfield and Kitchener; (b) Romanow and Kochegarov; (c) Vatolin *et al.*; (d) Saito and Watanabe; (e) Thiele. Reproduced with permission from T. Iida and R. I. L. Guthrie, *The Physical Properties of Liquid Metals*. © 1988 (Clarendon Press/Oxford University Press, NY-academic.permissions@oup.com)

Model Based on the Pair Distribution Function Theory

Using statistical mechanics, an expression for η can be derived if the pair distribution function $g(r)$ (pages 402–403) and the pair potential (page 404) are known. Born and Green derived an expression for the coefficient of viscosity, based on $g(r)$ and $\phi(r)$:

$$\eta = \frac{2\pi}{15}\left(\frac{m}{k_B T}\right)^{1/2} w_0^2 \int_0^\infty g(r)\frac{\partial\phi(r)}{\partial r}r^4 dr \tag{8.54}$$

where

m = mass of atoms
k_B = Boltzmann's constant
T = absolute temperature
w_0 = probability of finding another atom in addition to the reference atom in any randomly selected unit volume
$g(r)$ = the pair distribution function.

The calculated values are compared with experimental values in Table 8.11.

Table 8.11 Comparison between calculated [Equation (8.54)] and experimental values of viscosity of some metal melts.

Metal	Temperature (K)	Calculated values Born–Green (mPa s)	Experimental values (mPa s)
Na	387	0.70	0.68
	476	0.59	∼0.40
K	343	0.68	∼0.51
	618	0.44	0.25
Hg	273	1.78	1.68
	423	1.57	∼1.1
Al	973	0.95	2.9
	1123	0.88	∼1.3
Pb	623	1.84	∼2.2
	823	1.60	1.7

Hard Sphere Model

The hard sphere model (Longuet-Higgins and Pople) is valid for a dense liquid of noninteracting hard spheres. If the model is applied to metal melts at their melting points, it gives the following expression for the viscosity coefficient:

$$\eta_M = 3.8 \times 10^{-8}\frac{(MT_M)^{1/2}}{(V_m^M)^{2/3}}\frac{p^{4/3}\left(1-\dfrac{p}{2}\right)}{(1-p)^3} \tag{8.55}$$

where

η_M = coefficient of viscosity at melting point
M = molar weight
T_M = melting point
V_m^M = molar volume at the melting point
p = packing fraction (page 407).

Longuet-Higgins and Pople used the value 0.45 for the packing fraction and obtained

$$\eta_M = 0.6 \times 10^{-7}\frac{(MT_M)^{1/2}}{(V_m^M)^{2/3}} \tag{8.56}$$

A randomly packed 'liquid' of hard spheres of equal size has a packing fraction $p = 0.66$. If we replace the packing fraction p by this value, we obtain

$$\eta_M = 3.7 \times 10^{-7} \frac{(MT_M)^{1/2}}{(V_m^M)^{2/3}} \qquad (8.57)$$

It is reasonable to assume that the metal melt approximately can be regarded as a randomly packed liquid. Consequently, the value 0.45 of the packing fraction seems to be far too low. On the other hand, a comparison with experimental values of the viscosity coefficient at the melting point indicates that the real packing fraction probably is lower than 0.66.

Andrade's Quasi-crystalline Model
Andrade's model (1934) is based on the idea that the atoms in the liquid state vibrate about their equilibrium positions in random directions and with a variety of frequencies, in analogy with Einstein's oscillators in the solid state. According to his theory, the viscosity of pure (monoatomic) metal melts close to their melting points can be written as

$$\eta_M \approx \frac{4}{3} \frac{\nu m}{a} \qquad (8.58)$$

where
ν = characteristic frequency of vibration
m = atomic mass
a = average interatomic distance.

Andrade used Lindemann's melting rule (page 406) to eliminate the frequency ν and replaced the interatomic distance a by

$$a = \left(\frac{V_m^M}{N_A}\right)^{1/3} \qquad (8.59)$$

where
V_m^M = molar volume of the liquid
N_A = Avogadro's number.

In this way, he obtained (the original constant is slightly adjusted) the coefficient of viscosity at the melting point:

$$\eta_M \approx 1.8 \times 10^{-7} \frac{(MT_M)^{1/2}}{(V_m^M)^{2/3}} \qquad (8.60)$$

where
η_M = dynamic viscosity coefficient of the melt at the melting point
T_M = melting point
M = molar weight
V_m^M = molar volume at the melting point.

A comparison between Equations (8.56), (8.57) and (8.60) shows that the hard sphere model and Andrade's model, from different starting points, give the same result, apart from the value of the constant.

Table 8.12 shows that Andrade's equation in most cases agrees well with experimental data.

Table 8.12 Comparison between calculated [Equation (8.60)] and experimental values of viscosity at melting point of some metal melts.

Metal	T_M (K)	Calculated values of η_M. Andrade's model (mPa s)	Experimental values of η_M (mPa s)
Na	371	0.62	0.70
K	337	0.50	0.54
Al	933	1.79	1.2–4.2
Fe	1808	4.55	6.92
Cu	1356	4.20	4.34
Zn	693	2.63	3.5
Hg	234	2.06	2.04
Pb	600	2.78	2.61

Other Models

Eyring and co-workers in 1964 proposed the so-called *theory of significant structures* for viscosities in liquids. It is an extension of Eyring's vacancy model of a liquid. The viscosity of the liquid is supposed to be a linear function of solid-like and gas-like structures.

Macedo and Litovitz considered the *energy aspect of viscosity*. The layers in the fluid slide relative to each others. All atoms in one layer must pass periodic potential barriers between equilibrium positions when they move along the chain of atoms in the neighbouring layer. The coefficient of viscosity can be expressed as a function of the height of the barrier, called the *activation energy of viscous flow*.

In 1972, Pasternak suggested a theory of viscosity based on the so-called theorem of corresponding states for transport properties, i.e. viscosity and diffusion. At the melting point, his equation becomes similar to Andrade's equation [Equation (8.60)].

8.10.4 Temperature Dependence of Viscosity of Pure Metal Melts. Relationship Between D and η

Relationship Between the Diffusion and Viscosity Coefficients

Diffusion and viscosity are related in the sense that both phenomena deal with atoms and layers of atoms, which move relative to each other in a liquid. The relationship is derived in the box.

Relationship Between D and η

Consider a solution with a concentration gradient. The solute atoms, which are assumed to be spherical, diffuse under the influence of the concentration gradient. According to Fick's first law [Equation (8.31) on page 416], the flux J can be written as

$$J = -D\frac{dc}{dy} \qquad (1')$$

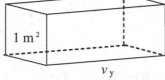

Alternatively, the flux is equal to the concentration c times the volume of the box in the figure:

$$J = c \times 1 \times v_y = cv_y \qquad (2')$$

Equation (2') is inserted into Equation (1'):

$$v_y = -\frac{D}{c}\frac{dc}{dy} = -D\frac{d\ln c}{dy} \qquad (3')$$

(Continued)

Each solute atom can be regarded as a sphere with radius r moving in a viscous solution. The force F_a, acting on an atom, which is required to maintain the motion is given by Stokes' law:

$$F_a = 6\pi\eta r v_y \tag{4'}$$

For 1 kmol the force will be

$$F = N_A F_a = N_A \times 6\pi\eta r v_y \tag{5'}$$

The work to move the N_A atoms a distance dy against the friction force F is equal to the change in the free energy $(-\Delta G)$ or the change $d\mu$ in the chemical potential:

$$d\mu = -F dy \tag{6'}$$

provided that the temperature and pressure are constant. The chemical potential is a function of the activity a of the solute:

$$\mu = \mu^0 + RT \ln a \tag{7'}$$

In a dilute solution, the activity is approximately equal to the concentration:

$$\mu = \mu^0 + RT \ln c \tag{8'}$$

Differentiation of Equation (8') gives

$$d\mu = RT d(\ln c) \tag{9'}$$

Combining Equations (6') and (9') gives

$$-F dy = RT d(\ln c)$$

or

$$-F = RT \frac{d(\ln c)}{dy} \tag{10'}$$

With the aid of Equation (5'), we obtain

$$-N_A \times 6\pi\eta r v_y = RT \frac{d(\ln c)}{dy} \tag{11'}$$

or with the aid of Equation (3')

$$-N_A \times 6\pi\eta r \left(-D \frac{d\ln c}{dy} \right) = RT \frac{d(\ln c)}{dy}$$

which can be reduced to

$$6\pi\eta r D = RT/N_A \tag{12'}$$

The relationship between D and η which is derived in the box can be written as

$$D = \frac{k_B T}{6\pi\eta r} \tag{8.61}$$

where k_B is Boltzmann's constant and r is the radius of the atoms. Equation (8.61) is called the Stokes–Einstein relationship.

Temperature Dependence of Viscosity

The temperature dependence of the diffusion coefficient is given on page 416 [Equation (8.32)]. Equation (8.61) shows that the viscosity coefficient is inversely proportional to the diffusion coefficient and the temperature dependence of η can be expressed as

$$\eta = \eta_0 \frac{T}{T_0} e^{\frac{\overline{U_{a\ act}^{visc}}}{k_B T}} \tag{8.62}$$

A comparison between Table 8.9 on page 417 and Table 8.13 below supports the assumption that $\overline{U_{a\ act}}$ is the same for D and η for the metals in question when the poor accuracy of the measurements is taken into consideration.

Empirical Relationship of η and T

On page 432, the concept of the activation energy of viscous flow was introduced. This quantity appears in a frequently used empirical function, which is often applied to describe the temperature dependence of viscosity:

$$\eta = \eta_0 e^{\frac{\overline{U_{a\ act}^{visc}}}{k_B T}} \tag{8.63}$$

Some values of $\overline{U_{act}^{visc}}$ derived from plots of the function $\ln \eta$ versus $1/T$ are given in Table 8.13.

Table 8.13 Activation energy of viscous flow.

Metal	$\overline{U_{act}^{visc}}$ (kJ/kmol)
Hg	5.2
Na	10.2
Sn	12.2
Pb	16.4
Ag	31.4
Fe 2.5%C	71

Examples of the temperature dependence of the dynamic viscosity in pure metal melts at constant pressure are shown in Figures 8.26 and 8.27. The figures show that most of the curves are straight lines.

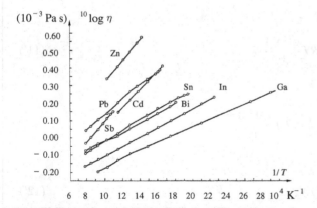

Figure 8.26 Temperature dependence of the dynamic coefficient of viscosity at constant pressure of some metals. Reproduced with permission from T. Iida and R. I. L. Guthrie, *The Physical Properties of Liquid Metals*. © 1988 (Clarendon Press/Oxford University Press, NY-academic. permissions@oup.com)

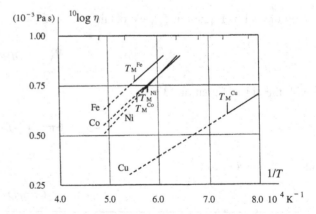

Figure 8.27 Viscosities of Fe, Co and Ni above and below their melting points as functions of temperature. Below the melting point the melts are supercooled (dashed curves). Reproduced with permission from F. D. Richardson, *Physical Chemistry of Melts in Metallurgy*, Vol. 1. © 1974 Academic Press Inc. (London), now by Elsevier.

Within a limited temperature interval, T can obviously be regarded as a constant because the exponential factor changes much more rapidly than T in Equation (8.62). Hence it may be acceptable to use Equation (8.63) for the derivation of $\overline{U}_{a\ act}^{visc}$ instead of Equation (8.62).

8.10.5 Viscosity of Alloy Melts

Dilute Alloys

The viscosities of binary alloys vary in most cases with the composition of the alloy. In Figure 8.28, the influence of the carbon concentration on the viscosity of Fe–C melts of various compositions and different temperatures is shown.

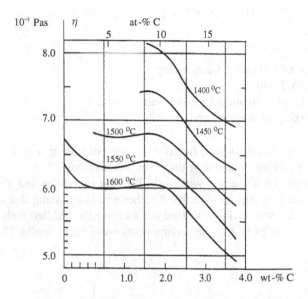

Figure 8.28 Viscosities of Fe–C melts as a function of composition at different temperatures. Reproduced with permission from J. F. Elliott, M. Coleiser and V. Ramakrishna, *Thermochemistry for Steelmaking – Thermodynamics and Transport Properties*, Vol. II. © 1963 Addison-Wesley Publishing Company, Inc. (now under Pearson Education).

The difference between the coefficient of viscosity of a dilute alloy and that of a pure metal melt is often comparatively small, especially at higher temperatures. It may amount to 1–5 % per at- % of solute.

So far, no reliable theoretical model for the viscosity of dilute liquid alloys exists as the knowledge of the forces between unlike atoms is too poor.

Thermodynamic Model of Viscosity for Binary Alloy Melts

Most viscosity models involve the atomic masses and the sizes of the component atoms but there are also models which are based entirely thermodynamic quantities. One of them is the following:

$$\eta = A e^{\frac{\Delta G_{act}^*}{RT}} \qquad\qquad A = \frac{h N_A \rho}{M} \qquad\qquad (8.64)$$

where

$h =$ Planck's constant
$N_A =$ Avogadro's number
$\rho =$ density of the melt
$M =$ molar mass of the alloy melt
$\Delta G_{act}^* =$ Gibbs molar free energy of activation
$R =$ the general gas constant
$T =$ absolute temperature.

ΔG_{act}^* is a function of both composition and temperature of the alloy melt. It has been successfully applied to various metallic and ionic systems.

The viscosity of liquids and melts depend on the motion of atoms, molecules or ions and on the mutual interaction, i.e. the binding between the particles in the liquid or melt and their configurations. The same is true for other quantities, including the heat of mixing of two components in a liquid or melt.

For this reason, it is reasonable to assume that there must exist a relationship between the Gibbs energy of activation and the molar heat of mixing.

Seetharaman and co-workers derived a viscosity model based on the Gibbs free energy of mixing. For binary alloys it is based on Equation (8.64) and the relationship

$$\Delta G^* = \sum_{1}^{2} x_i \Delta G_{i \ act}^* + \Delta G_{mix} + 3RTx_1 x_2 \tag{8.65}$$

where

$\Delta G_{act}^* =$ Gibbs molar free energy of activation for the alloy
$x_i =$ molar fraction of components 1 and 2
$\Delta G_{i \ act}^* =$ Gibbs molar free energy of activation for components 1 and 2
$\Delta G_{mix} =$ Gibbs molar heat of mixing of components 1 and 2.

The viscosities of substitutional metal alloys as a function of composition at a given temperature can be derived from Equation (8.65) if the molar heat of mixing is known at the given temperature.

The model has been tested on eight binary alloys with very different properties and phase diagrams. One of them shows complete miscibility in the solid state (the Ag–Au system). Another one is a system, that includes intermediate phases in the solid state (Ag–Sn). The system Fe–Co was included to find out whether the unfilled d electron shells in the atoms influence the viscosity. In some alloys the viscosities of the pure components were fairly similar (Sn–Bi), in other cases they differed considerably (Cu–Bi).

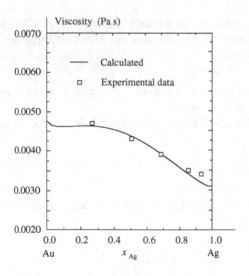

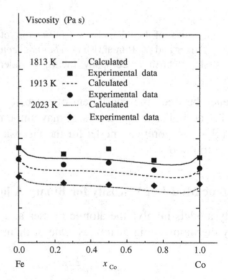

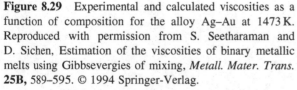

Figure 8.29 Experimental and calculated viscosities as a function of composition for the alloy Ag–Au at 1473 K. Reproduced with permission from S. Seetharaman and D. Sichen, Estimation of the viscosities of binary metallic melts using Gibbsevergies of mixing, *Metall. Mater. Trans.* **25B,** 589–595. © 1994 Springer-Verlag.

Figure 8.30 Experimental and calculated viscosities as a function of composition for the alloy Co–Fe at 1813, 1913 and 2023 K. Reproduced with permission from S. Seetharaman and D. Sichen, Estimation of the viscosities of binary metallic melts using Gibbsevergies of mixing, *Metall. Mater. Trans.* **25B,** 589–595. © 1994 Springer-Verlag.

A severe problem was to find reliable values of the viscosities of the components as the measured values differ considerably and are inconsistent. The experimental values of the Gibbs free energies of mixing were taken from the SGTE Solution

Database (Scientific Group Thermodata Europe) of Hultgren *et al.*: *Selected values of the Thermodynamic Properties of Binary Alloys*. The densities of the pure liquid metals were assumed to be the same as those at the melting point.

Overall, the agreement between the measured and calculated values of the viscosities was found to be good. Figures 8.29 and 8.30 show the viscosity as a function of composition for two of the eight alloys.

8.10.6 Experimental Methods of Viscosity Measurements on Metal Melts

Influence of Gravity Forces and Walls on Viscosity

A striking and unfortunate feature of different experimental viscosity results on metal melts is that they often differ widely. An example is given in Figure 8.25 on page 429, which shows some measurements of the viscosity of iron as a function of temperature. Another puzzling feature of the measured viscosities in liquids is that they depend on the shape and position of the flowing liquid.

Forces other than friction forces, which act on the atoms, are the *gravity forces*. They change the motions of the atoms after a collision, which depends on the impact but also on the direction of the liquid flow, i.e. whether it is parallel, anti-parallel or perpendicular to the gravity force. Consequently, the friction forces change, which affects the value of the viscosity coefficient.

Another factor that affects the viscosity coefficient is whether the melt flows through a narrow passage or not. If the walls are close to the flowing melt they will influence the result of the viscosity measurements. Experimental evidence shows that this effect appears at dimensions of 1–2 mm.

The latter phenomenon is likely to be the reason why viscosity measurements of metal melts differ considerably.

Mainly three different methods are used for such measurements, the capillary tube method ($r < 0.2$ mm) (Figure 8.31), the oscillating vessel method (Figure 8.32) and the oscillating plate method (Figure 8.33).

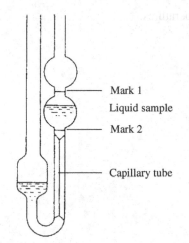

Figure 8.31 The time required for a known amount of the melt to pass a heat-resisting glass or a quartz tube ($r < 0.2$ mm) of given length ($l >$ 70 mm) is measured. η can be calculated from the measurements. Reproduced with permission from T. Iida and R. I. L. Guthrie, *The Physical Properties of Liquid Metals*. © 1988 Oxford University Press.

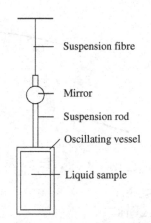

Figure 8.32 A vessel containing a viscous metal melt, hanging in a fibre with a mirror, is set into oscillation about its vertical axis. The motion is gradually damped. The time period and the damping of the decreasing oscillations are measured and η can be calculated. Reproduced with permission from T. Iida and R. I. L. Guthrie, *The Physical Properties of Liquid Metals*. © 1988 Oxford University Press.

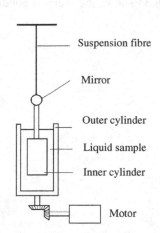

Figure 8.33 A planar plate is set into harmonic linear motion by a driving force. The amplitude of the motion is measured when the plate is surrounded by air or vacuum and a viscous metal melt. The amplitude is reduced in the latter case and η of the melt can be calculated. Reproduced with permission from T. Iida and R. I. L. Guthrie, *The Physical Properties of Liquid Metals*. © 1988 Oxford University Press.

Hence the experimental method of viscosity measurements influences the result strongly. It is difficult to obtain accurate values of η.

8.11 Thermal Conduction

Thermal conductivites of solids has been extensively discussed in Chapter 7.

In *pure metals*, the free electrons are responsible for the heat transport through the metal and the phonon contribution can be neglected.

In *ionic crystals* and *nonmetals with covalent bonds*, there are very few free electrons and the transport of heat is performed by phonons (lattice vibrations) which move through the crystal lattice.

In other nonmetals, for example alloys, both free electrons and phonons contribute to the heat transport in various proportions.

Nonmetals normally have very high melting points, which are hard to achieve. Hence they are of little practical interest and we will only study the thermal conductivity of metal melts below.

Thermal conduction in metal melts is of great importance, for example convection phenomena in furnaces and baths and in casting processes and other industrial processes.

Thermal Conduction in Metal Melts

When a metal melts, its structure will be broken and the electron bands will be strongly changed. Again, it is the complex conditions in the liquid concerning interatomic forces, potential energies and motion of the atoms and the free electrons which have made theoretical studies of the transport phenomena extremely difficult, not to say impossible, so far.

In addition, accurate measurements of heat flow are experimentally difficult, owing to losses to the environment. Relatively few experimental data on thermal conduction are available and there may be large discrepancies between different sets of measurements and different authors.

For these reasons, no consistent theory of thermal conduction is known so far. Some experimental data will be discussed below.

8.11.1 Thermal Conductivity of Pure Metal Melts

Figure 8.34 shows the thermal conductivities λ of some pure metal melts at various temperatures.

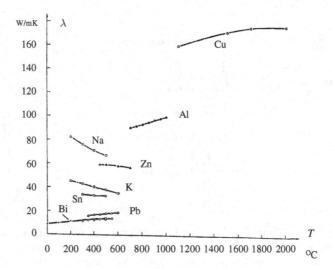

Figure 8.34 Thermal conductivities of some pure metal melts as functions of temperature. Reproduced with permission from T. Iida and R. I. L. Guthrie, *The Physical Properties of Liquid Metals*. © 1988 (Clarendon Press/Oxford University Press, NY-academic. permissions@oup.com).

From Figure 8.34, we can conclude that

- The curves start at different temperatures because the metals have different melting points.
- Copper is the outstanding liquid thermal conductor.

A comparison with Table 7.3 on page 347 shows that the thermal conductivities of solid metals are very much higher than those of the corresponding liquid metals.

Temperature Dependence of Thermal Conductivity

Figure 8.34 shows also the temperature dependence of the thermal conductivities of the metal melts.

- The temperature dependence is comparatively small for most metals.
- Some conductivities increase whereas others decrease with temperature.

As the complexity of the liquid state and the energy states of the free electrons in the melts are very high, no nearby explanation for the latter observation can be given at present.

8.11.2 Thermal Conductivity of Alloy Melts

Figure 8.35 shows the thermal conductivity of some alloy melts as a function of temperature. A comparison between Figures 8.34 and 8.35 indicates that

- The presence of solute atoms lowers the thermal conductivity of the alloy melt considerably compared with those of the two pure metal if their values differ.
- The temperature dependence of the thermal conductivity of alloy melts is comparatively small just as in pure metal melts.

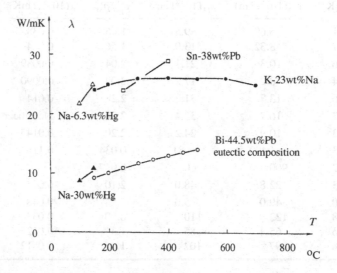

Figure 8.35 Thermal conductivities of some alloy melts as functions of temperature. Reproduced with permission from J. F. Elliott, M. Coleiser and V. Ramakrishna, *Thermochemistry for Steelmaking – Thermodynamics and Transport Properties*, Vol. II. © 1963 Addison-Wesley Publishing Company, Inc. (now under Pearson Education).

Foreign atoms change the average electron concentration of the free electrons in a solid metal. At special critical values this may cause a total change of the crystal structure. It is not surprising that foreign atoms in alloy melts makes it more difficult for the electrons to move through the alloy melt. This reduces the thermal conductivity as heat is transported by the free electrons in metals and alloys.

8.12 Electrical Conduction

The theory of electrical conductivities of solids has been extensively discussed in Chapter 7. Substances with very few free electrons, for example ionic crystals, are insulators at room temperature. When ionic crystals melt, the situation becomes different. The ions become mobile and are able to transport electrical charge through the melt.

Most solid *pure metals* conduct electric charge very well. They contain one or more free electrons per atom. These electrons are entirely responsible for the transport of electric charge through the metal. The theory of electrical conduction in solid metals, which has taken the band theory of electrons into account, agrees very well with experimental data.

Below we will discuss electrical conduction in metal melts, which is important in, for example, refining processes in metals, electrical heating of metals and the production of aluminium.

Electrical Conduction in Metal Melts

The structure of metal melts is very complex. These circumstances have consequences for both thermal and electrical conduction in metal melts. As electric currents are much easier to measure than heat flows, the accuracy of the available experimental data of electrical conduction in liquid metals is much better than that of thermal conduction.

So far, no general and satisfactory theory of electrical conduction for metal melts has been developed. Electrical conductivity is defined as the inverse of resistivity: $\sigma = 1/\rho$.

8.12.1 Electrical Conductivity of Pure Metal Melts

Electrical conduction is normally discussed in terms of resistivity instead of conductivity (page 348 in Chapter 7). The electrical resistivities of some pure solid and liquid metals and semiconductors at various temperatures are given in Table 8.14.

Table 8.14 Electrical resistivities of some liquid and solid metals and semiconductors at their melting points together with the temperature dependence of the metal melt. $\rho = \rho_M + \alpha(T - T_M)$.

Metal	T_M (K)	$\rho_M^s (10^{-8}\,\Omega\,m)$	$\rho_M^L (10^{-8}\,\Omega\,m)$	ρ_M^L/ρ_M^s	$\alpha(10^{-8}\,\Omega\,m\,K^{-1})$	$T_M{-}T$ (K)
Na	371	6.60	9.57	1.45	0.038	371–573
K	337	8.32	13.0	1.56	0.064	337–573
Cu	1356	10.3	21.1	2.04	0.0089	1356–1473
Ag	1234	8.2	17.2	2.09	0.0090	1234–1473
Au	1336	13.7	31.2	2.28	0.014	1336–1473
Zn	693	16.7	37.4	2.24	Not linear	
Al	933	10.9	24.2	2.20	0.0145	933–1473
Si	1713	~2400	~81	0.034	0.113	1713–1820
Ge	1232	900	60	0.067		
Sn	505	22.8	48.0	2.10	0.025	505–1473
Pb	600	49.0	95.0	1.94	0.048	600–1273
Fe	1808	122	110	0.90	0.033	1808–1973
Ni	1726	65.4	85	1.3	0.0127	1726–1973
Co	1768	97	102	1.05	0.0612	1768–1973

The following conclusions can be drawn from the table:

- The resistivities of most metals increase on melting. The mobilities of the lattice atoms increase, which makes it more difficult for the valence electrons to pass.
- The resistivities of the semiconductors Si and Ge decrease considerably on melting because they loose their special semiconductor character during the transformation.
- Ferromagnetic metals show little variation of resistivity on melting. The magnetic interaction between the atoms does not change during the melting process.

Temperature Dependence of Resistivity of Metal Melts

The temperature dependence of many metal melts is linear. In this case the resistivity of the melt is a linear function of the temperature:

$$\rho^L = \rho_M^L + \alpha\,(T - T_M) \tag{8.66}$$

The constant α in sixth column in Table 8.14 is identical with the temperature coefficient α in Equation (8.66). The datas are missing for some metals. The reason for this may be that the experimental data are insufficient or that the temperature dependence is not linear. Table 8.14 shows that

- The resistivities of most metal melts increase linearly with increasing temperature.

The increased kinetic motion of the atoms in a metal melt obstructs the motion of the free electrons, which results in an increase in resistivity. Figures 8.36 and 8.37 show the resistivities of some alloys and pure metals as functions of temperature.

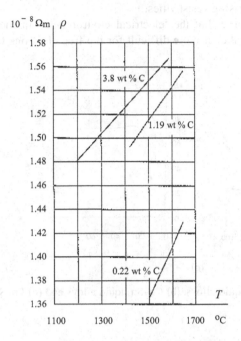

Figure 8.36   Resistivities of some liquid Fe–C alloys as functions of temperature. Reproduced with permission from T. Iida and R. I. L. Guthrie, *The Physical Properties of Liquid Metals*. © 1988 (Clarendon Press/Oxford University Press, NY- academic.permissions@oup.com).

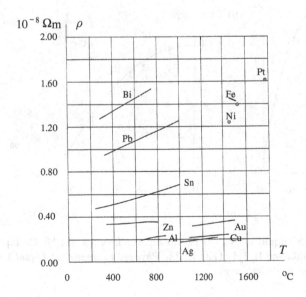

Figure 8.37   Resistivities of some metal melts as functions of temperature. Reproduced with permission from J. F. Elliott, M. Coleiser and V. Ramakrishna, *Thermochemistry for Steelmaking – Thermodynamics and Transport Properties*, Vol. II. © 1963 Addison-Wesley Publishing Company, Inc. (now under Pearson Education).

8.12.2 *Electrical Conductivities of Alloy Melts*

Metal products are often made of alloys to improve, for example, their mechanical strength, magnetic properties, wear resistance, temperature resistance, oxidation resistance, corrosion resistance (stainless steel) and other properties. Hence it is important to investigate the properties of alloys in addition to those of pure metals.

On page 154 in Chapter 3, the electron structures of *solid* alloys was discussed. The introduction of foreign atoms into a crystal lattice of a pure metal changes strongly the 'electrical environment' in the crystal lattice. It was shown that the change in electron concentration, i.e. changed composition of an alloy, may even result in a complete change in the crystal structure. Another consequence is that the levels of the energy bands become distorted and the motion of the free electrons becomes obstructed, which is manifested as increasing resistivities.

Analogously, it is reasonable to assume that the 'electrical environment' in metal melts also becomes distorted by the introduction of alloying elements and makes it more difficult for the free electrons to move through the liquid.

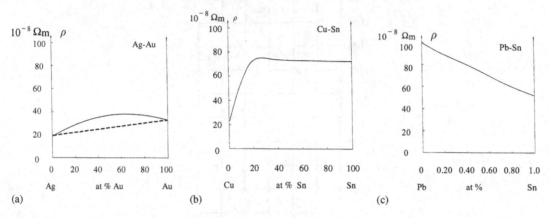

Figure 8.38   Resistivities of (a) Ag–Au liquid alloys, (b) Cu–Sn liquid alloys and (c) Pb–Sn liquid alloys as functions of composition.

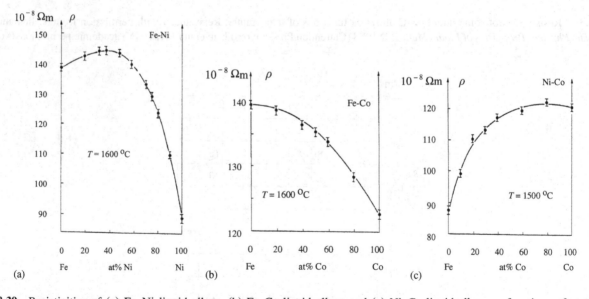

Figure 8.39   Resistivities of (a) Fe–Ni liquid alloys, (b) Fe–Co liquid alloys and (c) Ni–Co liquid alloys as functions of composition. Reproduced with permission from T. Iida and R. I. L. Guthrie, *The Physical Properties of Liquid Metals*. © 1988 (Clarendon Press/Oxford University Press, NY-academic.permissions@oup.com).

Figures 8.38 and 8.39 show the appearances of some typical binary alloy systems. From the figures and other experimental evidence, it can be concluded that

- The resistivity of an alloy, in most cases, *cannot* be obtained by linear interpolation (using mole fractions) of the resistivities of the components. The resistivity of the alloy normally exceeds the interpolated value (dashed line in Figures 8.38a).
- The resistivity–composition curve seems to have a smooth convex shape with a maximum resistivity value for an intermediate composition if the two component atoms have a similar electron configurations (Figure 8.38a and Figure 8.39a–c).

8.12.3 Ratio of Thermal and Electrical Conductivities in Pure Metal Melts

It is well known that the Wiedemann–Franz law (Section 7.4.5 on page 357) is valid for *pure solid metals*. It states that

- The ratio of the thermal and electrical conductivities of a pure metal is proportional to the absolute temperature.

$$\frac{\lambda}{\sigma} = \frac{\pi^2 k_B^2}{3e^2}T = constant \times T \tag{8.67}$$

The value of the constant is $2.45 \times 10^{-8}\,W\Omega/K^2$.

The Wiedemann–Franz law can be regarded as confirmation of the fact that the transport mechanisms behind both thermal and electrical conduction in a metal are the free electrons in metals.

Even if the theory of thermal and electrical conduction in metal melts is not developed, it is more than likely that the Wiedemann–Franz law is also true for metal melts.

To check this assumption, the ratio of the thermal and electrical conductivities, measured separately but at the same temperature, was calculated. The results were not very promising as deviations between 30 and 100% were obtained.

At the beginning of the 1970s, a new experimental method was developed and the measurements of the thermal and electrical conductivities of some metal melts were repeated. The difference was that the two conductivities were measured *simultaneously* and *with much higher accuracy* than in the earlier measurements. In this way, Busch *et al.* and Haller *et al.* could show that the Wiedemann–Franz law *is* valid for liquid Ga, Hg and Sn.

The earlier failure can be ascribed to the poor accuracy of the measurements of the thermal conductivity. The method for measuring the thermal and electrical conductivities simultaneously offers a possibility to obtain more accurate determinations of thermal conductivities than with direct measurements. By experimental determination of λ/σ and σ with good accuracies, λ can be calculated with satisfactory accuracy.

The method is analogous to the classical determination of the electron mass m. The charge e of the electron is determined by Millikan's oil drop method. The ratio e/m is determined by deviation of a beam of electrons in a magnetic field and the mass m can be calculated with high accuracy.

Summary

■ *Models of Metal Melts*

X-ray analysis shows that a short-range order exists in liquid metals.

Vacancy Model

The liquid is assumed to have the same structure as the corresponding solid, with the important distinction that the vacancy concentration in the liquid is much larger than in the solid. This difference in vacancy concentration between melts and solids is assumed to be responsible for the large difference in the properties of a melt compared with the corresponding solid.

The vacancy model of liquids can explain some of the properties of liquids in a semiquantitative way. If Eyring's vacancy model were true, the X-ray plots of melts should show the presence of longe-range order structures in the liquid states like those found in solids. A serious objection to the model is that no such effects have been found.

■ *Modern Theory of Liquids*

Liquids and melts have irregular structures and resemble amorphous solids, which can be regarded as an 'instant picture' of a liquid. It differs only slightly from the liquid structure we obtain when we consider *an average over time and over all atoms.*

Pair Distribution Function

The lack of a consistent and useful model of liquid structure and the necessity for a mathematical description of the structure have led to the introduction of the concept of the *pair distribution function*, which is of great importance for the description of the structures and properties of liquids and melts.

The pair distribution function $g(r)$ deals with the relationship between a pair of atoms, an arbitrary reference atom and another atom. It is defined as *the relative probability of finding another atom in a unit volume at a distance r from the reference atom*:

$$g(r) = \frac{w_r}{w_0}$$

The probability of finding the second atom at a distance r from the reference atom (origin), independent of direction:

Radial distribution function $= g(r)w_0 \times 4\pi r^2$.

Coordination number of an atom: $Z_{coord} \approx 2 \int\limits_{r_0}^{r_{max}} 4\pi r^2 g(r) dr$

Pair Potential

The pair potential ϕ_{ij} is defined as the potential energy between an atom and its surrounding neighbours. If we choose the atom as the reference atom, the pair potential is denoted by $\phi(r)$.

The pair distribution function and the pair potential are very important quantities as all the properties of pure liquids and melts are functions of $g(r)$ and $\phi(r)$. Pair potentials can, in principle, be derived from quantum mechanics, but this is practically impossible for all atoms except H and He so far. Instead, various models of $\phi(r)$ are used:

Hard sphere potential:

$$\phi(r) = +\infty \text{ for } r < r_0$$
$$\phi(r) = 0 \quad \text{for } r \geq r_0$$

Inverse power potential:

$$\phi(r) = E \left(\frac{r_0}{r} \right)^n$$

Lennard-Jones potential:

$$\phi(r) = 4E_e \left[\left(\frac{r_0}{r} \right)^n - \left(\frac{r_0}{r} \right)^m \right]$$

Ion–ion potential:

$$\phi = \frac{A}{r^3} \cos(2k_F r)$$

■ Melting Points of Solid Metals

The melting points of metals vary widely. They depend on the structure of the crystal lattice of the metal and the strength of the interatomic forces.

Lindemann's Melting Rule

$$\nu = C \left(\frac{T_M}{M V_m^{2/3}} \right)^{1/2}$$

■ Volume Change and Heat of Mixing on Mixing Metal Melts

When two liquids mix, their total volume changes. The mixing process is accompanied by a change in energy of the system. The volume change can be positive or negative. The heat of mixing can also be either positive or negative. All combinations of signs occur.

$$V_{m\ alloy}^{excess} = V_{end} - V_{beginning} = \frac{x_1 M_1 + x_2 M_2}{\rho_{alloy}} - \left(\frac{x_1 M_1}{\rho_1} + \frac{x_2 M_2}{\rho_2} \right)$$

Heat of Mixing

$$H_{m\ alloy}^{excess} = \frac{\alpha}{\kappa_T} T V_{m\ alloy}^{excess}$$

■ Thermal Expansion

Thermal Volume Expansion

$$\beta = \frac{1}{V} \left(\frac{\partial V}{\partial T} \right)_p$$

Melts normally have higher volume expansion coefficients than solids.

Experimental evidence shows that β_{liquid} is practically constant within the temperature range from the melting point T_M to the boiling point T_b.

$$V_m(T) = V_m^M [1 + \beta(T - T_M)]$$

Temperature Dependence of Density

$$\rho = \frac{M}{V_m} = \frac{M}{V_m^M [1 + \beta(T - T_M)]}$$

If β is small:

$$\rho = \rho_M - \frac{M\beta}{V_m^M}(T - T_M)$$

■ Heat Capacity

The motion of the atoms in the liquid state is extremely complex. It has not been possible to describe the motion of the atoms as a function of time. Therefore, it is very difficult to develop a satisfactory theory of heat capacity for liquids.

The only way to proceed is to use the theory of the pair distribution function and some approximate model of the pair potential and derive an expression for the internal energy expressed with the aid of these functions.

$$U = \frac{3}{2} RT + \overline{\Phi}$$

where

$$\overline{\Phi} = \frac{w_0 N_a}{2} \int_0^\infty g(r)\phi(r) \times 4\pi r^2 dr$$

or

$$U = \frac{3}{2} RT + \frac{2\pi (N_a)^2}{V_m} \int_0^\infty g(r)\phi(r) r^2 dr$$

Heat Capacity at Constant Volume

$$C_V = \left(\frac{\partial U}{\partial T}\right)_V = \frac{3}{2}R + \frac{2\pi (N_a)^2}{V_m} \int\limits_0^\infty \left[\frac{\partial g(r)}{\partial T}\right]_V \phi(r)\, r^2 \mathrm{d}r$$

Heat Capacity at Constant Pressure

The difference between C_p and C_V is described by the same relation for both liquids and solids:

$$C_p - C_V = \frac{\beta^2 V_A^{\theta_D}}{\kappa}(T - \theta_D) \quad \text{Grüneisen's rule}$$

The heat capacities of metal melts are of the same magnitude as those of the corresponding solids. The heat capacities at constant pressure of some metal melts seem to be constant over a considerable temperature range and a mean value can be used.

In other cases, when the C_p–T curve is inclined or bent, an empirical function is used to describe the temperature dependence of C_p:

$$C_p = a + bT + \frac{c}{T^2} + dT^2$$

■ Transport Properties of Liquids

Transport phenomenon	Involves transport of
Diffusion	Mass
Viscosity	Momentum
Thermal conduction	Energy
Electrical conduction	Charge

■ Diffusion

Basic Theory of Diffusion

The macroscopic laws of diffusion in liquids are analogous to those in solids:

Fick's first law: $J = -D\,\mathrm{grad}\ c$

In one dimension: $J = -D\dfrac{\mathrm{d}c}{\mathrm{d}y}$

Simple Model of Diffusion in Liquids

Diffusion constant as a function of activation energy and temperature:

$$D = D_0 e^{-\frac{\overline{U_{a\ act}^{diff}}}{k_B T}}$$

Theories of Diffusion in Liquids

Better theoretical models for the diffusion constant in disordered media than the simple one have been developed. Four models are discussed in the text.

- fluctuation theory
- significant structure theory
- random barrier theory
- Theory of dense gases.

Diffusion of Interstitially Solved Atoms in Metal Melts

The lighter the interstitial atoms are, the more easily can they move in the solvent and the larger will be their diffusion constants in the melt.

Influence of Gravity and Other Factors on Measurements of Diffusion Coefficients

Gravity and the influence of the walls of capillary tubes result in varying values of the diffusion coefficient when different experimental methods are used. Space experiments show that even the laws change.

Diffusion in Alloys

Experiments on liquid alloys indicate that the same final equations can be applied to liquid and solid alloys:

$$\tilde{D} = x_A \tilde{D}_B + x_B \tilde{D}_A$$

where

$$\tilde{D}_A = D_A \left(1 + \frac{\mathrm{d}\ln \gamma_A}{\mathrm{d}\ln x_A}\right) \quad \text{and} \quad \tilde{D}_B = D_B \left(1 + \frac{\mathrm{d}\ln \gamma_B}{\mathrm{d}\ln x_B}\right)$$

■ *Viscosity*

Basic Theory of Viscosity

$$F = -\eta A \frac{\mathrm{d}v}{\mathrm{d}y} \qquad \tau = -\eta \frac{\mathrm{d}v}{\mathrm{d}y}$$

The nature of the friction forces between the layers is entirely different for gases and liquids. In the case of liquids the intermolecular forces in the liquid are responsible for the frictional adhesion forces.

The lack of a consistent single model for liquid structure makes the theory of viscosity difficult. Approximate models have to be introduced. Three of them are discussed in the text:

- model based on the pair distribution function theory
- hard sphere model
- Andrade's quasi-crystalline model.

Andrade's equation:

$$\eta_M \approx 1.8 \times 10^{-7} \frac{(M T_M)^{1/2}}{(V_m^M)^{2/3}}$$

agrees well with experimental data in most cases.

Temperature Dependence of Viscosity of Pure Metal Melts

Viscosity decreases rapidly with increase in temperature. Empirical relationship:

$$\eta = \eta_0 \mathrm{e}^{\frac{\overline{U_{a\ act}^{visc}}}{k_B T}}$$

Influence of Gravity and Other Factors on Measurements of Viscosity Coefficients

Gravitation and influence of the walls result in varying values of the viscosity coefficient when different experimental methods are used.

Relationship Between D and η

$$D = \frac{k_B T}{6\pi\eta r}$$

which gives

$$\overline{U_{a\ act}^{diff}} = \overline{U_{a\ act}^{visc}} = \overline{U_{a\ act}}$$

and

$$\frac{\eta}{T} = constant \times e^{\frac{\overline{U_a}}{k_B T}}$$

Viscosities of Alloy Melts

The viscosities of binary alloys vary in most cases with the composition of the alloy. So far no reliable theoretical model for the viscosity of dilute liquid alloys exists as the knowledge of the forces between unlike atoms is too poor.

■ Thermal Conduction

The complex conditions in the liquid concerning interatomic forces, potential energy and motion of the atoms and the free electrons make theoretical studies of the transport phenomena extremely difficult. In addition, accurate measurements of heat flow are experimentally difficult, due to losses to the environment.

For these reasons, no consistent theory of thermal conductivity is known so far. Conclusions are based on experimental evidence.

Thermal Conductivity in Pure Metal Melts

Thermal conductivities of solid metals are very much higher than those of the corresponding liquid metals. Copper is the outstanding liquid thermal conductor.

Temperature Dependence of Thermal Conductivity

The temperature dependence is comparatively small for most liquid metals. Some conductivities increase and others decrease with temperature.

Thermal Conductivity in Alloy Melts

The presence of solute atoms lowers the thermal conductivities of alloy melts considerably compared with those of the pure metal. The temperature dependence of the thermal conductivities of alloy melts is comparatively small just as in pure metal melts.

■ Electrical Conduction

As electrical currents are much easier to measure than heat flows, the accuracy of the available experimental data on electrical conductivities in liquid metals is much better than those on thermal conductivities.

Owing to the complexity of a metal melt, no general and satisfactory theory of electrical conductivity for metal melts has been found. Conclusions are based on experimental evidence.

Electrical Conductivities of Pure Metal Melts

- The resistivities of most metals increase on melting. The mobilities of the atoms increase, which makes it more difficult for the valence electrons to pass.

- The resistivities of the semiconductors Si and Ge decrease remarkably on melting because they change from solid semi-conductors into metal melts.
- Ferromagnetic metals show little variation of resistivity on melting. The magnetic interaction between the atoms does not change during the melting process.

Temperature Dependence of Resistivity of Metal Melts

The resistivities of most metal melts increase linearly with increase in temperature.

Electrical Conductivities of Alloy Melts

The 'electrical environment' in metal melts becomes distorted by the introduction of alloying elements and makes it more difficult for the free electrons to move through the liquid. Therefore, the electrical conductivity is higher in alloys than in pure metals.

The resistivity of an alloy, in most cases, *cannot* be obtained by linear interpolation of the resistivities of the components. The resistivity of the alloy normally exceeds the interpolated value.

■ Wiedemann–Franz Law

The ratio of the thermal and electrical conductivities of a pure solid metal is proportional to the absolute temperature:

$$\frac{\lambda}{\sigma} = \frac{\pi^2 k_B^2}{3e^2} T = constant \times T$$

The Wiedemann–Franz law can be regarded as a confirmation of the fact that the carriers behind both thermal and electrical conduction in a metal are the free electrons in metals. It has been known for a long time that the law is valid for solid metals. With the aid of a new technique, accurate measurements on metal melts were performed in the 1970s, which showed that the Wiedemann–Franz law is also valid for pure metal melts.

Exercises

8.1 When metals solidify their volumes shrink. A measure of the shrinkage is the so-called solidification shrinkage β. It is defined as

$$\beta = \frac{\rho_s - \rho_L}{\rho_s}$$

where ρ_s and ρ_L are the densities of the solid and liquid metal, respectively.

Calculate β for Al, Cu and Fe from the diagrams shown.

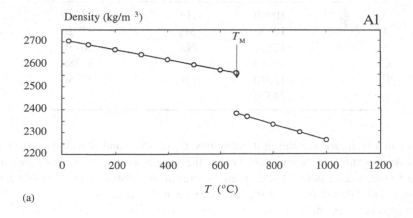

(a)

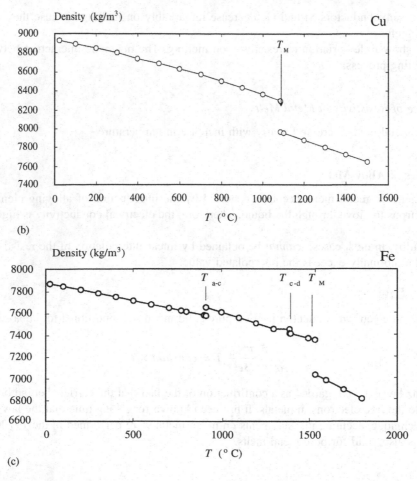

(b)

(c)

Figures (a), (b) and (c) are reproduced by permission of the Controller of HMSO (which is now The Office of Public Sector Information (OPSI)). © 2002 Crown.

8.2 There is an empirical rule called Richards's rule that says that the latent heat of fusion of an element divided by the gas constant R is equal to the melting point temperature (K) of the metal.

Test the rule on the metals in the table and compare the result with Table 8.3 on page 406.

Element	$-\Delta H_{\mathrm{m}}^{\mathrm{fusion}}$ (kJ/kmol)	Element	$-\Delta H_{\mathrm{m}}^{\mathrm{fusion}}$ (kJ/kmol)
Ag	10960	Hg	23240
Al	10560	Mg	87020
Au	12820	Na	26430
Cu	13010	Ni	17380
Fe	13670	Pb	49780
Ge	34870		

8.3 The curves shown in the figure are the heat capacities C_p of solid and liquid Al (upper curve) and C_V of liquid Al (lower curve) as functions of temperature within the temperature interval from just above absolute zero to a temperature close to the boiling point. The diagram is a result of Grimvall's and Forsblom's simulation calculations, based on Ercolessi's and Adams' model of the forces between the Al atoms.

The melting process is an example of a transformation process from an ordered to a disordered state of the system.

Use the diagram to answer the following questions:

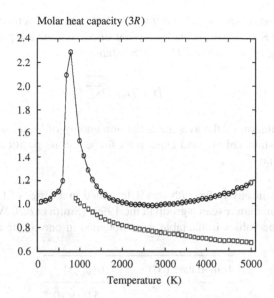

Molar heat capacity (3R)

After Grimvall and Forsblom, *Heat Capacity of Liquid Al.*

(a) Which are the degrees of freedom in Al just before and after the melting process? Express C_V in terms of R.
(b) Calculate an approximate value of the heat of fusion from the diagram. Compare your value with that in the table in Exercise 8.2.
(c) Derive an average value of $C_p - C_V$ for Al for the temperature interval 1500–2000 K. Check your result by calculation of a theoretical value of $C_p - C_V$. Some material constants for solid Al are as follows:

Linear expansion coefficient $\alpha = 2.4 \times 10^{-5}$ K^{-1} Density at 428 K = 2670 kg/m^3.
Debye temperature = 428 K.   Compressibility coefficient $\kappa = 15.7$ m^2/TN.

(d) Discuss the possible reasons for the changes in C_p and C_V for liquid Al within the temperature interval 1500–2750 K shown in the figure.

8.4

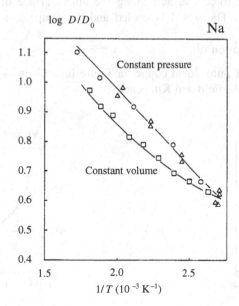

log D/D_0 Na

1/T (10^{-3} K^{-1})

The diagram shows $\log D/D_0$ as a function of the inverse of the absolute temperature $1/T$ for molten Na within the temperature interval from 573 K down to the melting point of Na at constant pressure and constant volume, where D is the self-diffusion coefficient of sodium and D_0 is a constant.

$$D = D_0 e^{-\frac{U_{a\ act}}{k_B T}}$$

(a) Use the diagram for calculation of the average activation energy of liquid sodium.
(b) Discuss the curve for constant volume and explain its shape, i.e. its position and the reason why the lower curve deviates from a straight line.

8.5 The diffusion coefficient of antimony in bismuth was determined in a series of low Earth orbital space experiments at different temperatures by a Canadian research group in the 1990s (Smith *et al.*). When the influence of the microgravity was reduced to a minimum, the values in the table were obtained in one of the experiments.

Temperature (°C)	$D(m^2/s)$
390	5.00×10^{-9}
592	8.36×10^{-9}
785	11.82×10^{-9}

(a) Plot the diffusion coefficient D as a function of temperature $T(°C)$.
(b) What type of function does the curve represent?
(c) Calculate the increase in the diffusion coefficient per °C.

8.6 (a) A thin layer of lubricating oil separates two large parallel plates at a distance of 0.060 mm from each other. The lower plate is a rest. A (tangential) shear stress of $4.8 \times 10^3 \, N/m^2$ is required to keep the upper plate moving with a constant velocity of 0.40 m/s relative to the lower plate. Calculate the coefficient of viscosity of the lubricating oil.

(b) The same lubricating oil is used as a lubricant in a bearing, which consists of a central axis at rest and a coaxial rotating wheel (height 40 mm and inner radius 20 mm). The thickness d of the oil interface between the axis and the wheel is 0.060 mm. The angular velocity of the wheel is 200 r.p.m.

Calculate the total friction force that acts along the inner surface of the wheel and the friction power that is developed at the interface. The wheel is cooled and the temperature can be assumed to be equal to room temperature.

(c) Describe the benefit of lubrication oil.

8.7 The figures show the viscosity of pure liquid copper and pure liquid iron as functions of temperature. The diagrams were derived experimentally by Barfield and Kitchener.

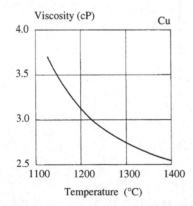

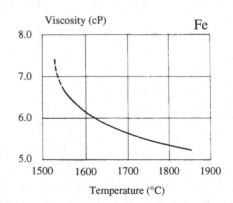

Both figures are reproduced with permission from J. F. Elliott, M. Coleiser and V. Ramakrishna, *Thermochemistry for Steelmaking – Thermodynamics and Transport Properties*, Vol. II. © 1963 Addison-Wesley Publishing Company, Inc. (now under Pearson Education).

(a) Assume that the simple empirical relationship

$$\eta = \eta_0 e^{\frac{\overline{U_a}}{k_B T}} \tag{1'}$$

between η and T is valid. Is this equation adequate for derivation of the average activation energy? If so, derive the values of $\overline{U_{a\ act}}$ for liquid copper and liquid iron.

(b) With the aid of the Stokes–Einstein relationship [Equation (8.61) on page 433], another, more accurate, relationship [Equation (8.62) on page 434]:

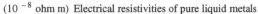

$$\eta = \eta_0 \frac{T}{T_0} e^{\frac{\overline{U_{a\ act}}}{k_B T}} \tag{2'}$$

can be derived. Test this relationship by plotting log η/T as a function of $1/T$ for liquid copper and liquid iron. Discuss the result and compare it with the result in (a).

8.8 Derive the thermal conductivity for the pure liquid and solid metals Cu, Ni and Fe at their melting points with the aid of the Wiedemann–Franz law.

The electrical conductivities, which are easier to determine experimentally than the thermal conductivities, can be derived from the figure. More accurate values of the melting points of the metals than those you can read from the figure can be found in Table 8.3 on page 406.

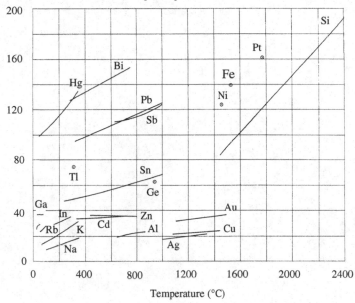

Reproduced with permission from J. F. Elliott, M. Coleiser and V. Ramakrishna, *Thermochemistry for Steelmaking – Thermodynamics and Transport Properties*, Vol. II. © 1963 Addison-Wesley Publishing Company, Inc. (now under Pearson Education).

Answers to Exercises

Chapter 1

1.1a A continuous spectrum, overlapped by a number of sharp discrete lines.

1.1b 11 kV.

1.1c No. Vacancies are formed in a number of Cu atoms, which results in K lines. All K lines appear simultaneously at constant intensity ratios. A filter, which transmits the K_α line but not the more energetic K lines, can be used if only the K_α line is wanted.

1.2 0.280 nm.

1.3 There is no stationary structure in liquids but there exists a permanent short-range order.

1.4 0.154 nm. The K_α line in the characteristic X-ray spectrum of copper.

1.5a ω_r = probability of finding another atom in a unit volume at a distance r from the origin; ω_0 = average probability of finding another atom in an arbitrary unit volume.

Curve 1.11 = curve 1.10 multiplied by $4\pi r^2$. Integration will be a lot simpler when you use a volume element equal to a thin shell $4\pi r^2 dr$ than a volume element in polar coordinates.

1.5b 0.22 nm (intersection between the curve and the r axis).

1.5c 0.28 nm (maximum of the first peak of the curve).

1.5d Coordination number = 12.
[The area under the curve (first peak) $\Lambda = bh/2 = (0.40 - 0.22) \times 13/2 = 11.7$.]

1.6a $\pi/6 \approx 0.52$.

1.6b $\pi\sqrt{3}/8 \approx 0.68$.

1.6c $\pi\sqrt{2}/6 \approx 0.74$.

1.7a *Hint:* Express a and c in terms of the radius R of the atoms.

1.7b $\pi\sqrt{2}/6 \approx 0.74$.

1.8 $\sim 10\%$.

1.9a $5.71 \times 10^3 \, \text{kg/m}^3$.

1.9b Stacking faults, composition deviations.
The structure is close to HCP but not exactly ideal.
$[(c/a)_{\text{ideal}} = \sqrt{8/3} \approx 1.63$ and $(c/a)_{\text{ZnO}} = 0.5195/0.3243 \approx 1.60$.]

1.9c 0.462 nm.

1.10 $\rho_{\text{Cu}} = 8.93 \times 10^3 \, \text{kg/m}^3$ and $M_{\text{Cu}} = 63.54 \, \text{kg/kmol}$.

1.10a 0.362 nm $[4/a^3 = N_A/(M/\rho)]$.

1.10b 0.256 nm $(a/\sqrt{2})$.

1.10c 12 (consider the unit cell).

1.11a $S(\text{hkl}) = f \sum_j e^{-2\pi i(u_j h + v_j k + w_j l)}$

where u_j, v_j and w_j are the components of a position vector r that successively describes the different basic positions of atoms in the unit cell of the crystal which contribute to the sum $S(\text{hkl})$ and f is the atomic scattering factor.

Condition for constructive reflection in a BCC crystal:

h + k + l = even integer

Condition for constructive reflection in an FCC crystal:

hkl are all integers of the same kind, odd or even.

If hkl are mixed integers (odd and even) the intensity of the diffracted X-ray is zero in an FCC structure.

1.11b In a BCC structure the following lines appear: (110), (200), (211), (220), (222), (310), (321).

1.11c Reflections occur in the following planes in an FCC crystal: (111), (200), (220), (222) and (311).

1.12 $\lambda_{\text{deB}} = 3.92 \times 10^{-12} \, \text{m}$. Relativistic calculations are necessary.

1.12a MgO shows FCC structure. The designations of the lattice planes are shown in the last column of the table.

Physics of Functional Materials   Hasse Fredriksson and Ulla Åkerlind
© 2008 John Wiley & Sons, Ltd

Exercise Table 1.12a

$2R_j$	$\dfrac{R_j}{\lambda_{deB}L}$	$\left(\dfrac{R_j}{\lambda_{deB}L}\right)^2$	$\left[h^2+k^2+l^2 = a^2\left(\dfrac{R_j}{\lambda_{deB}L}\right)^2\right]$	hkl
19.9×10^{-3}	4.169×10^9	17.38×10^{18}	3	111
22.9×10^{-3}	4.798×10^9	23.02×10^{18}	4	200
32.4×10^{-3}	6.790×10^9	46.10×10^{18}	8	220
37.8×10^{-3}	7.915×10^9	62.65×10^{18}	11	311
39.7×10^{-3}	8.315×10^9	69.14×10^{18}	12	222
45.8×10^{-3}	9.595×10^9	92.06×10^{18}	16	400
49.9×10^{-3}	10.455×10^9	109.31×10^{18}	19	331
51.1×10^{-3}	10.705×10^9	114.60×10^{18}	20	420
56.0×10^{-3}	11.730×10^9	137.59×10^{18}	24	422

hkl are all integers of the same kind, odd or even. Hence MgO shows FCC structure.

1.12b 0.417 nm. $\left[a^2 = (h^2+k^2+l^2)/\left(\dfrac{R_j}{\lambda_{deB}L}\right)^2\right]$

1.13 Vacancies, interstitials, substitutionals. (pages 24–25 in Chapter 1.)

1.14 Line defects: dislocations (pages 26–27 in Chapter 1).
Interfacial defects: stacking faults, grain boundaries, twinned crystals (pages 28–30 in Chapter 1).

1.15 There are substitutional solutions and interstitial solutions.

1.15a Substitutional solutions. The foreign atoms are too large to be dissolved as interstitials.

1.15b See pages 33–34 in Chapter 1.

1.15c Interstitial solutions contain only small foreign atoms such as H̲, C̲, N̲ and O̲ for space reasons.

1.16 When a solution is saturated and more foreign atoms are added a new phase is precipitated. The new phase is called intermediate phase or secondary phase.

1.17a A random solid solution is a substitutional solution where the substitutional atoms are distributed at random in the lattice sites.

1.17b An ordered solid solution is a substitutional solution where the substitutional atoms appear in a regular and repeated pattern.

1.17c A superlattice has a regular alternation of unlike atoms through the entire crystal.

1.17d Order is maintained locally but not at distant parts of the solution, i.e. there is no long-range order. This is often the situation in alloy melts.

1.17e $\alpha = (1 - P_A/x_A)$. See page 38 in Chapter 1.
α is a measure of the short-range order.
In a random solution $P_A = x_A$ and $\alpha = 0$.

1.17f A group of atoms or other particles (for example vacancies) held together by mutual internal forces.

Chapter 2

2.1 13.6 eV.

2.2 12.8 eV ($n = 1 \rightarrow n = 4$).

2.3a 540 nm.

2.3b 3.9 eV.

2.4 6.61×10^{-34} J s.

2.5 1.91×10^{-11} m for electrons close to the target.

2.6 $0.335 \times p_i^2$ eV, where p_i is a low integer.

2.7 Solutions of the Schrödinger wave equation, which includes a potential function $E_{pot}(x, y, z)$, are only possible for quantized eigenvalues. The solution is the wave function ψ. The wave function ψ is used for probability calculations. See page 53 in Chapter 2.

2.8a Set up an expression of the probability and differentiate it to find the maximum.

2.8b $<r> = 3a_0/2$ and $<r^2> = 3a_0^2$.

2.9a Principal quantum number $n = 1, 2, 3, \ldots$.
The azimuthal quantum number $l = 0, 1, 2, \ldots, (n-1) = $ (s, p, d electrons) represents the angular momentum l of the electron where $|l| = \hbar\sqrt{l(l+1)}$. Magnetic quantum number $m_l = 0, \pm 1, \pm 2, \ldots, \pm l$ gives the projection of the l vector in a special direction.
Electron spin $|s| = 1/2$. Two directions are possible.

2.9b All four quantum numbers (n, l, m_l, m_s) of two electrons within an atom cannot be equal. At least one of them must differ.

2.9c There is a resultant $|L| = \hbar\sqrt{L(L+1)}$ and a resultant $|S| = \hbar\sqrt{S(S+1)}$ of all the l and s vectors of the electrons in an atom. $|L|$ and $|S|$ have a resultant $|J| = \hbar\sqrt{J(J+1)}$. The resultants of filled shells are zero.
Nomenclature: the capital letter S, P, D, ... symbolizes the L values. The principle is the same as for orbital electrons. Superscript $= 2S + 1$. Subscript $=$ the J value.

2.9d No restrictions for Δn, $\Delta L = 0, \pm 1$, $\Delta M_L = 0, \pm 1$, $\Delta S = 0$, $\Delta J = 0, \pm 1$, with the exception that $J' = 0$ does not combine with $J'' = 0$.

2.9e Singlet states have $S = 0$. Triplet states have $S = 1$. This means that the resulting spin vector has different values, which is allowed and common. Transitions from singlet to triplet states are forbidden. Singlet–singlet and triplet–triplet transitions occur provided that the other selection rules besides $\Delta S = 0$ are fulfilled.

2.10a 254 nm.

2.10b 6.4×10^{16} photons/s.

2.11 The total energy required to remove both electrons in He is 79.0 eV.

Hint: The constant in Rydberg's equation can be written $Z^2 R$, where Z is the charge of the nucleus of the atom.

2.12

Exercise Table 2.12

Atom	Atomic number	Electron configuration	Electronic state
H	1	1s	^{2}S
He	2	$1s^2$	^{1}S
Li	3	$1s^2 2s$	^{2}S
Be	4	$1s^2 2s^2$	^{1}S
B	5	$1s^2 2s^2 2p$	^{2}P
F	9	$1s^2 2s^2 2p^5$	^{2}P
Ne	10	$1s^2 2s^2 2p^6$	^{1}S
Na	11	$1s^2 2s^2 2p^6 3s$	^{2}S
Cl	17	$1s^2 2s^2 2p^6 3s^2 3p^5$	^{2}P
Ar	18	$1s^2 2s^2 2p^6 3s^2 3p^6$	^{1}S
K	19	$1s^2 2s^2 2p^6 3s^2 3p^6 4s$	^{1}S

2.13a The selection rules are:

$$\Delta S = 0$$
$$\Delta L = 0, \pm 1$$
$$\Delta J = 0, \pm 1, \text{ except } J' = 0 \leftrightarrow J'' = 0.$$

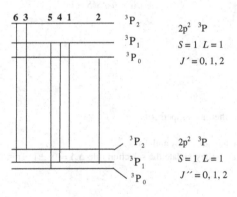

Exercise 2.13a

The lines shown in the figure agree with the selection rules.

2.13b The intervals between the sublevels in the lower triplet help in the analysis.

$$\alpha = 265.0454 \text{ nm} = \text{line 6}$$
$$\beta = 265.0619 \text{ nm} = \text{line 3}$$
$$\gamma = 265.0550 \text{ nm} = \text{line 5}$$
$$\delta = 265.0596 \text{ nm} = \text{line 4}$$
$$\varepsilon = 265.0760 \text{ nm} = \text{line 1}$$
$$\eta = 265.0694 \text{ nm} = \text{line 2}.$$

2.14 Spectral lines split up into components in a magnetic field (pages 69–71 in Chapter 2). For the normal Zeeman effect, $\Delta E_{\text{pot}} = \mu_B \Delta M_L B$, where $\mu_B = eh/(4\pi m)$.

2.15 15 530.00 cm^{-1} and 15 530.00 ± 0.47 cm^{-1}. Compare Figure 2.30 on page 71.

Exercise 2.15

2.16a $-16.25\,\text{eV}$.

2.16b $10.95\,\text{eV}$.

2.16c $15.43\,\text{eV}$.

2.17a There are similar designations and symbols for resulting vectors in atoms and molecules.

$\Lambda =$ the magnitude of the projection of the resulting L vector on the internuclear axis in a diatomic molecule (Figure 2.38 on page 75).

Λ is quantisized: $\Lambda = 0, \pm 1, \ldots \pm L$.

The resulting spin vector S is the same as for atoms. S is integer or half integer.

$$|S| = \hbar\sqrt{S(S+1)}$$

Nomenclature for states:

$\Lambda = 0 \rightarrow \Sigma$ state
$\Lambda = 1 \rightarrow \Pi$ state
$\Lambda = 2 \rightarrow \Delta$ state.

The multiplicity of the state $= 2S + 1$. It is written as a superscript.

2.17b $\Delta\Lambda = 0, \pm 1$ and $\Delta S = 0$.

2.17c $L = 2$, $S = 1/2$, $J = 5/2$ and $3/2$. The components are $^2\Delta_{5/2}$ and $^2\Delta_{3/2}$.

Transitions to the $^2\Sigma$ states do not occur because they violate the selection rule $\Delta\Lambda = 0, \pm 1$. Transitions to $^2\Pi$ and $^2\Delta$ states are possible as they fulfil the same selection rule.

2.18 $E_{\text{rot}} = \frac{I\omega^2}{2} = \frac{(I\omega)^2}{2I} = \frac{p^2}{2I} = \frac{\hbar^2}{2I}J(J+1)$.

2.19a Pure rotation (far-infrared or microwave region): $\Delta J = \pm 1$.

Rotation + vibration near-infrared region: $\Delta J = \pm 1$ and $\Delta v = \pm 1$.

Rotation + vibration + simultaneous electronic transition:

$\Delta\Lambda = 0, \pm 1$ and ΔS.

$\Delta J = \pm 1$ if $\Delta\Lambda = 0$ and no restriction on the vibration quantum number v.

$\Delta J = 0, \pm 1$ if $\Delta\Lambda = \pm 1$ and no restriction on the vibration quantum number v.

2.19b The intensities are determined by the transition integral:

$I = constant \times \int \psi(v')\psi(v'')^* dr$. See pages 81–82 and 84–85.

2.20a $6.43 \times 10^{13}\,\text{Hz}$.

2.20b $2.13 \times 10^{-20}\,\text{J}$ or $0.133\,\text{eV}$.

2.21a $3.75\,\text{eV}$.

2.21b $3.70\,\text{eV}$.

2.22a $2.64 \times 10^{-47}\,\text{kg m}^2$.

2.22b $0.128\,\text{nm}$.

2.23 $0.113\,\text{nm}$.

2.24 Polyatomic molecules have several vibrational frequencies and the moments of inertia for these molecules are $\neq 0$ for all axes. This increases the number of transitions which are allowed and occur in the spectrum. Rotations around three axes are possible.

2.25 $0.096\,\text{nm}$ and $105°$.

Chapter 3

3.1 Cohesive energy refers to separation into neutral free atoms, whereas lattice energy refers to separation into free ions. See page 100.

3.2a 8.97 eV.

3.2b 8.6.

3.3a *Ion bonds*: electrostatic attraction between ions. Example: KCl.

Covalent bonds = *homopolar bonds* = *electron pair bonds*: exchange energy, i.e. a quantum-mechanical effect based on the fact that electrons are indistinguishable. Example: H_2.

Metallic bonds: a lattice of metal ions, surrounded by a common 'electron cloud' which is a very stable system. Examples: all metals.

3.3b Four sp^3 wave functions combine to four symmetrical wave functions, which give four equivalent (half) electron pair bonds in tetrahedral directions. Examples: diamond structure and the CH_4 molecule.

sp^2 hybridization is another example, which occurs in graphite and benzene rings.

3.4a $3.51 \times 10^3 \, kg/m^3$.

3.4b $2.26 \times 10^3 \, kg/m^3$.

3.5a The Schrödinger equation is solved for an electron in a box (the metal) with constant potential energy $E_{pot} = 0$. The solution shows that the energy of the electron is quantized. $E = \hbar^2/2m^* \left(k_x^2 + k_y^2 + k_z^2\right)$ where $k = (k_x k_y k_z)$ is the wavevector of the matter wave of the electron. E and $k_x k_y k_z$ are quantized.

The number of occupied energy levels = number of atoms times the valence number.

The Pauli principle is valid for electrons. Hence much higher energy levels are occupied than that which a Boltzmann distribution would have indicated, because of the Pauli principle. The upper limit of the electron energy of the free electrons is the Fermi energy E_F. See Figure 3.18 on page 117. The work function (Figure 3.17 on page 116) is the energy required to release the most energetic free electrons from the metal surface.

3.5b $E = \dfrac{\hbar^2}{2m^*} k_x^2 \quad \psi = A \sin\left(\sqrt{\dfrac{2m^* E}{\hbar^2}} x\right) \quad k_x = \dfrac{1}{\hbar}\sqrt{2m^* E}$.

3.5c $E_{kin} = \dfrac{\hbar^2}{2m^*}\left(k_x^2 + k_y^2 + k_z^2\right)$

or in one dimension $E = \dfrac{\hbar^2}{2m^*} k_x^2$

The curve (Figure 3.21 on page 120) is *not* continuous. It consists of numerous close and discrete energy levels, because k_x is quantized.

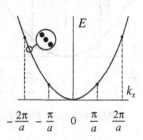

Exercise 3.5c Reproduced with permission from M. Alonso and E. Finn, *Fundamental University Physics*. © Addison-Wesley.

3.6 4.3 eV.

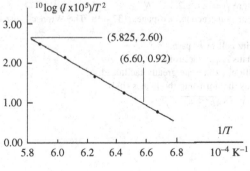

Exercise 3.6

3.7a $f_{FD} = \dfrac{1}{e^{\frac{E - E_F}{k_B T}} + 1}$

See Figure 3.19a and b on page 118.

3.7b The numbers of available and occupied electron energy states between E and $E + dE$ are:

Available energy states : $N(E)dE = \dfrac{(2m^*)^{3/2}}{4\pi^2 \hbar^3} E^{1/2} dE$

Occupied energy states : $N(E)f_{FD}dE = \dfrac{(2m^*)^{3/2}}{4\pi^2 \hbar^3} E^{1/2} f_{FD} dE$

See Figure 3.26 on page 123.

3.7c m^* = effective mass of an electron in the interior of the metal $\neq m$ for an electron outside the metal. See page 112.

3.7d $n_{\text{total}} = \int_0^{E_F} N(E)f_{FD}dE = \dfrac{(2m^*)^{3/2}}{3\pi^2 \hbar^3} E_F^{3/2}$

3.7e $E_F = \dfrac{\hbar^2}{2m^*}\left(3\pi^2 n_{\text{total}}\right)^{2/3}$.

3.8a 1.57×10^6 m/s.

3.8b 1.18×10^6 m/s.

3.9 -4.5×10^{-2} eV.

3.10 Na 3.2 eV, Li 4.7 eV and Al 11.6 eV.

3.11a A periodical potential energy E_{pot} instead of $E_{\text{pot}} = 0$.

3.11b No, both the k values and the kinetic energy of the matter wave of the electron are quantized.

3.11c $k_x = p \times \pi/a$, where p is a positive or negative integer.

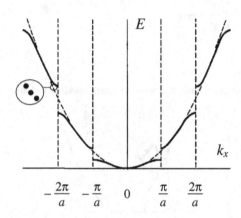

Exercise 3.11c   Reproduced with permission from M. Alonso and E. Finn, *Fundamental University Physics*. © Addison-Wesley.

3.12a Unit cell = the volume in a space lattice, which by translation movements can fill the whole lattice without overlapping or leaving hollow space inside the lattice.

  Primitive cell = unit cell, which contains only one lattice point per cell.

3.12b $a = L/N_{\text{total}}$. Condition: $k = p \times 2\pi/L$, where $p = \pm 1, \pm 2, \ldots \pm N_{\text{total}}$.

3.12c A Wigner–Seitz cell is a primitive cell. For construction, see pages 137–138. The Wigner–Seitz cell in reciprocal space is identical with the first Brillouin zone.

3.12d N_{total}. Each atom represents a Wiegner–Seitz cell in r space.

3.12e Each band (Brillouin zone) can accommodate $2N_{\text{total}}$ electrons.

3.13a The theoretical calculations to find the Brillouin zones are greatly facilitated.

3.13b The $G[\text{hkl}]$ vector has the same direction as the normal to the planes (hkl).

  $G(\text{hkl}) = \dfrac{2\pi}{d_{\text{hkl}}}\hat{n}$ and $|G(\text{hkl})| = \sqrt{(h\boldsymbol{a}^*)^2 + (k\boldsymbol{b}^*)^2 + (l\boldsymbol{c}^*)^2}$

3.13c $d_{\text{hkl}} = \dfrac{2\pi}{\sqrt{(h\boldsymbol{a}^*)^2 + (k\boldsymbol{b}^*)^2 + (l\boldsymbol{c}^*)^2}}$

3.13d $d_{\text{hkl}} = \dfrac{a}{\sqrt{h^2 + k^2 + l^2}}$

3.14a The diffraction order is included in Laue indices. Laue indices = $p \times$ Miller indices.

3.14b $\Delta k = (h\boldsymbol{a}^* + k\boldsymbol{b}^* + l\boldsymbol{c}^*) = G[\text{hkl}]$.

3.14c All lattice atoms must be included in the parallel planes (hkl) for constructive interference.

3.15 Volume density = area density × distance between parallel planes.

  Maximum area density → minimum distance d_{hkl} between the planes → maximum of $\sqrt{h^2 + k^2 + l^2}$. The only plane that includes all the lattice atoms and has a larger value of $\sqrt{h^2 + k^2 + l^2}$ than all other planes with low hkl numbers is the (111) plane.

3.16a The condition for Brillouin zone boundaries is $|k| = k_x = p \times \pi/a$, where $p = \pm 1, \pm 2, \ldots, \pm N_{\text{total}}$.

3.16b $k = 2\pi/\lambda = p \times \pi/a$, which can be transformed into $2a = p\lambda$. The latter condition can be interpreted as total reflection of the matter wave towards a set of parallel crystal planes. A standing matter wave is formed.

3.16c

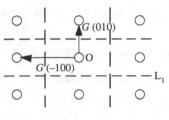

Exercise 3.16c

Bisectors are drawn to the shortest, next shortest and so on G vectors from a k point. The enclosed areas represent Brillouin zone 1, Brillouin zone $1 + 2$, Brillouin zone $1 + 2 + 3$, etc.

3.17

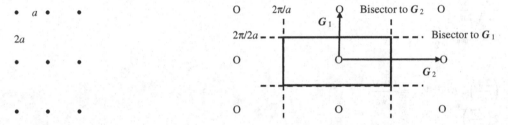

Exercise 3.17a   Crystal lattice in real space.   **Exercise 3.17b**   Reciprocal space. The first Brillouin zone is bounded by the bisectors to vectors of the type G_1 and G_2.

The first Brillouin zone is bounded by the bisectors to vectors of the type G_1 and G_2. The first Brillouin zone is shown in Figure Exercise 3.17b.

Figure Exercise 3.17c show the first and second Brillouin zones of the lattice in Figure Exercise 3.17a. The first Brillouin zone is bounded by bisectors to vectors of the type G_1 and G_2. The second Brillouin zone is bounded by bisectors to vectors of the type G_1, G_2, G_3 and G_4. The 'area' of each Brillouin zone equals $2(\pi/a)^2$.

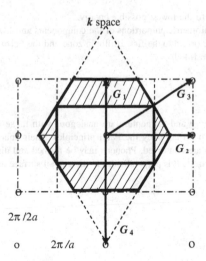

Exercise 3.17c   Reciprocal space. First and second Brillouin zones of the unit cell in real space in Figure Exercise 3.17a.

3.18a   1. $k = p \times \pi/a$, where $p = \pm 1, \pm 2, \ldots, \pm N_{\text{total}}$.
   2. All atoms must be included in the parallel planes in r space.

3.18b Parallel planes (hkl) in r space correspond to a point (hkl) in k space. When the reciprocal lattice has been derived, a Wigner–Seitz cell is constructed. It is the first Brillouin zone with 'volume V_1' and corresponds to some of the shortest G vectors. Similar constructions give a 'volume V_2'. The second Brillouin zone is 'V_2'- 'V_1'.

3.19a The first Brillouin zone of an SC crystal structure with the lattice constant a is a SC cube with side length = $2\pi/a$ (reciprocal space).

3.19b *Hint:* Consider the interior lattice points in k space and add those on the surfaces, the edges and the corners, which are shared with other cells.

3.20a Twelve {110} planes.

3.20b *Hint:* The {110} planes give reflections and define a G [110] vector. Construct the first Brillouin zone. Plot k [100], k[110] and k[111] and calculate their lengths in terms of $2\pi/a$.

3.21a Eight {111} planes and six {200} planes.

3.21b *Hint:* The {111} and {200} planes give reflections and define G[111] and G[200] vectors. Construct the first Brillouin zone and plot k[111] and k[200] and calculate their lengths in terms of $2\pi/a$.

3.22a Fermi sphere: the k points at the surface of the sphere have the kinetic energy E_F.

The definition equation of k_F is $E_F = \dfrac{\hbar^2}{2m^*}k_F^2$.

3.22b With the aid of Equation (3.67) on page 124, you obtain

$$k_F = \left(3\pi^2 \times \frac{\text{number of atoms per unit } r \text{ cell} \times \text{number of valence electrons per atom}}{\text{volume of a unit } r \text{ cell}}\right)^{1/3}.$$

3.22c $k_F^{BCC} = \dfrac{2\pi}{a}\sqrt[3]{\dfrac{3}{4\pi}}$.

3.22d $k_F^{FCC} = \dfrac{2\pi}{a}\sqrt[3]{\dfrac{3}{2\pi}}$.

3.23a $\dfrac{\pi}{a}\sqrt{2}$. *Hint* : $k_{\min} = \dfrac{G}{2}$.

3.23b $\dfrac{\pi}{a}\sqrt{3}$.

3.24a $'V_{Bz}' = \left(\dfrac{2\pi}{a}\right)^3$. *Hint* : $V_{WS}^k = \dfrac{(2\pi)^3}{V_{WS}^r}$.

3.24b $'V_{Bz}' = 2\left(\dfrac{2\pi}{a}\right)^3$.

3.24c $'V_{Bz}' = 4\left(\dfrac{2\pi}{a}\right)^3$.

3.25a $\dfrac{\pi}{a}\sqrt{2} = 0.85 \times 10^{10}\,\text{m}^{-1}$ (compare Exercise 3.22).

3.25b 88% (compare Exercise 3.21).

3.26a A Schottky defect is a vacancy which is not coupled to any interstitial in the crystal lattice. A Frenkel defect consists of a vacancy and a nearby interstitial in a crystal lattice. See Figures 3.64 and 3.65 on page 151.

3.26b Colour centres appear in ion crystals. They are crystal defects, which absorb characteristic frequency regions of visible light and reflect and/or transmit the complementary colours.

3.27a The alloy structure is the one that corresponds to the lowest possible energy.

3.27b An intermediate phase is a solution with stoichiometric proportions of the components and low ability to dissolve additional atoms of either kind.

3.28 *Hint:* Calculate the minimum distance from the origin to the first Brillouin zone and the Fermi radius (see Exercises 3.23 and 3.22).

The lowest energy band in the metal is far from filled.

3.29a 36 at-%. *Hint:* See Exercise 3.28.

3.29b *Hint:* See Exercise 3.28.

3.30a 5 at-%.

3.30b 52%.

3.31 A phonon is an elastic wave quantum. Its energy and momentum are analogous with those of the photons. Phonons obey Bose–Einstein statistics (identical and indistinguishable particles, which do not obey the Pauli principle). Their energy is quantized. They interact with other particles. In collisions the total energy and total momentum are conserved. Phonons may be formed and disappear in collisions.

3.32 The magnitude of the wave vector is $9.4 \times 10^6\,\text{m}^{-1}$. It is perpendicular to the crystal surface and directed towards the interior of the crystal.

Chapter 4

4.1 38 K.

4.2a $k_B T/2$.

4.2b $0.43 \times 10^{-2}\,\text{eV}$, in agreement with Figure 4.7.

The agreement between the figure and calculations is good.

4.2c 0.042 independent of temperature and molar weight.

4.3 $\sim 7.5\%$.

4.4a $c_A = \dfrac{x_A M_A \times 100}{x_A M_A + x_B M_B}$ $c_B = \dfrac{x_B M_B \times 100}{x_A M_A + x_B M_B}$ $(c_{N_2} = 78\%, \; c_{O_2} = 22\%)$.

4.4b $x_A = \dfrac{\dfrac{c_A}{M_A}}{\dfrac{c_A}{M_A} + \dfrac{c_B}{M_B}}$ $x_B = \dfrac{\dfrac{c_B}{M_B}}{\dfrac{c_A}{M_A} + \dfrac{c_B}{M_B}}$ $(x_{N_2} = 0.80, \; x_{O_2} = 0.20)$.

4.4c $p_A = x_A p$ $p_B = x_B p$.

4.4d $\overline{M} = x_A M_A + x_B M_B \, (\overline{M}_{air} = 28.8)$.

4.4e $0.179 \, kg/m^3$.

4.4f 0.269×10^{26} molecules/m^3.

4.5 $230\,°C\,(225\,°C)$.

4.6a   Gas and liquid have identical properties at the critical point. At $T > T_{cr}$ it is impossible to compress the gas into a liquid.

4.6b   Region III: the pressure of the gas increases.

     Region II: condensation occurs along the horizontal part of the curve. The condensation starts at point G and ends at point L (the amounts of gas and liquid are described by the lever rule).

     Region I: compression of the liquid (which requires much energy).

4.7 $94\,atm$. $p_{ideal} = 100\,atm$.

4.8a   Two molecules, both with zero electric dipole moment, e.g. diatomic molecules with equal atoms.

4.8b $E_{pot}^a = -4E_e \left(\dfrac{r_0}{r}\right)^6$ (van der Waals attraction). $F_a = -\dfrac{dE_{pot}^a}{dr} = -4E_e \dfrac{6 r_0^6}{r^7}$

     $E_{pot}^r = 4E_e \left(\dfrac{r_0}{r}\right)^{12}$ (electron shell repulsion). $F_r = -\dfrac{dE_{pot}^r}{dr} = 4E_e \dfrac{12 r_0^{12}}{r^{13}}$

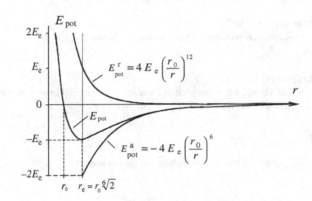

Exercise 4.8

where

     r_0 = the r value which corresponds to $E_{pot} = 0$.

     E_e = the depth of the potential well. The minimum of $E_{pot} = -E_e$.

     $r_e = r_0 \sqrt[6]{2}$ = equilibrium value of r corresponding to the minimum of E_{pot}.

4.8c   The Lennard-Jones potential of $H_2 = 0.0024\,eV$. The dissociation energy of the H_2 molecule $= 4.48\,eV$. The interaction between the H_2 molecules is much weaker than that between the H atoms.

4.9a   See pages 187–188.

4.9b   At low temperatures, very few molecules are excited up to higher rotational and vibrational energy levels and only the translation motion contributes to C_V which is $3R/2$. At increasing temperature, rotational excitation occurs and two additional degrees of freedom contribute to C_V which is $5R/2$. At still higher temperatures, vibrational excitation occurs, which contributes with two additional degrees of freedom, i.e. $C_V = 7R/2$. In many cases the temperature is not high enough for a fully developed vibrational contribution.

4.10a   Diatomic gas. Translation + rotation around two axes contribute to C_V.

4.10b   Polyatomic gas (nonlinear molecules). Translation + rotation around three axes contribute to C_V.

4.10c   Monoatomic gas. Translation contributes to C_V.

4.11a $0.18\,nm$.

4.11b $1.12 \times 10^{-15}\,m$ at $p = 10^6\,atm$ and $1.12 \times 10^{-3}\,m$ at $p = 10^{-6}\,atm$.

4.12 $0.046\,W/m\,K$. Comparison with a standard table indicates that the gas is neon.

4.13a 0.22.

4.13b 1.1. *Hint:* $k = \eta c_V$.

4.14a $\dfrac{dm}{dt} = -DA \dfrac{dc}{dy}$, where c is measured as number of atoms or mass per unit volume.

     A concentration gradient of radioactive tracer atoms is formed and the radiation as a function of position and time is measured.

4.14b   Compare Equation (4.62) on page 197 and Equation (4.77) on page 201.

4.14c Fick's first law: $\dfrac{\mathrm{d}m}{\mathrm{d}t} = -DA\dfrac{\mathrm{d}c}{\mathrm{d}y}$ (here in one dimension).

Fick's second law: $\dfrac{\partial c}{\partial t} = D\dfrac{\partial^2 c}{\partial y^2}$ (here in one dimension).

See pages 199 and 202.

4.15a *Hint:* Use $D = \dfrac{\overline{v_{kin}}\, l}{3}$ and find expressions for $\overline{v_{kin}}$ and l. $\dfrac{D_{H_2}}{D_{D_2}} = 1.41$ and $\dfrac{D^{235}UF_6}{D^{238}UF_6} = 1.004$.

4.15b It is very much easier to separate D_2 and H_2 than $^{235}UF_6$ and $^{238}UF_6$.

Chapter 5

5.1 20 W or 1.8×10^2 kWh/year. The real power is higher, mainly because the cooling device has a lower efficiency than the ideal Carnot cycle.

5.2a 2 kW or ~ 50 kWh/24 hours.

5.2b 5.5 kW or $\sim 1.3 \times 10^2$ kWh/24 hours.

5.2c Direct heating requires about three times as much electric power as the heat pump.

5.3 $\Delta S = 2mc \ln\left[(T_1 + T_2)/2\sqrt{T_1 T_2}\right] = 61\,\mathrm{J/K}$.

5.4 54 MJ. 1.5×10^2 kJ/K.

5.5 5.0 J/K.

5.6a $H = U + pV$.

5.6b An ideal solution has equal forces between like (A–A and B–B) and unlike (A–B) atoms, total solubility at all proportions and heat of mixing = zero. All other solutions are nonideal.

5.6c Molar heat of mixing ΔH_{mix} = the amount of net energy that has to be added to the system to mix two components into 1 kmol of the solution of given composition. If heat is formed, $\Delta H_{mix} < 0$. If heat is required, $\Delta H_{mix} > 0$.

5.7a 11.3 kJ/kmol K.

5.7b $P_L/P_s \approx 10^{3.6 \times 10^{26}}$ (the number of Al atoms is 6×10^{26}).

5.8a $G = H - TS$. Condition for equilibrium at constant pressure: G has a minimum, hence $\mathrm{d}G_{T,p} = 0$.

5.8b Driving force $= -\Delta G = -\int\limits_{\mathrm{initial}}^{\mathrm{final}} \mathrm{d}G$.

5.8c Chemical potential of a pure substance $\mu = G_A^0$.

5.8d G is a most useful instrument for studies of various processes such as chemical reactions and other transformation processes.

5.9a k = the fraction of the total number of particles that reach the final state per unit time.

$$k = \text{Constant} \times e^{-\frac{G_{a\ act}}{k_B T}} = Ae^{-\frac{U_{a\ act}}{k_B T}}$$

as $G_{a\ act} = H_{a\ act} - TS_{a\ act} \approx U_{a\ act} - TS_{a\ act}$ and the last term is included in the constant A.

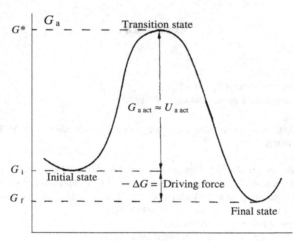

Exercise 5.9a

5.9b 1. The *frequency* with which the atoms have an opportunity to react.

2. The *fraction* of the total number of atoms which have enough energy to overcome the potential barrier (activation energy).

3. *A probability factor* of the process. See pages 242–243.

5.10a 1.3×10^{-2} kmol/m³s.

5.10b 2.0×10^{-19} J = 1.2 eV.

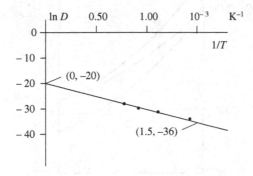

Exercise 5.10b

5.11a First order of A, second order of B, third overall order.

5.11b 2.3×10^2 m^6/s (kmol)^2.

5.11c 7.2 kmol/m^3/s.

5.12a *The interstitial mechanism*: in the crystal lattice there are interstitial positions available for *small* foreign atoms. The activation energy is low enough to be overcome by some *interstitial atoms*, which move from one site to another.

  The vacancy mechanism: vacancies in the crystal lattice may exchange positions with *lattice atoms*. The activation energy includes vacancy formation energy.

5.12b Fick's first law: $J = -D \times \mathrm{d}c/\mathrm{d}y$.

  Self-diffusion constant: D_A = diffusion constant of diffusion of atoms A in the pure substance A.

5.12c Jump frequency = the number of jumps per unit time.

  Jump distance = distance between two adjacent positions of the jumping atom in the diffusion direction. $D = d_j^2 f$.

5.12d $D = D_0 \mathrm{e}^{-\frac{U_{a\ act}}{k_B T}}$.

  Diffusion experiments at different temperatures give D as a function of T. $\ln D$ is plotted as a function of $1/T$ and the activation energy is derived from the slope of the line.

5.13a The possible positions for the interstitial atoms:

  1. Twelve positions at the middle of each edge and one position in the centre per unit cell or on average four positions per unit cell with a radius $\leq R(\sqrt{2}-1) \approx 0.414\,R$.

  2. Eight positions close to the corners per unit cell along the principal diagonals with a radius $\leq R\left(\sqrt{\frac{3}{2}}-1\right) \approx 0.225\,R$.

5.13b In case 1 each interstitial atom touches six lattice atoms (octahedral position).

  In case 2 each interstitial atom touches four lattice atoms. Each of them is enclosed in a tetrahedron of one 'corner atom' and three 'face centre atoms' (tetrahedral position).

5.14a $\sim 3 \times 10^{21}$ H atoms.

5.14b ~ 6 mg.

5.14c The pressure loss is of the magnitude 10^{-3} atm, which cannot be detected in the pressure measurement.

5.15 $D = 2 \times 10^{-9}$ m^2/s and $U_{a\ act} = 1.5 \times 10^{-19}$ J = 0.92 eV.

Exercise 5.15

5.16 5.5 length units.

5.17 1–2 MHz at 600 K.

5.18a The solution shows that

$$D = d_j^2 Z_{coord}\, \nu_{vibr} e^{\frac{\Delta S_{a\ form}}{k_B}} e^{\frac{\Delta S_{a\ barrier}}{k_B}} e^{-\frac{\Delta U_{a\ act}}{k_B T}} = D_0 e^{-\frac{\Delta U_{a\ act}}{k_B T}}$$

where

$$D_0 = d_j^2 Z_{coord}\, \nu_{vibr} e^{\frac{\Delta S_{a\ form}}{k_B}} e^{\frac{\Delta S_{a\ barrier}}{k_B}}$$

5.18b In substitutional diffusion, the concentration of vacancies is fundamental and can be shown more clearly:

$$D = d_j^2 Z_{coord}\, \nu_{vibr} x_{vac}^{eq} e^{\frac{\Delta S_{a\ barrier}}{k_B}} e^{-\frac{\Delta U_{a\ barrier}}{k_B T}} = D_0' e^{-\frac{\Delta U_{a\ barrier}}{k_B T}}$$

where

$$D_0' = d_j^2 Z_{coord}\, \nu_{vibr} x_{vac}^{eq} e^{\frac{\Delta S_{a\ barrier}}{k_B}}$$

The diffusion constant D_0' contains the factor x_{vac}^{eq}, which is temperature dependent.

5.18c No. The vacancy concentration increases strongly at the higher temperature. The vacancies become 'frozen' on quenching.

$$D = d_j^2 Z_{coord}\, \nu_{vibr} x_{vac} e^{\frac{\Delta S_{a\ barrier}}{k_B}} e^{-\frac{\Delta U_{a\ barrier}}{k_B T}} = D_0'' e^{-\frac{\Delta U_{a\ barrier}}{k_B^T}}$$

where

$$D_0'' = d_j^2 Z_{coord}\, \nu_{vibr} x_{vac} e^{\frac{\Delta S_{a\ barrier}}{k_B}} \quad \text{or} \quad D_0'' = \frac{x_{vac}}{x_{vac}^{eq}} D_0'$$

5.18d The interstitial sites are present in the crystal lattice from the beginning and the majority of them are empty. No formation energy is required. The concentration of the small interstitial atoms determines the number of diffusing atoms which is independent of the temperature. The diffusivity is influenced by the energy barrier.

$$D = d_j^2 Z_{coord}\, \nu_{vibr} e^{\frac{\Delta S_{a\ barrier}}{k_B}} e^{-\frac{\Delta U_{a\ barrier}}{k_B T}} = D_0 e^{-\frac{\Delta U_{a\ barrier}}{k_B T}}$$

where

$$D_0 = d_j^2 Z_{coord}\, \nu_{vibr} e^{\frac{\Delta S_{a\ barrier}}{k_B}}$$

5.19a If there are no sinks and no sources of vacancies, $J_A + J_B + J_{vac} = 0$.

The vacancy flux $J_{vac} = -[J_A + J_B]$. The net flux of atoms has the same magnitude as the vacancy flux but opposite sign. This leads to an uneven vacancy distribution, which does not accumulate but decays owing to annihilation or production of vacancies at edge dislocations.

5.19b $\tilde{D} = x_A \tilde{D}_B + x_B \tilde{D}_A$

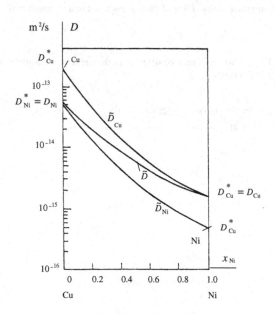

Exercise 5.19b

5.20a The net flux of atoms is directed towards the Cu side. The Zn atoms move faster into the Cu side than the Cu atoms into the alloy side. Hence there is a net flux of vacancies towards the alloy side. Excess of vacancies appears on the alloy side and a lack of vacancies on the Cu side. A steady state is developed. The vacancies annihilate on the alloy side and are produced on the Cu side at edge dislocations.

5.20b Markers are insoluble inclusions at the weld plane. They follow the displacement of the weld plane, owing to diffusion. The weld plane and the markers move towards the alloy side because the Cu side expands slightly.

5.21a

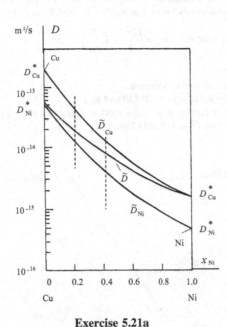

Exercise 5.21a

Choose for example $x_{Ni} = 0.2$ and 0.4 and read the quantities involved in Equation (5.176) from the diagram. Calculate the two $\tilde{D}$ values by means of Equation (5.176) and compare the result with the values, which you have read from the diagram.

The agreement between the calculated values and the values of $\tilde{D}$ read from the diagram is satisfactory. The vertical scale is logarithmic which makes readings more difficult than usual.

5.21b Along the line $x_{Ni} = 0$:
Upper point: Cu* diffuses in pure Cu (self-diffusion coefficient).
Lower point: Ni* diffuses in pure Cu.
Along the line $x_{Ni} = 1$:
Upper point: Cu* diffuses in pure Ni.
Lower point: Ni* diffuses in pure Ni (self-diffusion coefficient).

Chapter 6

6.1 $4.1 \times 10^8 \, N/m^2$.

6.2 0.30.

6.3a $5 \times 10^2 \, J$.

6.3b The potential energy is primarily transformed into kinetic energy (vibrational energy of the lattice atoms in the spring) and is successively transferred as heat to the surroundings.

6.3c Yes, the temperature initially increases by 0.1 K.

6.4 22 m. $\left[h = \dfrac{Y}{3(1-2\nu)} \dfrac{\ln(\rho/\rho_0)}{\rho_w g} \right.$ where ρ_w = density of water, ρ_0 = density of the body in air and ρ = density of the body at depth h. $\Big]$

6.5a r increases 5.4 nm.

6.5b r decreases 54 nm.

6.5c 8.2 J.

6.6 -300α %.

6.7a 1. Anharmonicity of the potential well of the lattice atoms, i.e. the lattice constant a increases with increasing vibrational energy (vibration numbers). See Figure 2.45 on page 83 in Chapter 2.

2. Increase of vacancy formation with increasing temperature.

6.7b The condition $dG = 0$ (equilibrium) and not primarily the increasing distances between the lattice atoms is the origin of thermal expansion of solids. The entropy plays an important role in the theory of expansion in solids.

6.8a Einstein regarded each atom as three perpendicular oscillators, vibrating around their equilibrium positions with quantized frequencies, a single basic frequency and multiples of this frequency up to infinity. Addition of all the oscillator energies gave the total energy U and C_V could be derived by taking the derivative of U with respect to temperature.

Debye modified the theory by introduction of a manifold of frequencies and an upper limit ν_D for the frequencies and introduced the Debye temperature $\theta_D = h\nu_D/k_B$. At temperatures considerably lower than θ_D the so-called T^3 law is valid:

$$C_v = 9R\left(\frac{T}{\theta_D}\right)^3 \int_0^{\frac{\theta_D}{T}} \frac{x^4 e^x}{(e^x - 1)^2} dx \quad \text{or} \quad \text{If } T \ll \theta_D, \quad C_v = \frac{12\pi^4}{5} R\left(\frac{T}{\theta_D}\right)^3$$

6.8b $C_p - C_v = \dfrac{\beta^2 V_A^{\theta_D}}{\kappa} (T - \theta_D)$ Grüneisen's rule

where $V_A^{\theta_D}$ = molar volume of crystal at the Debye temperature.

6.9 Dulong–Petit's law: $C_V = 3R$. $C_V = Mc_V = 3 \times 8.314 \times 10^3 = 25 \text{ kJ/kmol K}$.

Beryllium deviates strongly from Dulong–Petit's law because $\theta_D \gg$ room temperature. The value for Be agrees well with the Debye model.

6.10a Graphical verification: C_V plotted as a function of T^3 is a straight line.

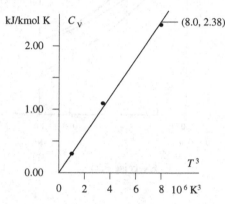

Exercise 6.10a

6.10b $\theta_D = 1870$ K, which is high compared with the Debye temperatures of other elements. A high value of θ_D corresponds to strong interatomic forces.

6.11 $\sim 1.5\%$.

6.12a The transition from an ordered to a disordered state requires energy. At $T = T_{cr}$, C_p has a narrow characteristic maximum (Figure 6.23 on page 316).

6.12b $S = (f_A - x_A)/(1 - x_A)$.

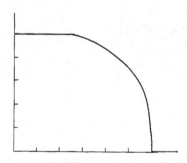

Exercise 6.12b

6.13a Magnetism is associated with the magnetic moments and the angular momentum of the electrons in their orbits around the nucleus.

The Bohr magneton comes from the relationship $\mu_l = \mu_B m_l$ where $\mu_B = \dfrac{e}{2m} \hbar$. See pages 70–71 in Chapter 2.

The resulting magnetic moment $\mu = \mu_L + \mu_S$ of the atoms is the origin of the magnetic properties of materials.

6.13b If the resultant magnetic moment $= 0$ the material is *diamagnetic*.
   If the atom has a resultant magnetic moment $\neq 0$ the material is *paramagnetic*.
   If the atom has a resulting magnetic moment $\neq 0$ and there is a strong coupling between the magnetic moments of the atoms leading to alignment, the material is *ferromagnetic*.

6.14 There is no real difference between paramagnetic and ferromagnetic materials when the magnetic moments of the atoms are considered. The difference in the strong alignment of the magnetic moments of a large number of atoms is due to the *strong coupling between their spin vectors*. A large number of atoms with aligned magnetic moments form stable clusters of atoms called Weiss domains.

   The observations have been confirmed by a quantum mechanical model and by the Ising model. The Ising model is statistical. It deals with atoms with spin up or spin down. An energy function is set up and the condition for minimum free energy is derived. The minimum corresponds to a strong coupling between parallel spins.

6.15a The Curie point is the transition temperature when ferromagnetic materials lose their ferromagnetic properties and become paramagnetic. The Curie point temperature is a material constant.

6.15b Curie point is a typical example of an order to disorder transformation, due to violent kinetic motion of the atoms in the crystal lattice.

6.16a A single atom has an integer number of electrons in an energy state. Some Ni atoms have two electrons in 4s and eight in 3d whereas other atoms may have one electron in 4s and nine in 3d. Hence the *average* number of electrons is not necessarily an integer.

6.16b The electron distribution in 4s and 3d is a consequence of *hybridization* (pages 109–110 in Chapter 3). The difference in energy between 3d↑ and 3d↓ in the Figure (a) is due to *exchange energy* (pages 73–74 in Chapter 2). In both figures the electron distribution corresponds to the lowest possible total energy.

6.16c Figure (a): the lowest 3d↑ sub-band is filled and the 4s and 3d↓ sub-bands share the remaining five electrons. This is done in such a way that that the electrons with the highest energy in each sub-band have *the same energy* as the highest level in the 3d↑ sub-band.

6.16d Figure (b): The exchange energy, which results in different energies of the 3d↑ and 3d↓ sub-bands, disappears at the Curie point. Then the electrons redistribute in such a way that the upper energy levels in 4s, 3d↑ and 3d↓ become equal and that the two 3d sub-bands accommodate an equal number of electrons.

6.17 See Figure 6.45 on page 331. It shows that a small H field is sufficient to magnetize the material nearly up to saturation in the direction [100]. Much stronger H fields are required to magnetize the material in other directions.

6.18 The periodical change of the magnetic field is in principle the same as that in the virgin curve. Consider Figure 6.47 on page 332.
   Region I: reversible displacement of domain wall boundaries.
   Region II: irreversible displacements of domain wall boundaries. The direction which is closest to the easiest crystallographic direction 'wins'.
   Region III: rotation of Weiss domains from the easiest direction to the direction of the H field until saturation.
   These processes are repeated periodically in the hysteresis loop when the material is exposed to an alternating magnetizing field. Each loop leads to heat losses.

Chapter 7

7.1a Insulators contain very few free electrons as $E_g \gg 0$. Conductors contain plenty of free electrons as $E_g = 0$.

7.1b In the absence of free electrons, heat is transported by phonons in insulators.

$$\lambda = \lambda_{\text{lattice}} = nc\overline{v_{\text{kin}}}l/3.$$

7.1c In pure metals the electron contribution is much greater than the phonon contribution, which can be neglected.

$$\lambda_e = n_e c_e v l/3 = (n_e v l/3 \times \pi^2 \lambda_B^2 T)/2E_F, \text{ which gives } \lambda_e = n_e \pi^2 k_B^2 T\tau)/3m^*.$$

   In alloys the phonon contribution cannot be neglected: $\lambda = \lambda_{\text{lattice}} + \lambda_e$.

7.2 $0.9\,\text{W/m K}$.

7.3a *Conductors*: valence and conduction bands often overlap. $E_g = 0$. Electron–hole pairs. *Semiconductors*: valence and conduction bands are separated by an energy gap of magnitude $1\,\text{eV}$. Thermal excitation is possible.
   Insulators: $E_g \gg 0$, hence very few electrons in the conduction band.

7.3b The valence band corresponds to the first Brillouin zone and the conduction band corresponds to the second Brillouin zone. The energy levels of the inner electron shells of the lattice ions are narrow. The valence and conduction bands are wide. The free electrons do not belong to any particular ions but to the whole lattice.

7.3c ρ_{metals} increase with T. $\rho_{\text{semiconductors}}$ decrease strongly with T. $\rho_{\text{insulators}}$ decrease somewhat with T, owing to thermal excitation, but is still very high.

7.4a *Classical theory* (pages 353–355):
   The conduction electrons collide frequently with lattice ions and move with a small constant net velocity in an electric field. The collisions are equivalent with a friction force that balances the electrical force: $mv_e/\tau = eE$. This law, in combination with Ohm's law, $j = \sigma E$, and the relationship $j = n_e e v_e$ gives the relationship below between σ and τ.
   Drift velocity = average velocity v_e of the free electrons in an electrical field.
   Relaxation time = average time τ between two successive collisions and also time constant in the function $v = v_e e^{-\frac{t}{\tau}}$.
   Mobility = drift velocity per unit electric field.

$$\sigma = \frac{n_e e^2 \tau}{m} \qquad \sigma = n_e e \mu \qquad \mu = \frac{v_e}{E}$$

Band theory:

The *band theory* of conductivity is a modification of the classical theory.

$$\sigma = ne^2\tau/m^*,$$

where m^* is the effective mass of the free electrons in the metal.

7.4b Wiedemann–Franz law: by introducing the expressions for the thermal and electrical conductivities for pure metals, the relationship $\lambda/\sigma = constant \times T$ can be derived. The conclusion is that the transport of heat and charge in both cases are performed entirely by the free electrons in pure metals.

7.5a 2.5×10^2 W/mK$\{[\rho(20\,°C) = 2.7 \times 10^2\,\Omega\,mm^2/m]; \alpha = 4.3 \times 10^{-3}\,K^{-1}\}$.

Hint: σ and λ must be compared *at the same temperature* (150°C).

7.5b λ increases strongly owing to a dominant phonon contribution.

7.6a 1.6×10^6 m/s.

7.6b 0.74 mm/s.

7.6c 4.3×10^{-3} m^2/Vs.

7.7a 3.8×10^{-14} s.

7.7b 53 nm.

7.8 At $T > 0$ some electrons are excited up to the conduction band and leave holes in the valence band. The electron concentrations in the bands are

$$n_c = \frac{2N}{e^{\frac{E_c - E_F}{k_B T}} + 1} \quad \text{and} \quad n_v = \frac{2N}{e^{\frac{E_v - E_F}{k_B T}} + 1}$$

where N is the number of atoms per unit volume.

The condition $n_v + n_c = 2N$ gives

$$E_F = \frac{E_v + E_c}{2}.$$

This value is introduced into the expressions for n_c and n_v, which gives

$$n_c = \frac{2N}{e^{\frac{E_g}{2k_B T}} + 1} \quad \text{and} \quad n_v = 2N - n_c = \frac{2N}{e^{-\frac{E_g}{2k_B T}} + 1}$$

$$\sigma = n_e e \mu_e + n_h e \mu_h,$$

where $n_c = n_e$ and $n_h = 1 - n_v = n_c$

More accurate expressions are

$$n_e = \frac{(2\pi m_e^* k_B T)^{3/2}}{4\pi^2 \hbar^3} e^{-\frac{E_c - E_F}{k_B T}} \quad \text{and} \quad n_h = \frac{(2\pi m_h^* k_B T)^{3/2}}{4\pi^2 \hbar^3} e^{-\frac{E_F - E_v}{k_B T}}$$

$$n_i = \sqrt{n_e n_h} = \frac{(2\pi m k_B T)^{3/2}}{4\pi^2 \hbar^3} \left(\frac{m_e^*}{m} \frac{m_h^*}{m} \right)^{3/4} e^{-\frac{E_g}{2k_B T}}$$

which gives

$$\sigma = n_i e(\mu_e + \mu_h)$$

7.9a $\sigma_{theor} \approx 4 \times 10^{-39}\,(\Omega\,m)^{-1} \ll \sigma_{exp} = 10^{-12}\,(\Omega\,m)^{-1}$.

7.9b No. The reason is presence of impurity atoms. The concentration is 3.5×10^7 atoms/m^3, which is a low impurity concentration.

7.10a The *conductivity of an intrinsic semiconductor* depends strongly on the *temperature* (increases with increase in T). Instead of n_e and n_h you use $n_i = \sqrt{n_e n_h}$ and Equation (7.71) on page 368 for calculation of the conductivity: $\sigma = n_i e\,(\mu_e + \mu_h)$.

The *conductivity of a doped semiconductor* depends strongly on the *concentration of the dopant* and the intrinsic conductivity can be neglected. n_e and n_h are not equal.

For calculation of σ you use the general relationship $\sigma = n_e e \mu_e + n_h e \mu_h$. Instead of n_e and n_h you use n_e and n_i^2/n_e or n_h and n_i^2/n_h, respectively. If $n_e \gg n_h$ or $n_h \gg n_e$, the minor term can be neglected and the general equation will be simplified to $\sigma_{dopant} = n_{dopant} e \mu_{dopant}$.

7.10b $E_{kin} = \frac{\hbar^2}{2m^*} k^2$ [Equation (3.55) on page 120 in Chapter 3]. Taking the derivative of $E_{kin}(k)$ twice with respect to k gives the desired expression.

Conduction band

Donor level

Valence band

Exercise 7.10c Donor level in an n-doped semiconductor, i.e. with impurity atoms from group 5 in the periodic table.

Conduction band

Acceptor level

Valence band

Exercise 7.10c Acceptor level in a p-doped semiconductor, i.e. with impurity atoms from group 3 in the periodic table.

7.11 0.24 eV.

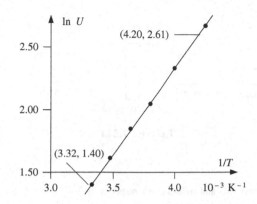

Exercise 7.11 $\ln U = \ln C - \ln \sigma = constant + E_g/2k_B T$.

7.12a $6.6 \times 10^{15}\,\mathrm{m}^{-3}$.
7.12b Fraction occupied energy levels in the conduction band $\approx 6 \times 10^{-14}$.
 7.13 $0.61\,\mathrm{eV}$.
7.14a $0.65\,\mathrm{eV}$ at $T = 0\,\mathrm{K}$; $0.69\,\mathrm{eV}$ at $T = 300\,\mathrm{K}$.
7.14b $n_e = n_h = 0$ at $T = 0\,\mathrm{K}$; $n_e = n_h \approx 3.2 \times 10^{13}\,\mathrm{m}^{-3}$ at $T = 300\,\mathrm{K}$.
 7.15 $1.6 \times 10^2\,\Omega$.
 7.16 $\sigma_e/\sigma_i = 4 \times 10^3$.
 7.17 $7.4 \times 10^{-6}\,\Omega\,\mathrm{m}$.
 7.18 $n_e = 1.5 \times 10^{18}\,\mathrm{m}^{-3}$; $n_h = 3.1 \times 10^{20}\,\mathrm{m}^{-3}$.
7.19a $E_B = 1.9 \times 10^{-2}\,\mathrm{eV}$ or about 2% of the band gap of Si.
7.19b $r_0 = 3.3\,\mathrm{nm}$ or about six times the lattice constant of Si.
7.20a $0.21\,\mathrm{eV}$ and $5.9 \times 10^{-6}\,\mathrm{m}$.
7.20b The absorption is zero. The semiconductor is transparent.
7.20c The exciton energy equals approximately $8 \times 10^{-3}\,\mathrm{eV}$.
7.21a The p-type semiconductor has its acceptor level $0.04\,\mathrm{eV}$ above the upper edge of the valence band.

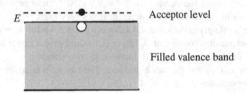

Exercise 7.21a

7.21b At room temperature, the thermal energy is of the magnitude $k_B T \approx 0.025\,\mathrm{eV}$. A considerable fraction of the free electrons have enough energy to become excited up to the acceptor level. In this case the minimum cannot be observed. At very low temperature no thermal excitation is possible and the minimum can be observed.
7.22a $I_0/2$.
7.22b Zero.
7.22c $I_0/8$.
 7.23 $1.7\,\mu\mathrm{m}$.

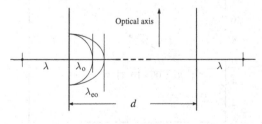

Exercise 7.23

7.24a *Without* the plate, the polarizer and the analyser are initially crossed, i.e. have perpendicular polarization planes. No light is transmitted.
With the plate, the analyser is rotated until no light can pass. This angle is the angle of rotation of the polarization plane.

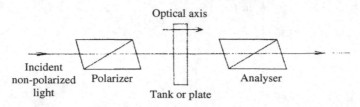

Optical axis

Incident non-polarized light   Polarizer   Tank or plate   Analyser

Exercise 7.24a

7.24b $22°/\text{mm}$.
7.24c 3.3 mm.
7.24d The only difference is that the rotation occurs left- and right-handed, respectively.

Chapter 8

8.1 $\beta_{Al} = 0.070$, $\beta_{Cu} = 0.039$ and $\beta_{Cu} = 0.043$.
8.2 On average, the calculated values of the melting points are 10% higher than the experimental values for the metals. Richard's rule is not valid for semiconductors.
8.3a Solid Al: no translation motion is possible. There are six degrees of freedom due to vibrations (potential and kinetic energy) in three perpendicular directions in the crystal lattice: $C_V = 2 \times 3 \times r/2 = 3R$.

Liquid Al: if the liquid is considered to lack shear resistance there are four degrees of freedom due to translational motion in three directions and one degree of freedom that refers to the potential energy of compression and expansion motion. However, close to the melting point the liquid still has a considerable shear resistance on the atomic level. This corresponds to two more degrees of freedom. The number of degrees of freedom will then be six, which gives

$$C_V = (3 + 1 + 2) \times R/2 = 3R.$$

With increasing temperature the shear resistance is gradually lost and the value of C_V decreases.
8.3b Assume that the curve is symmetrical around the line $T = T_{max}$ and calculate the area under the curve from $T = 0$ to 933 K. It is calculated as the sum of a base rectangle minus a fourth of an ellipse plus a top triangle. The heat of fusion is twice this value.

The heat of fusion is calculated as the area under the curve. $\Delta H^{fusion} = 9.9 \times 10^6 \text{J/kmol}$ or $\sim 370 \text{ kJ/kg}$. Correct value = 397 kJ/kg (391 kJ/kg in Exercise 8.2). The agreement is acceptable as the accuracy of the diagram is not very high.
8.3c $\overline{C_p - C_V} \approx 0.6R = 5.0 \text{ kJ/kmolK}$. Grüneisen's rule gives the value 4.4 kJ/kmolK. It is doubtful whether Grüneisen's rule is valid at temperatures higher than the melting point or not. The volume expansion coefficient $\beta_L (\beta_L > \beta_s)$ would give a better agreement than β_s.
8.3d The decrease in C_V is probably due to loss of shear resistance with increase in temperature.

C_p is influenced by shear resistance (just as C_V) and thermal expansion. The two effects counteract each other. The volume expansion dominates over the loss of shear resistance at high temperatures.
8.4a 0.020 eV.
8.4b At constant *pressure*, the average distance between the atoms is constant, which means that the average activation $\overline{U}_{a\ act}$ is constant. The upper curve is a straight line.

At constant *volume*, the average distance between the atoms decreases when the temperature decreases. At decreasing temperatures the curve becomes bent owing to gradual decrease in $\overline{U}_{a\ act}$.
8.5a See figure.

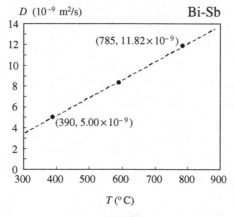

Exercise 8.5a

8.5b The curve is a straight line: $D = A + BT$.

8.5c $1.7 \times 10^{-11} \text{m}^2/\text{sK}$.

8.6a $0.72 \, \text{N s/m}^2 = 0.72 \, \text{Pa s}$.

8.6b 26 N and 11 W.

8.6c The oil reduces the friction, the friction losses and the damage on the metal surfaces considerably.

8.7a The simple relationship between viscosity η and temperature T in Figure (a) can be written as

$$\ln \eta = constant \times \frac{\overline{U_{a \, act}}}{k_B} \frac{1}{T}$$

In η is plotted here as a function of $1/T$ for copper and iron.

The diagrams show that the simple Equations (1′) and (2′) are not valid and no values of $\overline{U_{a \, act}}$ can be calculated.

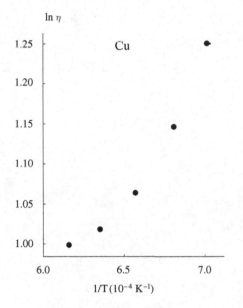

Exercise 8.7a   Copper

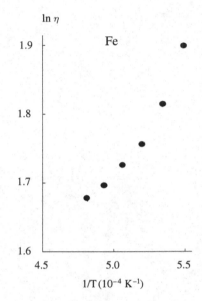

Exercise 8.7a   Iron

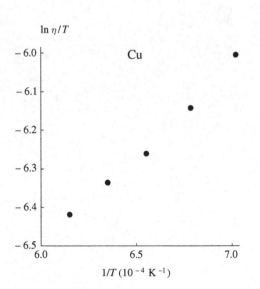

Exercise 8.7b   Copper

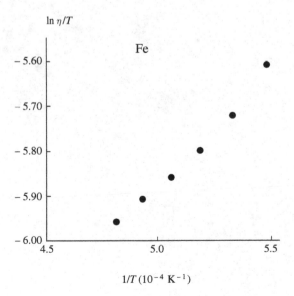

Exercise 8.7a   Iron

8.7b The relationship between η and T in Figure (b) can be written as

$$\ln \eta/T = constant \times \frac{\overline{U_{a\ act}}}{k_B} \frac{1}{T}.$$

$\ln \eta/T$ is plotted here as a function of $1/T$ for copper and iron.

The diagrams show that Stokes–Einstein relationship is a much better approximation for the viscosity than the simple exponential function, especially for Cu. Approximate values for $\overline{U_{a\ act}}$ can be calculated from the diagrams: $\overline{U_{a\ Cu}} \approx 0.34\,\mathrm{eV}$ and $\overline{U_{a\ Fe}} \approx 0.36\,\mathrm{eV}$.

8.8 $\lambda_{Fe} = 32\,\mathrm{W/mK}$, $\lambda_{Ni} = 34\,\mathrm{W/mK}$ and $\lambda_{Cu} = 1.5 \times 10^2$ W/mK at their melting points.

Index

Absolute temperature   179
Acceptor level   370, 371
Activation energy   238
Angular momentum of the electron in its orbital   58
Anharmonic oscillator   82
Arrhenius equation   243
Atomic distribution diagram   6
Atomic number Z   62
Atomic radius   10
Atomic scattering factor   18
Atomic size   10
Avogadros number   116, 230

B and H fields   330–1
Balmer series   46
Band intensities for vibrational transitions in molecules   84–5
Band theory of solids   125
   energy bands in solids   126
      eigenvalues and eigenfunctions of free electrons in a crystal
         lattice   126–8
      number of electrons per Brillouin zone   128
      number of energy states per band   129
   energy distribution of electrons in the energy states of Brillouin
      zones   140
   origin of energy bands of solids   125–6
   see also Brillouin zones
Basis of crystal structure   13
BCC   20
BCC structure   21
Bernals liquid model   11–12
Bloch functions   127–8
Bohr magneton   70–1, 318
Bohr model
   of atomic structure   46
   of many electron atoms   48
   of the hydrogen atom   46
Boltzmann   174
Boltzmanns constant   229
Bonds
   binding energy and dissociation energy   98–9
   cohesive energy   100
   ionization energy and electron affinity   99

   lattice energy   100
   sublimation energy and condensation energy   100
Bonds in molecules and non-metallic solids   100
   energy of a free ionic molecule   102
   ionic bonds   102
   ionic bonds in ionic crystals   103
   lattice energy of ionic crystals   104
   molecular bonds   101
   polarization energy   106
   residual covalent binding energy   106
   zero point vibrational energy   106
Born   51, 54
Bose-Einstein statistics   148–9
Boyles law   177
Braggs law   3
Bravais   13
Brillouin zones   128, 134
   Brillouin zones in three dimensions   130, 137–40
   Brillouin zones in two dimensions   130, 135–7
      construction of Brillouin zone boundaries   135–6
   Brillouin zones of a simple cubic crystal lattice   138
   Brillouin zones of FCC, BCC and HCP crystal
      lattices   139–40
   definition   128
Burgers vector   26

Carnot cycle   223
Charless law   179–80
Chemical potential   236
Classification of electronic states in diatomic molecules   75
Closed-packed metal structures   19–20
Colour centres in ion crystals   152
Coordination number   7
Coordination numbers of metals   21
Coordination shell   7–8
Covalent bonds   108
   covalent bonds in carbon   109
   covalent bonds in the H_2 molecule   108
C_p and C_V
   definition   187
   ratio of C_p and C_V 187, 191
   relationship between C_p and C_V in gases   187

Crystal defects in pure metals 24
  interfacial defects 24, 28
  line defects 24, 26
  point defects 24, 25
Crystal structure of solid metals 19
Crystal structure of solids 12
  types of crystal structures 13–15
Crystal systems 15
Curie temperature 323, 328, 329
Current density 353–4

Daltons law 205–6
De Broglie 50
De Broglie wave length 50, 54
Debye–Scherrers X-ray method 3–4
Degree of freedom 186, 191
Density and volume of liquids and
  melts 407–9
Diamagnetism 320, 322
  theory of diamagnetism 321
Diatomic molecules 75
Diffusion in gases 200
  diffusion coefficient D 199, 207
  Ficks laws 199, 201
  kinetic theory of diffusion in
    gases 200
  relationship between D and l 268
  relationship between D and η 432
Diffusion in liquids and melts 416
  determination of diffusion coefficients 418
  diffusion coefficient as a function of activation energy and
    temperature 416
  diffusion in alloys 421
  diffusion of interstitially solved atoms in
    metal melts 421
  experimental methods for measurements of diffusivities
    in liquid metals 423
  influence of gravity on measurements of diffusivities 424
  models of diffusion in liquids 417–21
  simple model of diffusion in liquids 416
Diffusion in solids 253
  basic theory of diffusion 254
    diffusion coefficient 254
  determination of diffusion coefficients by tracer methods
    273–5
  diffusion in alloys 265
    calculation of D as a function of diffusion coefficient
      L in binary alloys 266
  diffusion in binary alloys 268
    interstitial and substitutional diffusion in binary alloys 269
    theory of substitutional diffusion in binary alloys 269–73
      chemical diffusion coefficient 271–4
      Kirkendall effect 270
  diffusion in ternary alloys 274
  diffusion mechanisms 254–5
  theory of diffusion 255–6
    diffusion coefficient as a function of activation
      energy and temperature 264

diffusion coefficient of lattice atoms as a function
  of atomic quantities 260–4
driving force of diffusion 265
energy barrier of a jumping atoms 258
jump frequency of lattice atoms to vacancies and their
  activation energy 259
jump frequency of small interstitial atoms 260
self-diffusion 265
Diffusivity 199, 254
Dipole-dipole interaction 180
Dislocations 26
Dissociation energy of diatomic
  molecules 83
Donor level 371, 372
Drift velocity 353–4, 356
Dulong–Petits law 304

Eddy currents 332
Effective mass 112, 145–6
  determination of m* 311–12
Einstein 49
Elasticity 290
  compressibility 293
  deformation energy 295
  Hookes law 291–3
  modulus of elasticity 292
  Poissons ratio 292
  relationship between Y and κ 294
  stress and strain 291
Elastic vibrations in insulators 357
Elastic vibrations in solids 146
Electrical conduction in liquids and melts 440
  electrical conductivity in alloy melts 442
  electrical conductivity in pure metal melts 440
  ratio of thermal and electrical conductivities in pure
    metal melts 443
Electrical conduction in solids 347
  conductors, insulators and semiconductors 348–9
  resistivity and conductivity 347
  see also Metallic conductors, Insulators;
    Semiconductors
Electrical conductivity in solids 348, 357
  of doped semiconductors 369
  of insulators 361
  of metallic conductors 353–354
  of pure semiconductors 368
Electron configurations in atoms 62, 64
Electron density distribution in atoms 66
Electron structures in alloys 154–6
Energy distribution in particle systems 174
Energy distribution of electrons in the energy states of
  Brilloin zones 140
Energy gap 368, 376, 362
Energy gap of diamond 363
Energy levels and spectral lines 65, 82
Energy of metallic crystals 288
  electron energy 288
  phonon energy 290
Enthalpy 221

Entropy 224
   Boltzmanns relationship between entropy and probability 227
   entropy and probability 226–7
   entropy change in irreversible processes 225–8
   entropy change on mixing two components 229–30
   entropy changes in reversible processes 224, 227
Expansion 294
   length and volume expansion 296–9
   relationship between α and β 298
   theory of thermal expansion 301
   thermal expansion 299
   thermodynamic explanation of thermal
       expansion 302

FCC 20
FCC structure 22
Fermi–Dirac statistics 117, 147–9
Fermi distribution of energies in the electron gas 121–3
Fermi energies of some metals 124
Fermi factor 117, 123
Fermi level 116, 123, 124
Fermi radius 124, 405
Fermi surface 120, 124, 141
Ferromagnetic materials in magnetic fields 330
   hysteresis 331
   magnetisation of ferromagnetic materials 330
Ferromagnetism 323
   antiferromagnetism, ferrimagnetism and ferromagnetism 323
   electron configurations and magnetic moments of ferromagnetic
       atoms 324
   spin interaction 323
   the Ising model 327
   theory of ferromagnetism 326
   the quantum mechanical model 326
Fick's laws 199, 201
Frenkel defects 151–2
Fullerens 111–12

Gas laws 178
General law of ideal gases 178
Gibbs free energy 235
   chemical potential 236
Grain boundaries 24, 28–9
Grüneisens rule 299, 302–3,
       312, 413
G vector 132, 134, 135–6

H_2 molecule 74–5
H_2^+ molecule 72–4
Hard sphere model of atoms 10
Hard sphere model of liquids 11
Harmonic oscillator 80
HCP 20
HCP structure 23
Heat capacities of gases 186
   heat capacities of diatomic gases 188
   heat capacities of diatomic gases as a function of
       temperature 187, 191

heat capacities of ideal monoatomic
       gases 186
   heat capacities of polyatomic gases 192
Heat capacity of liquids and melts 413
   heat capacity at constant
       pressure 413
   heat capacity at constant volume 413
   theory of heat capacity of liquids 412
Heat capacity of solids 303
   classical model 304
   Debyes model 307–9
   Debye temperature 308
   Debye temperatures of some metals 309
   Einsteins model 305–7
   heat capacity of the free electrons 310
   molar heat capacity at constant pressure 312
   total molar heat capacity at low temperatures 311
Heat engines 221–3
Heisenbergs uncertainty principle 51, 54–5
Hume–Rotherys rules 33
Hybrid carbon bonds in graphite 111
Hybrid formation of sp^3 orbitals in CH_4 109
Hysteresis 331

Influence of lattice defects on electronic structures in crystals 151
Influence of lattice defects on electronic structures in metals and
       semiconductors 153
Influence of lattice defects on electronic structures in non-metallic
       crystals 151
Insulators 357
   conductivities of insulators 361–2
       band theory of electrical conductivity in insulators 361–2
       determination of the energy gap of an insulator 362
   simple model of electrical conduction in insulators 358–60
       calculation of the electron concentration in the valence and
           conduction bands 359
       calculation the position of the Fermi level 360
Intensities of X-ray diffractions in crystal planes 18–19
Interaction between molecules in gases 180
Interfacial defects 24, 28
Intermediate phases 31, 35, 156
Interpretation of atomic distribution diagrams 6
Interstitial atoms 25, 152, 156
Intrinsic diffusion coefficient 265

Kinetics of homogeneous reactions in gases 245
   activated complex theory 250
   collision theory of homogeneous chemical reactions 245
   driving force and reaction rate of homogeneous chemical
       reactions 248
   order of homogeneous chemical reactions 248
   temperature dependence of the rate constant 247
Kinetic theory of gases 170
Kirkendall effect 270
Kronig–Penney model of periodically varying potential
       energy 127
k space 115, 134
k vector 127, 134

Lattice   13
Lattice directions   15
Lattice energy, *see* Bonds
Lattice planes   16
Lattice symbols   14–15
Laue indices   18, 134
Laue's diffraction condition in k space   133
Laws of real gases   183
Laws of real gases based on Lennard-Jones potential   181–2
Lennard-Jones potential energy   180–3
Liquid structures, *see* Models of pure liquids and melts

Magnetic moment   59
Magnetism   317
   B and H fields   330
   Bohr magneton   70–1, 318
   Curie point   328–30
   eddy currents   332
   electron spin and intrinsic magnetic moment of electrons   319
   magnetic moments of atoms   317–19
   magnetising field   319–20
   magnetostriction   299
   types of magnetic materials   320, 323
   Weiss domains   328, 330
Matter waves   50
Maxwell–Boltzmanns distribution law   148–9, 172–3, 240
Maxwells velocity distribution law   172
   mean velocity   173
   most probable velocity   172
   root mean square velocity   172
Mean free path in a gas   193
Mean free path in a gas mixture   206–7
Mean free path of electrons in solids   356
Melting points of solid metals   406
   Lindemanns melting rule   406
Mendeleev   62
Metallic bonds   112
   classical model of the electron gas   112
   energy levels of free electrons in a metal   116
   free electron model of a metal   112
   quantum mechanical model of the electron gas   114
   the Schrödinger equation of free electrons in a metal   114
Metallic conductors   350
   band theory of electrical conduction in metals   356
   classical theory of electrical conduction in metals   353
      conductivity   355
      current density   353–4
      drift velocity   353–4, 356
      mean free path   356
      mobility   356
   ratio of thermal and electrical conductivities in pure metals   357
   resistivities of alloys   352
   resistivities of pure metals   350
      temperature dependence of resistivity   351
Methods of X-ray examination of solid materials   2
Miller indices   16–17
Millikan   112
Mobility   356

Models of pure liquids and melts   402
   pair distribution function   402
   pair potential models   402
   theory of liquid structure   402
Morse function   83
Multiplicity   62

Nearest neighbour distances   6, 20
   nearest neighbour distances of metals   21
   relationship between nearest neighbour distance and lattice parameter   20
Nernsts theorem   228
Nicol prism   382
Nomenclature of atomic electronic states   64
Nomenclature of electron orbitals   59
Nomenclature of molecular orbitals   72
Nomenclature of molecular states   76
Non-rigid rotator   84
Number of atoms per unit cell   15, 20
   number of electrons per Brillouin zone   129
   number of energy states per band   129

Octahedron   12
Optical properties of solids   375
   of insulators   378
   of metals   375
   of semiconductors   376
Order-disorder transformations
   degree of order, long range parameter   315
   influence of order-disorder transformations on heat capacity in solids   315
   order–disorder transformations in binary alloys   315
   order–disorder transformations in CuZn   316
   order–disorder transformations in ferromagnetic materials   329
Ordered solid solutions   37
Origin of energy bands in solids   125–6

Pair distribution function   402
Paramagnetism   319–20, 321
   theory of paramagnetism   321–3
Particle in a box   55–7
Partition function   240
Pauli   51
Pauli principle   62, 116
Periodic table of elements   62
Permeability
   permeability in vacuum   330
   relative permeability   330
Phonons   146
   angular frequency of phonons as a function of wave number   149–50
   phonon statistics   147–9
   properties of phonons   146
      phonon energies   147
      phonon velocities   147–8
Photoelectric effect   48–9
Photons   49
Physical interpretation of the wave function   54

Planck   48
Plancks quantum theory   48–9
Plancks radiation law   49
Plasma – the fourth state of matter   208
Polarized light   379
  applications   387–8
  dichroism   383
  double refraction   380–2
  Fresnels relationships for reflected light   379
  linearly polarized light   380
  methods to produce linearly polarized light   379
  optical activity   385–7
  plane-polarized light   380
  reflection   380
Polyatomic molecules   88
Potential energy of a magnet in a B field   60
Pressure   170–1
Primitive cell   13, 137

Quantum mechanical model of atomic structure   48
Quantum mechanical model of molecular structure   72
Quantum mechanics   48, 51
Quantum mechanics and probability   65
Quantum numbers and their interpretation   58
  azimuthal quantum number   58
  magnetic quantum number   59
  principal quantum number   58
  spin quantum number   61
Quantum numbers of many-electron atoms   62
  resulting vector J 62
  total angular momentum L 62
  total spin vector S 62
Quenching   246

Radial distribution function   8, 402
Reaction rates   242
  definition   242
  determination of reaction rates   243–5
  reaction rates of simple reactions and transformations   242–3
Real space   130
Reciprocal lattices of crystals   130–1
Reciprocal lattice vector   132, 134
Reciprocal space   130–1, 134
Reduced mass   77, 80
Relativistic mechanics   51
Relaxation time   355–6
Representation of ψ and E_{kin} in k space   118
  graphical representation of E_{kin} 120
  graphical representation of ψ 119
  kinetic energy of the free electron   120
Rigid rotator   76
Rutherford   44

Schottky defects   151–2
Schrödinger equation   52, 53
  solutions of the Schrödinger equation   55
  solutions of the Schrödinger equation for atoms   57
Selection rules for electronic transitions in atoms   66–9

Selection rules in molecules
  selection rules for electronic and vibrational transitions   84
  selection rules for electronic, rotational and vibrational transitions   87
  selection rules for electronic transitions   84
  selection rules for pure rotational transitions   87
  selection rules for pure vibrational transitions   84
Semiconductors   362
  doped semiconductors   369
    theory of electrical conduction in doped semiconductors   369
      electrical conductivity of doped semiconductors   369
      influence of temperature and dopant concentration on the number of charge carriers   369–70
  pure semiconductors   362–4
    theory of electrical conduction in pure semiconductors   364–8
      calculation of density distribution of electron energy states and electron concentration in the conduction band   364
      calculation of density distribution of hole energy states and hole concentration in the valence band   366
      calculation of the energy gap in intrinsic semiconductors   368
      calculation of the Fermi level in an intrinsic semiconductor   367
      electrical conductivities of intrinsic semiconductors   368
  types of doped semiconductors   370
    calculation of donor and acceptor levels in doped semiconductors   372, 374–5
    compound semiconductors   372
    n-doped and n-type semiconductors   370
    p-doped and p-type semiconductors   371
Short range order   6, 38
  degree of short range order   38
  degree of short range order coefficient   38
Solutions
  interstitial solutions   31, 34
  primary solid solutions   31
  random solid solutions   30
  secondary solid solution   31
  substitutional solid solutions   31, 33
Sommerfeld   114, 118, 125
Space lattice   13, 120
Space lattice terms   14
Spectra of diatomic molecules   82
Splitting of energy levels in a magnetic field   60
Stable and metastable states   237
Stacking faults   24, 28
Statistical weight   175
Stirlings equation   230
Structures of alloy melts and solids   30
Structures of liquid alloys   31
Structures of metal melts   4

Structures of solid alloys   33
  interstitial solid solutions   31
  substitutional solid solutions   31
  with interstitials atoms   25, 154
  with substitutional atoms   25, 154
Superlattices   37
Susceptibility   320, 322

Terminology   220
Tetrahedron   12
Thermal conduction in gases
  kinetic theory of thermal conduction in gases   198
  thermal conductivity in gases   199
Thermal conduction in solids   342
  thermal conductivities of alloys   347
  thermal conductivities of metals   345
  thermal conductivities of nonmetals   343
  thermal conductivities of pure metals   346
  thermal conductivity of diamond   345
Thermal conduction of liquids and melts   438
  thermal conduction in metal melts   438
  thermal conductivity in alloy melts   439
  thermal conductivity in pure metal melts   438
Thermal energy distribution in a gas   175–8, 241
Thermal energy distribution in particle systems   240
  application to reactions and transformations   241
Thermal expansion of liquids and melts   409
  temperature dependence of density   410
  thermal volume expansion   409
Thermal velocity distribution in a gas   172
Thermodynamics   220
  first law of thermodynamics   220
  second law of thermodynamics   221, 222–3
  third law of thermodynamics   224, 227–8
Thermodynamics of ideal and non-ideal solutions   231
  enthalpy change on mixing two components   231
  entropy change of ideal and nonideal solutions   232
  entropy change on mixing two components   232
  heat of mixing of ideal and nonideal solutions   231
  ideal and non-ideal solutions   231
Thermodynamics of phase transformations   232
  enthalpy and entropy changes in phase transformations   233–4
  survey of phase transformations   235
Thomson J. J.   112
Tracer diffusion in a two component gas mixture   208
Transformation kinetics   236–9
  activation energy   238–9
  driving force   239
  endothermic and exothermic reactions and transformations   239–40
  reactions and transformations   237

stable and metastable states   237
  thermodynamic condition for equilibrium   237
Transport properties of liquids   415
Twinned crystals   29

Unit cell   13, 129

Vacancies   25, 151, 157, 258
Van der Waals equation for real gases   183–5
Van der Waals forces   101, 112
Van der Waals interaction   180–1
Viscosity of gases   197
  dynamic coefficient of viscosity   196
  kinetic coefficient of viscosity in gases   425
Viscosity of liquids and melts   429
  basic theory of viscosity   426
  experimental methods of viscosity measurements on metal melts   437
  influence of gravity forces and walls on viscosity   437
  kinetic coefficient of viscosity   426
  models of viscosity in liquids   429
  newtonian and non-newtonian fluids   427
  Ostwald's law   428
  relationship between D and η 432–5
  temperature dependence of viscosity in pure metal melts   432
  thermodynamical model of viscosity for binary alloy melts   435
  viscosity in pure metal melts   429
  viscosity of alloy melts   435

Wave mechanics   51
Weiss domains   328, 330
Wiedemann–Franz law   357, 443
Wigner–Seitz cell   137
Work function   49, 116–17

X-ray analysis of liquids and melts
  temperature dependence of the atomic distribution in pure metal melts   401
  X-ray analysis of pure liquids and melts   400
  X-ray spectra of binary alloy melts   402
  X-ray spectra of liquids and melts   400
X-ray analysis of solids   2–4
X-ray diffraction   2–3
X-ray examination of metal melts   4–5
X-ray spectrometer   3

Youngs modulus   292

Zeeman effect   69, 318